TRENDS IN MATHEMATICS

Trends in Mathematics is a series devoted to the publication of volumes arising from conferences and lecture series focusing on a particular topic from any area of mathematics. Its aim is to make current developments available to the community as rapidly as possible without compromise to quality and to archive these for reference.

Proposals for volumes can be sent to the Mathematics Editor at either

Birkhäuser Verlag
P.O. Box 133
CH-4010 Basel
Switzerland

or

Birkhäuser Boston Inc.
675 Massachusetts Avenue
Cambridge, MA 02139
USA

Material submitted for publication must be screened and prepared as follows:

All contributions should undergo a reviewing process similar to that carried out by journals and be checked for correct use of language which, as a rule, is English. Articles without proofs, or which do not contain any significantly new results, should be rejected. High quality survey papers, however, are welcome.

We expect the organizers to deliver manuscripts in a form that is essentially ready for direct reproduction. Any version of TeX is acceptable, but the entire collection of files must be in one particular dialect of TeX and unified according to simple instructions available from Birkhäuser.

Furthermore, in order to guarantee the timely appearance of the proceedings it is essential that the final version of the entire material be submitted no later than one year after the conference. The total number of pages should not exceed 350. The first-mentioned author of each article will receive 25 free offprints. To the participants of the congress the book will be offered at a special rate.

Differential Equations with Symbolic Computation

Dongming Wang
Zhiming Zheng
Editors

Birkhäuser Verlag
Basel · Boston · Berlin

Editors:

Dongming Wang
School of Science
Beihang University
37 Xueyuan Road
Beijing 100083
China

and

Laboratoire d'Informatique de Paris 6
Université Pierre et Marie Curie - CNRS
8, rue due Capitaine Scott
75015 Paris
France

e-mail: Dongming.Wang@lip6.fr

Zhiming Zheng
School of Science
Beihang University
37 Xueyuan Road
Beijing 100083
China

e-mail: zzheng@pku.edu.cn

2000 Mathematical Subject Classification 34-06; 35-06; 68W30

A CIP catalogue record for this book is available from the
Library of Congress, Washington D.C., USA

Bibliographic information published by Die Deutsche Bibliothek
Die Deutsche Bibliothek lists this publication in the Deutsche Nationalbibliografie; detailed bibliographic data is available in the Internet at http://dnb.ddb.de

ISBN 3-7643-7368-7 Birkhäuser Verlag, Basel – Boston – Berlin

Part of Springer Science+Business Media
Printed on acid-free paper produced from chlorine-free pulp. TCF ∞
Printed in Germany
ISBN-10: 3-7643-7368-7
ISBN-13: 976-3-7643-7368-9
e-ISBN: 3-7643-7429-2

9 8 7 6 5 4 3 2 1

www.birkhauser.ch

Preface

This book provides a picture of what can be done in differential equations with advanced methods and software tools of symbolic computation. It focuses on the symbolic-computational aspect of three kinds of fundamental problems in differential equations: transforming the equations, solving the equations, and studying the structure and properties of their solutions. Modern research on these problems using symbolic computation, or more restrictively using computer algebra, has become increasingly active since the early 1980s when effective algorithms for symbolic solution of differential equations were proposed, and so were computer algebra systems successfully applied to perturbation, bifurcation, and other problems. Historically, symbolic integration, the simplest case of solving ordinary differential equations, was already the target of the first computer algebra package SAINT in the early 1960s.

With 20 chapters, the book is structured into three parts with both tutorial surveys and original research contributions: the first part is devoted to the qualitative study of differential systems with symbolic computation, including stability analysis, establishment of center conditions, and bifurcation of limit cycles, which are closely related to Hilbert's sixteenth problem. The second part is concerned with symbolic solutions of ordinary and partial differential equations, for which normal form methods, reduction and factorization techniques, and the computation of conservation laws are introduced and used to aid the search. The last part is concentrated on the transformation of differential equations into such forms that are better suited for further study and application. It includes symbolic elimination and triangular decomposition for systems of ordinary and partial differential polynomials. A 1991 paper by Wen-tsün Wu on the construction of Gröbner bases based on Riquier–Janet's theory, published in China and not widely available to the western readers, is reprinted as the last chapter. This book should reflect the current state of the art of research and development in differential equations with symbolic computation and is worth reading for researchers and students working on this interdisciplinary subject of mathematics and computational science. It may also serve as a reference for everyone interested in differential equations, symbolic computation, and their interaction.

The idea of compiling this volume grew out of the Seminar on Differential Equations with Symbolic Computation (DESC 2004), which was held in Beijing, China in April 2004 (see http://www-calfor.lip6.fr/~wang/DESC2004) to facilitate the interaction between the two disciplines. The seminar brought together active researchers and graduate students from both disciplines to present their work and to report on their new results and findings. It also provided a forum for over 50 participants to exchange ideas and views and to discuss future development and cooperation. Four invited talks were given by Michael Singer, Lan Wen, Wen-tsün Wu, and Zhifen Zhang. The enthusiastic support of the seminar speakers and the high quality of their presentations are some of the primary motivations for our endeavor to prepare a coherent and comprehensive volume with most recent advances on the subject for publication. In addition to the seminar speakers, several distinguished researchers who were invited to attend the seminar but could not make their trip have also contributed to the present book. Their contributions have helped enrich the contents of the book and make the book beyond a proceedings volume. All the papers accepted for publication in the book underwent a formal review-revision process.

DESC 2004 is the second in a series of seminars, organized in China, on various subjects interacted with symbolic computation. The first seminar, held in Hefei from April 24–26, 2002, was focused on geometric computation and a book on the same subject has been published by World Scientific. The third seminar planned for April 2006 will be on symbolic computation in education.

The editors gratefully acknowledge the support provided by the Schools of Science and Advanced Engineering at Beihang University and the Key Laboratory of Mathematics, Informatics and Behavioral Semantics of the Chinese Ministry of Education for DESC 2004 and the preparation of this book. Our sincere thanks go to the authors for their contributions and cooperation, to the referees for their expertise and timely help, and to all colleagues and students who helped for the organization of DESC 2004.

Beijing
May 2005

Dongming Wang
Zhiming Zheng

Contents

Trends in Mathematics: *Differential Equations with Symbolic Computation*, 1–22

Symbolic Computation of Lyapunov Quantities and the Second Part of Hilbert's Sixteenth Problem

Stephen Lynch

Abstract. This tutorial survey presents a method for computing the Lyapunov quantities for Liénard systems of differential equations using symbolic manipulation packages. The theory is given in detail and simple working MATLAB and Maple programs are listed in this chapter. In recent years, the author has been contacted by many researchers requiring more detail on the algorithmic method used to compute focal values and Lyapunov quantities. It is hoped that this article will address the needs of those and other researchers. Research results are also given here.

Mathematics Subject Classification (2000). Primary 34C07; Secondary 37M20.

Keywords. Bifurcation, Liénard equation, limit cycle, Maple, MATLAB, small-amplitude.

1. Introduction

Poincaré began investigating isolated periodic cycles of planar polynomial vector fields in the 1880s. However, the general problem of determining the maximum number and relative configurations of limit cycles in the plane has remained unresolved for over a century. In the engineering literature, limit cycles in the plane can correspond to steady-state behavior for a physical system (see [25], for example), so it is important to know how many possible steady states there are. There are applications in aircraft flight dynamics and surge in jet engines, for example.

In 1900, David Hilbert presented a list of 23 problems to the International Congress of Mathematicians in Paris. Most of the problems have been solved, either completely or partially. However, the second part of the sixteenth problem remains unsolved. Ilyashenko [37] presents a centennial history of Hilbert's 16th problem and Li [19] has recently written a review article.

The Second Part of Hilbert's Sixteenth Problem. Consider planar polynomial systems of the form

$$\dot{x} = P(x, y), \quad \dot{y} = Q(x, y), \tag{1.1}$$

where P and Q are polynomials in x and y. The question is to estimate the maximal number and relative positions of the limit cycles of system (1.1). Let H_n denote the maximum possible number of limit cycles that system (1.1) can have when P and Q are of degree n. More formally, the Hilbert numbers H_n are given by

$$H_n = \sup \{\pi(P, Q) : \partial P, \partial Q \leq n\},$$

where ∂ denotes "the degree of" and $\pi(P, Q)$ is the number of limit cycles of system (1.1).

Dulac's Theorem states that a given polynomial system cannot have infinitely many limit cycles. This theorem has only recently been proved independently by Ecalle et al. [13] and Ilyashenko [36], respectively. Unfortunately, this does not imply that the Hilbert numbers are finite.

Of the many attempts to make progress in this question, one of the more fruitful approaches has been to create vector fields with as many isolated periodic orbits as possible using both local and global bifurcations [3]. There are relatively few results in the case of general polynomial systems even when considering local bifurcations. Bautin [1] proved that no more than three small-amplitude limit cycles could bifurcate from a critical point for a quadratic system. For a homogeneous cubic system (no quadratic terms), Sibirskii [33] proved that no more than five small-amplitude limit cycles could be bifurcated from one critical point. Recently, Zoladek [39] found an example where 11 limit cycles could be bifurcated from the origin of a cubic system, but he was unable to prove that this was the maximum possible number.

Although easily stated, Hilbert's sixteenth problem remains almost completely unsolved. For quadratic systems, Songling Shi [32] has obtained a lower bound for the Hilbert number $H_2 \geq 4$. A possible global phase portrait is given in Figure 1. The line at infinity is included and the properties on this line are determined using Poincaré compactification, where a polynomial vector field in the plane is transformed into an analytic vector field on the 2-sphere. More detail on Poincaré compactification can be found in [27]. There are three small-amplitude limit cycles around the origin and at least one other surrounding another critical point. Some of the parameters used in this example are very small.

Blows and Rousseau [4] consider the bifurcation at infinity for polynomial vector fields and give examples of cubic systems having the following configurations:

$$\{(4), 1\}, \{(3), 2\}, \{(2), 5\}, \{(4), 2\}, \{(1), 5\} \text{ and } \{(2), 4\},$$

where $\{(l), L\}$ denotes the configuration of a vector field with l small-amplitude limit cycles bifurcated from a point in the plane and L large-amplitude limit cycles

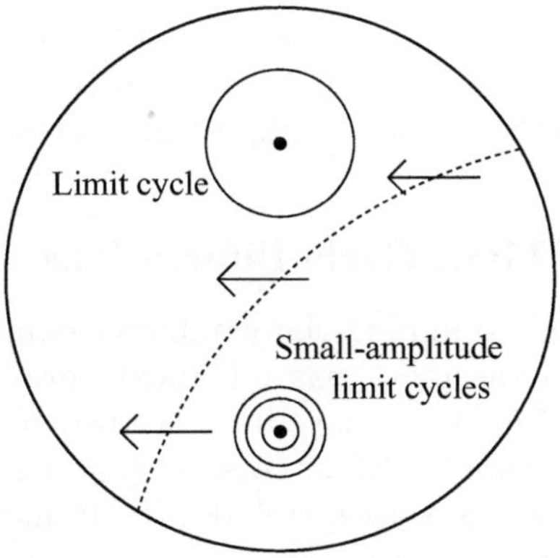

FIGURE 1. A possible configuration for a quadratic system with four limit cycles: one of large amplitude and three of small amplitude.

simultaneously bifurcated from infinity. There are many other configurations possible, some involving other critical points in the finite part of the plane as shown in Figure 2. Recall that a limit cycle must contain at least one critical point.

By considering cubic polynomial vector fields, in 1985, Jibin Li and Chunfu Li [18] produced an example showing that $H_3 \geq 11$ by bifurcating limit cycles out of homoclinic and heteroclinic orbits; see Figure 2.

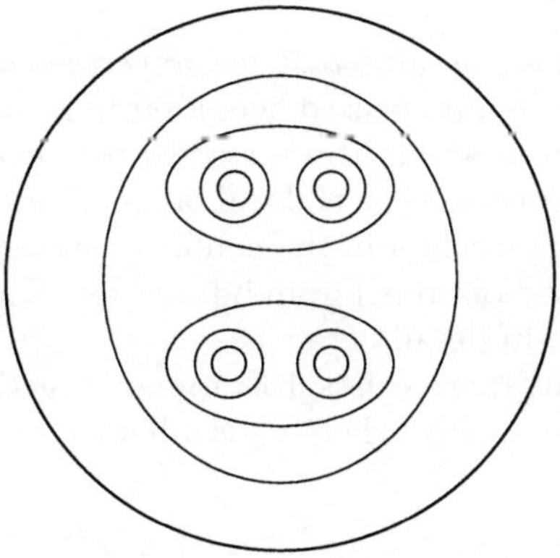

FIGURE 2. A possible configuration for a cubic system with 11 limit cycles.

Returning to the general problem, in 1995, Christopher and Lloyd [7] considered the rate of growth of H_n as n increases. They showed that H_n grows at least as rapidly as $n^2 \log n$.

In recent years, the focus of research in this area has been directed at a small number of classes of systems. Perhaps the most fruitful has been the Liénard system. A method for computing focal values and Lyapunov quantities for Liénard systems is given in detail in the next section. Liénard systems provide a very suitable starting point as they do have ubiquity for systems in the plane [14, 16, 28].

2. Small-Amplitude Limit Cycle Bifurcations

The general problem of determining the maximum number and relative configurations of limit cycles in the plane has remained unresolved for over a century. Both local and global bifurcations have been studied to create vector fields with as many limit cycles as possible. All of these techniques rely heavily on symbolic manipulation packages such as Maple, and MATLAB and its Symbolic Math Toolbox. Unfortunately, the results in the global case number relatively few. Only in recent years have many more results been found by restricting the analysis to *small-amplitude limit cycle* bifurcations.

It is well known that a nondegenerate critical point, say $\mathbf{x_0}$, of center or focus type can be moved to the origin by a linear change of coordinates, to give

$$\dot{x} = \lambda x - y + p(x, y), \quad \dot{y} = x + \lambda y + q(x, y), \tag{2.1}$$

where p and q are at least quadratic in x and y. If $\lambda \neq 0$, then the origin is structurally stable for all perturbations.

Definition 2.1. A critical point, say $\mathbf{x_0}$, is called a *fine focus* of system (1.1) if it is a center for the linearized system at $\mathbf{x_0}$. Equivalently, if $\lambda = 0$ in system (2.1), then the origin is a fine focus.

In the work to follow, *assume that the unperturbed system does not have a center at the origin.* The technique used here is entirely local; limit cycles bifurcate out of a fine focus when its stability is reversed by perturbing λ and the coefficients arising in p and q. These are said to be local or small-amplitude limit cycles. How close the origin is to being a center of the nonlinear system determines the number of limit cycles that may be obtained from bifurcation. The method for bifurcating limit cycles will be given in detail here.

By a classical result, there exists a Lyapunov function, $V(x, y) = V_2(x, y) + V_4(x, y) + \cdots + V_k(x, y) + \cdots$ say, where V_k is a homogeneous polynomial of degree k, such that

$$\frac{dV}{dt} = \eta_2 r^2 + \eta_4 r^4 + \cdots + \eta_{2i} r^{2i} + \cdots, \tag{2.2}$$

where $r^2 = x^2 + y^2$. The η_{2i} are polynomials in the coefficients of p and q and are called the *focal values*. The origin is said to be a fine focus of order k if $\eta_2 = \eta_4 = \cdots = \eta_{2k} = 0$ but $\eta_{2k+2} \neq 0$. Take an analytic transversal through the origin parameterized by some variable, say c. It is well known that the return map of (2.1), $c \mapsto h(c)$, is analytic if the critical point is nondegenerate. Limit cycles of system (2.1) then correspond to zeros of the *displacement function*, say $d(c) = h(c) - c$. Hence at most k limit cycles can bifurcate from the fine focus. The stability of the origin is clearly dependent on the sign of the first non-zero focal value, and the origin is a nonlinear center if and only if all of the focal values are zero. Consequently, it is the reduced values, or *Lyapunov quantities*, say $L(j)$, that are significant. One needs only to consider the value η_{2k} reduced modulo the ideal $(\eta_2, \eta_4, \ldots, \eta_{2k-2})$ to obtain the Lyapunov quantity $L(k-1)$. To bifurcate limit cycles from the origin, select the coefficients in the Lyapunov quantities such that

$$|L(m)| \ll |L(m+1)| \quad \text{and} \quad L(m)L(m+1) < 0,$$

for $m = 0, 1, \ldots, k-1$. At each stage, the origin reverses stability and a limit cycle bifurcates in a small region of the critical point. If all of these conditions are satisfied, then there are exactly k small-amplitude limit cycles. Conversely, if $L(k) \neq 0$, then at most k limit cycles can bifurcate. Sometimes it is not possible to bifurcate the full complement of limit cycles.

The algorithm for bifurcating small-amplitude limit cycles may be split into the following four steps:

1. computation of the focal values using a mathematical package;
2. reduction of the n-th focal value modulo a Gröbner basis of the ideal generated by the first $n-1$ focal values (or the first $n-1$ Lyapunov quantities);
3. checking that the origin is a center when all of the relevant Lyapunov quantities are zero;
4. bifurcation of the limit cycles by suitable perturbations.

Dongming Wang [34, 35] has developed software to deal with the reduction part of the algorithm for several differential systems. For some systems, the following theorems can be used to prove that the origin is a center.

The Divergence Test. Suppose that the origin of system (1.1) is a critical point of focus type. If

$$\operatorname{div}(\psi \mathbf{X}) = \frac{\partial(\psi P)}{\partial x} + \frac{\partial(\psi Q)}{\partial y} = 0,$$

where $\psi : \Re^2 \to \Re^2$, then the origin is a center.

The Classical Symmetry Argument. Suppose that $\lambda = 0$ in system (2.1) and that either

(i) $p(x, y) = -p(x, -y)$ and $q(x, y) = q(x, -y)$ or

(ii) $p(x, y) = p(-x, y)$ and $q(x, y) = -q(-x, y)$.

Then the origin is a center.

Adapting the classical symmetry argument, it is also possible to prove the following theorem.

Theorem 2.1. *The origin of the system*

$$\dot{x} = y - F(G(x)), \quad \dot{y} = -\frac{G'(x)}{2} H(G(x)),$$

where F and H are polynomials, $G(x) = \int_0^x g(s)ds$ with $g(0) = 0$ and $g(x)\operatorname{sgn}(x) > 0$ for $x \neq 0$, is a center.

To demonstrate the method for bifurcating small-amplitude limit cycles, consider Liénard equations of the form

$$\dot{x} = y - F(x), \quad \dot{y} = -g(x), \tag{2.3}$$

where $F(x) = a_1x + a_2x^2 + \cdots + a_ux^u$ and $g(x) = x + b_2x^2 + b_3x^3 + \cdots + b_vx^v$. This system has proved very useful in the investigation of limit cycles when showing existence, uniqueness, and hyperbolicity of a limit cycle. In recent years, there have also been many local results; see, for example, [9]. Therefore, it seems sensible to use this class of system to illustrate the method.

The computation of the first three focal values will be given. Write

$$V_k(x, y) = \sum_{i+j=k} V_{i,j}x^iy^j$$

and denote $V_{i,j}$ as being odd or even according to whether i is odd or even and that $V_{i,j}$ is 2-odd or 2-even according to whether j is odd or even, respectively. Solving equation (2.2), it is easily seen that $V_2 = \frac{1}{2}(x^2 + y^2)$ and $\eta_2 = -a_1$. Therefore, set $a_1 = 0$. The odd and even coefficients of V_3 are then given by the two pairs of equations

$$\begin{aligned} 3V_{3,0} - 2V_{1,2} &= b_2, \\ V_{1,2} &= 0 \end{aligned}$$

and

$$\begin{aligned} -V_{2,1} &= a_2, \\ 2V_{2,1} - 3V_{0,3} &= 0, \end{aligned}$$

respectively. Solve the equations to give

$$V_3 = \frac{1}{3}b_2x^3 - a_2x^2y - \frac{2}{3}a_2y^3.$$

Both η_4 and the odd coefficients of V_4 are determined by the equations

$$\begin{aligned} -\eta_4 - V_{3,1} &= a_3, \\ -2\eta_4 + 3V_{3,1} - 3V_{1,3} &= -2a_2b_2, \\ -\eta_4 + V_{1,3} &= 0. \end{aligned}$$

The even coefficients are determined by the equations

$$\begin{aligned} 4V_{4,0} - 2V_{2,2} &= b_3 - 2a_2^2, \\ 2V_{2,2} - 4V_{0,4} &= 0 \end{aligned}$$

and the supplementary condition $V_{2,2} = 0$. In fact, when computing subsequent coefficients for V_{4m}, it is convenient to require that $V_{2m,2m} = 0$. This ensures that there will always be a solution. Solving these equations gives

$$V_4 = \frac{1}{4}(b_3 - 2a_2^2)x^4 - (\eta_4 + a_3)x^3y + \eta_4xy^3$$

and

$$\eta_4 = \frac{1}{8}(2a_2b_2 - 3a_3).$$

Suppose that $\eta_4 = 0$ so that $a_3 = \frac{2}{3}a_2b_2$. It can be checked that the two sets of equations for the coefficients of V_5 give

$$V_5 = \left(\frac{b_4}{5} - \frac{2a_2^2b_2}{3}\right)x^5 + (2a_2^3 - a_4)x^4y + \left(\frac{8a_2^3}{3} - \frac{4a_4}{3} + \frac{2a_2b_3}{3}\right)x^2y^3$$
$$+ \left(\frac{16a_2^3}{15} - \frac{8a_4}{15} - \frac{4a_2b_3}{15}\right)y^5.$$

The coefficients of V_6 may be determined by inserting the extra condition $V_{4,2} + V_{2,4} = 0$. In fact, when computing subsequent even coefficients for V_{4m+2}, the extra condition $V_{2m,2m+2} + V_{2m+2,2m} = 0$, is applied, which guarantees a solution. The polynomial V_6 contains 27 terms and will not be listed here. However, η_6 leads to the Lyapunov quantity

$$L(2) = 6a_2b_4 - 10a_2b_2b_3 + 20a_4b_2 - 15a_5.$$

Lemma 2.1. *The first three Lyapunov quantities for system* (2.3) *are* $L(0) = -a_1$, $L(1) = 2a_2b_2 - 3a_3$, and $L(2) = 6a_2b_4 - 10a_2b_2b_3 + 20a_4b_2 - 15a_5$.

Example. Prove that

(i) there is at most one small-amplitude limit cycle when $\partial F = 3, \partial g = 2$ and

(ii) there are at most two small-amplitude limit cycles when $\partial F = 3, \partial g = 3$,

for system (2.3).

Solutions. (i) Now L(0)=0 if $a_1 = 0$ and $L(1) = 0$ if $a_3 = \frac{2}{3}a_2b_2$. Thus system (2.3) becomes

$$\dot{x} = y - a_2x^2 - \frac{2}{3}a_2b_2x^3, \quad \dot{y} = -x - b_2x^2,$$

and the origin is a center by Theorem 2.1. Therefore, the origin is a fine focus of order one if and only if $a_1 = 0$ and $2a_2b_2 - 3a_3 \neq 0$. The conditions are consistent. Select a_3 and a_1 such that

$$|L(0)| \ll |L(1)| \quad \text{and} \quad L(0)L(1) < 0.$$

The origin reverses stability once and a limit cycle bifurcates. The perturbations are chosen such that the origin reverses stability once and the limit cycles that bifurcate persist.

(ii) Now $L(0) = 0$ if $a_1 = 0$, $L(1) = 0$ if $a_3 = \frac{2}{3}a_2b_2$, and $L(2) = 0$ if $a_2b_2b_3 = 0$. Thus $L(2) = 0$ if

(a) $a_2 = 0$,

(b) $b_3 = 0$, or

(c) $b_2 = 0$.

If condition (a) holds, then $a_3 = 0$ and the origin is a center by the divergence test ($\mathrm{div}\mathbf{X} = 0$). If condition (b) holds, then the origin is a center from result (i) above. If condition (c) holds, then $a_3 = 0$ and system (2.3) becomes

$$\dot{x} = y - a_2x^2, \quad \dot{y} = -x - b_3x^3,$$

and the origin is a center by the classical symmetry argument. The origin is thus a fine focus of order two if and only if $a_1 = 0$ and $2a_2b_2 - 3a_3 = 0$ but $a_2b_2b_3 \neq 0$. The conditions are consistent. Select b_3, a_3, and a_1 such that

$$|L(1)| \ll |L(2)|, \; L(1)L(2) < 0 \quad \text{and} \quad |L(0)| \ll |L(1)|, \; L(0)L(1) < 0.$$

The origin has changed stability twice, and there are two small-amplitude limit cycles. The perturbations are chosen such that the origin reverses stability twice and the limit cycles that bifurcate persist.

3. Symbolic Computation

Readers can download the following program files from the Web. The MATLAB M-file lists all of the coefficients of the Lyapunov function up to and including degree six terms. The output is also included for completeness. The program was written using MATLAB version 7 and the program files can be downloaded at

http://www.mathworks.com/matlabcentral/fileexchange

under the links "Companion Software for Books" and "Mathematics".

```
% MATLAB Program - Determining the coefficients of the Lyapunov
% function for a quintic Lienard system.

% V3=[V30;V21;V12;V03], V4=[V40;V31;V22;V13;V04;eta4],
% V5=[V50;V41;V32;V23;V14;V05],
% V6=[V60;V51;V42;V33;V24;V15;V06;eta6]
% Symbolic Math toolbox required.
clear all

syms a1 a2 b2 a3 b3 a4 b4 a5 b5;
A=[3 0 -2 0;0 0 1 0;0 -1 0 0;0 2 0 -3];
B=[b2; 0; a2; 0];

V3=A\B

A=[0 -1 0 0  0 -1;0 3 0 -3 0 -2;0 0 0 1 0 -1;4 0 -2 0 0 0;
0 0 2 0 -4 0; 0 0 1 0 0 0];
B=[a3; -2*a2*b2; 0; b3-2*a2^2;0;0];

V4=A\B

A=[5 0 -2 0 0 0;0 0 3 0 -4 0;0 0 0 0 1 0;0 -1 0 0 0 0;0 4 0 -3 0 0;
```

```
0 0 0 2 0 -5];
B=[b4-10*a2^2*b2/3;0;0;a4-2*a2^3;-2*a2*b3;0];

V5=A\B

A=[6 0 -2 0 0 0 0 0;0 0 4 0 -4 0 0 0;0 0 0 0 2 0 -6 0;
0 0 1 0 1 0 0 0;0 -1 0 0 0 0 0 -1;0 5 0 -3 0 0 0 -3;
0 0 0 3 0 -5 0 -3;0 0 0 0 0 1 0 -1];
B=[b5-6*a2*a4-4*a2^2*b2^2/3+8*a2^4;16*a2^4/3+4*a2^2*b3/3-8*a2*a4/3;
0;0;a5-8*a2^3*b2/3;-2*a2*b4+8*a2^3*b2+2*a2*b2*b3-4*a4*b2;
16*a2^3*b2/3+4*a2*b2*b3/3-8*a4*b2/3;0];

V6=A\B

L0=-a1
eta4=V4(6,1)
L1=maple('numer(-3/8*a3+1/4*a2*b2)')
a3=2*a2*b2;
eta6=V6(8,1)
L2=maple('numer(-5/16*a5+1/8*a2*b4-5/24*a2*b2*b3+5/12*a4*b2)')
%End of MATLAB Program
```

```
V3 =

[  1/3*b2]
[     -a2]
[       0]
[ -2/3*a2]

V4 =

[   1/4*b3-1/2*a2^2]
[ -5/8*a3-1/4*a2*b2]
[                 0]
[ -3/8*a3+1/4*a2*b2]
[                 0]
[ -3/8*a3+1/4*a2*b2]

V5 =

[              1/5*b4-2/3*a2^2*b2]
[                     -a4+2*a2^3]
[                               0]
[     -4/3*a4+8/3*a2^3+2/3*a2*b3]
[                               0]
[ -8/15*a4+16/15*a2^3+4/15*a2*b3]

V6 =
```

```
[    14/9*a2^4+1/6*b5-10/9*a2*a4-2/9*a2^2*b2^2+1/18*a2^2*b3]
[ -11/16*a5-1/8*a2*b4+5/24*a2*b2*b3-5/12*a4*b2+8/3*a2^3*b2]
[                            2/3*a2^4+1/6*a2^2*b3-1/3*a2*a4]
[      2/9*a4*b2-5/6*a5+1/3*a2*b4-1/9*a2*b2*b3+16/9*a2^3*b2]
[                          -2/3*a2^4-1/6*a2^2*b3+1/3*a2*a4]
[              -5/16*a5+1/8*a2*b4-5/24*a2*b2*b3+5/12*a4*b2]
[                         -2/9*a2^4-1/18*a2^2*b3+1/9*a2*a4]
[              -5/16*a5+1/8*a2*b4-5/24*a2*b2*b3+5/12*a4*b2]


L0 =-a1

L1 =-3*a3+2*a2*b2

L2 =-15*a5+6*a2*b4-10*a2*b2*b3+20*a4*b2
```

The Maple 9 program files can be found at
http://www.maplesoft.com/books/.

```
> # MAPLE program to compute the first two Lyapunov quantities for
> # a quintic Lienard system.
> restart:
> kstart:=2:kend:=5:
> pp:=array(1..20):qq:=array(1..20):
> vv:=array(1..20):vx:=array(0..20):
> vy:=array(0..20):xx:=array(0..20,0..20):
> yy:=array(0..20,0..20):uu:=array(0..20,0..20):
> z:=array(0..20):ETA:=array(1..20):
> pp[1]:=y:qq[1]:=-x:vv[2]:=(x^2+y^2)/2:vx[2]:=x:vy[2]:=y:
>
> for j1 from 0 to 20 do
> for j2 from 0 to 20 do
> xx[j1,j2]:=0:yy[j1,j2]:=0:
> od:od:
>
> # Insert the coefficients for a quintic Lienard system.
> xx[0,1]:=1:xx[2,0]:=-a2:xx[3,0]:=-a3:xx[4,0]:=-a4:xx[5,0]:=-a5:
> yy[1,0]:=-1:yy[2,0]:=-b2:yy[3,0]:=-b3:yy[4,0]:=-b4:yy[5,0]:=-b5:
>
> for kloop from kstart to kend do
> kk:=kloop:
> dd1:=sum(pp[i]*vx[kk+2-i]+qq[i]*vy[kk+2-i],i=2..kk-1):
> pp[kk]:=sum(xx[kk-i,i]*x^(kk-i)*y^i,i=0..kk):
> qq[kk]:=sum(yy[kk-i,i]*x^(kk-i)*y^i,i=0..kk):
> vv[kk+1]:=sum(uu[kk+1,i]*x^(kk+1-i)*y^i,i=0..kk+1):
> d1:=y*diff(vv[kk+1],x)-x*diff(vv[kk+1],y)+pp[kk]*vx[2]+
```

```
      qq[kk]*vy[2]+dd1:
> dd:=expand(d1):
>
> if irem(kk,2)=1 then dd:=dd-ETA[kk+1]*(x^2+y^2)^((kk+1)/2):
> fi:
> dd:=numer(dd):x:=1:
> for i from 0 to kk+1 do z[i]:=coeff(dd,y,i);
> od:
>
> if kk=2 then
> seqn:=solve({z[0],z[1],z[2],z[3]},
            {uu[3,0],uu[3,1],uu[3,2],uu[3,3]}):
> elif kk=3 then
> seqn:=solve({z[0],z[1],z[2],uu[4,2],z[3],z[4]},
            {uu[4,0],uu[4,1],uu[4,2],uu[4,3],uu[4,4],ETA[4]}):
> elif kk=4 then
> seqn:=solve({z[0],z[1],z[2],z[3],z[4],z[5]},
            {uu[5,0],uu[5,1],uu[5,2],uu[5,3],uu[5,4],uu[5,5]}):
> elif kk=5 then
> seqn:=solve({z[0],z[1],z[2],uu[6,2]+uu[6,4],z[3],z[4],z[5],z[6]},
            {uu[6,0],uu[6,1],uu[6,2],uu[6,3],uu[6,4],uu[6,5],
             uu[6,6],ETA[6]}):
> fi:
>
> assign(seqn):x:='x':i:='i':
> vv[kk+1]:=sum(uu[kk+1,i]*x^(kk+1-i)*y^i,i=0..kk+1):
> vx[kk+1]:=diff(vv[kk+1],x):vy[kk+1]:=diff(vv[kk+1],y):
> ETA[kk+1]:=ETA[kk+1]:
> od:
>
> print(L1=numer(ETA[4])):a3:=2*a2*b2/3:print(L2=numer(ETA[6]));

                    L1 = -3 a3 + 2 b2 a2

       L2 = 20 b2 a4 - 10 b3 b2 a2 + 6 b4 a2 - 15 a5
```

The programs can be extended to compute further focal values. The algorithm in the context of Liénard systems will now be described. Consider system (2.3); the linearization at the origin is already in canonical form. Write D_k for the collection of terms of degree k in $\dot{V}$. Hence

$$D_k = y\frac{\partial V_k}{\partial x} - x\frac{\partial V_k}{\partial y} - \left\{\sum_{r=2}^{k-1}\left(a_r\frac{\partial V_{k-r+1}}{\partial x} + b_r\frac{\partial V_{k-r+1}}{\partial y}\right)x^r\right\}. \tag{3.1}$$

Choose V_k and η_{2k} $(k = 2, 3, \ldots)$ such that $D_{2k} = \eta_{2k}r^k$ and $D_{2k-1} = 0$. The focal values are calculated recursively, in a two-stage procedure. Having determined V_ℓ

with $\ell \leq 2k$, V_{2k+1} is found by setting $D_{2k+1} = 0$, and then V_{2k+2} and η_{2k+2} are computed from the relation $D_{2k+2} = \eta_{2k+2}\left(x^2+y^2\right)^{k+1}$. Setting $D_{2k+1} = 0$ gives $2k+2$ linear equations for the coefficients of V_{2k+1} in terms of those of V_ℓ with $\ell \leq 2k$. These uncouple into two sets of $k+1$ equations, one of which determines the odd coefficients of V_{2k+1} and the other the even coefficients. For system (2.3) the two sets are as follows:

$$\begin{aligned}
(2k+1)V_{2k+1,0} - 2V_{2k-1,2} &= \sum_{i=2}^{2k-1} iV_{i,1}a_{2k+1-i} + \sum_{i=0}^{2k-2} 2V_{i,2}b_{2k-i} \\
(2k-1)V_{2k-1,2} - 4V_{2k-3,4} &= \sum_{i=1}^{2k-3} iV_{i,3}a_{2k-1-i} + \sum_{i=0}^{2k-4} 4V_{i,4}b_{2k-2-i} \\
\vdots \quad &= \quad \vdots \\
3V_{3,2k-2} - 2kV_{1,2k} &= V_{1,2k-1}a_2 + 2kV_{0,2k}b_2 \\
V_{1,2k} &= 0,
\end{aligned}$$

and

$$\begin{aligned}
-V_{2k,1} &= \sum_{i=2}^{2k} iV_{i,0}a_{2k+2-i} + \sum_{i=1}^{2k-1} V_{i,1}b_{2k+1-i} \\
2kV_{2k,1} - 3V_{2k-2,3} &= \sum_{i=1}^{2k-2} iV_{i,2}a_{2k-i} + \sum_{i=0}^{2k-3} 3V_{i,3}b_{2k-1-i} \\
\vdots \quad &= \quad \vdots \\
4V_{4,2k-3} - (2k-1)V_{2,2k-1} &= \sum_{i=1}^{2} a_{4-i} + \sum_{i=0}^{1}(2k-1)V_{i,2k-1}b_{3-i} \\
2V_{2,2k-1} - (2k+1)V_{0,2k+1} &= 0.
\end{aligned}$$

The coefficients of V_{2k+2} are obtained by setting $D_{2k+2} = \eta_{2k+2}\left(x^2+y^2\right)^{k+1}$. The odd and even coefficients are given by the following two sets of equations:

$$\begin{aligned}
-\eta_{2k+2} - V_{2k+1,1} &= \sum_{i=2}^{2k+1} iV_{i,0}a_{2k+3-i} + \sum_{i=1}^{2k} V_{i,1}b_{2k+2-i} \\
-k\eta_{2k+2} + (2k+1)V_{2k+1,1} - 3V_{2k-1,3} &= \sum_{i=1}^{2k-1} iV_{i,2}a_{2k+1-i} + \sum_{i=0}^{2k-2} 3V_{i,3}b_{2k-i} \\
\vdots \quad &= \quad \vdots \\
-k\eta_{2k+2} + 3V_{3,2k-1} - (2k+1)V_{1,2k+1} &= V_{1,2k}a_2 + (2k+1)V_{0,2k+1}b_2 \\
-\eta_{2k+2} + V_{1,2k+1} &= 0,
\end{aligned}$$

and

$$\begin{aligned}
(2k+2)V_{2k+2,0} - 2V_{2k,2} &= \sum_{i=1}^{2k} iV_{i,1}a_{2k+2-i} + \sum_{i=0}^{2k-1} 2V_{i,2}b_{2k+1-i} \\
2kV_{2k,2} - 4V_{2k-2,4} &= \sum_{i=1}^{2k-2} iV_{i,3}a_{2k-i} + \sum_{i=0}^{2k-3} 4V_{i,4}b_{2k-1-i} \\
\vdots \quad &= \quad \vdots \\
4V_{4,2k-2} - 2kV_{2,2k} &= \sum_{i=1}^{2} iV_{i,2k-1}a_{4-i} + \sum_{i=0}^{1} 2kV_{i,2k}b_{3-i} \\
2V_{2,2k} - (2k+2)V_{0,2k+2} &= 0.
\end{aligned} \tag{3.2}$$

To simplify the calculations for system (3.3) set $V_{2r,2r+2}+V_{2r+2,2r}=0$ if $k=4r+2$, and $V_{2r,2r}=0$ if $k=4r$.

4. Results

Liénard systems have proved very useful in the investigation of multiple limit cycles and also when proving existence, uniqueness, and hyperbolicity of a limit cycle. Let ∂ denote the degree of a polynomial, and let $H(m,n)$ denote the maximum number of global limit cycles, where m is the degree of f and n is the degree of g for the Liénard equation

$$\ddot{x} + f(x)\dot{x} + g(x) = 0. \tag{4.1}$$

The results in the global case are listed below:

1. In 1928, Liénard [17] proved that when $\partial g = 1$ and F is a continuous odd function, which has a unique root at $x = a$ and is monotone increasing for $x \geq a$, then (2.3) has a unique limit cycle.
2. In 1975, Rychkov [30] proved that if $\partial g = 1$ and F is an odd polynomial of degree five, then (2.3) has at most two limit cycles.
3. In 1976, Cherkas [5] gave conditions in order for a Liénard equation to have a center.
4. In 1977, Lins, de Melo, and Pugh [20] proved that $H(2,1) = 1$. They also conjectured that $H(2i,1) = H(2i+1,1) = i$, where i is a natural number.
5. In 1988, Coppel [10] proved that $H(1,2) = 1$.
6. In 1992, Zhifen Zhang [38] proved that a certain generalised Liénard system has a unique limit cycle.
7. In 1996, Dumortier and Chengzhi Li [11] proved that $H(1,3) = 1$.
8. In 1997, Dumortier and Chengzhi Li [12] proved that $H(2,2) = 1$.

More recently, Giacomini and Neukirch [15] introduced a new method to investigate the limit cycles of Liénard systems when $\partial g = 1$ and $F(x)$ is an odd polynomial. They are able to give algebraic approximations to the limit cycles and obtain information on the number and bifurcation sets of the periodic solutions even when the parameters are not small. Sabatini [31] has constructed Liénard systems with coexisting limit cycles and centers.

Although the Liénard equation (4.1) appears simple enough, the known global results on the maximum number of limit cycles are scant. By contrast, if the analysis is restricted to local bifurcations, then many more results may be obtained. Consider the Liénard system

$$\dot{x} = y, \quad \dot{y} = -g(x) - f(x)y, \tag{4.2}$$

where $f(x) = a_0 + a_1x + a_2x^2 + \cdots + a_ix^i$ and $g(x) = x + b_2x^2 + b_3x^3 + \cdots + b_jx^j$; i and j are natural numbers. Let $\hat{H}(i,j)$ denote the maximum number of small-amplitude limit cycles that can be bifurcated from the origin for system (4.2) when the unperturbed system does not have a center at the origin, where i is the degree of f and j is the degree of g. The following results have been proved by induction using the algorithm presented in Section 2.

1. If $\partial f = m = 2i$ or $2i+1$, then $\hat{H}(m,1) = i$.
2. If g is odd and $\partial f = m = 2i$ or $2i+1$, then $\hat{H}(m,n) = i$.
3. If $\partial g = n = 2j$ or $2j+1$, then $\hat{H}(1,n) = j$.
4. If f is even, $\partial f = 2i$, then $\hat{H}(2i,n) = i$.
5. If f is odd, $\partial f = 2i+1$ and $\partial g = n = 2j+2$ or $2j+3$; then $\hat{H}(2i+1,n) = i+j$.
6. If $\partial f = 2$, $g(x) = x + g_e(x)$, where g_e is even and $\partial g = 2j$; then $\hat{H}(2,2j) = j$.

Note that the first result seems to support the conjecture of Lins, de Melo, and Pugh [20] for global limit cycles. Results 1 and 2 were proven by Blows and Lloyd [2], and the results 3 to 5 were proven by Lloyd and Lynch [21]. An example illustrating the result in case 5 is given below.

Example. Use the algorithm in Section 2 to prove that at most four limit cycles can be bifurcated from the origin for the system

$$\dot{x} = y - \left(a_2x^2 + a_4x^4 + a_6x^6\right), \quad \dot{y} = -\left(b_2x^2 + b_3x^3 + b_4x^4 + b_5x^5 + b_6x^6 + b_7x^7\right).$$

Solution. Note in this case that $\partial f = 2i+1 = 5$ and $\partial g = 2j+3 = 7$, therefore in this case $i = 2$ and $j = 2$. That $\eta_4 = L(1) = 2a_2b_2$, follows directly from the computation of the focal values in the second section of the chapter. The computer programs given earlier can be extended to compute further focal values. This is left as an exercise for the reader. Let us assume that the reader has computed the focal values correctly.

Suppose that $b_2 = 0$, then $L(2) = a_2b_4$. Select $a_2 = 0$, then $L(3) = 7a_4b_4$. Next, select $b_4 = 0$, then $L(4) = 9a_4b_6$. Finally, select $a_4 = 0$, then $L(5) = 99a_6b_6$. A similar argument is used if a_2 is chosen to be zero from the equation $L(1) = 0$. The first five Lyapunov quantities are as follows:

$$L(1) = a_2b_2, L(2) = a_2b_4, L(3) = 7a_4b_4, L(4) = 9a_4b_6, L(5) = 99a_6b_6.$$

If $a_2 = a_4 = a_6 = 0$ with $b_2, b_4, b_6 \neq 0$, then the origin is a center by the divergence test. If $b_2 = b_4 = b_6 = 0$ with $a_2, a_4, a_6 \neq 0$, then the origin is a center by the classical symmetry argument.

From the above, the origin is a fine focus of order four if and only if

$$a_2b_2 = 0, \quad a_2b_4 = 0, \quad a_4b_4 = 0, \quad a_4b_6 = 0,$$

but

$$a_6b_6 \neq 0.$$

The conditions are consistent: for example, let $b_2 = a_2 = a_4 = b_4 = 0$ and $a_6 = b_6 = 1$. Select a_6, b_6, a_4, b_4, a_2 and b_2 such that

$$|a_4b_6| \ll 1 \quad \text{and} \quad a_4a_6 < 0,$$

$$|a_4b_4| \ll 1 \quad \text{and} \quad b_4b_6 < 0,$$

$$|a_2b_4| \ll 1 \quad \text{and} \quad a_2a_4 < 0,$$

$$|a_2b_2| \ll 1 \quad \text{and} \quad b_2b_4 < 0,$$

respectively. The origin has reversed stability four times and so four small-amplitude limit cycles have bifurcated.

5. A New Algorithm

The author [22] considered the generalized Liénard equation

$$\dot{x} = h(y) - F(x), \quad \dot{y} = -g(x), \tag{5.1}$$

where $h(y)$ is analytic with $h'(y) > 0$. It is not difficult to show that the above results 1. – 6. listed above also hold for the generalized system.

In [23], the author gives explicit formulae for the Lyapunov quantities of generalized quadratic Liénard equations. This work along with the results of Christopher and Lloyd [8] has led to a new algorithmic method for computing Lyapunov quantities for Liénard systems. An outline of the method is given below and is taken from [9].

Define

$$u = \sqrt{2G(x)}\operatorname{sgn}(x), \quad u(0) = 0, \quad u'(0) > 0, \tag{5.2}$$

where

$$G(x) = \int_0^x g(\xi)d\xi = \frac{x^2}{2} + O(x^3).$$

The function u is analytic and invertible. Denote its inverse by $x(u)$ and let $F^*(u) = f(x(u))$, then (5.1) becomes

$$\dot{u} = h(y) - F^*(u), \quad \dot{y} = -u, \tag{5.3}$$

after scaling time by $u/g(x(u)) = 1 + O(u)$.

It turns out that the Lyapunov quantities can be expressed very simply in terms of the coefficients of $F^*(u)$, as shown in [23] and stated in the following theorem:

Theorem 5.1. *Let $F^*(u) = \sum_1^\infty a_i u^i$, and suppose that $a_{2i+1} = 0$ for all $i < k$ and $a_{2k+1} \neq 0$, $k > 0$. Then system* (5.3) *has a fine focus of order k at the origin. Furthermore, for k greater than zero, $L(k) = C_k a_{2k+1}$, where $C_k < 0$ is some non-zero constant, depending only on k.*

A proof to this theorem can be found in [9].

The following corollary is then immediate from the above theorem:

Corollary 5.1. Let F and G be as above. If there exists a polynomial H so that

$$F(x) - H(G(x)) = c_k x^{2k+1} + O(x^{2k+2}),$$

then the system (5.1) or (4.2) has a fine focus of order k and $L(k)$ is proportional to c_k, the constant of proportionality depending only on k.

From this corollary, we can show that the calculation of Lyapunov quantities is entirely symmetric if we swap f and g, provided that $f'(0) \neq 0$.

Theorem 5.2. *Suppose $f(0) = 0$. If the origin is a fine focus and $F''(0) = f'(0)$ is non-zero then the order of the fine focus of* (5.1) *or* (4.2) *given above is the same when F is replaced by G (f replaced by g) and g replaced by $f/f'(0)$. Furthermore, the Lyapunov quantities of each system are constant multiples of each other (modulo the lower order Lyapunov quantities).*

Proof. From the hypothesis, we have $F(0) = F'(0) = 0$, $F''(0) \neq 0$ and

$$F(x) - H(G(x)) = c_k x^{2k+1} + O(x^{2k+2}),$$

for some $k > 0$ and some constant c_k; thus $H'(0) \neq 0$ too, and

$$G(x) = H^{-1}(F(x) - c_k x^{2k+1} + O(x^{2k+2}))$$

or

$$G(x) = H^{-1}F(x) - c_k x^{2k+1} H'(0)^{-1} + O(x^{2k+1}).$$

The result follows directly. The form $f/f'(0)$ is used in the statement of the theorem just to guarantee that the conditions on g are satisfied when the functions are swapped. □

We now apply these results to calculate the Lyapunov quantities of the system (4.2) or (5.1) with $\deg(g) = 2$. We shall show that the same results hold for $\deg(f) = 2$ at the end of this section.

We write F and G in the form

$$F = \sum_{i=2}^{n+1} c_i x^i, \qquad 2G = x^2 + ax^3.$$

We shall always assume that g is fully quadratic, that is $a \neq 0$. If not, we can apply the results of [2] to conclude that at most $\lfloor n/2 \rfloor$ limit cycles can bifurcate. By a simultaneous scaling of the x and y axes, we can also assume that $a = 1$. As before, such a scaling respects the weights of the Lyapunov quantities, and therefore will have no effect on the dynamics.

We now use the transformation (5.2) to obtain x as a function of u:

$$u^2 = x^2 + x^3. \tag{5.4}$$

The return map depends on the odd terms of the function $F(x(u))$; however, since $x(u)$ satisfies the algebraic identity (5.4), we can write this as

$$F(x(u)) = A(u^2) + B(u^2)x(u) + C(u^2)x(u)^2, \tag{5.5}$$

where A, B and C are polynomials of degree at most

$$\left\lfloor \frac{n+1}{3} \right\rfloor, \quad \left\lfloor \frac{n}{3} \right\rfloor, \quad \left\lfloor \frac{n-1}{3} \right\rfloor,$$

respectively, whose coefficients are linear in the c_i. There is no constant term in A, so the total number of parameters in A, B and C is equal to the total number of parameters in F. It is clear to see that we can transform between the two sets of coefficients by a linear transformation.

In order to pick out the odd degree terms of $F(x(u))$, consider the function

$$F(x(u)) - F(x(-u)) = (x(u) - x(-u))\left(B(u^2) + C(u^2)[x(u) + x(-u)]\right). \tag{5.6}$$

Let $\zeta(u)$ be the third root of the algebraic equation (5.4). Then $x(u) + x(-u) + \zeta(u) = -1$. Thus, ζ is even in u. Since $x(u) - x(-u) = 2u + O(u^2)$, the first non-zero term of (5.6) will be of order

$$\operatorname{ord}_u(B(u^2) + C(u^2)[-1 - \zeta(u)]) + 1.$$

From the results above, rewriting $v = u^2$ and $\xi(v) = 1 + \zeta(u)$, the order of fine focus at the origin will be given by

$$\operatorname{ord}_v(B(v) - C(v)\xi(v)), \qquad \xi(\xi - 1)^2 = v, \qquad \xi(0) = 0. \tag{5.7}$$

Since the coefficients of B and C are linear in the c_i, the coefficients a_{2i} above will also be linear in the c_i, and hence $L(k)$ will be a multiple of the coefficient of v^k in the polynomial

$$B(v) - C(v)\xi(v). \tag{5.8}$$

Note that the coefficients of A play no part in the calculations, and so the number of parameters remaining is

$$\left\lfloor \frac{n}{3} \right\rfloor + \left\lfloor \frac{n-1}{3} \right\rfloor + 2 = \left\lfloor \frac{2n+1}{3} \right\rfloor + 1.$$

Since the coefficients (and hence the Lyapunov quantities $L(i)$) are linear in the c_i, we need only show that the coefficients for the first $\lfloor \frac{2n+1}{3} \rfloor + 1$ terms of (5.8) are linearly independent. It is then clear that we can have at most a fine focus of order $\lfloor \frac{2n+1}{3} \rfloor$. Furthermore, we can choose the coefficient of the $\lfloor \frac{2n+1}{3} \rfloor$ term to be non-zero with the other terms zero, and bifurcate $\lfloor \frac{2n+1}{3} \rfloor$ limit cycles.

It turns out that to show that these coefficients are linearly independent is quite hard (even though an explicit matrix for this problem may be written down). We shall therefore proceed on a different path. We first show that the parameters of B and C are all effective, that is there are no non-trivial values of the parameters for which the coefficients of (5.8) all vanish, and then establish that the maximum order of vanishing of (5.8) is $\lfloor \frac{2n+1}{3} \rfloor$.

Suppose, therefore, that the expression (5.8) vanishes for some polynomials B and C. Thus $\xi(v)$ is a rational function of v, and we write $\xi(v) = r(v)/s(v)$, where r and s have no common factors. Thus (5.7) gives

$$r(v)(r(v) - s(v))^2 = s(v)^3 v.$$

Any linear factor of $s(v)$ cannot divide $r(v)$ and so cannot divide $r(v) - s(v)$. Thus, s is a constant and $\deg(r) = 1$, which is certainly not the case.

We now wish to show that there are no non-trivial expressions of the form (5.8) with order greater than $\lfloor \frac{2n+1}{3} \rfloor$. The proof of this assertion is more tricky. We use a counting argument reminiscent to that used by Petrov [29] to investigate the bifurcation of limit cycles from the Hamiltonian system

$$\dot{x} = y, \qquad \dot{y} = x - x^2.$$

This of course is of a similar form to our system. It is interesting to speculate on whether some stronger connection can be obtained between these local and global results.

Theorem 5.3. *Let $\xi(v)$ be the solution of*

$$\xi(\xi - 1)^2 = v.$$

Suppose B and C are polynomials of degree at most n and m respectively one of which is non-zero. Then the multiplicity of the zero of

$$H(v) = B(v) - C(v)\xi(v), \qquad \deg(C) \le m, \quad \deg(B) \le n$$

at $v = 0$ is at most $m + n + 1$.

Proof. By induction on $m + n$ we can assume that $\deg(C) = m$ and $\deg(B) = n$. From the argument above, H cannot vanish identically. We now consider the function ξ over the complex plane with branch points at $v = 4/27$, $\xi = 1/3$ and $v = 0$, $\xi = 1$. If we take a cut along $[4/27, \infty]$ then the branch of ξ with $\xi(0) = 0$ is well defined and single-valued over the complex plane. Clearly for a zero at $v = 0$ we need $B(0) = 0$, which we shall assume from now on.

Let Γ_r denote the closed curve $re^{i\theta} + 4/27$, $\theta \in [0, 2\pi]$. We measure the change in the argument of H as v describes the contour

$$[4/27 + \rho, 4/27 + R] + \Gamma_R - [4/27 + \rho, 4/27 + R] - \Gamma_\rho.$$

Along $(4/27, \infty)$, ξ is complex and so there will be no zeros of H, and for ρ sufficiently small and R sufficiently large, these curves will not pass through a zero of H either.

At $v = 4/27$, $\xi(v) \approx 1/3 - (4/27 - v)^{1/2}$, and so the contribution to the argument of H as we move around $-\Gamma_\rho$, will tend to a negative number or zero as $\rho \to 0$. On the other hand, about Γ_R we have $\xi \approx v^{1/3}$ and so

$$H(v) \approx \alpha v^{m+1/3} \text{ or } \beta v^n.$$

Finally, we consider the change in argument of ξ along the two sides of the cut. We only consider the upper half of the cut, as the contribution on the lower half is the same by conjugation. Note that $\operatorname{Im}\xi \neq 0$ along $[\rho, R]$, and therefore, if the argument of H is to be increased by more than $(k+1)\pi$, $C(v)$ must change sign k times. Furthermore, as v tends from 0 to ∞, the argument of ξ above the cut changes from 0 to $2\pi/3$. If the argument of H increases by more than $(k+5/3)\pi$, then H must be a real multiple of ξ at least k times. At each of these points $B(v)$ must have a root. However, one root of B is already at the origin.

Thus, the maximum change in the argument of H is given by

$$2\pi\big(\min(n+2/3, m+1) + \max(n, m+1/3)\big) = 2\pi(n+m+1),$$

as $\rho \to 0$ and $R \to \infty$. This proves the theorem. □

Theorem 5.4. *No more than* $\lfloor \frac{2n+1}{3} \rfloor$ *limit cycles can appear from the class of systems* (4.2) *with*

(i) $\deg(f) \leq n$ *and* $\deg(g) = 2$;
(ii) $\deg(g) \leq n$ *and* $\deg(f) = 2$.

Proof. We have demonstrated (i) above, so it only remains to show how (ii) follows from Theorem 5.2. If $f(0) \neq 0$ then there is a focus which is structurally stable and no limit cycles are produced. Hence we may assume that $f(0) = 0$. If $f'(0) = 0$ also then $F = ax^3$, for some constant a. Using Corollary 5.1, we find that there must be a fine focus of order one unless $a = 0$, in which case the system has a centre at the origin (a Hamiltonian system). If $f'(0) \neq 0$, then we can apply Theorem 5.2 to show that this case is entirely symmetric to the case (i). □

Using similar methods to those above and expanding on the work in [24], Christopher and the author [9] were able to prove the following result:

Theorem 5.5. *For system* (4.2) *with* $m = k, n = 3$ *or with* $m = 3, n = k$ *and* $1 < k \leq 50$, *the maximum number of bifurcating limit cycles from the origin as a fine focus is equal to the maximum order of the fine focus. In the real case this number is*

$$\left\lfloor \frac{3(k+2)}{8} \right\rfloor.$$

Table 1 shows the results for system (4.2) using the theorems appearing in this chapter, along with computed results by the author and Christopher.

6. Conclusions and Further Work

Table 1 is symmetric, it remains an open question whether the completed table will also be symmetric. The relationship between global and local results is still to be addressed as is the question of simultaneous bifurcations when there is more than one fine focus for these systems. The ultimate aim, however, is to establish a general formula for $\hat{H}(i,j)$ as a function of the degrees of f and g, for Liénard

degree of f \ degree of g	**1**	**2**	**3**	**4**	**5**	**6**	**7**	**8**	**9**	**10**	**11**	**12**	**13**	**⋯**	**48**	**49**	**50**
50	↑	↑	38														
49	24	33	38														
48	24	32	36														
⋮	⋮	⋮	⋮														
13	6	9	10														
12	6	8	10														
11	5	7	8														
10	5	7	8														
9	4	6	8	9													
8	4	5	6	9													
7	3	5	6	8													
6	3	4	6	7													
5	2	3	4	6	6												
4	2	3	4	4	6	7	8	9	9								
3	1	2	2	4	4	6	6	6	8	8	8	10	10	⋯	36	38	38
2	1	1	2	3	3	4	5	5	6	7	7	8	9	⋯	32	33	→
1	0	1	1	2	2	3	3	4	4	5	5	6	6	⋯	24	24	→

TABLE 1. The maximum number of small-amplitude limit cycles that can be bifurcated from the origin of the Liénard system (4.2) for varying degrees of f and g.

systems. It is hoped that future results might also give some indication on how to provide a solution to the second part of Hilbert's sixteenth problem. Symbolic manipulation packages will undoubtedly play a significant role in future work on these problems, but ultimately it will be the mathematicians who will prevail.

References

[1] N. Bautin, *On the number of limit cycles which appear with the variation of the coefficients from an equilibrium point of focus or center type.* Amer. Math. Soc. Trans. **5** (1962), 396–414.

[2] T. R. Blows and N. G. Lloyd, *The number of small-amplitude limit cycles of Liénard equations.* Math. Proc. Camb. Philos. Soc. **95** (1984), 359–366.

[3] T. R. Blows and L. M. Perko, *Bifurcation of limit cycles from centers and separatrix cycles of planar analytic systems.* SIAM Review **36** (1994), 341–376.

[4] T. R. Blows and C. Rousseau, *Bifurcation at infinity in polynomial vector fields.* J. Diff. Eqns. **104** (1993), 215–242.

[5] L. A. Cherkas, *Conditions for a Liénard equation to have a center.* Differentsial'nye Uravneniya **12** (1976), 201–206.

[6] C. J. Christopher and J. Devlin, *On the classification of Lienard systems with amplitude-independent periods.* J. Diff. Eqns. **200**(1) (2004), 1–17.

[7] C. J. Christopher and N. G. Lloyd, *Polynomial systems: a lower bound for the Hilbert numbers.* Proc. Roy. Soc. Lond. A **450** (1995), 219–224.

[8] C. J. Christopher and N. G. Lloyd, *Small-amplitude limit cycles of polynomial Liénard systems.* Nonlinear Diff. Eqns. Appl. **3** (1996), 183–190.

[9] C. J. Christopher and S. Lynch, *Small-amplitude limit cycle bifurcations for Liénard systems with quadratic or cubic damping or restoring forces.* Nonlinearity **12** (1999), 1099–1112.

[10] W. A. Coppel, *Some quadratic systems with at most one limit cycle.* Dynamics Reported, New York **2** (1988), 61–68.

[11] F. Dumortier and Chengzhi Li, *On the uniqueness of limit cycles surrounding one or more singularities for Liénard equations.* Nonlinearity **9** (1996), 1489–1500.

[12] F. Dumortier and Chengzhi Li, *Quadratic Liénard equations with quadratic damping.* J. Diff. Eqns. **139** (1997), 41–59.

[13] J. Ecalle, J. Martinet, J. Moussu and J. P. Ramis, *Non-accumulation des cycles-limites I.* C.R. Acad. Sci. Paris Sér. I Math. **304** (1987), 375–377.

[14] Z. S. Feng, *Exact solutions to the Liénard equation and its applications.* Chaos Solitons Fractals **21**(2) (2004), 343–348.

[15] H. Giacomini and S. Neukirch, *Improving a method for the study of limit cycles of the Liénard equation.* Phys. Rev. E **57** (1998), 6573–6576.

[16] M. A. Han, P. Bi and D. M. Xiao, *Bifurcation of limit cycles and separatrix loops in singular Lienard systems.* Chaos Solitons & Fractals **20**(3) (2004), 529–546.

[17] A. Liénard, *Étude des oscillations entrenues.* Revue Génerale de Électricité **23** (1928), 946–954.

[18] Jibin Li and Chunfu Li, *Global bifurcation of planar disturbed Hamiltonian systems and distributions of limit cycles of cubic systems.* Acta. Math. Sinica **28** (1985), 509–521.

[19] Jibin Li, *Hilbert's 16th problem and bifurcations of planar polynomial vector fields.* Int. J. Bifurcation and Chaos **13**(1) (2003), 47–106.

[20] A. Lins, W. de Melo and C. Pugh, *On Liénards equation with linear damping.* Lecture Notes in Mathematics **597** (edited by J. Palis and M. do Carno), Springer-Verlag, Berlin, 1977, 335–357.

[21] N.G. Lloyd and S. Lynch, *Small-amplitude limit cycles of certain Liénard systems.* Proc. Roy. Soc. Lond. Ser. A **418** (1988), 199–208.

[22] S. Lynch, *Limit cycles of generalized Liénard equations.* Appl. Math. Lett. **8**(6) (1995), 15–17.

[23] S. Lynch, *Generalized quadratic Liénard equations.* Appl. Math. Lett. **11**(3) (1998), 7–10.

[24] S. Lynch, *Generalized cubic Liénard equations.* Appl. Math. Lett. **12**(2) (1999), 1–6.

[25] S. Lynch and C. J. Christopher, *Limit cycles in highly nonlinear differential equations.* J. Sound and Vibration **224**(3) (1999), 505–517.

[26] S. Lynch, *Dynamical Systems with Applications Using Maple*, Birkhäuser, Basel, 2001.

[27] S. Lynch, *Dynamical Systems with Applications Using MATLAB*, Birkhäuser, Basel, 2004.

[28] H. N. Moreira, *Liénard-type equations and the epidemiology of maleria.* Ecological Modelling **60** (1992), 139–150.

[29] G. S. Petrov, *Number of zeros of complete elliptic integrals.* Funct. Anal. Appl. **18** (1984), 148–149.

[30] G. S. Rychkov, *The maximum number of limit cycles of the system* $\dot{x} = y - a_0x - a_1x^3 - a_2x^5, \dot{y} = -x$ *is two.* Differentsial'nye Uravneniya **11** (1973), 380–391.

[31] M. Sabatini, *Liénard limit cycles enclosing priod annuli, or enclosed by period annuli.* Rocky Mount. J. Math. **35**(1) (2005), 253–266.

[32] Songling Shi, *A concrete example of the existence of four limit cycles for plane quadratic systems.* Sci. Sinica A **23** (1980), 153–158.

[33] K. S. Sibirskii, *The number of limit cycles in the neighborhood of a critical point.* Differential Equations **1** (1965), 36–47.

[34] Dongming Wang, *Mechanical manipulation for a class of differential systems.* J. Symb. Comp. **12** (1991), 233–254.

[35] Dongming Wang, *Elimination Practice: Software Tools and Applications.* Imperial College Press, London, 2004.

[36] Yu Ilyashenko, *Finiteness Theorems for Limit Cycles.* Translations of Mathematical Monographs 94, American Mathematical Society, Providence, RI, 1991.

[37] Yu Ilyashenko, *Centennial history of Hilbert's 16th problem.* Bull. Amer. Math. Soc. **39**(3) (2002), 301–354.

[38] Zhifen Zhang, Tongren Ding, Wenzao Huang and Zhenxi Dong, *Qualitative Theory of Differential Equations*, Translation of Mathematical Monographs 102, American Mathematical Society, Providence, RI, 1992.

[39] H. Zoladek, *Eleven small limit cycles in a cubic vector field.* Nonlinearity **8** (1995), 843–860.

Stephen Lynch
Department of Computing and Mathematics
Manchester Metropolitan University
Chester Street
Manchester M1 5GD
UK
e-mail: `s.lynch@mmu.ac.uk`

Trends in Mathematics: *Differential Equations with Symbolic Computation*, 23–35

Estimating Limit Cycle Bifurcations from Centers

Colin Christopher

Abstract. We consider a simple computational approach to estimating the cyclicity of centers in various classes of planar polynomial systems. Among the results we establish are confirmation of Żołądek's result that at least 11 limit cycles can bifurcate from a cubic center, a quartic system with 17 limit cycles bifurcating from a non-degenerate center, and another quartic system with at least 22 limit cycles globally.

Mathematics Subject Classification (2000). 34C07.

Keywords. Limit cycle, center, multiple bifurcation.

1. Introduction

The use of multiple Hopf bifurcations of limit cycles from critical points is now a well-established technique in the analysis of planar dynamical systems. For many small classes of systems, the maximum number, or *cyclicity*, of bifurcating limit cycles is known and has been used to obtain important estimates on the general behavior of these systems. In particular, quadratic systems can have at most three such limit cycles [1]; symmetric cubic systems (those without quadratic terms) and projective quadratic systems at most five [11, 15, 8]. Results are also known explicitly for several large classes of Liénard systems [3].

The idea behind the method is to calculate the successive coefficients α_i in the return map for the vector field about a non-degenerate monodromic critical point. That is, we choose a one-sided analytic transversal at the critical point with local analytic parameter c, and represent the return map by an expansion

$$h(c) = c + \sum_{i\geq 0} \alpha_i\, c^i.$$

The cyclicity can then be found from examining these coefficients and their common zeros. The terms α_{2k} are merely analytic functions of the previous α_i, so the only interesting functions are the ones of the form α_{2i+1}. If α_{2k+1} is the first non-zero one of these, then at most k limit cycles can bifurcate from the origin, and, provided we have sufficient choice in the coefficients α_i, we can also obtain that many limit cycles in a simultaneous bifurcation from the critical point.

We call the functions α_{2i+1} the *Liapunov quantities* of the system. If all the α_{2i+1} vanish then the critical point is a *center.* It is possible to analyze this case also, but to do fully requires a more intimate knowledge, not only of the common zeros of the polynomials α_i, but also of the ideal they generate in the ring of coefficients. The papers [1, 15] cover the case of a center also. We call the set of coefficients for which all the α_i vanish the *center variety.*

In the cases we consider here, when $\alpha_0 = 0$, the remaining coefficients are polynomials in the parameters of the system. By the Hilbert Basis Theorem, the center variety is then an algebraic set.

Unfortunately, although the calculation of the Liapunov quantities is straight forward, the computational complexity of finding their common zeros grows very quickly. The result is that some very simple systems have remained intractable (to direct calculation at least) at present; for example, the system of Kukles' [9]:

$$\dot{x} = y, \qquad y = -x + a_1x^2 + a_2xy + a_3y^2 + a_4x^3 + a_5x^2y + a_6xy^2 + a_7y^3.$$

For higher degree systems it seems that a more realistic approach would be to restrict our attention to finding good lower bounds to the cyclicity by carefully selecting subclasses of the systems investigated. For example, the Kukles' system above has cyclicity 6 when $a_2 = 0$ [10], and this is expected to be the maximum number. In the same way, Lloyd and James [6] found examples of cubic systems with cyclicity 8. Recently, a cubic system with 12 limit cycles has been found by generating two symmetric nests of 6 limit cycles [13].

The disadvantage of such an approach is that there is no clear geometry of the order of cyclicity, and so we must find suitable classes of systems on a rather ad hoc basis. The higher the cyclicity desired, the more parameters we need in our model, and the less likely it is that we will be able to complete the calculations due to the inevitable expression swell.

In contrast, the classification of centers in polynomial systems is much more accessible to an a-priori geometric approach. Żołądek and Sokulski have enumerated a great number of known classes of cubic centers of the two main types conjectured to comprise all non-degenerate centers [16, 17, 12]. Furthermore, the analysis of global bifurcations of limit cycles from integrable systems has yielded nice estimates of the number of limit cycles in such systems. For example, Li and Huang's proof that a cubic system has 11 limit cycles [7], and recent estimates on the growth of the Hilbert numbers [4].

A natural approach therefore would be to use center bifurcations rather than multiple Hopf bifurcations to estimate the cyclicity of a system. Using such a technique, Żołądek has shown that there are cubic systems with 11 limit cycles bifurcating from a single critical point [17]. However the proof is quite technical and in general such methods are hard to apply to systems of higher degree.

Naïvely, we would expect the number of limit cycles to be estimated by one less than the maximum codimension of a component of the center variety. A comparison of the cyclicities of the Liénard systems computed in [4] with the codimensions of their center varieties, using the results of [2], shows that this is indeed the case for Liénard systems of low degree. Making this observation rigorous, however, would be much harder.

Our aim here is to describe a simple computational technique which will allow us to estimate the generic cyclicity of a family of centers. It can also be used to check whether we have found the whole of an irreducible component of the center variety. One nice aspect of this work is that it removes on one hand the necessity of lengthy calculations or complex independence arguments, and on the other hand gives room for a more creative approach to estimating cyclicity, using the latent geometry of the centers of the system. We give several examples to prove the effectiveness of our technique, including a quartic system with 17 limit cycles bifurcating from a center, and another quartic system with at least 22 limit cycles. We also confirm Żołądek's result that 11 limit cycles can bifurcate from a center in a cubic system.

Throughout the paper, we have tried to keep the details of the individual calculations to a minimum. This is because, once an initial system is given, the intermediate calculations themselves are entirely automatic, and do not appear to be of any independent interest. The method has been implemented in REDUCE, but it should be a straight-forward matter to be able to write similar routines to work in any of the standard Computer Algebra systems. Copies of the REDUCE programs used and a detailed summary of the calculations can be obtained from the author via e-mail.

2. The Basic Technique

The idea of the method is very simple. We choose a point on the center variety, and linearize the Liapunov quantities about this point. In the nicest cases, we would hope that the point is chosen on a component of the center variety of codimension r, then the first r linear terms of the Liapunov quantities should be independent. If this is the case, we will show below that the cyclicity is equal to $r-1$. That is, there exist perturbations which can produce $r-1$ limit cycles, and this number is the maximum possible.

Sometimes it is possible that we have found a particular class of centers, and want to check whether the set comprises the whole of a component of the center variety. Again, in nice cases, a simple computation of the linear terms of the Liapunov quantities can establish that the codimension is in fact maximal. It would be an interesting task to go through the known families of centers found by Żołądek and Sokulski [16, 17, 12] to see how many of these families of centers form complete components of the center variety for cubic systems.

Of course, we cannot know a-priori whether the method will work for a given component of the center variety. We may not have chosen a good point on the variety or, worse still, the presence of symmetries, for example, might have forced the ideal generated by the Liapunov quantities to be non-radical. Furthermore, even in the nicest cases, this method will only determine the cyclicity of a generic point on that component of the center variety.

However, although these are serious shortfalls, there are also great advantages to this method. Since we choose the starting system explicitly, the computations involved are essential linear and therefore extremely fast. It is hard to see how some of the cyclicities given here could have been obtained by more standard approaches without a lot of hard effort.

We now explain the technique in more detail for the cases we are particularly interested in. Modifications to more general situations (analytic vector fields, analytic dependence on parameters etc.) should be clear.

Consider a critical point of focal or center type in a family of polynomial systems. After an affine change of coordinates, we can assume that the members of the family are of the form

$$\dot{x} = \lambda x - y + p(x,y), \qquad \dot{y} = \lambda y + x + q(x,y). \tag{2.1}$$

Where p and q are some polynomials of fixed degree. We let Λ denote the set of parameters, $\lambda_1, \ldots, \lambda_N$ of p and q where $\lambda_1 = \lambda$. We shall assume that the coefficients of p and q are polynomials in the parameters, and we let $K \equiv R^N$ denote the corresponding parameter space. That is, we identify each point in K with its corresponding system (2.1).

We choose a transversal at the origin and calculate the return map $h(c)$ as in the introduction. Standard theory shows this to be analytic in c and Λ. The limit cycles of the system are locally given by the roots of the expression

$$P(c) = h(c) - c = \alpha_1 c + \sum_{i=2}^{\infty} \alpha_k c^k,$$

where the α_i are analytic functions of Λ.

We are interested in a fixed point of the parameter space K, which we can choose to be the origin without loss of generality (for we know that λ must be zero for any bifurcations to take place, and the other parameters can be translated to zero).

More detailed calculations show that $\alpha_1 = e^{2\pi\lambda} - 1 = 2\pi\lambda\,(1 + O(\lambda))$ and that

$$\alpha_k = \beta_k + \sum_{i=1}^{k-1} \beta_i w_{ik}, \quad (k > 1)$$

where the β_i are polynomials in the coefficients of p and q. The w_{ik} are analytic functions of Λ. We set $\beta_1 = 2\pi\lambda$. Furthermore, β_{2k} always lies in the ideal generated by the previous β_i $(1 \le i \le 2k-1)$ in the polynomial ring generated by the coefficients in Λ. This means that in most of the calculations below the β_{2i} are effectively redundant. We call the functions β_{2i+1} the *i-th Liapunov quantity* and denote it by $L(i)$.

If at the origin of K, we have $L(i) = 0$ for $i < k$ and $L(k) \neq 0$, then $P(c)$ has order $2k+1$. In this case $P(c)$ can have at most $2k+1$ zeros in a neighborhood of the origin for small perturbations. It follows that at most k limit cycles can bifurcate from this point under perturbation (each limit cycle counts for two zeros of the return map, one of each sign). The number k is called the *order* of the fine focus.

If we can choose the $L(i)$ $(1 \leq i \leq k-1)$ independently in a neighborhood of $0 \in K$, for example when the Jacobian matrix of the $L(i)$'s with respect to the parameters Λ has rank $k-1$ then we can produce $k-1$ limit cycles one by one by choosing successively

$$|L(i-1)| \ll |L(i)|, \qquad L(i-1)L(i) < 0,$$

working from $L(k-1)$ down to $L(0)$. At each stage the lower terms remain zero. With a little more analysis we can show that this bifurcation can be made simultaneously.

Suppose now that at the origin of K, we have $L(i) = 0$ for all i, then the critical point is a center. Let $\mathbb{R}[\Lambda]$ denote the coordinate ring generated by the parameters Λ, and I the ideal generated in this ring by the Liapunov quantities. By the Hilbert Basis Theorem, there is some number n for which the first n of the $L(i)$ generate I. Thus, the set of all centers is in fact an algebraic set, which we call the *center variety.*

Since all the β_{2k}'s lie in the ideal generated by the $L(i)$ with $i < k$, we can write

$$P(c) = \sum_{i=0}^{n} b_{2i+1} c^{2i+1} (1 + \Psi_{2i+1}(c, \lambda_1, \ldots, \lambda_N)), \tag{2.2}$$

where the functions Ψ_{2i+1} are analytic in their arguments and $\Psi(0,0) = 0$. A standard argument from [1] shows that at most n limit cycles can bifurcate.

To find the cyclicity of the whole of the center variety, not only is it necessary to know about the zeros of the $L(i)$, but also the ideal that they generate. It is no surprise therefore that few examples are known of center bifurcations [1, 15].

However, if we are working about one point on the center variety, we can simplify these calculations greatly. Instead of taking the polynomial ring generated by the $L(i)$, we can take the ideal generate by the $L(i)$ in $\mathbb{R}\{\{\Lambda\}\}$, the power series ring of Λ about $0 \in K$ instead. This also has a finite basis, by the equivalent Noetherian properties of power series rings.

What makes this latter approach so powerful, however, is that in many cases this ideal will be generated by just the linear terms of the $L(i)$. In which case we have the following theorem.

Theorem 2.1. *Suppose that $s \in K$ is a point on the center variety and that the first k of the $L(i)$ have independent linear parts (with respect to the expansion of $L(i)$ about s), then s lies on a component of the center variety of codimension at* *least k and there are bifurcations which produce $k-1$ limit cycles locally from the center corresponding to the parameter value s.*

If, furthermore, we know that s lies on a component of the center variety of codimension k, then s is smooth point of the variety, and the cyclicity of the center for the parameter value s is exactly $k-1$.

In the latter case, $k-1$ is also the cyclicity of a generic point on this component of the center variety.

Proof. The first statement is obvious. As above we can choose s to be the origin without loss of generality. Since the theorem is local about the origin of K, we can perform a change of coordinates so that the first k of the $L(i)$ are given by λ_i.

Now since we can choose the λ_i independently, we can take $\lambda_i = m_i \epsilon^{2(k-i)}$ for some fixed values m_i $(0 \leq i \leq k-1)$, and $m_k = 1$. The return map will therefore be an analytic function of ϵ and c. From the (2.2) above, we see that

$$P(c)/c = \sum_{i=0}^{k} m_i c^{2i} \epsilon^{2(k-i)} + \Phi(c, \epsilon).$$

Here Φ contains only terms of order greater than $2k$ in c and ϵ. For appropriate choices of the m_i, the linear factors of $\sum_{i=0}^{r} m_i c^{2i} \epsilon^{2(k-i)}$ can be chosen to be distinct and real, and none tangent to $\epsilon = 0$; whence $P(c)/c$ has an ordinary $2k$-fold point at the origin as an analytic function of c and ϵ. Now it is well known that in this case each of the linear factors $c - v_i\epsilon$ of the terms of degree $2k$ can be extended to an analytic solution branch $c = v_i\epsilon + O(\epsilon^2)$ of $P(c)/c = 0$. This gives $2k$ distinct zeros for small ϵ, and the second statement follows.

The third statement follows from noticing that the first k of the $L(i)$ must form a defining set of equations for the component of the center variety. Any $L(i)$ for $i > k$ must therefore lie in the ideal of the $L(i)$ if we work over $\mathbb{R}\{\{\Lambda\}\}$. The results follows from Bautin's argument mentioned above [1].

The last statement follows from the fact that the points where the center variety is not smooth or where the linear terms of the first k Liapunov quantities are dependent form a closed subset of the component of the center variety we are on. □

In practice, the computation of the Liapunov quantities from the return map $P(c)$ is not the most efficient way to proceed. Instead we use a method which turns out to be equivalent. Recall that we only need to calculate the Liapunov quantities $L(k)$ modulo the previous $L(i)$, $i < k$. In particular, $L(1)$ is a multiple of λ and so we can assume that $\lambda = 0$ when we calculate the $L(k)$ for $k > 0$.

We seek a function $V = x^2 + y^2 + \cdots$ such that for our vector field X,

$$X(V) = \lambda\eta_4 y^4 + \eta_6 y^6 + \cdots, \tag{2.3}$$

for some polynomials η_{2k}. The calculation is purely formal, and the choice of V can be made uniquely if we specify that $V(x,0) - x^2$ is an odd function for example. It turns out that the polynomials η_{2k} are equivalent to $L(k)/2\pi$ modulo the previous $L(i)$ with $i < k$.

This is the method we shall adopt here, though there are many other methods of calculating equivalent sets of Liapunov quantities. In particular, it is more common to replace the quantities y^{2i} in right hand side of (2.3) by $(x^2+y^2)^i$. The two give equivalent sets of Liapunov quantities, in the sense explained above, however the version in y^{2i} is slightly easier to work with computationally.

If the linear parts of the system are not quite in the form of (2.1), then rather than transform the system to (2.1), we can replace the terms x^2+y^2 in expansion of V by the equivalent positive definite quadratic form which is annihilated by the linear parts of X.

Now suppose once again that our center corresponds to $0 \in K$. We can write the general vector field in the family as $X = X_0 + X_1 + X_2 + \cdots$, where X_i contains the terms of degree i in Λ (again, we can take $\lambda = 0$ if we only want to calculate the higher Liapunov quantities). Let $\eta_{2k,i}$ denote the terms of degree i in η_{2k}, and similarly let V_i denote the terms of degree i in Λ in V, then (2.3) gives

$$X_0 V_0 = 0, \qquad X_0 V_1 + X_1 V_0 = \eta_{4,1}\, y^4 + \eta_{6,1}\, y^6 + \cdots, \tag{2.4}$$

and, more generally

$$X_0 V_i + \cdots + X_i V_0 = \eta_{4,i}\, y^4 + \eta_{4,i}\, y^6 + \cdots. \tag{2.5}$$

We can then solve the two equations of (2.4) by linear algebra to find the linear terms of the $L(k)$ (modulo the $L(i)$, $(i < k)$). The algorithm can be implemented in a straight forward manner in a computer algebra system and is extremely fast (the author used REDUCE here). Higher order terms in the expansion of the $L(i)$ (considered later) can be generated using (2.5), but the calculations are no longer linear, and soon become unmanageable.

Now we give the main result of this section.

Theorem 2.2. *There exists a class of cubic systems with 11 limit cycles bifurcating from a critical point. There exists a class of quartic systems with 15 limit cycles bifurcating from a critical point.*

Proof. We first consider the family of cubic systems C_{31} in Żołądek's most recent classification [17]. These systems have a Darboux first integral of the form

$$\phi = \frac{(xy^2+x+1)^5}{x^3(xy^5+5xy^3/2+5y^3/2+15xy/8+15y/4+a)^2}.$$

There is a critical point at

$$x = \frac{6(8a^2+25)}{(32a^2-75)}, \qquad y = \frac{70a}{(32a^2-75)}.$$

If we translate this point to the origin and put $a = 2$ we find we have the system,

$$\begin{aligned}\dot{x} &= 10(342+53x)(289x-2112y+159x^2-848xy+636y^2),\\ \dot{y} &= 605788x-988380y+432745xy-755568y^2+89888xy^2-168540y^3,\end{aligned}$$

whose linear parts give a center.

We consider the general perturbation of this system in the class of cubic vector fields. That is, we take a parameter for each quadratic and cubic term and also a parameter to represent λ above, when the system is brought to the normal form (2.1).

From the discussion above, we know that $L(1)$ is just a multiple of λ and can be effectively ignored in the rest of the calculations. Furthermore, we do not need to bring the system to (2.1) to calculate the Liapunov quantities, as we use the alternative method described above, computing V starting from the more general quadratic form, $302894x^2/2 - 988380xy + 3611520y^2/2$. As this term can also be generated automatically, we do not mention it again in the examples which follow.

Automatic computations now show that the linear parts of $L(2), \dots, L(12)$ are independent in the parameters and therefore 11 limit cycles can bifurcate from this center.

For the quartic result, we look at a system whose first integral is given by

$$\phi = \frac{(x^5 + 5x^3 + y)^6}{(x^6 + 6x^4 + 6/5xy + 3x^2 + a)^5}.$$

The form is inspired by Żołądek's system C45 in [16]. We take $a = -8$ which gives a center at $x = 2$, $y = -50$, which we move to the origin. This gives a system

$$\begin{aligned}\dot{x} &= -510x - 6y - 405x^2 - 3xy - 120x^3 - 15x^4,\\ \dot{y} &= 49950x + 510y + 22500x^2 - 1335xy - 15y^2\\ &\quad + 2850x^3 - 630x^2y - 300x^4 - 105x^3y.\end{aligned}$$

This time we take a general quartic bifurcation and find that, assuming $L(1) = 0$ as above, we have $L(2)$ to $L(16)$ linearly independent in the parameters. Hence this center can produce 15 limit cycles by bifurcation. □

Remark 2.3. We note that an immediate corollary of the work is that there are components of the center variety of the class of all cubic systems which have codimension 12.

The result for cubic systems was first shown by Żołądek. However, the system he considers is different from ours. This is because, as noted in his paper, it is not possible to generate 11 limit cycles from his system by considering the linear terms only. The nice thing about the result here is that it depends on only the simplest arguments and a direct calculation.

We will improve the quartic bound in the next section.

3. Higher Order Perturbations

Of course, it will often happen that the linear terms of the Liapunov quantities are not independent. Several reasons for this are discussed in Żołądek [18].

Loosely speaking, we may be at an intersection point of two components of the center variety. Alternatively, the existence of a symmetry can sometimes imply that the parameters only appear to certain powers in the expansion of the

Liapunov quantities. Finally, it can also happen that the parameter space can be embedded in a larger parameter space where it is tangent to the center variety.

This last possibility occurs in the paper of Żołądek [18], and he must consider second order terms.

It is still possible in this case to obtain cyclicities by considering the higher order terms of the Liapunov quantities. These can be calculated as in (2.5). However, the procedure becomes much slower as the degree of the terms increases.

In general, as soon as higher order terms are taken into account, the situation becomes much more complex. However, we shall give one result here where we can say something concrete under some generic assumptions.

We apply this result to the quartic system considered in the previous section and show that in fact 17 limit cycles can bifurcate from this center when we consider the quadratic terms. We also show that the strata of symmetric centers C_{46} can generate 11 limit cycles under cubic perturbations.

This latter result dates back to an earlier attempt by Żołądek [14] to find 11 limit cycles, but has not been established until now.

Suppose that $L(1),\dots,L(r)$ have independent linear parts. Since we are interested only in the cyclicity in a neighborhood $0 \in K$, we can perform an analytic change of coordinates in parameter space and assume $L(i) = \lambda_i$ for $i = 2,\dots,r$ (recall $L(1) = 2\pi\lambda$ already).

Now, suppose we have expanded the Liapunov quantities $L(r+1),\dots,L(k)$ in terms of the parameters $\lambda_{r+1},\dots,\lambda_k$, and that the order of the first non-zero terms of each of these Liapunov quantities is the same, m say. In this case, we can write the Liapunov quantities as $L(i) = h_i(\lambda_{r+1},\dots,\lambda_k) + \cdots$ where h_i is a homogeneous polynomial of degree m. Here, we have reduced the $L(i)$, $(i > r)$, modulo the $L(i)$, $(i = 1,\dots r)$, so that they have no dependence on $\lambda_1,\dots,\lambda_r$.

Theorem 3.1. *Suppose the h_i are given as above, and consider the equations $h_i = 0$ as defining hypersurfaces in $S = \mathbb{R}^{k-r} \setminus \{0\}$. If there exists a line ℓ in S such that $h_i(\ell) = 0$ and the hypersurfaces $h_i = 0$ intersect transversally along ℓ for $i = r+1,\dots,k-1$, and such that $h_k(\ell) \neq 0$, then there are perturbations of the center which can produce $k-1$ limit cycles.*

Proof. In this case, there exists an analytic curve C in a neighborhood of $0 \in R$ given by $L(i) = h_i + \cdots = 0$, $i = r+1,\dots,k-1$, which is tangent to ℓ at $0 \in R$. We now move the parameters $\lambda_{r+1},\dots,\lambda_k$ along C, keeping $\lambda_1 = \cdots = \lambda_r = 0$. For a sufficiently small perturbation along C we shall have $L(1) = \cdots = L(k-1) = 0$ and $L(k) \neq 0$. Thus we have a weak focus of order $k-1$. Furthermore, the rank of the the other $L(i)$ will be equal to $k-1$, by hypothesis. Thus we can move away from this curve in a direction which produces $k-1$ limit cycles. □

Theorem 3.2. *There is a class of quartic system with 17 limit cycles bifurcating from a critical point.*

Proof. We calculate the linear and quadratic terms of the first 18 Liapunov quantities with respect to a general perturbation of a quartic system which has no quadratic or cubic terms in the perturbation of $\dot{y}$, so that we have 18 parameters (including λ). The first 16 Liapunov quantities are independent and we perform a change of variable to express $L(17)$ and $L(18)$ as quadratic terms in the remaining two parameters. In these new coordinates, $L(17)$ and $L(18)$ have only one linear factor in common and hence the remaining linear factor in $L(17)$ gives the line ℓ in the theorem above, and we obtain 17 limit cycles. □

We now consider the symmetric cubic systems

$$\dot{x} = -8ax + 4y - 4x^3 + xy^2, \qquad \dot{y} = -16x + 8ay + 32ax^3 - 28x^2y + 3y^3,$$

which are based on the first integral

$$\phi = \frac{(x^4 + xy + 1)^3}{(x^6 + 3x^3y/2 + 3x^2/2 + 3y^2/8 + a)^2}.$$

When $a = 1/2$ the origin is a center and the linear parts of $L(1)$ to $L(6)$ are independent. From the symmetry of the system, the six parameters which correspond to the quadratic perturbation terms always appear in monomials of even degree. The remaining Liapunov quantities $L(7)$ to $L(12)$ turn out to have leading terms which are quadratic in these six parameters.

From the above theorem, it is only necessary to verify that the first five of these quadratic terms have a common zero at which the sixth doesn't vanish, and for which the Jacobian of the first five terms with respect to five of the remaining parameters is non-zero. The computations are very hard in this case, and recourse was had to the specialist commutative algebra package SINGULAR [5] for some of the Gröbner basis calculations described below.

The original conjecture of Żołądek was that C46 centers formed their own component in the center variety of all cubic systems. We were not able to verify this here, as there exist non trivial solutions of the first five quadratic terms which are shared by the sixth. However, it is testimony to the original intuition of Żołądek (an investigation done without the aid of any computer algebra) that the Liapunov quantities are sufficiently independent that we can indeed obtain the number of limit cycles hoped for.

Theorem 3.3. *The symmetric cubic centers C46 can generate 11 limit cycles.*

Proof. Ideally, to obtain the common solutions for the intersections of the first five quadratic terms, we would like to compute their Gröbner basis with respect to a lexicographic ordering. Unfortunately, the computations are sufficiently difficult that such an approach does not work directly. However, we can compute a Gröbner basis with a total degree ordering, which is much more efficient computationally, but less useful for obtaining explicit equations for the zeros of the polynomials.

If the ideal was zero dimensional, then we could use the FGLM algorithm to return efficiently to a lexicographic basis, but in our case, it turns out that there is a non-trivial one dimensional solution (also shared by the sixth quadratic term). However, our Gröbner basis calculation shows that this solution lies on a quadratic hypersurface $h = 0$.

[6] E. M. James and N. G. Lloyd, A cubic system with eight small-amplitude limit cycles. *IMA J. Appl. Math.* **47** no. 2 (1991), 163–171.

[7] Ji Bin Li and Qi Ming Huang, Bifurcations of limit cycles forming compound eyes in the cubic system. *Chinese Ann. Math.* Ser. B **8** no. 4 (1987), 391–403.

[8] N. G. Lloyd, C. Christopher, J. Devlin, J. M. Pearson, and N. Yasmin, Quadratic-like cubic systems. Planar nonlinear dynamical systems (Delft, 1995). *Differential Equations Dynam. Systems* **5** no. 3-4 (1997), 329–345.

[9] N. G. Lloyd and J. M. Pearson, Computing centre conditions for certain cubic systems. *J. Comput. Appl. Math.* **40** no. 3 (1992), 323–336.

[10] N. G. Lloyd and J. M. Pearson, Conditions for a centre and the bifurcation of limit cycles in a class of cubic systems. Bifurcations of planar vector fields (Luminy, 1989), Lecture Notes in Math. **1455**, Springer, Berlin, 1990, 230–242.

[11] K. S. Sibirskii, On the number of limit cycles arising from a singular point of focus or center type. *Dokl. Akad. Nauk SSSR* **161** (1965), 304–307 (Russian). *Soviet Math. Dokl.* **6** (1965), 428–431.

[12] J. Sokulski, The beginning of classification of Darboux integrals for cubic systems with center. *Differential Integral Equations*, to appear.

[13] P. Yu and H. Maoan, Twelve limit cycles in a cubic order planar system with $\mathbb{Z}_2$-symmetry. *Communications on Pure and Applied Analysis* **3** no. 3 (2004), 515–526.

[14] H. Żołądek, The solution of the center-focus problem. Preprint, University of Warsaw, 1992.

[15] H. Żołądek, On a certain generalization of Bautin's theorem. *Nonlinearity* **7** no. 1 (1994), 273–279.

[16] H. Żołądek, The classification of reversible cubic systems with center. *Topol. Methods Nonlinear Anal.* **4** no. 1 (1994), 79–136.

[17] H. Żołądek, Remarks on: "The classification of reversible cubic systems with center" *Topol. Methods Nonlinear Anal.* **8** no. 2 (1996), 335–342.

[18] H. Żołądek, Eleven small limit cycles in a cubic vector field. *Nonlinearity* **8** no. 5 (1995), 843–860.

Colin Christopher
School of Mathematics and Statistics
University of Plymouth
Plymouth PL4 8AA
United Kingdom
e-mail: c.christopher@plymouth.ac.uk

Trends in Mathematics: *Differential Equations with Symbolic Computation*, 37–53

Conditions of Infinity to be an Isochronous Center for a Class of Differential Systems

Wentao Huang and Yirong Liu

Abstract. In this article, the definition of isochronous center at infinity is given and the center conditions and the isochronous center conditions at infinity for a class of differential systems are investigated. By a transformation, infinity is taken to the origin and therefore properties at infinity can be studied with the methods developed for finite critical points. Using the computations of singular point values and period constants at the origin, the problem of conditions for infinity to be a center or to be an isochronous center has been solved for complex vector fields in this case.

Mathematics Subject Classification (2000). 34C05.

Keywords. Infinity, center, isochronous center, singular point value, period constant.

1. Introduction

In the qualitative theory of polynomial differential systems, the problem of determining center conditions and isochronous center conditions are two interesting and difficult topics. In the case of a finite critical point a lot of work has been done. The case of a center is considered in [3, 4, 12, 13]. Several classes of isochronous systems have also been studied: quadratic isochronous centers [14]; isochronous centers of a linear center perturbed by third, fourth and fifth degree homogeneous polynomials [16, 17, 18]; complex polynomial systems [20]; reversible systems [19, 21]; isochronous centers of cubic systems with degenerate infinity [22, 23]. For the case

---

The authors thank the referees for many useful suggestions that helped improve the paper. This research is supported by the Nature Science Foundation of China (60464001) and the Nature Science Foundation of Guangxi Province.

of infinity, research has concentrated on the following $2n+1$ degree system

$$\frac{dx}{dt} = \sum_{k=0}^{2n} X_k(x,y) - y(x^2+y^2)^n,$$
$$\frac{dy}{dt} = \sum_{k=0}^{2n} Y_k(x,y) + x(x^2+y^2)^n,$$

where $X_k(x,y), Y_k(x,y)$ are homogeneous polynomials of degree k of x and y. For this system, the equator Γ_∞ on the Poincaré closed sphere is a trajectory of the system, having no real critical point. Γ_∞ is called *infinity* in this point of view. As far as center conditions at infinity are concerned, several special systems have been studied: cubic systems in [6, 9]; fifth degree systems in [10]. But for the problem concerning the conditions of infinity to be a isochronous center, there are very few results. In [11] the conditions of infinity to be an isochronous center for a real rational system are discussed. In this paper, we study the center conditions and the isochronous center conditions at infinity for a class of differential system of the form

$$\begin{aligned}\frac{dx}{dt} &= \frac{1}{(x^2+y^2)^2}(-\lambda y + A_{30}x^3 + A_{21}x^2y + A_{12}xy^2 + A_{03}y^3) - y,\\ \frac{dy}{dt} &= \frac{1}{(x^2+y^2)^2}(\lambda x + B_{30}x^3 + B_{21}x^2y + B_{12}xy^2 + B_{03}y^3) + x.\end{aligned} \tag{1.1}$$

By a time transformation, system (1.1) can be transferred into the following polynomial differential system

$$\begin{aligned}\frac{dx}{dt} &= -\lambda y + A_{30}x^3 + A_{21}x^2y + A_{12}xy^2 + A_{03}y^3 - y(x^2+y^2)^2,\\ \frac{dy}{dt} &= \lambda x + B_{30}x^3 + B_{21}x^2y + B_{12}xy^2 + B_{03}y^3 + x(x^2+y^2)^2.\end{aligned} \tag{1.2}$$

Systems (1.1) and (1.2) have the same of phase portraits in the phase plane, so their center conditions at infinity are identical.

This paper is organized as follows. In Section 2 we define isochronous center of infinity and restate some known results which are necessary for investigating center and isochronous center conditions. In Section 3 we give the singular point values and center conditions. In Section 4, we compute period constants and discuss isochronous center conditions of the system.

2. The Isochronous Center at Infinity and Some Preliminary Results

Definition 2.1. For a real planar differential system, infinity Γ_∞ is said to be an isochronous center if the trajectories of the system in the neighborhoods of infinity (the equator) are all closed and their periods are constant.

The following theorem is obvious.

Theorem 2.2. *For two differential systems (A) and (B), if there exists a homeomorphism taking infinity of system (A) to the origin of system (B), then the infinity of system (A) is a center (an isochronous center) if and only if the origin of system (B) is a center (an isochronous center).*

By means of transformation

$$x = \frac{\xi}{(\xi^2+\eta^2)^3}, \quad y = \frac{\eta}{(\xi^2+\eta^2)^3},$$

system(1.1) becomes the following polynomial system

$$\begin{aligned}
\frac{d\xi}{dt} &= -\eta + (\xi^2+\eta^2)^3\,((\tfrac{-\xi^2}{5}+\eta^2)\,(A_{30}\,\xi^3 + A_{21}\,\xi^2\,\eta + A_{12}\,\xi\,\eta^2 + A_{03}\,\eta^3) \\
&\quad - \frac{6}{5}\,\xi\,\eta\,(B_{30}\,\xi^3 + B_{21}\,\xi^2\,\eta + B_{12}\,\xi\,\eta^2 + B_{03}\,\eta^3)) - \eta\,(\xi^2+\eta^2)^{10}\,\lambda, \\
\frac{d\eta}{dt} &= \xi + (\xi^2+\eta^2)^3\,((\xi^2 - \frac{\eta^2}{5})\,(B_{30}\,\xi^3 + B_{21}\,\xi^2\,\eta + B_{12}\,\xi\,\eta^2 + B_{03}\,\eta^3) \\
&\quad + \frac{-6}{5}\,\xi\,\eta\,(A_{30}\,\xi^3 + A_{21}\,\xi^2\,\eta + A_{12}\,\xi\,\eta^2 + A_{03}\,\eta^3)) + \xi\,(\xi^2+\eta^2)^{10}\,\lambda.
\end{aligned} \tag{2.1}$$

From Theorem 2.2, the center (isochronous center) conditions at infinity of system (1.1) are the same as the center (isochronous center) conditions at the origin of system (2.1).

By means of transformation

$$z = \xi + \eta i, \quad w = \xi - \eta i, \quad T = it, \quad i = \sqrt{-1}, \tag{2.2}$$

system (2.1) can be transformed into the following complex system

$$\begin{aligned}
\frac{dz}{dT} &= z + \frac{2\,a_{03}\,w^7\,z^4}{5} + (\frac{2\,a_{12}}{5} + \frac{3\,b_{30}}{5})\,w^6\,z^5 + (\frac{2\,a_{21}}{5} + \frac{3\,b_{21}}{5})\,w^5\,z^6 \\
&\quad + (\frac{2\,a_{30}}{5} + \frac{3\,b_{12}}{5})\,w^4\,z^7 + \frac{3\,b_{03}\,w^3\,z^8}{5} + w^{10}\,z^{11}\,\lambda, \\
\frac{dw}{dT} &= -[w + + \frac{2\,b_{03}\,w^4\,z^7}{5} + (\frac{3\,a_{30}}{5} + \frac{2\,b_{12}}{5})\,w^5\,z^6 + (\frac{3\,a_{21}}{5} + \frac{2\,b_{21}}{5})\,w^6\,z^5 \\
&\quad + (\frac{3\,a_{12}}{5} + \frac{2\,b_{30}}{5})\,w^7\,z^4 + \frac{3\,a_{03}\,w^8\,z^3}{5} + w^{11}\,z^{10}\,\lambda],
\end{aligned} \tag{2.3}$$

where

$$\begin{aligned}
a_{30} &= \frac{1}{8}(A_{03} + iA_{12} - A_{21} - iA_{30} + iB_{03} - B_{12} - iB_{21} + B_{30}),\\
b_{30} &= \frac{1}{8}(A_{03} - iA_{12} - A_{21} + iA_{30} - iB_{03} - B_{12} + iB_{21} + B_{30}),\\
a_{03} &= \frac{1}{8}(-A_{03} + iA_{12} + A_{21} - iA_{30} - iB_{03} - B_{12} + iB_{21} + B_{30}),\\
b_{03} &= \frac{1}{8}(-A_{03} - iA_{12} + A_{21} + iA_{30} + iB_{03} - B_{12} - iB_{21} + B_{30}),\\
a_{21} &= \frac{1}{8}(-3A_{03} - iA_{12} - A_{21} - 3iA_{30} - 3iB_{03} + B_{12} - iB_{21} + 3B_{30}),\\
b_{21} &= \frac{1}{8}(-3A_{03} + iA_{12} - A_{21} + 3iA_{30} + 3iB_{03} + B_{12} + iB_{21} + 3B_{30}),\\
a_{12} &= \frac{1}{8}(3A_{03} - iA_{12} + A_{21} - 3iA_{30} + 3iB_{03} + B_{12} + iB_{21} + 3B_{30}),\\
b_{12} &= \frac{1}{8}(3A_{03} + iA_{12} + A_{21} + 3iA_{30} - 3iB_{03} + B_{12} - iB_{21} + 3B_{30}),
\end{aligned} \tag{2.4}$$

and z, w, T are complex variables. We say that systems (2.1) and (2.3) are *concomitant*. Evidently, the coefficients of system (2.1) are real if and only if the coefficients of system (2.3) satisfy the conjugate condition, i.e.,

$$\overline{a_{\alpha\beta}} = b_{\alpha\beta}, \quad \alpha \geq 0, \quad \beta \geq 0, \quad \alpha + \beta = 3. \tag{2.5}$$

In [1, 7, 8], the authors defined the complex center and complex isochronous center for the following complex systems

$$\begin{aligned}
\frac{dz}{dT} &= z + \sum_{k=2}^{\infty} Z_k(z, w) = Z(z, w),\\
\frac{dw}{dT} &= -w - \sum_{k=2}^{\infty} W_k(z, w) = -W(z, w),
\end{aligned} \tag{2.6}$$

with

$$Z_k(z, w) = \sum_{\alpha+\beta=k} a_{\alpha\beta} z^{\alpha} w^{\beta}, \quad W_k(z, w) = \sum_{\alpha+\beta=k} b_{\alpha\beta} w^{\alpha} z^{\beta},$$

and gave a recursive algorithm to determine the necessary conditions for a center. We now restate the definitions and the algorithm.

By means of the transformation

$$z = re^{i\theta}, \quad w = re^{-i\theta}, \quad T = it, \tag{2.7}$$

where r, θ are complex numbers, system (2.6) can be transformed into

$$\begin{aligned} \frac{dr}{dt} &= i\frac{wZ - zW}{2r} = ir\sum_{k=1}^{\infty}\frac{wZ_{k+1} - zW_{k+1}}{2zw} \\ &= \frac{ir}{2}\sum_{m=1}^{\infty}\sum_{\alpha+\beta=m+2}(a_{\alpha,\beta-1} - b_{\beta,\alpha-1})e^{i(\alpha-\beta)\theta}r^m, \\ \frac{d\theta}{dt} &= \frac{wZ + zW}{2zw} = 1 + \sum_{k=1}^{\infty}\frac{wZ_{k+1} + zW_{k+1}}{2zw} \\ &= 1 + \frac{1}{2}\sum_{m=1}^{\infty}\sum_{\alpha+\beta=m+2}(a_{\alpha,\beta-1} + b_{\beta,\alpha-1})e^{i(\alpha-\beta)\theta}r^m. \end{aligned} \tag{2.8}$$

For the complex constant h, $|h| \ll 1$, we write the solution of (2.8) satisfying the initial condition $r|_{\theta=0} = h$ as

$$r = \tilde{r}(\theta, h) = h + \sum_{k=2}^{\infty} v_k(\theta)h^k$$

and denote

$$\begin{aligned} \tau(\varphi, h) &= \int_0^{\varphi}\frac{dt}{d\theta}d\theta \\ &= \int_0^{\varphi}[1 + \frac{1}{2}\sum_{m=1}^{\infty}\sum_{\alpha+\beta=m+2}(a_{\alpha,\beta-1} + b_{\beta,\alpha-1})e^{i(\alpha-\beta)\theta}\tilde{r}(\theta, h)^m]^{-1}d\theta. \end{aligned}$$

Definition 2.3 ([1]). For a sufficiently small complex constant h, the origin of system (2.6) is called a *complex center* if $\tilde{r}(2\pi, h) \equiv h$. The origin is a *complex isochronous center* if

$$\tilde{r}(2\pi, h) \equiv h, \quad \tau(2\pi, h) \equiv 2\pi.$$

Lemma 2.4 ([5]). *For system (2.6), we can derive uniquely the following formal series*

$$\xi = z + \sum_{k+j=2}^{\infty} c_{kj}z^k w^j, \quad \eta = w + \sum_{k+j=2}^{\infty} d_{k,j}w^k z^j, \tag{2.9}$$

where $c_{k+1,k} = d_{k+1,k} = 0$, $k = 1, 2, \ldots$, *such that*

$$\frac{d\xi}{dT} = \xi + \sum_{j=1}^{\infty} p_j\, \xi^{j+1}\eta^j, \quad \frac{d\eta}{dT} = -\eta - \sum_{j=1}^{\infty} q_j\, \eta^{j+1}\xi^j. \tag{2.10}$$

Definition 2.5 ([7, 1]). Let $\mu_0 = 0, \mu_k = p_k - q_k$, $\tau(k) = p_k + q_k$, $k = 1, 2, \ldots$. μ_k is called the kth *singular point value* of the origin of system (2.6) and $\tau(k)$ is called the kth *period constant* of the origin of system (2.6).

Theorem A ([1]). For system (2.6), the origin is a complex center if and only if $\mu_k = 0$, $k = 1, 2, \ldots$. The origin is a complex isochronous center if and only if $\mu_k = \tau_k = 0$, $k = 1, 2, \ldots$.

Theorem B ([8]). For system (2.6), we can derive successively the terms of the following formal series

$$M(z,w) = \sum_{\alpha+\beta=0}^{\infty} c_{\alpha\beta} z^{\alpha} w^{\beta} \tag{2.11}$$

such that

$$\frac{\partial(MZ)}{\partial z} - \frac{\partial(MW)}{\partial w} = \sum_{m=1}^{\infty} (m+1)\mu_m (zw)^k,$$

where $c_{00} = 1, \forall c_{kk} \in R, k = 1, 2, \ldots$, and for any integer m, μ_m is determined by the following recursion formulas:

$$\begin{aligned}
&c_{0,0} = 1,\\
&\text{if } (\alpha = \beta > 0) \text{ or } \alpha < 0, \text{ or } \beta < 0, \text{ then } c_{\alpha,\beta} = 0\\
&\text{else}\\
&c_{\alpha,\beta} = \frac{1}{\beta - \alpha} \sum_{k+j=3}^{\alpha+\beta+2} [(\alpha+1)a_{k,j-1} - (\beta+1)b_{j,k-1}] c_{\alpha-k+1,\beta-j+1};
\end{aligned} \tag{2.12}$$

$$\mu_m = \sum_{k+j=3}^{2m+2} (a_{k,j-1} - b_{j,k-1}) c_{m-k+1,m-j+1}. \tag{2.13}$$

Theorem C ([1]). For system (2.6), we can derive uniquely the following formal series

$$f(z,w) = z + \sum_{k+j=2}^{\infty} c'_{kj} z^k w^j, \quad g(z,w) = w + \sum_{k+j=2}^{\infty} d'_{k,j} w^k z^j,$$

where $c'_{k+1,k} = d'_{k+1,k} = 0,\ k = 1, 2, \ldots$, is such that

$$\frac{df}{dT} = f(z,w) + \sum_{j=1}^{\infty} p'_j z^{j+1} w^j, \quad \frac{dg}{dT} = -g(z,w) - \sum_{j=1}^{\infty} q'_j w^{j+1} z^j, \tag{2.14}$$

and when $k - j - 1 \neq 0$, c'_{kj} and d'_{kj} are determined by the following recursive formulas:

$$\begin{aligned}
c'_{kj} &= \frac{1}{j+1-k} \sum_{\alpha+\beta=3}^{k+j+1} [(k-\alpha+1)a_{\alpha,\beta-1} - (j-\beta+1)b_{\beta,\alpha-1}] c'_{k-\alpha+1,j-\beta+1},\\
d'_{kj} &= \frac{1}{j+1-k} \sum_{\alpha+\beta=3}^{k+j+1} [(k-\alpha+1)b_{\alpha,\beta-1} - (j-\beta+1)a_{\beta,\alpha-1}] d'_{k-\alpha+1,j-\beta+1},
\end{aligned} \tag{2.15}$$

and for any positive integer j, p'_j and q'_j are determined by the following recursive formulas:

$$\begin{aligned} p'_j &= \sum_{\alpha+\beta=3}^{2j+2} [(j-\alpha+2)a_{\alpha,\beta-1} - (j-\beta+1)b_{\beta,\alpha-1}]c'_{j-\alpha+2,j-\beta+1}, \\ q'_j &= \sum_{\alpha+\beta=3}^{2j+2} [(j-\alpha+2)b_{\alpha,\beta-1} - (j-\beta+1)a_{\beta,\alpha-1}]d'_{j-\alpha+2,j-\beta+1}. \end{aligned} \tag{2.16}$$

In the expressions (2.15) and (2.16), we take $c'_{1,0} = d'_{1,0} = 1, c'_{0,1} = d'_{0,1} = 0$, and if $\alpha < 0$ or $\beta < 0$, let $a_{\alpha\beta} = b_{\alpha\beta} = c'_{\alpha\beta} = d'_{\alpha\beta} = 0$.

The relations between p_j, q_j and p'_j, q'_j $(j = 1, 2, \ldots)$ are as follows.

Theorem D. Let $p_0 = q_0 = p'_0 = q'_0 = 0$. If there exists a positive integer m, such that

$$p_0 = q_0 = p_1 = q_1 = \cdots = p_{m-1} = q_{m-1} = 0, \tag{2.17}$$

then

$$p'_0 = q'_0 = p'_1 = q'_1 = \cdots = p'_{m-1} = q'_{m-1} = 0,\ p_m = p'_m,\ q_m = q'_m, \tag{2.18}$$

and the corresponding statement with p and q swapped with p' and q' also holds.

Theorem B gives a recursive algorithm to compute singular point values μ_k and Theorems C and D give a method to compute the period constants τ_m. However, we cannot use Theorems C and D to calculate μ_k (see [1, Remark 3.1]). The computation of μ_k and τ_m gives necessary conditions for the origin to be a center or an isochronous center respectively, but to prove the sufficiency of the conditions other methods are needed. In the following we give a sufficient condition for the origin to be an isochronous center. The authors of [7] gave the definition of Lie invariants of system (2.6) and a proof of the constructive theorem of singular point values (see [7, Theorem 2.5] or [2, Theorem 4.15]). Similarly, in analogy to the constructive theorem of singular point values, we have

Theorem 2.6. *For system* (2.6), *the kth period constant at the origin τ_k is a Lie invariant of order k .*

From Theorem 2.6, we can introduce a sufficient condition for the origin to be a complex isochronous center.

Theorem 2.7. *If all the elementary Lie invariants of system* (2.6) *are zero, then the origin of system* (2.6) *is a complex isochronous center.*

Evidently, the isochronous center in the real number field is a special case of the complex isochronous center, i.e., the coefficients of the complex system satisfy an additional conjugate condition. So, in the following we only discuss the conditions for a complex center or complex isochronous center of the origin for system (2.3). In order to get more extended results, we assume that the coefficients of the system (2.3) $a_{\alpha\beta}, b_{\alpha\beta}$ are independent.

3. Singular Point Values and Center Conditions

We now discuss the computation of the singular point values and center conditions at the origin for system (2.3). Applying the recursive formula in Theorem B, we compute the singular point values of the origin of system (2.3) and simplify them (the detailed recursive formulas are given in Appendix A); then we have

Theorem 3.1. *The first 45 singular point values at the origin of system* (2.3) *are as follows:*

$$\mu_5 = \frac{1}{5}(-a_{21} + b_{21})$$

$$\mu_{10} = \frac{1}{5}(a_{12}a_{30} - b_{12}b_{30})$$

$$\mu_{15} = \frac{1}{40}(-9a_{03}a_{30}^2 + 9b_{03}b_{30}^2 + a_{03}b_{12}^2 - b_{03}a_{12}^2)$$

$$\mu_{20} = \frac{1}{20}(a_{21} + b_{21})(3a_{03}a_{30}^2 - 3b_{03}b_{30}^2 - a_{03}a_{30}b_{12} + b_{03}b_{30}a_{12})$$

$$\mu_{25} = -\frac{1}{120}(3a_{03}a_{30}^2 - 3b_{03}b_{30}^2 - a_{03}a_{30}b_{12} + b_{03}b_{30}a_{12})\cdot (3a_{03}b_{03} - 16a_{30}b_{30} - 24\lambda)$$

$$\mu_{30} = 0$$

$$\mu_{35} = -\frac{1}{9600}(3a_{03}a_{30}^2 - 3b_{03}b_{30}^2 - a_{03}a_{30}b_{12} + b_{03}b_{30}a_{12})\cdot (5a_{03}^2b_{03}^2 - 992a_{03}b_{03}a_{30}b_{30} + 6400a_{30}^2b_{30}^2)$$

$$\mu_{40} = \frac{7}{1440}(3a_{03}a_{30}^2 - 3b_{03}b_{30}^2 - a_{03}a_{30}b_{12} + b_{03}b_{30}a_{12})(a_{03}a_{30}^2 + b_{03}b_{30}^2)\cdot (5a_{03}b_{03} - 32a_{30}b_{30})$$

$$\mu_{45} = \frac{11}{31500}a_{30}^2b_{30}^2(3a_{03}a_{30}^2 - 3b_{03}b_{30}^2 - a_{03}a_{30}b_{12} + b_{03}b_{30}a_{12})\cdot (24909a_{03}b_{03} - 164000a_{30}b_{30})$$

and $\mu_k = 0,\ k \neq 5i, i < 9, i \in \mathbf{N}$. *In the above expression of* μ_k, *we have already let* $\mu_1 = \mu_2 = \cdots = \mu_{k-1} = 0,\ k = 2, 3, \ldots, 45$.

From Theorem 3.1, we get

Theorem 3.2. *For system* (2.3), *the first 45 singular point values are zero if and only if one of the following five conditions holds:*

$$\text{(i)}\quad a_{21} = b_{21},\ a_{12}a_{30} = b_{12}b_{30},\ a_{03}a_{30}^2 = b_{03}b_{30}^2,\ a_{30} \neq 0 \text{ or } b_{30} \neq 0; \tag{3.1}$$

$$\text{(ii)}\quad a_{21} = b_{21},\ a_{12} = 3b_{30},\ b_{12} = 3a_{30}, a_{03}a_{30}^2 \neq b_{03}b_{30}^2,\ a_{30} \neq 0 \text{ or } b_{30} \neq 0; \tag{3.2}$$

$$\text{(iii)}\quad \lambda = a_{21} = b_{21} = a_{30} = a_{03} = b_{12} = 0,\ a_{12} = -3b_{30}, b_{30}b_{03} \neq 0; \tag{3.3}$$

$$\text{(iv)}\quad \lambda = a_{21} = b_{21} = b_{30} = b_{03} = a_{12} = 0,\ b_{12} = -3a_{30}, a_{30}a_{03} \neq 0; \tag{3.4}$$

$$\text{(v)}\quad a_{21} = b_{21},\ a_{30} = b_{30} = 0,\ a_{03}b_{12}^2 = b_{03}a_{12}^2. \tag{3.5}$$

Proof. Let us prove the necessity. If $a_{30} = b_{30} = 0$, then from $\mu_{15} = 0$ we have $a_{03}b_{12}^2 = b_{03}a_{12}^2$, so condition (v) holds. If $a_{30} \neq 0$ or $b_{30} \neq 0$, then by $\mu_{10} = (a_{12}a_{30} - b_{12}b_{30})/5$ there is a constant p, such that

$$a_{12} = p\, b_{30}, \quad b_{12} = p\, a_{30}. \tag{3.6}$$

By substituting (3.6) into every expression of μ_k and simplifying them it is easy to complete the proof. □

In order to make use of the constructive theorem of singular point values ([7, Theorem 2.5] or [2, Theorem 4.15]) to get the center conditions of the system, we need find out all the elementary Lie invariants of the system. From the technique used in [7], we have the following lemma.

Lemma 3.3. *All the elementary Lie invariants of system* (2.3) *are as follows:*

$$\begin{aligned}
&a_{21}, b_{21}, \lambda \ (i.e., a_{10}, b_{10})\\
&a_{30}b_{30}, a_{12}b_{12}, a_{03}b_{03},\\
&a_{30}a_{12}, b_{30}b_{12},\\
&a_{30}^2a_{03}, b_{30}^2b_{03}, a_{30}b_{12}a_{03}, b_{30}a_{12}b_{03}, b_{12}^2a_{03}, a_{12}^2b_{03}.
\end{aligned}$$

In Theorem 3.2:

(1) If condition (i) holds, then from $a_{12}a_{30} = b_{12}b_{30}$, $a_{30} \neq 0$ or $b_{30} \neq 0$ there is a constant p, such that $a_{12} = p\, b_{30}, b_{12} = p\, a_{30}$. Hence, from $a_{03}a_{30}^2 = b_{03}b_{30}^2$ we see that $b_{30}a_{12}b_{03} = b_{30}a_{12}b_{03}$, $a_{12}^2b_{03} = b_{12}^2a_{03}$. Let $g = f(a_{\alpha\beta}, b_{\alpha\beta})$, $g* = f(b_{\alpha\beta}, a_{\alpha\beta})$. Under the condition (i), if g is any Lie invariant of system (2.3), then $g = g*$. According to the constructive theorem of singular point values (see [7]), we get that all $\mu_k = 0, k = 1, 2, \ldots$. So, the origin of the system is a center of the system (2.3). Similarly, if condition (v) holds, then the origin of the system is also a center of system (2.3).

(2) If condition (ii) holds, then an analytic first integral for the system is

$$\begin{aligned}
F(z,w) = (zw)^5(4 &+ 12\lambda w^{10}z^{10} + 6b_{21}w^5z^5 + 12b_{30}w^6z^4 + 12a_{30}w^4z^6\\
&+ 3b_{03}w^3z^7 + 3a_{03}w^7z^3)^{-\frac{1}{3}}.
\end{aligned}$$

(3) If condition (iii) holds, then the system has an integrating factor

$$J(z,w) = (zw)^{-16}(-1 + 3b_{30}w^6z^4).$$

(4) If condition (iv) holds, then the system has an integrating factor

$$J(z,w) = (zw)^{-16}(-1 + 3a_{30}w^4z^6).$$

Thus we have the following theorem.

Theorem 3.4. *For system* (2.3), *all the singular point values of the origin are zero if and only if the first 45 singular point values of the origin are zero, i.e., one of the five conditions in Theorem* 3.2 *holds. Thus, the five conditions in Theorem* 3.2 *are the center conditions of the origin.*

4. Period Constants and Isochronous Center Conditions

We now discuss the isochronous center conditions.

For system (2.3), according to Theorems C and D, we get recursive formulas to compute period constants (see Appendix B). From the center conditions, we investigate the following four cases.

Case 1: Expression (3.1) holds.

Since $a_{30} \neq 0$ or $b_{30} \neq 0$, from expression (3.1) we can take

$$a_{12} = A\, b_{30},\ b_{12} = A\, a_{30},\ a_{03} = B\, b_{30}^2,\ b_{03} = B\, a_{30}^2,\ a_{21} = b_{21} = r_{21}, \tag{4.1}$$

where A and B are any complex constants. Putting expression (4.1) into the recursive formulas in Appendix B, we see that the first 30 period constants of the origin of system (2.3) are as follows, after careful computations:

$$\begin{aligned}
\tau_5 &= 2r_{21},\\
\tau_{10} &= \frac{1}{2}(-4a_{30}b_{30} - 4Aa_{30}b_{30} - a_{30}^2 b_{30}^2 B^2 + 4\lambda),\\
\tau_{15} &= -\frac{1}{4}a_{30}^2 b_{30}^2 B(A-3)(A+3),\\
\tau_{20} &= -\frac{1}{96}{a_{30}}^2\, {b_{30}}^2\,(-384 - 192\,A + 192\,A^2 - 288\,a_{30}\,B^2\,b_{30}\\
&\qquad - 32\,A\,a_{30}\,B^2\,b_{30} + 3\,{a_{30}}^2\,B^4\,{b_{30}}^2),\\
\tau_{25} &= -\frac{1}{24}a_{30}^3 b_{30}^3 B(A+11)(-32+3B^2 a_{30}b_{30}),\\
\tau_{30} &= \frac{1}{600}(-3+A){a_{30}}^3{b_{30}}^3(-10056 - 12720A + 1032A^2 + 3696A^3\\
&\qquad + 957B^2 a_{30}b_{30} - 1301AB^2 a_{30}b_{30}),
\end{aligned} \tag{4.2}$$

and $\tau_k = 0,\ k \neq 5i,\ i < 6,\ i \in \mathbf{N}$.

In the above expression of τ_k, we have already let $\tau_1 = \cdots = \tau_{k-1} = 0$, $k = 2, 3, \ldots, 30$.

From (4.1) and (4.2), we have

Theorem 4.1. *Under the condition* (3.1)*, the first 30 period constants of the origin of system* (2.3) *are zero if and only if one of the following expressions holds:*

$$\lambda = a_{21} = b_{21} = a_{30} = b_{12} = b_{03} = 0,\ b_{30} \neq 0; \tag{4.3}$$

$$\lambda = a_{21} = b_{21} = b_{30} = a_{12} = a_{03} = 0,\ a_{30} \neq 0; \tag{4.4}$$

$$\lambda = a_{21} = b_{21} = a_{03} = b_{03} = 0,\ a_{12} = -b_{30},\ b_{12} = -a_{30},\ a_{30} \neq 0 \text{ or } b_{30} \neq 0. \tag{4.5}$$

Proof. According to $\tau_{15} = -\frac{1}{4}a_{30}^2 b_{30}^2 B(A-3)(A+3) = 0$, we get four cases with $A = 3, A = -3, B = 0$ and $a_{30}b_{30} = 0$. By taking the four cases separately and considering expression (4.1) it is easy to complete the proof of the theorem. □

If (4.3) or (4.4) holds, then all the elementary Lie invariants of system (2.3) are zero. According to Theorem 2.7, the origin is an isochronous center.

If expression (4.5) holds, then system (2.3) becomes

$$\begin{aligned} \frac{dz}{dT} &= z - \frac{1}{5}a_{30}z^7w^4 + \frac{1}{5}b_{30}z^5w^6, \\ \frac{dw}{dT} &= -(w - \frac{1}{5}b_{30}w^7z^4 + \frac{1}{5}a_{30}w^5z^6). \end{aligned} \tag{4.6}$$

From formula (2.8), we get

$$\frac{d\theta}{dt} = \frac{1}{2zw}(w(z - \frac{1}{5}a_{30}z^7w^4 + \frac{1}{5}b_{30}z^5w^6) + z(w - \frac{1}{5}b_{30}w^7z^4 + \frac{1}{5}a_{30}w^5z^6)) = 1.$$

System (4.6) therefore has an isochronous center at the origin.

Thus, we have the following result.

Theorem 4.2. *Under the center conditions* (3.1), *the origin of system* (2.3) *is an isochronous center if and only if one of the three conditions in Theorem* 4.1 *holds.*

Case 2: Expression (3.2) holds.

Putting expression (3.2) into the recursive formulas in Appendix B, we find that the first forty period constants are as follows:

$$\begin{aligned}
\tau_5 &= 2b_{21}, \\
\tau_{10} &= -\frac{1}{2}(a_{03}b_{03} + 16a_{30}b_{30} - 4\lambda), \\
\tau_{15} &= 0, \\
\tau_{20} &= \frac{1}{32}(-{a_{03}}^2\,{b_{03}}^2 + 128\,a_{03}\,a_{30}\,b_{03}\,b_{30} - 256\,{a_{30}}^2\,{b_{30}}^2), \\
\tau_{25} &= -\frac{7}{24}\,(3\,a_{03}\,b_{03} - 32\,a_{30}\,b_{30})\,(a_{03}\,{a_{30}}^2 + b_{03}\,{b_{30}}^2), \\
\tau_{30} &= 0, \\
\tau_{35} &= \frac{56933}{270}\,{a_{30}}^2\,{b_{30}}^2\,(a_{03}\,{a_{30}}^2 + b_{03}\,{b_{30}}^2), \\
\tau_{40} &= -\frac{376751200}{46080000}\,({a_{03}}^3\,{a_{30}}^4\,b_{03} - 1811085\,{a_{03}}^4\,{b_{03}}^4 + 3630412800\,{a_{03}}^2\,{a_{30}}^5\,b_{30} \\
&\quad + 198975544\,{a_{03}}^3\,a_{30}\,{b_{03}}^3\,b_{30} + 4255128088\,{a_{03}}^2\,{a_{30}}^2\,{b_{03}}^2\,{b_{30}}^2 \\
&\quad + 42511012864\,a_{03}\,{a_{30}}^3\,b_{03}\,{b_{30}}^3 - 101293762560\,{a_{30}}^4\,{b_{30}}^4 \\
&\quad - 376751200\,a_{03}\,{b_{03}}^3\,{b_{30}}^4 + 3630412800\,a_{30}\,{b_{03}}^2\,{b_{30}}^5), \\
\tau_k &= 0, \quad k \neq 5i,\ i < 8,\ i \in \mathbf{N}.
\end{aligned} \tag{4.7}$$

From $\tau_{35} = 0$, we get $a_{30}b_{30} = 0$ or $a_{03}\,{a_{30}}^2 + b_{03}\,{b_{30}}^2 = 0$. If $a_{30}b_{30} = 0$, then from $\tau_k = 0$, $k <= 40$, we get $a_{03}b_{03} = \lambda = a_{21} = b_{21} = 0$. If $a_{30}b_{30} \neq 0$, then let $a_{03} = hb_{30}^2$, $b_{03} = -ha_{30}^3$ with h being an arbitrary complex constant and it is easy to see that $\tau_{20} = 0$ is in contradiction with $\tau_{40} = 0$. Considering condition (3.2), we have

Theorem 4.3. *Under the condition* (3.2), *the first 40 period constants of the origin of system* (2.3) *are zero if and only if one of the following expressions holds:*

$$\lambda = a_{21} = b_{21} = a_{30} = a_{03} = b_{12} = 0,\ a_{12} = 3b_{30},\ b_{30} \neq 0; \tag{4.8}$$

$$\lambda = a_{21} = b_{21} = b_{30} = b_{03} = a_{12} = 0,\ b_{12} = 3a_{30},\ a_{30} \neq 0. \tag{4.9}$$

If expression (4.8) holds, then system (2.3) becomes

$$\begin{aligned} \frac{dz}{dT} &= z + \frac{9}{5} b_{30}\, w^6\, z^5 + \frac{3}{5} b_{03}\, w^3\, z^8, \\ \frac{dw}{dT} &= -(w + \frac{11}{5} b_{30}\, w^7\, z^4 + \frac{2}{5} b_{03}\, w^4\, z^7). \end{aligned} \tag{4.10}$$

By the transformation

$$u = \frac{\sqrt{1 + 3\, b_{30}\, w^6\, z^4}}{w^{\frac{5}{2}}\, z^{\frac{3}{2}}}, \qquad v = \frac{w^{\frac{7}{2}}\, z^{\frac{5}{2}}}{\sqrt{1 + 3\, b_{30}\, w^6\, z^4}},$$

system (4.10) becomes

$$\begin{aligned} \frac{du}{dT} &= \frac{u\,(10\,u - 32\,b_{30}\,v + b_{03}\,u^{15}\,v^{13}\,(u - 3\,b_{30}\,v)^3)}{10\,(u - 3\,b_{30}\,v)} = U(u,v), \\ \frac{dv}{dT} &= \frac{v\,(-10\,u + 28\,b_{30}\,v + b_{03}\,u^{15}\,v^{13}\,(u - 3\,b_{30}\,v)^3)}{10\,(u - 3\,b_{30}\,v)} = -V(u,v). \end{aligned} \tag{4.11}$$

From formula (2.8) we get

$$\frac{d\theta}{dt} = \frac{1}{2uv}(vU + uV) = 1, \tag{4.12}$$

so the system has an isochronous center at the origin.

By the symmetry of (4.8) and (4.9), it is easy to see that the system also has an isochronous center at the origin under the condition (4.9). Thus we have the following theorem.

Theorem 4.4. *Under the center condition* (3.2), *the origin of system* (2.3) *is an isochronous center if and only if one of the two conditions in Theorem* 4.3 *holds.*

Case 3: Expression (3.3) or (3.4) holds.

Putting expression (3.3) into the recursive formulas in Appendix B, we get that the first 45 period constants are all zero. It implies that the expression (3.3) is a necessary condition for the origin to be an isochronous center of system (2.3). Under this condition, system (2.3) becomes

$$\begin{aligned} \frac{dz}{dT} &= z - \frac{3}{5} b_{30}\, w^6\, z^5 + \frac{3}{5} b_{03}\, w^3\, z^8, \\ \frac{dw}{dT} &= -(w - \frac{7}{5} b_{30}\, w^7\, z^4 + \frac{2}{5} b_{03}\, w^4\, z^7). \end{aligned} \tag{4.13}$$

By the transformation

$$u = \frac{\sqrt{-1 + 3\, b_{30}\, w^6\, z^4}}{w^{\frac{5}{2}}\, z^{\frac{3}{2}}}, \qquad v = \frac{w^{\frac{7}{2}}\, z^{\frac{5}{2}}}{\sqrt{-1 + 3\, b_{30}\, w^6\, z^4}},$$

system (4.13) becomes

$$\begin{aligned}\frac{du}{dT} &= \frac{u}{10}\,(10 + b_{03}\,u^{15}\,v^{13}\,(u - 3\,b_{30}\,v)^2 + \frac{4\,b_{30}\,v}{-u + 3\,b_{30}\,v}) = U,\\ \frac{dv}{dT} &= \frac{v}{10}\,(-10 + b_{03}\,u^{15}\,v^{13}\,(u - 3\,b_{30}\,v)^2 + \frac{4\,b_{30}\,v}{-u + 3\,b_{30}\,v}) = -V.\end{aligned} \tag{4.14}$$

Because

$$\frac{d\theta}{dt} = \frac{1}{2uv}(vU + uV) = 1,$$

the system has an isochronous center at the origin. From the symmetry of (3.3) and (3.4), we have the following theorem.

Theorem 4.5. *For the origin of system* (2.3), *the center conditions* (3.3) *and* (3.4) *are also isochronous center conditions.*

Case 4: Expression (3.5) holds.

According to (3.5) and the recursive formulas in Appendix B, we have, after careful computations,

$$\begin{aligned}\tau_5 &= 2b_{21},\\ \tau_{10} &= \frac{-(a_{03}\,b_{03}) + 4\,\lambda}{2},\\ \tau_{15} &= \frac{-({a_{12}}^2\,b_{03}) - a_{03}\,{b_{12}}^2}{8},\\ \tau_{20} &= \frac{-({a_{03}}^2\,{b_{03}}^2)}{32}.\end{aligned} \tag{4.15}$$

From $\tau_5 = \tau_{10} = \tau_{15} = \tau_{20} = 0$ and (3.5), it is easy to get the following result.

Theorem 4.6. *Under the center condition* (3.5), *the necessary conditions for the origin of system* (2.3) *to be an isochronous center is that one of the following three expressions holds:*

$$\lambda = a_{21} = b_{21} = a_{30} = b_{30} = a_{03} = a_{12} = 0,\ b_{03} \neq 0; \tag{4.16}$$

$$\lambda = a_{21} = b_{21} = a_{30} = b_{30} = b_{03} = b_{12} = 0,\ a_{03} \neq 0; \tag{4.17}$$

$$\lambda = a_{21} = b_{21} = a_{30} = b_{30} = a_{03} = b_{03} = 0. \tag{4.18}$$

If (4.16) or (4.17) holds, then all the elementary Lie invariants of system (2.3) are zero. According to Theorem 2.7, the origin is an isochronous center. If expression (4.18) holds, then system (2.3) becomes

$$\begin{aligned}\frac{dz}{dT} &= z + \frac{2\,a_{12}\,w^6\,z^5}{5} + \frac{3\,b_{12}\,w^4\,z^7}{5},\\ \frac{dw}{dT} &= -(w + \frac{3\,a_{12}\,w^7\,z^4}{5} + \frac{2\,b_{12}\,w^5\,z^6}{5}).\end{aligned} \tag{4.19}$$

By the transformation

$$u = \frac{\sqrt{1 + a_{12}\, w^6\, z^4}}{w^{\frac{5}{2}}\, z^{\frac{3}{2}}}, \qquad v = \frac{w^{\frac{7}{2}}\, z^{\frac{5}{2}}}{\sqrt{1 + a_{12}\, w^6\, z^4}},$$

system (4.19) becomes

$$\begin{aligned}\frac{du}{dT} &= \frac{u\,(10\,u - 11\,a_{12}\,v + b_{12}\,u^{12}\,v^9 - 2\,a_{12}\,b_{12}\,u^{11}\,v^{10} + {a_{12}}^2\,b_{12}\,u^{10}\,v^{11})}{10\,(u - a_{12}\,v)} = U,\\ \frac{dv}{dT} &= \frac{v\,(-10\,u + 9\,a_{12}\,v + b_{12}\,u^{12}\,v^9 - 2\,a_{12}\,b_{12}\,u^{11}\,v^{10} + {a_{12}}^2\,b_{12}\,u^{10}\,v^{11})}{10\,(u - a_{12}\,v)} = -V.\end{aligned} \tag{4.20}$$

Since

$$\frac{d\theta}{dt} = \frac{1}{2uv}(vU + uV) = 1,$$

we see that system has an isochronous center at the origin. From the discussion above we have the following theorem.

Theorem 4.7. *Under the center condition* (3.5), *the origin of system* (2.3) *is an isochronous center if and only if one of the three conditions in Theorem* 4.6 *holds.*

Because all of the above isochronous center conditions have included the corresponding center conditions, we can summarize the results in the following theorem.

Theorem 4.8. *System* (2.3) *has an isochronous center at the origin if and only if one of the conditions* (3.3), (3.4), (4.3), (4.4), (4.5), (4.8), (4.9), (4.16), (4.17), *and* (4.18) *holds.*

If system (1.1) is a real system, then the coefficients of system (2.3) satisfy the conjugate condition. Hence we have the following corollary.

Corollary 4.9. *A real system* (1.1) *or* (1.2) *has a center at the origin if and only if one of the conditions* (3.1), (3.2), *and* (3.5) *holds. Real system* (1.1) *has an isochronous center at infinity if and only if one of the conditions* (4.5) *and* (4.18) *holds.*

Appendix A

The recursive formulas to compute singular values of the origin of system (2.3):

$c(0,0) = 1$;

if $(u = v > 0)$, or $u < 0$, or $v < 0$ then $c(u,v) = 0$ else

$c(u,v) = \frac{1}{-u+v}((1+u)\,\lambda - (1+v)\,\lambda)\,c(-10+u,-10+v) + (\frac{3\,b_{03}\,(1+u)}{5} - \frac{2\,b_{03}\,(1+v)}{5})\,c(-7+u,-3+v) + (\frac{(2\,a_{30}+3\,b_{12})\,(1+u)}{5} - \frac{(3\,a_{30}+2\,b_{12})\,(1+v)}{5})\,c(-6+u,-4+v) + (\frac{(2\,a_{21}+3\,b_{21})\,(1+u)}{5} - \frac{(3\,a_{21}+2\,b_{21})\,(1+v)}{5})\,c(-5+u,-5+v) + (\frac{(2\,a_{12}+3\,b_{30})\,(1+u)}{5} - \frac{(3\,a_{12}+2\,b_{30})\,(1+v)}{5})\,c(-4+u,-6+v) + (\frac{2\,a_{03}\,(1+u)}{5} - \frac{3\,a_{03}\,(1+v)}{5})\,c(-3+u,-7+v))$.

$\mu_m = \frac{b_{03}\,c(-7+m,-3+m)}{5} + (\frac{-3\,a_{30}-2\,b_{12}}{5} + \frac{2\,a_{30}+3\,b_{12}}{5})\,c(-6+m,-4+m) + (\frac{-3\,a_{21}-2\,b_{21}}{5} + \frac{2\,a_{21}+3\,b_{21}}{5})\,c(-5+m,-5+m) + (\frac{-3\,a_{12}-2\,b_{30}}{5} + \frac{2\,a_{12}+3\,b_{30}}{5})\,c(-4+m,-6+m) - \frac{a_{03}\,c(-3+m,-7+m)}{5}$.

Appendix B

The recursive formulas to compute the period constants of the origin of system (2.3):

$c'(1,0) = d'(1,0) = 1; c'(0,1) = d'(0,1) = 0;$

if $k < 0$, or $j < 0$, or $(j > 0$ and $k = j+1)$ then $c'(k,j) = 0$, $d'(k,j) = 0$ else

$c'(k,j) = -\frac{1}{5\,(1+j-k)}(5\,j\,\lambda\,c'(-10+k,-10+j) - 5\,k\,\lambda\,c'(-10+k,-10+j) + 15\,b_{03}\,c'(-7+k,-3+j) + 2\,b_{03}\,j\,c'(-7+k,-3+j) - 3\,b_{03}\,k\,c'(-7+k,-3+j) + 10\,b_{12}\,c'(-6+k,-4+j) + 3\,a_{30}\,j\,c'(-6+k,-4+j) + 2\,b_{12}\,j\,c'(-6+k,-4+j) - 2\,a_{30}\,k\,c'(-6+k,-4+j) - 3\,b_{12}\,k\,c'(-6+k,-4+j) - 5\,a_{21}\,c'(-5+k,-5+j) + 5\,b_{21}\,c'(-5+k,-5+j) + 3\,a_{21}\,j\,c'(-5+k,-5+j) + 2\,b_{21}\,j\,c'(-5+k,-5+j) - 2\,a_{21}\,k\,c'(-5+k,-5+j) - 3\,b_{21}\,k\,c'(-5+k,-5+j) - 10\,a_{12}\,c'(-4+k,-6+j) + 3\,a_{12}\,j\,c'(-4+k,-6+j) + 2\,b_{30}\,j\,c'(-4+k,-6+j) - 2\,a_{12}\,k\,c'(-4+k,-6+j) - 3\,b_{30}\,k\,c'(-4+k,-6+j) - 15\,a_{03}\,c'(-3+k,-7+j) + 3\,a_{03}\,j\,c'(-3+k,-7+j) - 2\,a_{03}\,k\,c'(-3+k,-7+j));$

$d'(k,j) = \frac{1}{5\,(1+j-k)}(5\,j\,\lambda\,d'(-10+k,-10+j) - 5\,k\,\lambda\,d'(-10+k,-10+j) + 15\,a_{03}\,d'(-7+k,-3+j) + 2\,a_{03}\,j\,d'(-7+k,-3+j) - 3\,a_{03}\,k\,d'(-7+k,-3+j) + 10\,a_{12}\,d'(-6+k,-4+j) + 2\,a_{12}\,j\,d'(-6+k,-4+j) + 3\,b_{30}\,j\,d'(-6+k,-4+j) - 3\,a_{12}\,k\,d'(-6+k,-4+j) - 2\,b_{30}\,k\,d'(-6+k,-4+j) + 5\,a_{21}\,d'(-5+k,-5+j) - 5\,b_{21}\,d'(-5+k,-5+j) + 2\,a_{21}\,j\,d'(-5+k,-5+j) + 3\,b_{21}\,j\,d'(-5+k,-5+j) - 3\,a_{21}\,k\,d'(-5+k,-5+j) - 2\,b_{21}\,k\,d'(-5+k,-5+j) - 10\,b_{12}\,d'(-4+k,-6+j) + 2\,a_{30}\,j\,d'(-4+k,-6+j) + 3\,b_{12}\,j\,d'(-4+k,-6+j) - 3\,a_{30}\,k\,d'(-4+k,-6+j) - 2\,b_{12}\,k\,d'(-4+k,-6+j) - 15\,b_{03}\,d'(-3+k,-7+j) + 3\,b_{03}\,j\,d'(-3+k,-7+j) - 2\,b_{03}\,k\,d'(-3+k,-7+j)).$

$p'(j) = \frac{1}{5}(5\,\lambda\,c'(-9+j,-10+j) - 12\,b_{03}\,c'(-6+j,-3+j) + b_{03}\,j\,c'(-6+j,-3+j) + 2\,a_{30}\,c'(-5+j,-4+j) - 7\,b_{12}\,c'(-5+j,-4+j) - a_{30}\,j\,c'(-5+j,-4+j) + b_{12}\,j\,c'(-5+j,-4+j) + 7\,a_{21}\,c'(-4+j,-5+j) - 2\,b_{21}\,c'(-4+j,-5+j) - a_{21}\,j\,c'(-4+j,-5+j) + b_{21}\,j\,c'(-4+j,-5+j) + 12\,a_{12}\,c'(-3+j,-6+j) + 3\,b_{30}\,c'(-3+j,-6+j) - a_{12}\,j\,c'(-3+j,-6+j) + b_{30}\,j\,c'(-3+j,-6+j) + 17\,a_{03}\,c'(-2+j,-7+j) - a_{03}\,j\,c'(-2+j,-7+j));$

$q'(j) = \frac{1}{5}(5\,\lambda\,d'(-9+j,-10+j) - 12\,a_{03}\,d'(-6+j,-3+j) + a_{03}\,j\,d'(-6+j,-3+j) - 7\,a_{12}\,d'(-5+j,-4+j) + 2\,b_{30}\,d'(-5+j,-4+j) + a_{12}\,j\,d'(-5+j,-4+j) - b_{30}\,j\,d'(-5+j,-4+j) - 2\,a_{21}\,d'(-4+j,-5+j) + 7\,b_{21}\,d'(-4+j,-5+j) + a_{21}\,j\,d'(-4+j,-5+j) - b_{21}\,j\,d'(-4+j,-5+j) + 3\,a_{30}\,d'(-3+j,-6+j) + 12\,b_{12}\,d'(-3+j,-6+j) + a_{30}\,j\,d'(-3+j,-6+j) - b_{12}\,j\,d'(-3+j,-6+j) + 17\,b_{03}\,d'(-2+j,-7+j) - b_{03}\,j\,d'(-2+j,-7+j));$

$\tau_m = p(m) + q(m) = p'(m) + q'(m).$

References

[1] Y. Liu, W. Huang, A new method to determine isochronous center conditions for polynomial differential systems, *Bull. Sci. Math.* 127 (2003), 133–148.

[2] J. Li, Hilbert's 16th problem and bifurcations of planar polynomial vector fields, *Int. J. Bifurcation and Chaos* 13 (2003), 47–106.

[3] J. Chavarriga, Integrable systems in the plane with a center type linear part, *Applicationes Mathematicae* 22 (1994), 285–309.

[4] Y. Ye, Qualitative Theory of Polynomial Differential Systems, Shanghai Sci. Tech. Publ., Shanghai, 1995 (in Chinese).

[5] B. B. Amelbkin, H. A. Lukasevnky, A. N. Catovcki, Nonlinear vibration, *B Γ Y Lenin Publ.*, 1982, 19–21 (in Russian).

[6] T. R. Blows, C. Rousseau, Bifurcation at infinity in polynomial vector fields, *J. Diff. Eqns.* 104 (1993), 215–242.

[7] Y. Liu, J. Li, Theory of values of singular point in complex autonomous differential system, *Sci. China Ser. A* 33 (1990), 10–24.

[8] Y. Liu, H. Chen, Formulas of singular point values and the first 10 saddle values for a class of cubic system, *Acta. Math. Appl. Sin.* 25 (2002), 295–302 (in Chinese).

[9] Y. Liu, Theory of center-focus for a class of higher-degree critical points and infinite points, *Sci. China Ser. A* 44 (2001), 37–48.

[10] Y. Liu, M. Zhao, Stability and bifurcation of limit cycles of the equator in a class of fifth polynomial systems, *Chinese J. Contemp. Math.*, 23(1), Allerton Press, Inc. New York.

[11] Y. Liu, W. Huang, Center and isochronous center at infinity for differential systems, *Bull. Sci. Math.* 128 (2004), 77–89.

[12] N. Salih, R. Pons, Center conditions for a lopsided quartic polynomial vector field, *Bull. Sci. Math.* 126 (2002), 369–378.

[13] J. Giné, Conditions for the existence of a center for the Kukles homogeneous systems, *Comput. Math. Appl.* 43 (2002), 1261–1269.

[14] W. S. Loud, Behavior of the period of solutions of certain plane autonomous systems near centers, *Contributions to Differential Equations* 3 (1964), 21–36.

[15] C. J. Christopher, J. Devlin, Isochronous centers in planar polynomial systems, *SIAM J. Math. Anal.* 28 (1997), 162–177.

[16] I. Pleshkan, A new method of investigating the isochronicity of a system of two differential equations, *Differential Equations* 5 (1969), 796–802.

[17] J. Chavarriga, J. Giné, I. García, Isochronous centers of a linear center perturbed by fourth degree homogrneous polynomial, *Bull. Sci. Math.* 123 (1999),77–96.

[18] J. Chavarriga, J. Giné, I. García, Isochronous centers of a linear center perturbed by fifth degree homogrneous polynomial, *J. Comput. Appl. Math.* 126 (2000), 351–368.

[19] J. Chavarriga, J. Giné, I. García, Isochronicity into a family of time-reversible cubic vector fields, *Appl. Math. Comput.* 121 (2001), 129–145.

[20] Y. Lin, J. Li, Normal form and critical points of the period of closed orbits for planar autonomous systems, *Acta Math. Sin.* 34 (1991), 490–501 (in Chinese).

[21] L. Cairó, J. Chavrriga, J. Giné, J. Llibre, A class of reversible cubic systems with an isochronous center, *Comput. Math. Appl.* 38 (1999), 39–53.

[22] N.G. Lloyd, J. Christopher, J. Devlin, J.M. Pearson, N. Yasmin, Quadratic like cubic systems, *Differential Equations Dynamical Systems* 5(3-4) (1997), 329–345.

[23] J. Chavarriga, J. Giné, I. García, Isochronous centers of cubic systems with degenerate infinity, *Differential Equations Dynamical Systems* 7(2) (1999), 221–238.

Wentao Huang
Department of Computing Science and Mathematics
Guilin University of Electronic Technology
Guilin 541004, P.R. China
e-mail: huangwt@gliet.edu.cn

Yirong Liu
Department of Mathematics
Zhejiang Normal University
Jinhua 321004, P.R. China
e-mail: liuyirong@163.com

Trends in Mathematics: *Differential Equations with Symbolic Computation*, 55–65

Darboux Integrability and Limit Cycles for a Class of Polynomial Differential Systems

Jaume Giné and Jaume Llibre

Abstract. We consider the class of polynomial differential equations $\dot{x} = P_n(x,y)+P_{n+m}(x,y)+P_{n+2m}(x,y)+P_{n+3m}(x,y)$, $\dot{y} = Q_n(x,y)+Q_{n+m}(x,y)+Q_{n+2m}(x,y)+Q_{n+3m}(x,y)$, for $n, m \geq 1$ and where P_i and Q_i are homogeneous polynomials of degree i. Inside this class we identify new subclasses of Darboux integrable systems. Moreover, under additional conditions such Darboux integrable systems can have at most three limit cycles. We provide the explicit expression of these limit cycles.

Mathematics Subject Classification (2000). Primary 34C05; Secondary 34C07.

Keywords. Integrability, algebraic limit cycle.

1. Introduction and Statement of the Results

In 1900 Hilbert [14] in the second part of his 16th problem proposed to find an estimation of the uniform upper bound for the number of limit cycles of all polynomial vector fields of a given degree, and also to study their distribution or configuration in the plane $\mathbb{R}^2$. This has been one of the main problems in the qualitative theory of planar differential equations in the 20th century. The contributions of Bamón [2] for the particular case of quadratic vector fields, and mainly of Écalle [6] and Ilyashenko [15] proving that any polynomial vector field has finitely many limit cycles have been the best results in this area. But until now the existence of a uniform upper bound is not proved. This problem remains open even for quadratic polynomial vector fields.

Another main problem in the qualitative theory of planar differential systems is the determination of first integrals. This paper deals with both problems for the

The first author is partially supported by a DGICYT grant number BFM 2002-04236-C02-01 and by a DURSI grant of Government of Catalonia "Distinció de la Generalitat de Catalunya per a la promoció de la recerca universitària". The second author is partially supported by a DGICYT grant number BFM 2002-04236-C02-02 and by a CICYT grant number 2001SGR00173.

following class of real polynomial differential systems

$$\begin{aligned} \dot{x} &= P_n(x,y) + P_{n+m}(x,y) + P_{n+2m}(x,y) + P_{n+3m}(x,y), \\ \dot{y} &= Q_n(x,y) + Q_{n+m}(x,y) + Q_{n+2m}(x,y) + Q_{n+3m}(x,y), \end{aligned} \tag{1.1}$$

where P_i and Q_i are homogeneous polynomials of degree i.

A *limit cycle* of system (1.1) is a periodic orbit isolated in the set of all periodic orbits of system (1.1). Let $W \subset \mathbb{R}^2$ be the domain of definition of a C^1 vector field (P,Q), and let U be an open subset of W. A function $V : U \to \mathbb{R}$ satisfying the linear partial differential equation

$$P\frac{\partial V}{\partial x} + Q\frac{\partial V}{\partial y} = \left(\frac{\partial P}{\partial x} + \frac{\partial Q}{\partial y}\right)V \tag{1.2}$$

is called an *inverse integrating factor* of the vector field (P,Q) on U. We note that $\{V=0\}$ is formed by orbits of the vector field (P,Q). This function V is very important because $R = 1/V$ defines on $U \setminus \{V=0\}$ an integrating factor of system (1) (which allows to compute a first integral of the system on $U \setminus \{V=0\}$), and $\{V=0\}$ contains the limit cycles of system (1) which are in U, see [8]. This fact allows to study the limit cycles which bifurcate from periodic orbits of a center (Hamiltonian or not) and compute their shape, see [9] and [10]. For doing that, the function V in power series of the small perturbation parameter is developed. A remarkable fact is that the first term of this expansion coincides with the first non-identically zero Melnikov function, see [20].

A function of the form $f_1^{\lambda_1} \cdots f_p^{\lambda_p} \exp(h/g)$, where f_i, g and h are polynomials of $\mathbb{C}[x,y]$ and the λ_i's are complex numbers, is called a *Darboux function.* System (1.1) is called *Darboux integrable* if it has a first integral or an integrating factor which is a Darboux function (for a definition of a first integral and of an integrating factor, see for instance [3, 5]). The problem of determining when a polynomial differential system (1.1) has a Darboux first integral is, in general, open.

We shall study a subclass of real polynomial differential systems (1.1) which will be Darboux integrable, and such that under an additional assumption over their inverse integrating factors they can have at most 3 limit cycles. Moreover, we shall prove that this upper bound is reached.

In order to present our results we need some preliminary notation and results. Thus, in polar coordinates (r,θ), defined by

$$x = r\cos\theta, \qquad y = r\sin\theta, \tag{1.3}$$

system (1.1) becomes

$$\begin{aligned} \dot{r} &= f_{n+1}(\theta)r^n + f_{n+m+1}(\theta)r^{n+m} \\ &\quad + f_{n+2m+1}(\theta)r^{n+2m} + f_{n+3m+1}(\theta)r^{n+3m}, \\ \dot{\theta} &= g_{n+1}(\theta)r^{n-1} + g_{n+m+1}(\theta)r^{n+m-1} \\ &\quad + g_{n+2m+1}(\theta)r^{n+2m-1} + g_{n+3m+1}(\theta)r^{n+3m-1}, \end{aligned} \tag{1.4}$$

where

$$f_i(\theta) = \cos\theta P_{i-1}(\cos\theta, \sin\theta) + \sin\theta Q_{i-1}(\cos\theta, \sin\theta),$$
$$g_i(\theta) = \cos\theta Q_{i-1}(\cos\theta, \sin\theta) - \sin\theta P_{i-1}(\cos\theta, \sin\theta).$$

We remark that f_i and g_i are homogeneous trigonometric polynomials in the variables $\cos\theta$ and $\sin\theta$ having degree in the set $\{i, i-2, i-4, \ldots\} \cap \mathbf{N}$, where $\mathbf{N}$ is the set of non-negative integers. This is due to the fact that $f_i(\theta)$ can be of the form $(\cos^2\theta + \sin^2\theta)^s f_{i-2s}$ with f_{i-2s} a trigonometric polynomial of degree $i - 2s \geq 0$. A similar situation occurs for $g_i(\theta)$.

If we impose $g_{n+1}(\theta) = g_{n+2m+1}(\theta) = g_{n+3m+1}(\theta) = 0$ and $g_{n+m+1}(\theta)$ either > 0 or < 0 for all θ, and do the change of variable $R = r^m$, then system (1.4) becomes the Abel differential equation

$$\frac{dR}{d\theta} = \frac{m}{g_{n+m+1}(\theta)} \left[f_{n+1}(\theta) + f_{n+m+1}(\theta)\, R + f_{n+2m+1}(\theta)\, R^2 + f_{n+3m+1}(\theta)\, R^3 \right]. \tag{1.5}$$

This kind of differential equations appeared in the studies of Abel on the theory of elliptic functions. For more details on Abel differential equations, see [16], [4] or [7].

We say that all polynomial differential systems (1.1) with

$$g_{n+1}(\theta) = g_{n+2m+1}(\theta) = g_{n+3m+1}(\theta) = 0$$

and $g_{n+m+1}(\theta)$ either > 0 or < 0 for all θ define the class $\mathcal{F}$ if they satisfy the equality

$$\begin{aligned} & g_{n+m+1}(\theta)\left(f'_{n+3m+1}(\theta) f_{n+2m+1}(\theta) - f_{n+3m+1}(\theta) f'_{n+2m+1}(\theta)\right) \\ & \quad + 3\,a^3 w^3(\theta) f^3_{n+3m+1}(\theta) \\ & = \frac{2}{9} f^3_{n+2m+1}(\theta) - f_{n+m+1}(\theta) f_{n+2m+1}(\theta) f_{n+3m+1}(\theta) + 3 f_{n+1}(\theta) f^2_{n+3m+1}(\theta) \end{aligned} \tag{1.6}$$

for some $a \in \mathbb{R}$, where

$$w(\theta) = \exp\left(\int \left[\frac{f_{n+m+1}(\theta)}{g_{n+m+1}(\theta)} - \frac{f^2_{n+2m+1}(\theta)}{3 f_{n+3m+1}(\theta) g_{n+m+1}(\theta)} \right] d\theta \right). \tag{1.7}$$

Here $f^n(\theta)$ means $[f(\theta)]^n$, and $' = d/d\theta$.

Since $g_{n+m+1}(\theta)$ either > 0 or < 0 for all θ, it follows that the polynomial differential systems (1.1) in the class $\mathcal{F}$ must satisfy that $n + m + 1$ is even.

We have found the subclass $\mathcal{F}$ thanks to case (g) of Abel differential equations studied on page 26 of the book of Kamke [16] where a complete prescription for the integrability of Abel differential equations having constant invariant is given. These equations are referred to in the literature as *constant invariant Abel differential equations* which can always and systematically be transformed into separable equations.

Our main results are the following two theorems. In the first, it is proved that all polynomial differential systems (1.1) in the class $\mathcal{F}$ are Darboux integrable, and in the second, it is proved that under an additional assumption over the inverse integrating factor they can have at most 3 limit cycles. Moreover, this upper bound is reached.

Theorem 1.1. *For a polynomial differential system* (1.1) *in the class* $\mathcal{F}$ *the following statements hold.*

(a) *If* $f_{n+m+1}(\theta)f_{n+2m+1}(\theta)f_{n+3m+1}(\theta) \neq 0$, *then the system is Darboux integrable, see Proposition* 2.1 *for the explicit expression of the first integral.*
(b) *If* $f_{n+m+1}(\theta) \neq 0$, $a = 0$ *and* $f_{n+2m+1}(\theta) = 0$, *then the system is Darboux integrable, see Proposition* 2.2 *for the explicit expression of the first integral.*

Theorem 1.1 will be proved in Section 2.

Theorem 1.2. *Assume that* $f_{n+m+1}(\theta)f_{n+2m+1}(\theta)f_{n+3m+1}(\theta) \neq 0$ *for a polynomial differential system* (1.1) *in the class* $\mathcal{F}$. *If* $g(R,\theta) = R + f_{n+2m+1}(\theta)/(3f_{n+3m+1}(\theta))$, *then the following statements hold.*

(a) *The maximum number of its limit cycles contained in the domain of definition of the inverse integrating factor*

$$V(R,\theta) = \frac{1}{w(\theta)^2}\, g(R,\theta)^3 + a^3\, w(\theta)$$

is one.

(b) *The maximum number of its limit cycles contained in the domain of definition of the inverse integrating factor* $V(R,\theta)$ *equal to*

$$\frac{1}{2}\, g(R,\theta) + \frac{1}{w(\theta)^2}\, g(R,\theta)^3 \int \frac{f_{n+3m+1}(\theta) w^2(\theta)}{g_{n+m+1}(\theta)}\, d\theta \tag{1.8}$$

is three.

Moreover, these upper bounds are reached.

We remark that in the proof of Theorem 1.2 in Section 3 we provide the explicit expression for the limit cycles.

It is easy to check that the polynomial differential systems (1.1) in the class $\mathcal{F}$ with $n = 1$ have always a node at the origin. Moreover, systems (1.1) in the class $\mathcal{F}$ have no monodromic singular points at the origin because $g_{n+1}(\theta) = 0$, see Theorem 3.3 in the book [22]. A singular point is *monodromic* for system (1.1) if there is no *characteristic orbit* associated to it; i.e., there is no orbit tending to the singular point with a defined tangent at this point. When the vector field is analytic, a monodromic singular point is either a center or a focus, see [1]. In short, the limit cycles given by Theorem 1.2 cannot surround a focus; in general they will surround a node. Using similar techniques to the ones for proving Theorems 1.1 and 1.2, in the papers [11], [12] and [13] we found different families of Darboux integrable polynomial differential systems with at most two limit cycles surrounding foci.

2. Proof of Theorem 1.1

Statement (a) of Theorem 1.1 follows from the next proposition which is a consequence of the case (g) presented in [16] for the integrability of the Abel differential equations.

Proposition 2.1. *If $f_{n+m+1}(\theta)f_{n+2m+1}(\theta)f_{n+3m+1}(\theta) \neq 0$ for a polynomial differential system* (1.1) *in the class $\mathcal{F}$, then it is Darboux integrable with the first integral $\tilde{H}(x,y)$ obtained from*

$$H(\eta,\xi) = \exp\left[\xi - \frac{1}{\sqrt{3}\,a^2}\arctan\left[\frac{2\eta - a}{\sqrt{3}\,a}\right]\right](\eta+a)^{-\frac{1}{3a^2}}(\eta^2-a\eta+a^2)^{\frac{1}{6a^2}}, \quad \textit{if } a \neq 0;$$

and

$$H(\eta,\xi) = \xi + 1/(2\eta^2), \quad \textit{if } a = 0,$$

where

$$\eta = \frac{1}{w(\theta)}g(R,\theta), \quad \textit{and} \quad \xi = \int \frac{f_{n+3m+1}(\theta)w^2(\theta)}{g_{n+m+1}(\theta)}\,d\theta,$$

with $w(\theta)$ given by (1.7), *through the changes of variables* (1.3) *and with $R = r^m$.*

Proof. Following the case (a) of Abel differential equation studied on page 24 of the book [16], which just defines the transformation of an Abel differential equation to its normal form, we do the change of variables $(R,\theta) \to (\eta,\xi)$ defined by $R = w(\theta)\eta(\xi) - f_{n+2m+1}(\theta)/(3f_{n+3m+1}(\theta))$. This transformation writes the Abel equation (1.5) into the normal form

$$\eta'(\xi) = [\eta(\xi)]^3 + I(\theta), \tag{2.1}$$

where

$$I(\theta) = \frac{g_{n+m+1}(\theta)}{f_{n+3m+1}(\theta)w^3(\theta)}\left[\frac{f_{n+1}(\theta)}{g_{n+m+1}(\theta)} + \frac{d}{d\theta}\left(\frac{f_{n+2m+1}(\theta)}{3f_{n+3m+1}(\theta)}\right)\right.$$
$$\left. - \frac{f_{n+m+1}(\theta)f_{n+2m+1}(\theta)}{3f_{n+3m+1}(\theta)g_{n+m+1}(\theta)} + \frac{2f^3_{n+2m+1}(\theta)}{27f^2_{n+3m+1}(\theta)g_{n+m+1}(\theta)}\right].$$

It is easy to see that for the case $I(\theta) = a^3$ with $a \in \mathbb{R}$, the differential equation (2.1) is of separable variables and we can obtain a first integral for it. For $a \neq 0$ a first integral is the one given in the statement of the proposition, and for $a = 0$ the first integral is $H(\eta,\xi) = \xi + 1/(2\eta^2)$. The condition $I(\theta) = a^3$ is equivalent to (1.6) which defines the class $\mathcal{F}$. We must mention that cases (b) and (c) of Abel differential equation studied on page 25 of the book [16] provide again the case studied for $a = 0$.

Now, we are going to prove that systems of statement (a) are Darboux integrable. For systems (1.1) in the class $\mathcal{F}$ with $f_{n+m+1}(\theta)f_{n+2m+1}(\theta)f_{n+3m+1}(\theta) \neq 0$, it is easy to check that an inverse integrating factor for its associated Abel equation (2.1) with $I(\theta) = a^3$ is $\overline{V}(\xi,\eta) = \eta^3 + a^3$. Consequently, an inverse integrating factor for the Abel differential equation (1.5) is

$$V(R,\theta) = w(\theta)(\eta^3 + a^3) = \frac{1}{w(\theta)^2}\, g(R,\theta)^3 + a^3\ w(\theta), \tag{2.2}$$

for all $a \in \mathbb{R}$. Taking into account that an indefinite integral of the form

$$\int \frac{P(\sin\theta, \cos\theta)}{Q(\sin\theta, \cos\theta)}\, d\theta\ ,$$

where P and Q are polynomials, can be transformed to a rational indefinite integral which always give elementary functions, we have that $w(\theta)$ is an elementary function. So, this inverse integrating factor $V(R,\theta)$ is an elementary function in Cartesian coordinates (see [18, 19] for more details and a definition of elementary function), then systems (1.1) in the class $\mathcal{F}$ with $f_{n+m+1}(\theta) f_{n+2m+1}(\theta) f_{n+3m+1}(\theta) \neq 0$ have a Liouvillian first integral according with the results of Singer [19]. □

Statement (b) of Theorem 1.1 is a direct consequence of the following proposition.

Proposition 2.2. *If $f_{n+m+1}(\theta) \neq 0$, $a = 0$ and $f_{n+2m+1}(\theta) = 0$ for a polynomial differential system (1.1) in the class $\mathcal{F}$, then condition (1.6) implies $f_{n+1}(\theta)$ $f_{n+3m+1}(\theta) = 0$ and the system is Darboux integrable with the first integral $\tilde{H}(x,y)$ obtained from*

$$H(R,\theta) = \frac{\exp\left(2\int \frac{f_{n+m+1}(\theta)}{g_{n+m+1}(\theta)}\, d\theta\right)}{R^2} + 2\int \frac{\exp\left(2\int \frac{f_{n+m+1}(\theta)}{g_{n+m+1}(\theta)}\, d\theta\right) f_{n+3m+1}(\theta)}{g_{n+m+1}(\theta)}\, d\theta$$

if $f_{n+1}(\theta) = 0$, and

$$H(R,\theta) = \exp\left(-\int \frac{f_{n+m+1}(\theta)}{g_{n+m+1}(\theta)}\, d\theta\right) R - \int \frac{\exp\left(-\int \frac{f_{n+m+1}(\theta)}{g_{n+m+1}(\theta)}\, d\theta\right) f_{n+1}(\theta)}{g_{n+m+1}(\theta)}\, d\theta$$

if $f_{n+3m+1}(\theta) = 0$, through the change of variables (1.3) and with $R = r^m$.

Proof. Under the assumptions condition (1.6) implies $f_{n+1}(\theta) f_{n+3m+1}(\theta) = 0$. If $f_{n+1}(\theta) = 0$, then the Abel differential equation (1.5) is the Bernoulli differential equation $dR/d\theta = m(f_{n+m+1}(\theta)R + f_{n+3m+1}(\theta)R^3)/g_{n+m+1}(\theta)$. If $f_{n+3m+1}(\theta) = 0$, then the Abel differential equation (1.5) is the linear differential equation $dR/d\theta = m(f_{n+1}(\theta) + f_{n+m+1}(\theta)R)/g_{n+m+1}(\theta)$. Solving these differential equations we obtain the first integrals of the proposition.

The systems of the proposition are Darboux integrable because their first integrals are obtained by integrating elementary functions, see [19] for more details. □

3. Existence of Limit Cycles in the Class $\mathcal{F}$

It is known that the polynomial systems of degree 1 have no limit cycles, and that if a polynomial differential system of degree 2 has a limit cycle, then in its interior the system can only have a unique singular point, which must be a focus. But this is not true for polynomial differential systems of higher degree. The next proposition provides probably the easiest well-known example of a polynomial differential system with a node at the origin. The example is a family of cubic polynomial differential systems depending on two parameters. This family was considered by Vorobev [21] for showing that around a node, a cubic polynomial system can have limit cycles.

Proposition 3.1. *The differential system*

$$\dot{x} = -y + ax(x^2+y^2-1), \qquad \dot{y} = x + by(x^2+y^2-1),$$

with $ab > -1$ *and* $(a-b)^2 > 4$ *has the algebraic solution* $x^2+y^2-1=0$ *as a limit cycle surrounding a node, namely the origin.*

Proof. See [21]. □

In order to study the existence and non-existence of limit cycles for system (1.1) we use the following result.

Theorem 3.2. *Let* (P,Q) *be a* C^1 *vector field defined in the open subset* U *of* $\mathbb{R}^2$. *Let* $V = V(x,y)$ *be a* C^1 *solution of the linear partial differential equation* (1.2) *defined in* U. *If* γ *is a limit cycle of* (P,Q) *contained in the domain of definition* U, *then* γ *is contained in* $\{(x,y) \in U : V(x,y) = 0\}$.

Proof. See Theorem 9 in [8], or [17]. □

We recall that under the assumptions of Theorem 3.2 the function $1/V$ is an integrating factor in $U \setminus \{V(x,y)=0\}$. Again for more details, see [3, 5]. As we have seen, the function V is called an inverse integrating factor. In fact, using this notion, recently it is proved that any topological finite configuration of limit cycles is realizable by algebraic limit cycles of a Darboux integrable polynomial differential system, see [17].

3.1. Proof of Theorem 1.2

Proof. As we have seen in the proof of Proposition 2.1, systems (1.1) in the class $\mathcal{F}$ with $f_{n+m+1}(\theta)f_{n+2m+1}(\theta)f_{n+3m+1}(\theta) \neq 0$ have an inverse integrating factor for its associated Abel differential equation (1.5) given by expression (2.2) for all $a \in \mathbb{R}$. By Theorem 3.2, if system (1.1) and consequently its associated Abel equation (1.5) have limit cycles in the domain of definition of (2.2), such limit cycles for the Abel equation must be contained into the set $\{V(R,\theta)=0\}$. From the expression of the inverse integrating factor, the possible limit cycles must be given by

$$R = \begin{cases} -\dfrac{f_{n+2m+1}(\theta)}{3f_{n+3m+1}(\theta)} - a\ w(\theta), \\ -\dfrac{f_{n+2m+1}(\theta)}{3f_{n+3m+1}(\theta)} + \dfrac{1}{2}(1 \pm i\sqrt{3}\ )\ a\ w(\theta). \end{cases}$$

In order that these expressions of $R(\theta)$ define limit cycles, we must have $R(\theta) > 0$ for all θ. Since, from equality (1.6), $w(\theta)$ is a real function, we must take $a = 0$ in the above second expression for the possible limit cycles. So, the maximum number of possible limit cycles in the domain of definition of $V(R, \theta)$ is at most 1 for all $a \in \mathbb{R}$.

We can find other inverse integrating factors multiplying $V(R, \theta)$ by an arbitrary function of the first integral given in Proposition 2.1. Therefore, for $a = 0$ another inverse integrating factor for its associated Abel equation (2.1) is $\overline{V}(\xi, \eta) = \eta/2 + \eta^3\xi$. Consequently, an inverse integrating factor for the Abel differential equation (1.5) is $V(R, \theta) = w(\theta)[\eta(R, \theta)/2 + \eta^3(R, \theta)\ \xi(\theta)]$, or equivalently (1.8). From this expression of the inverse integrating factor, the possible limit cycles must be

$$R = \begin{cases} -\dfrac{f_{n+2m+1}(\theta)}{3f_{n+3m+1}(\theta)}, \\ -\dfrac{f_{n+2m+1}(\theta)}{3f_{n+3m+1}(\theta)} \pm \dfrac{w(\theta)}{\sqrt{2}\sqrt{-\displaystyle\int \frac{f_{n+3m+1}(\theta) w(\theta)^2}{g_{n+m+1}(\theta)}\, d\theta}}. \end{cases}$$

In conclusion, in the domain of definition of the inverse integrating factor (1.8) such systems can have at most three limit cycles.

Now it only remains to prove that the upper bounds for the number of the limit cycles in the statement of Theorem 1.2 are reached.

Systems (1.1) with $n = 1$ and $m = 2$ satisfying $g_2(\theta) = g_6(\theta) = g_8(\theta) = 0$ for all θ can be written into the form

$$\begin{aligned} \dot{x} &= a_{10}x + P_3(x, y) + x(A_4(x, y) + A_6(x, y)), \\ \dot{y} &= a_{10}y + Q_3(x, y) + y(A_4(x, y) + A_6(x, y)), \end{aligned} \tag{3.1}$$

where a_{ij} are arbitrary constants and $P_i(x, y)$, $Q_i(x, y)$ and $A_i(x, y)$ are homogeneous polynomials of degree i. Consider the following particular system of (3.1)

$$\begin{aligned} \dot{x} = &\ -12x + (13x - y)(x^2 + y^2) - x\left(9(x^2 + y^2)^2 - (x^2 + y^2)^3\right)/2, \\ \dot{y} = &\ -12y + (x + 13y)(x^2 + y^2) - y\left(9(x^2 + y^2)^2 - (x^2 + y^2)^3\right)/2; \end{aligned} \tag{3.2}$$

for $a = 0$ it is in the class $\mathcal{F}$ and satisfies $f_{n+m+1}(\theta)f_{n+2m+1}(\theta)f_{n+3m+1}(\theta) \neq 0$. It is easy to check that this system has exactly three limit cycles given by the circles $x^2 + y^2 - 2 = 0$, $x^2 + y^2 - 3 = 0$ and $x^2 + y^2 - 4 = 0$. Moreover it has a node at the origin and therefore this example generalizes the example of Proposition 3.1

given by Vorobev. This example has the Darboux first integral

$$H(x,y) = \exp\left(-2\ \arg(x+iy)\right) \frac{(x^2+y^2-2)(x^2+y^2-4)}{x^2+y^2-3}$$

and the polynomial inverse integrating factor

$$V(x,y) = (x^2+y^2)(x^2+y^2-2)(x^2+y^2-3)(x^2+y^2-4).$$

As system (3.2) has a unique singular point at the origin which is a node and a polynomial inverse integrating factor defined in the whole plane, applying Theorem 3.2, one sees that system (3.2) has exactly 3 limit cycles which are algebraic.

Consider now the following particular system of (3.1)

$$\begin{aligned} \dot{x} = &\ (1+a^3)x + 3(x-2y)(x^2+y^2) + 3x(x^2+y^2)^2 + x(x^2+y^2)^3, \\ \dot{y} = &\ (1+a^3)y + 3(2x+y)(x^2+y^2) + 3y(x^2+y^2)^2 + y(x^2+y^2)^3; \end{aligned} \tag{3.3}$$

for $a \neq 0$ it is in the class $\mathcal{F}$ and satisfies $f_{n+m+1}(\theta)f_{n+2m+1}(\theta)f_{n+3m+1}(\theta) \neq 0$. It is easy to check that for $a < 1$ this system has exactly one limit cycle given by the circle $x^2+y^2+1+a=0$. This system has the Darboux first integral

$$\begin{aligned} H(x,y) = \exp&\left[\frac{1}{3}\ \arg(x+iy) - \frac{1}{\sqrt{3}\ a^2} \arctan\left[\frac{2(1+x^2+y^2)-a}{\sqrt{3}\ a}\right]\right] \\ &\cdot(1+a+x^2+y^2)^{-1/(3a^2)}(a^2-a(1+x^2+y^2)+(1+x^2+y^2)^2)^{1/(6a^2)} \end{aligned}$$

and the polynomial inverse integrating factor

$$V(x,y) = (x^2+y^2)[(x^2+y^2+1)^3+a^3].$$

As system (3.3) has this polynomial inverse integrating factor defined in the whole plane, applying Theorem 3.2, one sees that system (3.3) has exactly one limit cycle which is algebraic. This completes the proof of Theorem 1.2. □

4. The Appendix

We denote the polynomials appearing in system (3.1) as follows:

$$\begin{aligned} P_3(x,y) &= a_{30}x^3 + a_{21}x^2y + a_{12}xy^2 + a_{03}x^3, \\ Q_3(x,y) &= b_{30}x^3 + b_{21}x^2y + b_{12}xy^2 + b_{03}x^3, \\ A_4(x,y) &= \alpha x^4 + \beta x^3y + \gamma x^2y^2 + \delta x^3y + \epsilon y^4, \\ A_6(x,y) &= Ax^6 + Bx^5y + Cx^4y^2 + Dx^3y^3 + Ex^2y^4 + Fxy^5 + Gy^6. \end{aligned}$$

Then, the following proposition provides some explicit systems which belong to the class $\mathcal{F}$.

Proposition 4.1. *System (3.1) belongs to the class $\mathcal{F}$ if one of the following statements holds.*

(a) $a_{10} = \alpha = \beta = \gamma = \delta = \epsilon = 0$ *and* $a = 0$.

(b) $b_{30} = b_{12}$, $b_{03} = b_{21}$, $a_{30} = a_{12}$, $a_{03} = a_{21}$, $b_{21} = a_{12}$, $b_{12} = -a_{21}$, $a = e = -a_{21}k_2$, $c = -2a_{21}k_2$, $b = d = 0$, $A = G = -a_{21}k_3$, $E = -3a_{21}k_3$, $B = D = F = 0$, $a_{12} = -(2a_{21}k_2^3 - 27a_{10}k_3^2)/(9k_2k_3)$ *and* $a = 0$.

(c) $a_{10} = (k_1^3 + a^3k_2^3)/k_1$, $a_{30} = a_{12} = b_{21} = b_{03} = 3k_1k_2$, $a_{21} = a_{03} = -b_{30} = -b_{12} = -6k_2k_3$, $\beta = \delta = 0$, $\gamma = 2\alpha = 2\epsilon = 6k_2^2$, $B = D = F = 0$, $C = E = 3A = 3G = 3k_2^3/k_1$ *and* $a \neq 0$.

(d) $\alpha = \beta = \gamma = \delta = \epsilon = A = B = C = D = E = F = G = 0$ *and* $a = 0$.

Systems provided by statements (a) *and* (d) *are Darboux integrable with the first integral given by Proposition* 2.2 *with* $n = 1$ *and* $m = 2$ *and where* $f_2(\theta) = 0$ *and* $f_8(\theta) = 0$, *respectively. Systems provided by statements* (b) *and* (c) *are Darboux integrable with the first integral given by Proposition* 2.1 *with* $n = 1$ *and* $m = 2$.

The proof of this proposition follows from doing tedious computations and using Propositions 2.1 and 2.2 when $n = 1$ and $m = 2$.

References

[1] A. A. Andronov, E. A. Leontovich, I. I. Gordon and A. G. Maier, *Qualitative Theory of Second-order Dynamic Systems.* John Wiley & Sons, New York-Toronto, Israel Program for Scientific Translations, Jerusalem-London, 1973.

[2] R. Bamón, *Quadratic Vector Fields in the Plane Have a Finite Number of Limit Cycles.* Inst. Hautes Études Sci. Publ. Math. **64** (1986), 111–142.

[3] J. Chavarriga, H. Giacomini, J. Giné and J. Llibre, *On the Integrability of Two-Dimensional Flows.* J. Differential Equations **157** (1999), 163–182.

[4] E. S. Cheb-Terrab and A. D. Roche, *An Abel ODE Class Generalizing Known Integrable Classes.* Eur. J. Appl. Math. **14** (2003), 217–229.

[5] C. J. Christopher and J. Llibre, *Integrability via Invariant Algebraic Curves for Planar Polynomials Differential Systems.* Annals of Differential Equations **16** (2000), 5–19.

[6] J. Écalle, *Introduction aux Fonctions Analysables et Preuve Constructive de la Conjecture de Dulac.* Hermann, 1992.

[7] A. Gasull and J. Llibre, *Limit Cycles for a Class of Abel Equations.* SIAM J. Math. Anal. **21** (1990), 1235–1244.

[8] H. Giacomini, J. Llibre and M. Viano, *On the Nonexistence, Existence, and Uniquennes of Limit Cycles.* Nonlinearity **9** (1996), 501–516.

[9] H. Giacomini, J. Llibre and M. Viano, *On the Shape of Limit Cycles that Bifurcate from Hamiltonian Centers.* Nonlinear Anal. **41** (2000), no. 3-4, Ser. A: Theory Methods, 523–537.

[10] H. Giacomini, J. Llibre and M. Viano, *On the Shape of Limit Cycles that Bifurcate from Non-Hamiltonian Centers.* Nonlinear Anal. **43** (2001), no. 7, Ser. A: Theory Methods, 837–859.

[11] J. Giné and J. Llibre, *Integrability and Algebraic Limit Cycles for Polynomial Differential Systems with Homogeneous Nonlinearities.* J. Differential Equations **197** (2004), 147–161.

[12] J. Giné and J. Llibre, *A Family of Isochronous Foci with Darbouxian First Integral.* Pacific J. Math. **218** (2005), to appear.

[13] J. Giné and J. Llibre, *Integrability, Degenerate Centers and Limit Cycles for a Class of Polynomial Differential Systems.* Preprint, Universitat Autònoma de Barcelona, 2004.

[14] D. Hilbert, *Mathematische Problem (Lecture).* Second International Congress on Mathematics, Paris, 1900, Nachr. Ges. Wiss. Göttingen Math.Phys. Kl. pp. 253–297, 1900.

[15] Yu. Ilyashenko, *Finiteness Theorems for Limit Cycles.* Translations of Mathematical Monographs 94, Amer. Math. Soc., 1991.

[16] E. Kamke, *Differentialgleichungen "Losungsmethoden und Losungen".* Col. Mathematik und ihre anwendungen, **18**, Akademische Verlagsgesellschaft Becker und Erler Kom-Ges., Leipzig, 1943.

[17] J. Llibre and G. Rodriguez, *Finite Limit Cycles Configurations and Polynomial Vector Fields.* J. Differential Equations **198** (2004), 374–380.

[18] M. J. Prelle and M. F. Singer, *Elementary First Integrals of Differential Equations.* Trans. Amer. Math. Soc. **279** (1983), 215–229.

[19] M. F. Singer, *Liouvillian First Integrals of Differential Equations.* Trans. Amer. Math. Soc. **333** (1992), 673–688.

[20] M. Viano, J. Llibre and H. Giacomini, *Arbitrary Order Bifurcations for Perturbed Hamiltonian Planar Systems via the Reciprocal of an Integrating Factor.* Nonlinear Anal. **48** (2002), no. 1, Ser. A: Theory Methods, 117–136.

[21] A. P. Vorobev, *Cycles Around a Singular Point of the Nodal Type.* (Russian), Dokl. Akad. Nauk. B.S.S.R. IV **9** (1960), 369–371.

[22] Zhi-Fen Zhang, Tong-Ren Ding, Wen-Zao Huang, and Zhen-Xi Dong, *Qualitative Theory of Differential Equations.* Translations of Mathematical Monographs, 101. Amer. Math. Soc., Providence, RI, 1992.

Jaume Giné
Departament de Matemàtica
Universitat de Lleida
Av. Jaume II, 69
25001 Lleida
Spain
e-mail: gine@eps.udl.es

Jaume Llibre
Departament de Matemàtiques
Universitat Autònoma de Barcelona
08193 Bellaterra, Barcelona
Spain
e-mail: jllibre@mat.uab.es

Trends in Mathematics: *Differential Equations with Symbolic Computation*, 67–83

Time-Reversibility in Two-Dimensional Polynomial Systems

Valery G. Romanovski and Douglas S. Shafer

Abstract. We characterize the set of all time-reversible systems within a particular family of complex polynomial differential equations in two complex dimensions. These results are a generalization to the complex case of theorems of Sibirsky for real systems. We also give an efficient computational algorithm for finding this set. An interconnection of time-reversibility and the center problem is discussed as well.

Mathematics Subject Classification (2000). Primary 34C14; Secondary 34C05.
Keywords. Time-reversibility, center problem.

1. Introduction

Consider a real system of the form

$$\dot{u} = v + U(u, v), \quad \dot{v} = -u + V(u, v), \tag{1.1}$$

where U and V are real analytic functions whose series expansions in a neighborhood of the origin start with terms of the second degree or higher. Conversion to polar coordinates shows that near the origin either all non-stationary trajectories of (1.1) are ovals (in which case the origin is called a *center*) or they are all spirals (in which case the origin is called a *focus*). The problem of distinguishing between centers and foci is known as the *Poincaré center problem.*

A *first integral* for system (1.1) on a neighborhood of the origin is a non-constant function that is constant on trajectories of (1.1). According to the Poincaré–Lyapunov theorem the origin is a center for (1.1) if and only if system (1.1) admits a formal first integral of the form

$$\Phi(u, v) = u^2 + v^2 + \sum_{k=3}^{\infty} \phi_k(u, v), \tag{1.2}$$

where for each k $\phi_k(u, v)$ is a homogeneous polynomial in u and v of degree k. (Actually it is known that if there exists a formal integral (1.2) then the series must in fact converge). However, there are no regular methods for constructing such an integral and this is one of the reasons why, despite its hundred-year history, the center problem has been solved for only a few relatively simple non-trivial subfamilies of system (1.1) (see, for example, [1, 3, 12, 13, 17, 18] and the references therein), most notably for the family of quadratic systems, those systems for which $\max\{\deg U, \deg V\} = 2$.

Sometimes it is possible to distinguish between a center and a focus in system (1.1) using a geometric argument, which we illustrate by the following example. Consider a system of the form

$$\dot{u} = v + vf(u, v^2), \qquad \dot{v} = -u + g(u, v^2), \tag{1.3}$$

where f is a real analytic function whose series expansion starts with terms of order at least one and g is a real analytic function whose series expansion starts with terms of order at least two. As already noted, in a sufficiently small neighborhood Ω of the origin either all non-stationary trajectories of (1.3) are spirals or all are closed curves. We observe, however, that the transformation $u \to u$, $v \to -v$, $t \to -t$ leaves the system unchanged. This implies that the u-axis is a line of symmetry for the orbits (as point-sets); hence no trajectory in Ω can be a spiral.

System (1.3) is an example of a *time-reversible* system, i.e. a system which is invariant under reflection with respect to a line and a change in the direction of time. To study the center problem in conjunction with reversibility properties it is convenient to introduce a complex structure on the phase plane $\{(u, v) \mid u, v \in \mathbb{R}\}$ by setting $x = u + iv$ ($i = \sqrt{-1}$). This can be done for any real system

$$\dot{u} = \widetilde{U}(u, v), \qquad \dot{v} = \widetilde{V}(u, v), \tag{1.4}$$

which is thereby transformed into the complex differential equation

$$\frac{dx}{dt} = P(x, \bar{x}); \tag{1.5}$$

one simply uses $\dot{x} = \dot{u} + i\dot{v}$ and the identities $u = \frac{1}{2}(x + \bar{x})$ and $v = \frac{1}{2i}(x - \bar{x})$ in $\widetilde{U}(u, v)$ and $\widetilde{V}(u, v)$. We will limit our study to the case that U and V in (1.1) and $\widetilde{U}$ and $\widetilde{V}$ in (1.4) are polynomial functions of u and v, so that P is a polynomial as well.

We say that a straight line L is an *axis of symmetry* of system (1.4) if as point-sets (ignoring the sense of the parametrization by time t) the orbits of the system are symmetric with respect to the line L.

The proof of the following lemma is straightforward.

Lemma 1.1 ([15]). *Let a denote the vector of coefficients of the polynomial $P(x, \bar{x})$ in* (1.5), *arising from the real system* (1.4) *by setting $x = u + iv$. If $a = \pm\bar{a}$ (meaning that either all the coefficients are real or all are pure imaginary), then the u-axis is an axis of symmetry of the real system* (1.4) *and of the corresponding complex differential equation* (1.5).

By the lemma the u-axis is an axis of symmetry for (1.5) if

$$P(\bar{x}, x) = -\overline{P(x, \bar{x})} \tag{1.6}$$

(the case $a = -\bar{a}$), or if

$$P(\bar{x}, x) = \overline{P(x, \bar{x})} \tag{1.7}$$

(the case $a = \bar{a}$). We now observe that if condition (1.6) is satisfied then under the change

$$x \to \bar{x}, \qquad \bar{x} \to x, \tag{1.8}$$

(1.5) is transformed to its negative,

$$\dot{x} = -P(x, \bar{x}), \tag{1.9}$$

and if condition (1.7) holds then (1.5) is unchanged. Thus condition (1.7) means that the system is reversible with respect to reflection across the u-axis (i.e., the transformation does not change the system) while condition (1.6) corresponds to time-reversibility with respect to the same transformation.

It is clear that if the origin is known to be a singular point of center or focus type and we are interested in a symmetry with respect to a line passing through the origin, then only time-reversibility is possible. So if (1.5) is obtained from (1.1) and the transformation (1.8) yields the system (1.9), then the real system (1.1) is time-reversible, and hence has a center at the origin.

If the line of reflection is not the u-axis but a distinct line L then we can apply the rotation $x_1 = e^{-i\varphi}x$ through an appropriate angle φ to make L the u-axis. In the new coordinates we have

$$\dot{x}_1 = e^{-i\varphi}P(e^{i\varphi}x_1, e^{-i\varphi}\bar{x}_1)\,.$$

By the discussion in the paragraph following (1.9) this system is time-reversible with respect to the line $\operatorname{Im} x_1 = 0$ if (1.6) holds, meaning that

$$e^{i\varphi}\overline{P(e^{i\varphi}x_1, e^{-i\varphi}\bar{x}_1)} = -e^{-i\varphi}P(e^{i\varphi}\bar{x}_1, e^{-i\varphi}x_1).$$

Hence, reverting to the variable x, (1.5) is time-reversible when there exists a φ such that

$$e^{2i\varphi}\overline{P(x, \bar{x})} = -P(e^{2i\varphi}\bar{x}, e^{-2i\varphi}x). \tag{1.10}$$

This suggests the following natural generalization of the notion of time-reversibility to the case of two-dimensional complex systems.

Definition 1.2. Let $\mathbf{z} = (x, y) \in \mathbb{C}^2$. We say that the system

$$\frac{d\mathbf{z}}{dt} = F(\mathbf{z}) \tag{1.11}$$

is *time-reversible* if there is a linear transformation T,

$$x \mapsto \alpha y, \ y \mapsto \alpha^{-1}x \tag{1.12}$$

($\alpha \in \mathbb{C}$), such that

$$\frac{d(T\mathbf{z})}{dt} = -F(T\mathbf{z}). \tag{1.13}$$

For more details about time-reversible systems and their applications the reader can consult the survey [9].

Direct calculation shows that the system

$$\dot{x} = P(x,y), \qquad \dot{y} = Q(x,y) \tag{1.14}$$

is time-reversible (in the sense of Definition 1.2) if and only if for some α

$$\alpha Q(\alpha y, x/\alpha) = -P(x,y), \qquad \alpha Q(x,y) = -P(\alpha y, x/\alpha)\,. \tag{1.15}$$

Thus we see that (1.10) is a particular case of (1.15): setting $\alpha = e^{2i\varphi}$, $y = \bar{x}$, and $Q = \bar{P}$ in (1.15) we obtain (1.10).

As mentioned above a real polynomial system that has a critical point at the origin of center or focus type and which is time-reversible with respect to a line passing through the origin must in fact have a center at the origin. When the notion of a center is properly generalized to the complex two-dimensional system displayed by equations (1.17) below, this fact is also true for them. A natural generalization is provided by the Poincaré–Lyapunov theorem. After the complexification $x = u+iv$ we obtain from system (1.4) the equation (1.5). Adjoining to the latter equation its complex conjugate we obtain the system

$$\frac{dx}{dt} = P(x,\bar{x}), \quad \frac{d\bar{x}}{dt} = \overline{P(x,\bar{x})},$$

which we also can write as

$$\frac{dx}{dt} = P(x,\bar{x}), \quad \frac{d\bar{x}}{dt} = \bar{P}(\bar{x},x)$$

(where only the coefficients in P are conjugated to form $\bar{P}$). It is convenient to consider $\bar{x}$ as a new variable y and $\bar{P}$ as a new function Q. Then from the latter system we obtain a system of two complex differential equations of the form (1.14). Without loss of generality we can write it in the form

$$\begin{aligned} \frac{dx}{dt} &= -\sum_{(p,q)\in\widetilde{S}} a_{p,q} x^{p+1} y^q = \widetilde{P}(x,y), \\ \frac{dy}{dt} &= \sum_{(p,q)\in\widetilde{S}} b_{q,p} x^q y^{p+1} = \widetilde{Q}(x,y), \end{aligned} \tag{1.16}$$

where $\widetilde{S}$ is the set

$$\widetilde{S} = \{(p_i,q_i) \mid p_i + q_i \geq 0, i = 1,\ldots,\ell\} \subset (\{-1\} \cup \mathbb{N}_+) \times \mathbb{N}_+,$$

and $\mathbb{N}_+$ denotes the set of nonnegative integers. The notation (1.16) simply emphasizes that we take into account only non-zero coefficients of the polynomials.

In the case that system (1.16) is obtained from system (1.1) it can be written in the form

$$\begin{aligned} i\frac{dx}{dt} &= x - \sum_{(p,q)\in S} a_{pq}x^{p+1}y^q = P(x,y), \\ i\frac{dy}{dt} &= -y + \sum_{(p,q)\in S} b_{qp}x^q y^{p+1} = Q(x,y), \end{aligned} \tag{1.17}$$

where

$$S = \{(p_i, q_i) \mid p_i + q_i \geq 1,\ i = 1, \ldots, \ell\} \subset (\{-1\} \cup \mathbb{N}_+) \times \mathbb{N}_+.$$

Definition 1.3. System (1.17) has a center at the origin if there is an analytic function of the form

$$\Psi(x,y) = xy + \sum_{s=3}^{\infty}\sum_{j=0}^{s} v_{j,s-j}x^j y^{s-j}, \tag{1.18}$$

where the $v_{j,s-j}$ are polynomials in the coefficients of P and Q, such that

$$\frac{\partial \Psi}{\partial x}P(x,y) + \frac{\partial \Psi}{\partial y}Q(x,y) = 0.$$

In particular if system (1.17) is obtained from (1.1) then setting $y = \bar{x}$ we obtain the first integral (1.2) of system (1.1).

We denote by $(a,b) = (a_{p_1,q_1}, a_{p_2,q_2}, \ldots, b_{q_1,p_1})$ the ordered vector of coefficients of system (1.16), by $\widetilde{E}(a,b) = \mathbb{C}^{2\ell}$ the parameter space of (1.16) (for the parameter space of (1.17) we use the notation $E(a,b)$), and by $\mathbb{C}[a,b]$ the polynomial ring in the variables a_{pq}, b_{qp}. It is clear that there is a one-to-one correspondence between points of $E(a,b)$ and systems of the form (1.17). Thus to solve the center problem, we need to identify all points (a,b) such that their corresponding systems have a center at the origin.

For system (1.17) one can always find (see, for example, [7, 12]) a Lyapunov function Ψ of the form (1.18) such that

$$\frac{\partial \Psi}{\partial x}P(x,y) + \frac{\partial \Psi}{\partial y}Q(x,y) = g_{11}\cdot(xy)^2 + g_{22}\cdot(xy)^3 + g_{33}\cdot(xy)^4 + \cdots, \tag{1.19}$$

where the g_{ii} are polynomials in the coefficients of (1.17) called *focus quantities*. Thus system (1.17) with coefficients (a^*,b^*) has a center at the origin if and only if $g_{ii}(a^*,b^*) = 0$ for all $i = 1, 2, \ldots,$ and Ψ is then a first integral. In general, for polynomials $f_1, \ldots, f_s$, we let $I = \langle f_1, \ldots, f_s\rangle$ denote the ideal in $\mathbb{C}[a,b]$ that they generate, and $\mathbf{V}(\langle f_1, \ldots, f_s\rangle) = \mathbf{V}(I)$ the variety of I (the set on which every element of I vanishes). Then a necessary and sufficient condition that system (1.17) with coefficients (a^*,b^*) has a center at the origin is that $(a^*,b^*) \in \mathbf{V}(\langle g_{11}, g_{22}, \ldots, g_{ii}, \ldots\rangle)$. This motivates the following definition.

Definition 1.4. The ideal

$$\mathcal{B} := \langle g_{11}, g_{22}, \ldots, g_{ii}, \ldots\rangle \subseteq \mathbb{C}[a,b]$$

is called the *Bautin ideal* of system (1.17) and its variety $\mathbf{V}(\mathcal{B})$ is called the *center variety* of system (1.17).

Since $\mathbb{C}[a,b]$ is a Noetherian ring, by the Hilbert basis theorem the ideal $\mathcal{B}$ is finitely generated. It is known that for quadratic systems $\mathcal{B}$ is generated by the first three focus quantities. At present there are no general methods for finding a basis of the Bautin ideal for an arbitrary polynomial system. There do exist, however, various methods for computing the variety of an ideal. But even for the general cubic system of the form (1.17), attempts to compute the center variety have failed because of the fact that the focus quantities are such enormous expressions that as of yet no computer algebra system can perform the computations needed.

For a fixed collection $(p_1,q_1),\ldots,(p_\ell,q_\ell)$ of elements of $(\{-1\}\cup\mathbb{N}_+)\times\mathbb{N}_+$, and letting ν denote the element $(\nu_1,\ldots,\nu_{2\ell})$ of $\mathbb{N}_+^{2\ell}$, let L be the map from $\mathbb{N}_+^{2\ell}$ to $\mathbb{N}_+^2$ (the elements of the latter written as column vectors) defined by

$$L(\nu)=\begin{pmatrix}L^1(\nu)\\L^2(\nu)\end{pmatrix}=\begin{pmatrix}p_1\\q_1\end{pmatrix}\nu_1+\cdots+\begin{pmatrix}p_\ell\\q_\ell\end{pmatrix}\nu_\ell+\begin{pmatrix}q_\ell\\p_\ell\end{pmatrix}\nu_{\ell+1}+\cdots+\begin{pmatrix}q_1\\p_1\end{pmatrix}\nu_{2\ell}. \quad (1.20)$$

Let $\mathcal{M}$ denote the set of all solutions $\nu=(\nu_1,\nu_2,\ldots,\nu_{2l})$ with non-negative components of the equation

$$L(\nu)=\begin{pmatrix}k\\k\end{pmatrix} \quad (1.21)$$

as k runs through $\mathbb{N}_+$, and the pairs (p_i,q_i) determining $L(\nu)$ come from system (1.17). Similarly, let $\widetilde{\mathcal{M}}$ denote the set of such solutions corresponding to (1.16). Obviously, $\mathcal{M}$ and $\widetilde{\mathcal{M}}$ are Abelian monoids. Let $\mathbb{C}[\mathcal{M}]$ (respectively, $\mathbb{C}[\widetilde{\mathcal{M}}]$) denote the subalgebra of $\mathbb{C}[a,b]$ generated by all monomials of the form

$$a_{p_1q_1}^{\nu_1}a_{p_2q_2}^{\nu_2}\cdots a_{p_\ell q_\ell}^{\nu_\ell}b_{q_\ell p_\ell}^{\nu_{\ell+1}}b_{q_{\ell-1}p_{\ell-1}}^{\nu_{\ell+2}}\cdots b_{q_1p_1}^{\nu_{2\ell}},$$

for all $\nu\in\mathcal{M}$ (respectively, $\nu\in\widetilde{\mathcal{M}}$). In order to simplify the notation we will abbreviate such a monomial by $[\nu]=[\nu_1,\ldots,\nu_{2\ell}]$. For ν in either $\mathcal{M}$ or $\widetilde{\mathcal{M}}$, let $\hat{\nu}$ denote the involution of the vector ν:

$$\hat{\nu}=(\nu_{2\ell},\nu_{2\ell-1},\ldots,\nu_1).$$

It is shown in [12] that the focus quantities of system (1.17) belong to $\mathbb{C}[\mathcal{M}]$ and have the form

$$g_{kk}=\sum_{L(\nu)=(k,k)^T}g_{(\nu)}([\nu]-[\hat{\nu}]), \quad (1.22)$$

with $g_{(\nu)}\in\mathbb{Q}$, $k=1,2,\ldots$. (Similar properties of the focus quantities were also obtained in [3, 11].)

Consider the ideals

$$I_{sym}=\langle[\nu]-[\hat{\nu}]\mid\nu\in\mathcal{M}\rangle\subset\mathbb{C}[\mathcal{M}]$$

and

$$\widetilde{I}_{sym}=\langle[\nu]-[\hat{\nu}]\mid\nu\in\widetilde{\mathcal{M}}\rangle\subset\mathbb{C}[\widetilde{\mathcal{M}}].$$

From (1.22) and Definition 1.4 it is clear that $\mathcal{B}\subseteq I_{sym}$; hence $\mathbf{V}(I_{sym})\subseteq\mathbf{V}(\mathcal{B})$.

Definition 1.5. For system (1.17) the variety $\mathbf{V}(I_{sym})$ is called the *Sibirsky (or symmetry) subvariety* of the center variety, and the ideal I_{sym} is called the *Sibirsky ideal*.

As already stated it is geometrically obvious that every time-reversible system of the form (1.1) has a center at the origin. It is easily seen that this property is transferred to systems of the form (1.17): *every time-reversible system* (1.17) *has a center at the origin.* Indeed, (1.15) immediately yields that systems (1.17) and (1.16) are time-reversible if and only if

$$b_{qp} = \alpha^{p-q} a_{pq}, \qquad a_{pq} = b_{qp}\alpha^{q-p}. \tag{1.23}$$

Hence in the case that (1.17) is time-reversible, using (1.23) we see that for $\nu \in \mathcal{M}$

$$[\hat{\nu}] = \alpha^{(L^1(\nu)-L^2(\nu))}[\nu] = [\nu] \tag{1.24}$$

and thus from (1.22) we obtain $g_{kk} \equiv 0$ for all k, which implies that the system has a center.

By (1.24) every time-reversible system $(a,b) \in \widetilde{E}(a,b)$ belongs to $\mathbf{V}(\widetilde{I}_{sym})$. The converse is false. A simple counterexample is the system

$$\dot{x} = x(1 - a_{10}x - a_{01}y), \qquad \dot{y} = -y(1 - b_{10}x - b_{01}y).$$

In this case $\widetilde{I}_{sym} = \langle a_{10}a_{01} - b_{10}b_{01}\rangle$. The system

$$\dot{x} = x(1 - a_{10}x), \qquad \dot{y} = -y(1 - b_{10}x) \tag{1.25}$$

arises from $\mathbf{V}(I_{sym})$ but the conditions (1.23) are not fulfilled, so (1.25) is not time-reversible.

The correct statement is the following theorem, announced in [8].

Theorem 1.6. *Let $\mathcal{R} \subset E(a,b)$ be the set of all time-reversible systems in the family* (1.16). *Then:*

(1) $\mathcal{R} \subset \mathbf{V}(\widetilde{I}_{sym})$;

(2) $\mathbf{V}(\widetilde{I}_{sym}) \setminus \mathcal{R} = \{(a,b) \mid \exists (p,q) \in S$ *such that* $a_{pq}b_{qp} = 0$ *but* $a_{pq} + b_{qp} \neq 0\}$.

We will present the proof of Theorem 1.6 in Section 2. Actually, the theorem is an easy generalization of the results of Sibirsky for real systems to the case of complex systems and we closely follow his reasoning. However, since the work of Sibirsky is often not well enough known to be used in the Western literature (one example is the recent paper [5] in which results of Sibirsky are re-derived using different methods), we believe that it is worthwhile to present detailed proof of the theorem (so, in a sense, the present paper can be considered as mainly a survey paper).

2. Invariants of the Rotation Group

We recall the notion of a polynomial invariant.

Definition 2.1. Let G be a matrix group acting on $\mathbf{x} = (x_1, \ldots, x_n)$ and let k be any field. A polynomial $f(\mathbf{x}) \in k[\mathbf{x}]$ is *invariant under* G if $f(\mathbf{x}) = f(A \cdot \mathbf{x})$ for every $A \in G$.

Example. Let $B = \left(\begin{smallmatrix} 0 & -1 \\ 1 & 0 \end{smallmatrix}\right)$. Then the set $C_4 = \{I_2, B, B^2, B^3\}$ is a group, and for the polynomial $p(\mathbf{x}) = p(x_1, x_2) = \frac{1}{2}(x_1^2 + x_2^2)$ we have that $p(\mathbf{x}) = p(B\mathbf{x})$, $p(\mathbf{x}) = p(B^2\mathbf{x})$, and $p(\mathbf{x}) = p(B^3\mathbf{x})$. Thus $p(\mathbf{x})$ is an invariant of the group C_4.

Consider the group of rotations of the phase space

$$x' = e^{-i\phi}x, \quad y' = e^{i\phi}y. \tag{2.1}$$

In (x', y') coordinates system (1.16) has the form

$$\dot{x}' = \sum_{(p,q)\in\widetilde{S}} a(\phi)_{(p,q)} x'^{p+1} y'^{q}, \quad \dot{y}' = \sum_{(p,q)\in\widetilde{S}} b(\phi)_{(q,p)} x'^{q} y'^{p+1},$$

and the coefficients of the transformed system are

$$a(\phi)_{p_k q_k} = a_{p_k q_k} e^{i(p_k - q_k)\phi}, \quad b(\phi)_{q_k p_k} = b_{q_k p_k} e^{i(q_k - p_k))\phi}, \tag{2.2}$$

where $k = 1, \ldots, \ell$. Let U_ϕ denote the transformation (2.2), so that U_ϕ is the diagonal $2\ell \times 2\ell$ matrix

$$U_\phi = \begin{pmatrix} U_\phi^{(a)} & 0 \\ 0 & U_\phi^{(b)} \end{pmatrix},$$

where $U_\phi^{(a)}$ and $U_\phi^{(b)}$ are diagonal matrices that act on the coordinates a and b, respectively. For example, for the system

$$\dot{x} = a_{00}x + a_{-1,1}y + a_{20}x^3, \quad \dot{y} = b_{-1,1}x + b_{00}y + b_{02}y^3$$

we have

$$U_\phi(a,b) = \begin{pmatrix} U_\phi^{(a)} & 0 \\ 0 & U_\phi^{(b)} \end{pmatrix} (a,b)^T =$$

$$\begin{pmatrix} 1 & 0 & 0 & 0 & 0 & 0 \\ 0 & e^{-i2\phi} & 0 & 0 & 0 & 0 \\ 0 & 0 & e^{i2\phi} & 0 & 0 & 0 \\ 0 & 0 & 0 & e^{-i2\phi} & 0 & 0 \\ 0 & 0 & 0 & 0 & e^{i2\phi} & 0 \\ 0 & 0 & 0 & 0 & 0 & 1 \end{pmatrix} \begin{pmatrix} a_{00} \\ a_{-1,1} \\ a_{20} \\ b_{02} \\ b_{1,-1} \\ b_{00} \end{pmatrix} = \begin{pmatrix} a_{00} \\ a_{-1,1}e^{-i2\phi} \\ a_{20}e^{i2\phi} \\ b_{02}e^{-i2\phi} \\ b_{1,-1}e^{i2\phi} \\ b_{00} \end{pmatrix},$$

so here

$$U_\phi^{(a)} = \begin{pmatrix} 1 & 0 & 0 \\ 0 & e^{-i2\phi} & 0 \\ 0 & 0 & e^{i2\phi} \end{pmatrix} \quad \text{and} \quad U_\phi^{(b)} = \begin{pmatrix} e^{-i2\phi} & 0 & 0 \\ 0 & e^{i2\phi} & 0 \\ 0 & 0 & 1 \end{pmatrix}.$$

Note that $U_\phi^{(a)}$ and $U_\phi^{(b)}$ do not really depend on a and b; rather, the notation simply indicates that $U_\phi^{(a)}$ acts on the vector of coefficients of the first equation of (1.16) and $U_\phi^{(b)}$ acts on the vector of coefficients of the second equation of (1.16).

We will write (2.2) in the short form

$$(a(\phi), b(\phi)) = U_\phi(a, b) = (U_\phi^{(a)} a, U_\phi^{(b)} b).$$

It follows from (2.2) that $\mathcal{U} = \{U_\phi\}$ is a group. We call $\mathcal{U}$ the *rotation group* of the system (1.16) (not to be confused with the group (2.1), which is the group of rotations of the phase plane, and is not associated with any particular system of differential equations). Here we are studying polynomial invariants of the group $\mathcal{U}$, which we call the *invariants of the rotation group*, or more simply the *invariants*. It is straightforward to see that a polynomial $f(a, b) \in \mathbb{C}[a, b]$ is an invariant of the group $\mathcal{U}$ if and only if each of its terms is an invariant.

We associate to system (1.16) the linear operator $L(\nu)$ defined by (1.20) (with $(p_m, q_m) \in \widetilde{S}$) and as above we denote by $\widetilde{\mathcal{M}}$ the monoid of all integer solutions $\nu = (\nu_1, \nu_2, \ldots, \nu_{2\ell})$ with non-negative components of the equation (1.21).

It is easily seen that there is a slightly different way to define this monoid. Consider the Diophantine equation

$$(p_1-q_1)\nu_1+(p_2-q_2)\nu_2+\cdots+(p_\ell-q_\ell)\nu_\ell+(q_\ell-p_\ell)\nu_{\ell+1}+\cdots+(q_1-p_1)\nu_{2\ell} = 0 \tag{2.3}$$

which is obtained by subtracting the second equation of system (1.21) from the first one.

Proposition 2.2. *The set of all non-negative integer solutions of equation* (2.3) *coincides with the monoid* $\widetilde{\mathcal{M}}$ *defined by equation* (1.21).

Proof. Obviously every solution of (1.21) is also a solution of (2.3). Conversely, let ν be a solution of (2.3) and let $e_i = (0, \ldots, 0, 1, 0, \ldots, 0)$ be the ith basis vector of $\mathbb{C}^{2\ell}$. Then

$$L^1(\nu) = L^2(\nu) = k, \tag{2.4}$$

yielding

$$L^1(\nu) + L^2(\nu) = 2k. \tag{2.5}$$

Note that

$$L^1(e_i) + L^2(e_i) = L^1(e_{2\ell-i}) + L^2(e_{2\ell-i}) = p_i + q_i \geq 0 \tag{2.6}$$

for $i = 1, \ldots, \ell$. Taking into account the fact that $L(\nu)$ is a linear operator, we conclude from (2.5) and (2.6) that the number k on the right-hand side of (2.4) is non-negative. □

For $\nu = (\nu_1, \ldots, \nu_{2\ell}) \in \mathbb{N}^{2\ell}$ define

$$(a, b)^{(\nu)} := [\nu] = a_{p_1 q_1}^{\nu_1} \cdots a_{p_\ell q_\ell}^{\nu_\ell} \, b_{q_\ell p_\ell}^{\nu_{\ell+1}} \cdots b_{q_1 p_1}^{\nu_{2\ell}}.$$

Then the action of U_ϕ on the coefficients a_{ij}, b_{ji} of the system of differential equations (1.16) yields the following transformation of the monomial $(a, b)^{(\nu)}$:

$$\begin{aligned}(a(\phi), b(\phi))^{(\nu)} &:= a(\phi)_{p_1 q_1}^{\nu_1} \cdots a(\phi)_{p_\ell q_\ell}^{\nu_\ell} \, b(\phi)_{q_\ell p_\ell}^{\nu_{\ell+1}} \cdots b(\phi)_{q_1 p_1}^{\nu_{2\ell}} \\ &= e^{i\phi(L^1(\nu)-L^2(\nu))} a_{p_1 q_1}^{\nu_1} \cdots a_{p_\ell q_\ell}^{\nu_\ell} \, b_{q_\ell p_\ell}^{\nu_{\ell+1}} \cdots b_{q_1 p_1}^{\nu_{2\ell}} \\ &= e^{i\phi(L^1(\nu)-L^2(\nu))} (a, b)^{(\nu)}.\end{aligned} \tag{2.7}$$

Thus we have the following characterization of the invariants.

Lemma 2.3. *The monomial* $[\nu] = (a, b)^{(\nu)}$ *is invariant under* $\mathcal{U}$ *if and only if* $L^1(\nu) = L^2(\nu)$, *i.e., if and only if* $\nu \in \widetilde{\mathcal{M}}$.

Definition 2.4. The vector

$$\zeta = (p_1 - q_1, \ldots, p_\ell - q_\ell, q_\ell - p_\ell, \ldots, p_1 - q_1)$$

is the *characteristic vector* of system (1.16). The non-negative integer

$$\{\zeta\} = GCD(\zeta) = GCD(p_1 - q_1, \ldots, p_\ell - q_\ell)$$

is the *characteristic number* of system (1.16). The vector $\kappa = \frac{\zeta}{\{\zeta\}}$ is the *reduced characteristic vector* of system (1.16).

Consider the equation

$$L^1(\nu) - L^2(\nu) = (p_1 - q_1)\nu_1 + \cdots + (p_\ell - q_\ell)\nu_\ell + (q_\ell - p_\ell)\nu_{\ell+1} + \cdots + (q_1 - p_1)\nu_{2\ell} = \{\zeta\}, \tag{2.8}$$

which can be written in the equivalent form

$$\zeta_1\nu_1 + \cdots + \zeta_\ell\nu_\ell + \zeta_{\ell+1}\nu_{\ell+1} + \cdots + \zeta_{2\ell}\nu_{2\ell} = \{\zeta\}.$$

Lemma 2.5. *If* $\zeta \neq 0$ *then equation* (2.8) *has a solution* $\nu \in \mathbb{N}_+^{2\ell}$ *for which* $\nu_i \geq 0$ *for all* i *and* $\nu_i > 0$ *for at least one* i.

Proof. The equation $(p_1 - q_1)t_1 + \cdots + (p_\ell - q_\ell)t_\ell = \{\zeta\}$ has a solution $t \in \mathbb{Z}^\ell$ because $\{\zeta\}$ is a generator of the ideal $\langle p_1 - q_1, \ldots, p_\ell - q_\ell\rangle$. Write $t' = t'^+ - t'^-$, where $t_i'^+ \geq 0$ and $t_i'^- \geq 0$ for all i. Then $\nu = (t'^+, \hat{t}'^-) \in \mathbb{N}_+^{2\ell}$ (where $\hat{t}'^-$ is the involution of the vector t'^- defined by $\hat{t}_j'^- = t_{\ell-j}'^-$) is a solution of (1.21) with non-negative coordinates and such that $|\nu| > 0$. □

We will write $z = (z_1, z_2, \ldots, z_{2\ell})$ for the vector (a, b), so that

$$z_i = \begin{cases} a_{p_i, q_i} & \text{if } 1 \leq i \leq \ell \\ b_{q_{2\ell+1-i}, p_{2\ell+1-i}} & \text{if } \ell + 1 \leq i \leq 2\ell. \end{cases}$$

A pair of variables z_i and z_j in z are *conjugate variables* provided $z_i = z_s$ and $z_j = z_{2\ell - s}$ for some s, so that $(z_i, z_j) = (a_{p_s q_s}, b_{q_s p_s})$. Invariants that depend on only one pair of conjugate variables are *unary* invariants; invariants that depend on two pairs of conjugate variables are *binary* invariants. Thus for example for system (4.1) considered below, a_{11} and $a_{02}b_{20}$ are the unary invariants and $a_{20}a_{02}$ and $b_{20}^4 a_{-13}^2$ are the binary invariants.

Lemma 2.6. (1) *The unary irreducible invariants of system* (1.16) *have the form* a_{pp}, b_{pp}, $a_{pq}b_{qp}$.

(2) *The binary irreducible invariants of system* (1.16) *are*

$$z_r^{\frac{|\zeta_s|}{GCD(\zeta_r, \zeta_s)}} z_s^{\frac{|\zeta_r|}{GCD(\zeta_r, \zeta_s)}}, \tag{2.9}$$

where $\zeta_r\zeta_s < 0$.

Proof. (1) From Lemma 2.3 we have that z_s is a unary invariant if and only if the corresponding coordinate of the characteristic vector ζ is equal to zero, that is, if and only if the corresponding coefficient of system (1.16), $a_{p_s q_s}$ or $b_{q_s p_s}$, satisfies $p_s = q_s$. Similarly, a monomial

$$a_{pq}^{\mu_1} b_{qp}^{\mu_2} \tag{2.10}$$

is a unary invariant if and only if (μ_1, μ_2) is a positive solution to the equation $\mu_1(p-q) + \mu_2(q-p) = 0$. Therefore $\mu_1 = \mu_2$, and the only irreducible invariant of the form (2.10) is $a_{pq} b_{qp}$.

(2) Again, by Lemma 2.3 and Proposition 2.2 the monomial $z_r^{\mu_1} z_s^{\mu_2}$ is an irreducible invariant if and only if $\mu = (\mu_r, \mu_s)$ is the minimal non-negative solution to $\mu_r \zeta_r + \mu_s \zeta_s = 0$ such that $|\mu| > 0$. This equation has such a solution if and only if $\zeta_r \zeta_s < 0$, and in such a case the solution is

$$\mu_r = \frac{|\zeta_s|}{GCD(\zeta_r, \zeta_s)}, \quad \mu_s = \frac{|\zeta_r|}{GCD(\zeta_r, \zeta_s)}.$$

□

Remark. In terms of the original variables $a_{p_k q_k}$ and $b_{q_k p_k}$ the second statement of the lemma is that the binary irreducible invariants are

$$a_{p_i q_i}^{\frac{|p_j - q_j|}{GCD(q_j - p_j, p_i - q_i)}} a_{p_j q_j}^{\frac{|p_i - q_i|}{GCD(q_j - p_j, p_i - q_i)}} \quad \text{and} \quad a_{p_i q_i}^{\frac{|p_j - q_j|}{GCD(q_j - p_j, p_i - q_i)}} b_{p_j q_j}^{\frac{|p_i - q_i|}{GCD(q_j - p_j, p_i - q_i)}}$$

according as $(p_i - q_i)(p_j - q_j) < 0$ and $(p_i - q_i)(p_j - q_j) > 0$, respectively.

3. Invariants and Time-Reversibility

For $h \in \mathbb{C}^{2\ell}$ let $R(h)$ be the set of indices of the non-zero coordinates of the vector h.

Lemma 3.1. *If $g, h \in E(a,b)$, $R(g) = R(h)$, and for all unary and binary invariants $J(a,b)$ the condition*

$$J(g) = J(h) \tag{3.1}$$

holds, then for any $r, s \in R(h) = R(g)$

$$\left(\frac{h_r}{g_r}\right)^{\kappa_s} = \left(\frac{h_s}{g_s}\right)^{\kappa_r}. \tag{3.2}$$

Proof. For $r, s \in R(h)$, none of h_r, h_s, g_r, and g_s is zero. If $\kappa_s = 0$ (that is, $p_s - q_s = 0$) then z_s (which is equal to $a_{p_s p_s}$ or to $b_{2\ell+1-s, 2\ell+1-s}$) is a unary invariant. By (3.1), $h_s = g_s$ and therefore (3.2) holds. The case $\kappa_r = 0$ is similar.

Now suppose $\kappa_r \kappa_s < 0$. If $r = s$ then certainly (3.2) holds. If $r \neq s$ then the binary invariants are given by (2.9). Therefore using (3.1) we obtain

$$h_r^{|\kappa_s|} h_s^{|\kappa_r|} = g_r^{|\kappa_s|} g_s^{|\kappa_r|} \tag{3.3}$$

(here we have raised both parts of (3.1) to the power $GCD(\zeta_s, \zeta_r)/\{\zeta\}$). Considering the two possibilities $\kappa_r < 0 < \kappa_s$ and $\kappa_r > 0 > \kappa_s$ we see that (3.3) yields (3.2).

Finally, consider the case $\kappa_r\kappa_s > 0$. Suppose κ_r and κ_s are positive, and note that $\kappa_i = -\kappa_{2\ell+1-i}$ for all $i = 1, \ldots, 2\ell$. From (2.9) we see that $z_r^{\kappa_s} z_{2\ell+1-s}^{\kappa_r}$ is a binary invariant. Using (3.1) we get

$$h_r^{\kappa_s} h_{2\ell+1-s}^{\kappa_r} = g_r^{\kappa_s} g_{2\ell+1-s}^{\kappa_r}.$$

Multiplying both sides of this equation by $h_s^{-\kappa_r} h_{2\ell+1-s}^{-\kappa_r} = g_s^{-\kappa_r} g_{2\ell+1-s}^{-\kappa_r}$ (which follows from (3.1) because $z_s z_{2\ell+1-s}$ is a unary invariant) we obtain (3.2). The case that κ_r and κ_s are both negative is similar. □

For $\sigma \in \mathbb{R}$ we denote by $\sigma \cdot h$ the vector $(\sigma^{-\zeta_1} h_1, \ldots, \sigma^{-\zeta_{2\ell}} h_{2\ell})$. The next theorem shows that if the values of all unary and binary invariants are the same on the ordered vectors g and h of the coefficients of two systems (1.16), and if $R(g) = R(h)$, then these systems can be transformed to each other by a transformation of the form (1.12).

Theorem 3.2. *The equation*

$$\sigma \cdot h = U_\phi g \tag{3.4}$$

has a solution with respect to $\sigma \neq 0$ and ϕ if and only if $R(h) = R(g)$ and for all unary and binary invariants $J(a, b)$ the condition $J(g) = J(h)$ holds.

Proof. If for some ϕ_0 and non-zero σ_0 the equality $\sigma_0 \cdot h = U_{\phi_0} g$ holds, then from (2.2) we see that $R(h) = R(g)$ and (3.1) holds not only for the unary and binary invariants but in fact for all invariants (by Lemma 2.3).

To prove the converse we must show that if for all unitary and binary invariants (3.1) holds, and if $R(g) = R(h)$, then the system

$$h_r = g_r \sigma^{\zeta_r} e^{i\zeta_r \phi} \quad (i = 1, \ldots, 2\ell) \tag{3.5}$$

has a solution, that is, that there are $\sigma_0 \neq 0$ and $\phi_0 \in [0, 2\pi)$ such that

$$h_r = g_r \sigma_0^{\zeta_r} e^{i\zeta_r \phi_0} \quad (i = 1, \ldots, 2\ell). \tag{3.6}$$

If $\zeta = 0$ then all the irreducible invariants are $z_1, \ldots, z_{2\ell}$ (that is, $a_{p_s p_s}, b_{p_s p_s}$, where $s = 1, \ldots, \ell$), and, in this case, for the unary invariants (3.1) is just the equality $h_r = g_r$, so that (3.6) holds with any σ and ϕ_0.

Hence suppose $\zeta \neq 0$. We consider two separate cases:

1. All the coordinates of the vectors g and h are different from zero;
2. Some coordinates of h and g are equal to zero.

Case 1. Let μ be a non-negative solution to (2.8) as provided by Lemma 2.5. For $s \in \mathbb{C}^{2\ell}$ we write

$$s^{(\mu)} := s_1^{\mu_1} s_2^{\mu_2} \cdots s_{2\ell}^{\mu_{2\ell}}.$$

By our assumption $g^{(\mu)} \neq 0$ and $\{\zeta\} \neq 0$, so the equation $h^{(\mu)} = g^{(\mu)} \sigma^{\{\zeta\}} e^{i\{\zeta\}\phi}$ has a solution $\sigma = \sigma_0$, $\phi = \phi_0$:

$$h^{(\mu)} = g^{(\mu)} \sigma_0^{\{\zeta\}} e^{i\{\zeta\}\phi_0}. \tag{3.7}$$

We will now show that in fact (3.6) is equivalent to (3.7). Indeed, if $\phi = \phi_0$, $\sigma = \sigma_0$ satisfy (3.6) then (3.7) holds because μ is a solution to (2.8). Conversely, suppose (3.7) holds. For any $r \in R(h)$

$$\left(h^{(\mu)}\right)^{\kappa_r} = \left(g^{(\mu)}\right)^{\kappa_r} \sigma_0^{\zeta_r} e^{i\zeta_r\phi_0}. \tag{3.8}$$

According to Lemma 3.1, for any r and s in $R(h)$ we have equality (3.2). Hence

$$\prod_{s=1}^{2\ell} h_r^{\kappa_s\mu_s} (h_s^{\mu_s})^{-\kappa_r} = \prod_{s=1}^{2\ell} g_r^{\kappa_s\mu_s} (g_s^{\mu_s})^{-\kappa_r} \tag{3.9}$$

which yields

$$h_r^{\sum_{s=1}^{2\ell} \kappa_s\mu_s} \left(h^{(\mu)}\right)^{-\kappa_r} = g_r^{\sum_{s=1}^{2\ell} \kappa_s\mu_s} \left(g^{(\mu)}\right)^{-\kappa_r}$$

and, therefore,

$$\left(h^{(\mu)}\right)^{\kappa_r} = \frac{h_r}{g_r} \left(g^{(\mu)}\right)^{\kappa_r}$$

(here we have used the fact that $\sum_{s=1}^{2\ell} \kappa_s\mu_s = \sum_{s=1}^{2\ell} \zeta_s\mu_s/\{\zeta\} = 1$). Substituting this expression for $\left(h^{(\mu)}\right)^{\kappa_r}$ into (3.8) we see that σ_0 and ϕ_0 satisfy (3.6).

Case 2. Assume now that some coordinates of the vector g are equal to zero. By our assumption that $R(h) = R(g)$ the same coordinates of h are also equal to zero.

Consider the system of equations (3.5) corresponding to the coordinates of g such that $g_r \neq 0$. If all ζ_r corresponding to these g_r are equal to zero then system (3.5) is satisfied for any ϕ and $\sigma \neq 0$ (in this case the corresponding polynomials z_r are unary invariants). Otherwise, in order to see that there is a solution of the system composed of the remaining equations of (3.5),

$$h_r = g_r \sigma^{\zeta_r} e^{i\zeta_r\phi} \quad (r \in R(g)),$$

we proceed as above, with the only difference that in (3.7) and (3.9) we take the product over all non-zero coordinates of the vector g, that is, over $R(g)$, and instead of $\{\zeta\}$ we use the number $\{\zeta(g)\}$, which is the *greatest common divisor of the coordinates ζ_i corresponding to the non-zero entries of the vector g.* (Another possible point of view for this case is that, in the set $\widetilde{S}$, we drop the indices corresponding to zero entries of g and then obtain from (1.16) a system of the same form but with the vector of coefficients containing no zero entry.) □

Remark. It is obvious that if (3.1) holds for all unary and binary invariants from a basis of the monoid $\widetilde{\mathcal{M}}$ then it holds for all unary and binary invariants (since every unary and binary invariant is a product of unary and binary invariants from a basis of $\widetilde{\mathcal{M}}$). Hence in the statements of Theorem 3.2 it is sufficient to require the fulfillment of condition (3.1) not for *all* unary and binary invariants but only for unary and binary invariants from a basis of $\widetilde{\mathcal{M}}$.

It is also easy to see that if (3.1) is fulfilled for unary and binary invariants then it is fulfilled for all invariants.

Theorem 3.3. *If $R(g) = R(h)$, and if $J(g) = J(h)$ holds for all unary and binary invariants (or, equivalently, for unary and binary invariants from a basis of $\widetilde{\mathcal{M}}$) then $J(g) = J(h)$ holds for all invariants $J(z)$.*

Proof. It is sufficient to consider only monomial invariants. Let $z^{(\theta)}$ be such an invariant and let σ_0 and ϕ_0 be solutions to (3.7) provided by Theorem 3.2. Then

$$h^{(\theta)} = g^{(\theta)} \sigma_0^{\zeta_1\theta_1+\cdots+\zeta_{2\ell}\theta_{2\ell}} e^{i(\zeta_1\theta_1+\cdots+\zeta_{2\ell}\theta_{2\ell})\phi_0}.$$

However, $z^{(\theta)}$ is an invariant. Therefore by Proposition 2.2 and Lemma 2.3 $\zeta_1\theta_1 + \cdots + \zeta_{2\ell}\theta_{2\ell} = 0$. Hence $h^{(\theta)} = g^{(\theta)}$. □

We let $[\nu]|_g$ denote the value of the monomial $[\nu] = z_1^{\nu_1} \cdots z_{2\ell}^{\nu_{2\ell}}$ at the vector g; that is, $[\nu]|_g = g^{(\nu)}$. We let (b, a) denote the involution of (a, b); that is, $(b, a) = (b_{q_1 p_1}, \ldots, b_{q_l p_l}, a_{p_l q l}, \ldots, a_{p_1 q_1})$.

We can now prove Theorem 1.6.

Proof of Theorem 1.6. Conclusion (1) of the theorem follows from formula (1.23) and the definition of $\widetilde{I}_{sym}$. As to conclusion (2), we begin by noting that, for fixed (a, b), the existence of (p, q) in S for which $a_{pq} b_{qp} = 0$ but $a_{pq} + b_{qp} \neq 0$ means the same thing as $R(a, b) \neq R(b, a)$. Thus for systems (a, b) from $\mathbf{V}(\widetilde{I}_{sym}) \setminus \mathcal{R}$ we have that $R(a, b) = R(b, a)$ and

$$[\nu]|_{(a,b)} = [\hat{\nu}]|_{(a,b)} \tag{3.10}$$

for all $\nu \in \widetilde{\mathcal{M}}$. However, the set $\{[\nu] \mid \nu \in \widetilde{\mathcal{M}}\}$ contains all unary and binary invariants of the system (1.16). Thus (3.10) means that $J(a, b) = J(b, a)$ for all such invariants. Thus by Theorem 3.2 the system $\sigma \cdot (b, a) = U_\phi(a, b)$ has a solution $\{U_{\phi_0}, \sigma_0\}$. Then $\alpha = \sigma_0 e^{i\phi_0}$ satisfies the equation (1.23), which implies that the system (a, b) is a time-reversible one. □

4. A Computational Algorithm

Theorem 1.6 implies that in order to find all time-reversible systems within a given polynomial family (1.16) it is sufficient to find a basis of the Sibirsky ideal of the system, or, equivalently, a Hilbert basis (i.e., a minimal generating set of the corresponding monoid $\widetilde{\mathcal{M}}$, which in [11] is called the set of "elementary Lie invariants"). Such a basis can be found by examining solutions of (1.21) through a step-by-step sorting ([11, 14]). An algorithmic method for computing $\widetilde{I}_{sym}$ and a basis of $\widetilde{\mathcal{M}}$ was proposed in [8]. It yields the following algorithm for finding all time-reversible systems within the family (1.16).

The correctness of the first two steps of the algorithm is proven in [8]; the correctness of step 3 follows from Theorem 3 (of course, the set $\{G \cap \mathbb{C}[a, b]\}$ also defines the variety of $\widetilde{I}_{sym}$, but as rule the set K is essentially smaller).

Algorithm (for finding time-reversible systems).

Input: Two sequences of integers $p_1, \dots, p_\ell$ ($p_i \geq -1$) and $q_1, \dots, q_\ell$ ($q_i \geq 0$). (These are the coefficient labels for a system of the form (1.16).)

Output: A finite set of generators for the Sibirsky ideal $\widetilde{I}_{sym}$ of (1.16) and a set K of polynomials defining the variety of $\widetilde{I}_{sym}$.

1. Compute a reduced Gröbner basis G for the ideal

$$\mathcal{J} = \langle a_{p_i q_i} - y_i t_1^{p_i} t_2^{q_i}, b_{q_i p_i} - y_{\ell-i+1} t_1^{q_{\ell-i+1}} t_2^{p_{\ell-i+1}} \mid i = 1, \dots, \ell \rangle \subset \mathbb{C}[a, b, y_1, \dots, y_\ell, t_1^{\pm}, t_2^{\pm}]$$

with respect to any elimination ordering for which

$$\{t_1, t_2\} > \{y_1, \dots, y_d\} > \{a_{p_1 q_1}, \dots, b_{q_1 p_1}\}.$$

2. $\widetilde{I}_{sym} = \langle G \cap \mathbb{C}[a, b] \rangle$.

3. K is formed by the generators of $\widetilde{I}_{sym}$ whose terms are unary or binary invariants.

We demonstrate how the algorithm works using as an example the system studied in [10]:

$$\begin{aligned} \dot{x} &= x(1 - a_{20}x^2 - a_{11}xy - a_{02}y^2 - a_{-13}x^{-1}y^3 - a_{22}x^2y^2), \\ \dot{y} &= -y(1 - b_{3,-1}x^3y^{-1} - b_{20}x^2 - b_{11}xy - b_{02}y^2 - b_{22}x^2y^2)\,. \end{aligned} \tag{4.1}$$

Computing a Gröbner basis of the ideal

$$\mathcal{J} = \langle a_{11} - t_1 t_2 y_1, b_{11} - t_1 t_2 y_1, a_{20} - t_1^2 y_2, b_{02} - t_2^2 y_2, a_{02} - t_2^2 y_3, b_{20} - t_1^2 y_3, a_{-13} - \frac{t_2^3 y_4}{t_1}, b_{3,-1} - \frac{t_1^3 y_4}{t_2}, a_{22} - t_1^2 t_2^2 y_5, b_{22} - t_1^2 t_2^2 y_5 \rangle$$

with respect to lexicographic order with

$$t_1 > t_2 > y_1 > y_2 > y_3 > y_4 > y_5 > a_{11} > b_{11} > a_{20} > b_{20} > a_{02} > b_{02} > a_{-13} > b_{3,-1} > a_{22} > b_{22}$$

we obtain a list of polynomials. (The list is too long to be presented here, but is easily computed using any standard computer algebra system.) According to step 2 of the algorithm above, in order to get a basis of $\widetilde{I}_{sym}$ we just have to pick up the polynomials that do not depend on $t_1, t_2, y_1, y_2, y_3, y_4, y_5$. In fact,

$$\widetilde{I}_{sym} = \langle a_{-13}\, b_{20}^2 - a_{02}^2\, b_{3,-1}, a_{02}\, a_{20} - b_{02}\, b_{20}, a_{-13}\, a_{20}\, b_{20} - a_{02}\, b_{02}\, b_{3,-1}, a_{-13}\, a_{20}^2 - b_{02}^2\, b_{3,-1}, a_{11} - b_{11}, a_{22} - b_{22} \rangle.$$

Now the third step of the algorithm yields

Theorem 4.1. *System* (4.1) *is time-reversible if and only if its coefficients satisfy the system of simultaneous equations*

$$a_{-13}\,b_{20}^2 - a_{02}^2\,b_{3,-1} = a_{02}\,a_{20} - b_{02}\,b_{20}$$
$$= a_{-13}\,a_{20}^2 - b_{02}^2\,b_{3,-1} = a_{11} - b_{11} = a_{22} - b_{22} = 0$$

and a_{ks} *and* b_{sk} *vanish simultaneously.*

References

[1] V. V. Amelkin, N. A. Lukashevich, A. P. Sadovskii, *Nonlinear Oscillations in Second Order Systems.* Minsk: BSU, 1982.

[2] Y. N. Bibikov, *Local Theory of Nonlinear Analytic Ordinary Differential Equations.* Lecture Notes in Math. vol. 702. Springer-Verlag, 1979.

[3] A. Cima, A. Gasull, V. Mañosa, F. Manosas, Algebraic properties of the Liapunov and period constants. *Rocky Mountain J. Math.* **27** (1997), 471–501.

[4] C. Christopher, J. Llibre, Algebraic aspects of integrability for polynomial systems. *Qual. Theory Dyn. Syst.* **1** (1999), 71–95.

[5] C. B. Collins, Poincaré's reversibility condition. *J. Math. Anal. Appl.* **259** (2001), 168–187.

[6] D. Cox, J. Little, D. O'Shea, *Ideals, Varieties, and Algorithms,* Springer-Verlag, 1992.

[7] H. Dulac, Détermination et intégration d'une certaine classe d'équations différentielles ayant pour point singulier un centre. *Bull. Sci. Math.* **32** (1908), 230–252.

[8] A. Jarrah, R. Laubenbacher, V. G. Romanovski, The Sibirsky component of the center variety of polynomial systems. *J. Symb. Comput.* **35** (2003), 577–589.

[9] J. S. W. Lamb, J. A. G. Roberts, Time-reversal symmetry in dynamical systems: a survey. *Phys. D* **112** (1998), 1–39.

[10] Y.-R. Liu, W.-T. Huang, A new method to determine isochronous center conditions for polynomial differential systems. *Bull. Sci. Math.* **127** (2003), 133–148.

[11] Y.-R. Lui, J.-B. Li, Theory of values of singular point in complex autonomous differential systems. *Sci. China Ser. A* **33** (1989), 10–23.

[12] V. G. Romanovski, M. Robnik, The center and isochronicity problems for some cubic systems. *J. Phys. A: Math. Gen.* **34** (2001), 10267–10292.

[13] C. Rousseau, D. Schlomiuk, Cubic vector fields symmetric with respect to a center. *J. Differential Equations* **123** (1995), 388–436.

[14] K. S. Sibirskii, *Algebraic Invariants of Differential Equations and Matrices.* Shtiintsa, 1976 (in Russian).

[15] K. S. Sibirsky, *Introduction to the Algebraic Theory of Invariants of Differential Equations.* Shtiintsa, 1982 (in Russian). English translation: Manchester University Press, 1988.

[16] B. Sturmfels, *Algorithms in Invariant Theory.* Springer-Verlag, 1993.

[17] D. Wang, Polynomial systems from certain differential equations. *J. Symb. Comput.* **28** (1999), 303–315.

[18] H. Żołądek, The classification of reversible cubic systems with center. *Topol. Methods Nonlinear Anal.* **4** (1994), 79–136.

Acknowledgment

The first author acknowledges the support of the work by the Ministry of Higher Education, Science and Technology of the Republic of Slovenia and by the Nova Kreditna Banka Maribor.

Valery G. Romanovski
Center for Applied Mathematics and Theoretical Physics
University of Maribor, Krekova 2
SI-2000 Maribor
Slovenia
e-mail: valery.romanovsky@uni-mb.si

Douglas S. Shafer
Department of Mathematics
University of North Carolina at Charlotte
Charlotte, NC 28223
USA
e-mail: dsshafer@uncc.edu

Trends in Mathematics: *Differential Equations with Symbolic Computation*, 85–107

On Symbolic Computation of the LCE of N-Dimensional Dynamical Systems

Shucheng Ning and Zhiming Zheng

Abstract. In the present paper, based on our earlier work, we propose a systematic method for symbolically computing the Lyapunov characteristic exponents, briefly LCE, of n-dimensional dynamical systems. First, we analyze in mathematics the LCE of n-dimensional dynamical systems. In particular, as an example, we discuss the LCE of the Lorenz systems. Then, to do the above task, a framework on representation and manipulation of a class of non-algebraic objects using non-standard analysis (NSA) is established. In this framework, an algorithm can be developed for deriving some unknown relations on some objects involving limit processes. Finally, applying this algorithm to n-dimensional dynamical systems, we can show that their maximal LCE can be derived mechanically; particularly, for the Lorenz systems, we obtain an important result on the maximal LCE of the chaotic attractors of these systems — their dependence on the systems parameters.

Mathematics Subject Classification (2000). Primary 58F13; Secondary 68Q40.

Keywords. Dynamical system, LCE, symbolic computation, non-algebraic object.

1. Introduction

Using computer algebra systems and symbolic computation techniques in dynamical systems research is an important recent development [8, 16, 13, 12, 11]. In the present paper the authors want to employ the techniques to establish a systematic method to compute the maximal LCE, say $h(\alpha)$, of n-dimensional chaotic systems, where $\alpha = (\alpha_1, \ldots, \alpha_m)$ are the systems parameters. As an example, we apply the method to compute the LCE of Lorenz systems. As we know, one of the most important methods in dynamical systems theory, used to analyze the behavior of chaotic dynamical systems, is to discuss their Lyapunov characteristic exponents, of which the maximal one is of special interest for us. Although the

Research supported by the National 863 Project and the NSF of China.

concept of LCE seems simple, it is never easy to get accurate LCE in practice when we meet a chaotic system. Thus, numerical methods have to be admitted. Such methods are used in many papers for studying the LCE of dynamical systems; unfortunately, the numerical results cannot satisfy many of the requirements in theoretical analysis. For example, how do the LCE depend on the corresponding systems parameters? Instead of numerically computing the LCE of attractors of chaotic dynamical systems, in the present paper we try to give a systematic method for finding LCE dependence on the systems parameters by symbolic computation for n-dimensional dynamical systems, and then apply it to investigate LCE of Lorenz systems of the following form:

$$\begin{aligned}\dot{x}_1 &= -\sigma x_1 + \sigma x_2,\\ \dot{x}_2 &= -x_1 x_3 + r x_1 - x_2,\\ \dot{x}_3 &= x_1 x_2 - b x_3,\end{aligned}$$

where σ, b and r are the systems parameters. One of the main results in the present paper is the following.

Main Result 1. For n-dimensional dynamical systems, some symbolic expressions satisfied by the systems parameters and the maximal LCE of their attractors can be derived mechanically.

Since computing the LCE of a dynamical system is a non-algebraic problem, which involves limit process, the algorithm constructed in this paper is beyond, in principle, the conventional symbolic computation methods (for example, the Gröbner basis method and Ritt–Wu's method), which are in nature algebraic and not appropriate for dealing with mathematical problems with non-algebraic background. Towards solving the above problem, our idea, based on our earlier work [13, 12, 11, 5, 18], is to extend the methods of algebraic equation solving to a class of non-algebraic objects. More precisely, our approach is first to introduce infinitesimals to describe in mathematics a class of non-algebraic objects, called *analytic equalities*, and the problem of computing LCE can be reduced into the category of the above non-algebraic objects; then we enlarge traditional algebraic axiom system by introducing infinitesimals such that we can develop a computing method for the analytic equalities, which leads to an algorithm for constructing necessary conditions for a set of analytic equalities to hold. The algorithm can be applied to mechanically derive unknown relations (unknown mathematical theorems). Armed with it, we try to approach some problems related to dynamical systems; in particular, we obtain another result on the LCE of Lorenz systems, which can be briefly described as follows.

Main Result 2. For the Lorenz systems, there exists an algorithm with inputs σ_0 and b_0, which can generate an analytic equality $\phi(\sigma_0, b_0, r, \widehat{h}) = 0$ automatically, where $\widehat{h} \in \widetilde{R}$ and $\widetilde{R}$ is the field of real numbers of non-standard analysis (NSA), and the maximal LCE, say h, of the Lorenz systems is the standard part of $\widehat{h}$, i.e.,

$h = \langle \widehat{h} \rangle \in R$, and R is the field of real numbers. Moreover, $\widehat{h}$ is one of the roots of $\phi(\sigma_0, b_0, r, y) = 0$.

We remark that when we investigate some problems of dynamical systems with the algorithm, we may meet some computing obstructions due to the complexity of the corresponding symbolic computation. However, the difficulties can be overcome from our understanding for some qualitative properties of the concerned dynamical systems or by combining the algorithm with other computing methods (see Section 3 and [11, 18, 15]).

This paper is arranged as follows. In Section 2, we show that a system of polynomial equations associating with the LCE of n-dimensional dynamical systems can be obtained and, as an example, the LCE of the Lorenz systems are discussed. In Section 3, we describe how to represent and manipulate the non-algebraic objects of our interest using non-standard analysis. In Section 4, two results on n-dimensional dynamical systems and the LCE of the Lorenz systems are obtained symbolically by applying our algorithm described in Section 3. The last section gives some remarks on our work.

2. On the LCE of N-Dimensional Dynamical Systems

2.1. The General Case

First, we consider in general the following n-dimensional dynamical systems

$$\begin{cases} \dot{x}_1 = f_1(\alpha_1, \ldots, \alpha_m, x_1, \ldots, x_n), \\ \dot{x}_2 = f_2(\alpha_1, \ldots, \alpha_m, x_1, \ldots, x_n), \\ \quad \vdots \\ \dot{x}_n = f_n(\alpha_1, \ldots, \alpha_m, x_1, \ldots, x_n), \end{cases} \tag{2.1}$$

where $f_1, \ldots, f_n$ are polynomials of $x_1, \ldots, x_n$, and $\alpha_1, \ldots, \alpha_m$ are the systems parameters.

Suppose that $x_1(t), \ldots, x_n(t)$ are the solutions with bounded initial values and $x_1(t), \ldots, x_n(t)$ are also bounded. The LCE h of an invariant set of (2.1) can be defined as follows:

$$h = \lim_{t \to \infty} \frac{1}{t} \ln \|D_{(x_1^0, \ldots, x_n^0)}(x_1(t), \ldots, x_n(t)) \circ \vec{v}\|,$$

where $x_1^0, \ldots, x_n^0$ are initial values of (2.1), and $\vec{v} = (a_1, \ldots, a_n) \in R^n$. Without loss of generality, we can assume that $\vec{v} \in S^{n-1}$, where S^{n-1} is a unit sphere in R^n. Roughly speaking, h describes the degree of sensitive dependence of solutions of (2.1) on the initial conditions. By parametrizing the initial values, i.e., letting

$x_1^0 = x_1^0(s), \ldots, x_n^0 = x_n^0(s)$, where s is the introduced parameter, we have

$$\begin{aligned}\frac{\partial x_1}{\partial s} &= \frac{\partial x_1}{\partial x_1^0}\frac{dx_1^0}{ds} + \cdots + \frac{\partial x_1}{\partial x_n^0}\frac{dx_n^0}{ds} = \frac{\partial x_1}{\partial x_1^0}a_1 + \cdots + \frac{\partial x_1}{\partial x_n^0}a_n,\\ &\vdots\\ \frac{\partial x_n}{\partial s} &= \frac{\partial x_n}{\partial x_1^0}\frac{dx_1^0}{ds} + \cdots + \frac{\partial x_n}{\partial x_n^0}\frac{dx_n^0}{ds} = \frac{\partial x_n}{\partial x_1^0}a_1 + \cdots + \frac{\partial x_n}{\partial x_n^0}a_n.\end{aligned} \tag{2.2}$$

We introduce $u_1 = \dfrac{\partial x_1}{\partial s}, \ldots, u_n = \dfrac{\partial x_n}{\partial s}$, and then have

$$\|D_{(x_1^0,\ldots,x_n^0)}(x_1(t), \ldots, x_n(t)) \circ \vec{v}\| = \sqrt{u_1^2 + \cdots + u_n^2},$$

and the following variational systems of (2.1):

$$\begin{pmatrix} \dot{u}_1 \\ \vdots \\ \dot{u}_n \end{pmatrix} = \begin{pmatrix} \dfrac{\partial f_1}{\partial x_1} & \cdots & \dfrac{\partial f_1}{\partial x_n} \\ \vdots & \ddots & \vdots \\ \dfrac{\partial f_n}{\partial x_1} & \cdots & \dfrac{\partial f_n}{\partial x_n} \end{pmatrix} \begin{pmatrix} u_1 \\ \vdots \\ u_n \end{pmatrix}, \tag{2.3}$$

or $\dot{u} = Df(x)u$, where $u = (u_1, \ldots, u_n)^T$, $x = (x_1, \ldots, x_n)$, and

$$f(x) = \begin{pmatrix} f_1(x_1, \ldots, x_n) \\ \vdots \\ f_n(x_1, \ldots, x_n) \end{pmatrix}, \quad Df(x) = \begin{pmatrix} \dfrac{\partial f_1}{\partial x_1} & \cdots & \dfrac{\partial f_1}{\partial x_n} \\ \vdots & \ddots & \vdots \\ \dfrac{\partial f_n}{\partial x_1} & \cdots & \dfrac{\partial f_n}{\partial x_n} \end{pmatrix}.$$

The above variational systems (2.3) are obtained by differentiation of (2.1) by s.

When attractors are strange, we let initial values $(x_1^0, \ldots, x_n^0)$ be in the chaotic attractors, and want to compute the maximal LCE h of the attractors. Note that there exists a $\vec{v}_0 \in S^{n-1}$, such that

$$\lim_{t\to\infty} \frac{1}{t} \ln \|D_{(x_1^0,\ldots,x_n^0)}(x_1(t), \ldots, x_n(t)) \circ \vec{v}_0\| = h,$$

and for sufficiently large t and any $\vec{v} \in S^{n-1}$,

$$\|D_{(x_1^0,\ldots,x_n^0)}(x_1(t), \ldots, x_n(t)) \circ \vec{v}\| \le C\|D_{(x_1^0,\ldots,x_n^0)}(x_1(t), \ldots, x_n(t)) \circ \vec{v}_0\|, \tag{2.4}$$

where C (≥ 1) is a positive constant independent of $\vec{v}$.

Let

$$\alpha(t, \vec{v}) = \frac{1}{t} \ln \|D_{(x_1^0,\ldots,x_n^0)}(x_1(t), \ldots, x_n(t)) \circ \vec{v}\| - h.$$

For $t \ge 0$, from (2.4), we have $\alpha(t, \vec{v}) \le \alpha(t, \vec{v}_0) + \frac{1}{t}\ln C$, for any $\vec{v} \in S^{n-1}$.

Since the maximal LCE h can be defined as

$$h = \lim_{t\to\infty} \frac{1}{t} \ln \|D_{(x_1^0,\ldots,x_n^0)}(x) \circ \vec{v}\|, \quad \text{for any } \vec{v} \in S^{n-1} \setminus S^{n-2},$$

where S^{n-2} is a unit sphere in R^{n-1} [1], by writing

$$h + \alpha(t, \vec{v}) = \frac{1}{t} \ln \|D_{(x_1^0,\ldots,x_n^0)}(x) \circ \vec{v}\| = \frac{1}{t} \ln |U|,$$

we can obtain

$$th + t\alpha(t, \vec{v}) = \ln |U|.$$

Differentiating the two sides of the above formula, we get

$$h + (t\alpha(t, \vec{v}))' = \frac{U^T \dot{U}}{U^T U} = \frac{U^T Df(x) U}{U^T U}. \tag{2.5}$$

Noting that (2.5) can be rewritten as

$$h = \frac{1}{u_1^2 + \cdots + u_n^2}(u_1, \ldots, u_n) \begin{pmatrix} \frac{\partial f_1}{\partial x_1} & \cdots & \frac{\partial f_1}{\partial x_n} \\ \vdots & \ddots & \vdots \\ \frac{\partial f_n}{\partial x_1} & \cdots & \frac{\partial f_n}{\partial x_n} \end{pmatrix} \begin{pmatrix} u_1 \\ \vdots \\ u_n \end{pmatrix} - (t\alpha(t, \vec{v}))',$$

we can let $z_1 = \frac{u_2}{u_1}, z_2 = \frac{u_3}{u_1}, \ldots, z_{n-1} = \frac{u_n}{u_1}$, and then have

$$h = \frac{1}{1 + z_1^2 + \cdots + z_{n-1}^2}(1, z_1, \ldots, z_{n-1}) \begin{pmatrix} \frac{\partial f_1}{\partial x_1} & \cdots & \frac{\partial f_1}{\partial x_n} \\ \vdots & \ddots & \vdots \\ \frac{\partial f_n}{\partial x_1} & \cdots & \frac{\partial f_n}{\partial x_n} \end{pmatrix} \begin{pmatrix} 1 \\ z_1 \\ \vdots \\ z_{n-1} \end{pmatrix} - (t\alpha(t, \vec{v}))'. \tag{2.6}$$

We also have

$$\dot{z}_1 = \frac{\dot{u}_2 u_1 - \dot{u}_1 u_2}{u_1^2},$$
$$\vdots$$
$$\dot{z}_{n-1} = \frac{\dot{u}_n u_1 - \dot{u}_1 u_n}{u_1^2},$$

or

$$\begin{pmatrix} \dot{z}_1 \\ \dot{z}_2 \\ \vdots \\ \dot{z}_{n-1} \end{pmatrix} = \begin{pmatrix} \frac{\dot{u}_2}{u_1} & -\frac{\dot{u}_1}{u_1} & 0 & & \cdots & 0 & 0 \\ \frac{\dot{u}_3}{u_1} & 0 & -\frac{\dot{u}_1}{u_1} & 0 & \cdots & 0 & 0 \\ \vdots & \vdots & \vdots & \vdots & & \vdots & \vdots \\ \frac{\dot{u}_n}{u_1} & 0 & 0 & 0 & \ldots & 0 & -\frac{\dot{u}_1}{u_1} \end{pmatrix} \begin{pmatrix} 1 \\ z_1 \\ \vdots \\ z_{n-1} \end{pmatrix}. \tag{2.7}$$

Theorem 2.1. There exist vectors $\vec{v}_1, \ldots, \vec{v}_n \in S^{n-1}$ and two polynomials F_1 and F_2 such that for $\vec{v} \in \{\vec{v}_1, \ldots, \vec{v}_n\}$, we have

$$\begin{cases} F_1(\alpha_1, \ldots, \alpha_m, h, x_1, \ldots, x_n, (t\alpha(t,\vec{v}))') = 0, \\ F_2(\alpha_1, \ldots, \alpha_m, h, x_1, \ldots, x_n, z_1, \ldots, z_{n-1}, (t\alpha(t,\vec{v}))') = 0. \end{cases}$$

Proof. From (2.5), we can obtain

$$U^T(Df(x) - (h + (t\alpha(t,\vec{v}))')I)U = 0, \quad \text{for any } t.$$

The above formula can be rewritten as

$$U^T((Df(x) + Df(x)^T)/2 - (h + (t\alpha(t,\vec{v}))')I)U = 0, \quad \text{for any } t.$$

Note that $(Df(x) + Df(x)^T)/2 - (h + (t\alpha(t,\vec{v}))')I)$ is a symmetric matrix. Let $\widetilde{U} = U/|U|$; then the above equation can be transformed into

$$\widetilde{U}^T((Df(x) + Df(x)^T)/2 - (h + (t\alpha(t,\vec{v}))')I)\widetilde{U} = 0,$$

where $|\widetilde{U}| = 1$. Note that $(Df(x)+Df(x)^T)/2$ is independent of $\vec{v}$ and, thus, there is an orthogonal matrix $|T(t)| = 1$, where $T(t) = T(t, x_1^0, \ldots, x_n^0, x_1(t), \ldots, x_n(t))$, independent of $\vec{v}$, such that

$$T(t)^T\ ((Df(x) + Df(x)^T)/2)\ T(t) = \begin{pmatrix} \beta_1(t) & 0 & \cdots & 0 \\ 0 & \beta_2(t) & \cdots & 0 \\ \vdots & \vdots & \ddots & \vdots \\ 0 & 0 & \cdots & \beta_n(t) \end{pmatrix},$$

where $\beta_i(t)$ is independent of $\vec{v}$, for $1 \le i \le n$.

Let $\widetilde{U} = T\widehat{U}$ and note that $T^T = T^{-1}$; then we have

$$\begin{aligned} h + (t\alpha(t,\vec{v}))' &= \widehat{U}^T \begin{pmatrix} \beta_1(t) & 0 & \cdots & 0 \\ 0 & \beta_2(t) & \cdots & 0 \\ \vdots & \vdots & \ddots & \vdots \\ 0 & 0 & \cdots & \beta_n(t) \end{pmatrix} \widehat{U} \\ &= \beta_1(t)\widehat{u}_1^2 + \beta_2(t)\widehat{u}_2^2 + \cdots + \beta_n(t)\widehat{u}_n^2, \end{aligned}$$

where $\widehat{U} = (\widehat{u}_1, \widehat{u}_2, \ldots, \widehat{u}_n)^T$.

As $\widetilde{U} = T\widehat{U}$ and $\widetilde{U}$ is a diffeomorphism, $\widehat{U}(t,\cdot) : R^n \to R^n$ is also a diffemorphism. Choose $\vec{v}_1, \vec{v}_2, \ldots, \vec{v}_n \in S^{n-1}$, such that $\widehat{U}(t, \vec{v}_i) = e_i$, where $(e_i, e_j) = \delta_{ij}$, $i, j = 1, 2, \ldots, n$. Thus, we have

$$h + (t\alpha(t,\vec{v}_1))' = \beta_1(t), \ \ldots, \ h + (t\alpha(t,\vec{v}_n))' = \beta_n(t).$$

Now, noting that $T(t)^T = T(t)^{-1}$, we have got the following important fact

$$\begin{aligned}
&|(Df(x) + Df(x)^T)/2 - (h + (t\alpha(t,\vec{v}))')I| \\
&\quad = |T(t)^T((Df(x) + Df(x)^T)/2 - (h + (t\alpha(t,\vec{v}))')I)T(t)| \\
&\quad = \left| \begin{pmatrix} \beta_1(t) & 0 & \cdots & 0 \\ 0 & \beta_2(t) & \cdots & 0 \\ \vdots & \vdots & \ddots & \vdots \\ 0 & 0 & \cdots & \beta_n(t) \end{pmatrix} - (h + (t\alpha(t,\vec{v}))')I \right| \\
&\quad = (\beta_1(t) - (h + (t\alpha(t,\vec{v}))')) \cdots (\beta_n(t) - (h + (t\alpha(t,\vec{v}))')) \\
&\quad = 0, \quad \text{for } \vec{v} \in \{\vec{v}_1, \vec{v}_2, \ldots, \vec{v}_n\}.
\end{aligned}$$

We write

$$F_1 = |(Df(x) + Df(x)^T)/2 - (h + (t\alpha(t,\vec{v}))')I|.$$

By rewriting (2.6) into

$$(h+(t\alpha(t,\vec{v}))')(1+z_1^2+\cdots+z_{n-1}^2)-(1,z_1,\ldots,z_{n-1})Df(x)\begin{pmatrix} 1 \\ z_1 \\ \vdots \\ z_{n-1} \end{pmatrix} = 0, \quad (2.8)$$

for any $\vec{v}$, we have another polynomial

$$F_2 = (h + (t\alpha(t,\vec{v}))')(1 + z_1^2 + \cdots + z_{n-1}^2) - (1, z_1, \ldots, z_{n-1})Df(x)\begin{pmatrix} 1 \\ z_1 \\ \vdots \\ z_{n-1} \end{pmatrix}.$$

Thus, for $\vec{v} \in \{\vec{v}_1, \ldots, \vec{v}_n\}$, we obtain the following system of polynomial equations

$$\begin{cases} F_1(\alpha_1, \ldots, \alpha_m, h, x_1, \ldots, x_n, (t\alpha(t,\vec{v}))') = 0, \\ F_2(\alpha_1, \ldots, \alpha_m, h, x_1, \ldots, x_n, z_1, \ldots, z_{n-1}, (t\alpha(t,\vec{v}))') = 0. \end{cases}$$

□

By differentiating F_1 and F_2, and substituting the systems (2.1) and (2.8), we can obtain

$$F_1'(\alpha_1, \ldots, \alpha_m, h, x_1, \ldots, x_n, (t\alpha(t,\vec{v}))', (t\alpha(t,\vec{v}))'') = 0,$$
$$F_2'(\alpha_1, \ldots, \alpha_m, h, x_1, \ldots, x_n, z_1, \ldots, z_{n-1}, (t\alpha(t,\vec{v}))', (t\alpha(t,\vec{v}))'') = 0.$$

Continuing in this fashion, we get

$$F_1^{(2n-1)}(\alpha_1, \ldots, \alpha_m, h, x_1, \ldots, x_n, (t\alpha(t,\vec{v}))', \ldots, (t\alpha(t,\vec{v}))^{(2n)}) = 0,$$
$$F_2^{(2n-1)}(\alpha_1, \ldots, \alpha_m, h, x_1, \ldots, x_n, z_1, \ldots, z_{n-1}, (t\alpha(t,\vec{v}))', \ldots, (t\alpha(t,\vec{v}))^{(2n)}) = 0.$$

So, we have the following.

Corollary 2.2. For n-dimensional dynamical systems, there is a system of polynomial equations associating with the LCE h as follows:

$$\begin{cases} F_1(\alpha_1,\ldots,\alpha_m,h,x_1,\ldots,x_n,(t\alpha(t,\vec{v}))')=0, \\ F_2(\alpha_1,\ldots,\alpha_m,h,x_1,\ldots,x_n,z_1,\ldots,z_{n-1},(t\alpha(t,\vec{v}))')=0, \\ F_1'(\alpha_1,\ldots,\alpha_m,h,x_1,\ldots,x_n,(t\alpha(t,\vec{v}))',(t\alpha(t,\vec{v}))'')=0, \\ F_2'(\alpha_1,\ldots,\alpha_m,h,x_1,\ldots,x_n,z_1,\ldots,z_{n-1},(t\alpha(t,\vec{v}))',(t\alpha(t,\vec{v}))'')=0, \\ \qquad\vdots \\ F_1^{(2n-1)}(\alpha_1,\ldots,\alpha_m,h,x_1,\ldots,x_n,(t\alpha(t,\vec{v}))',\ldots,(t\alpha(t,\vec{v}))^{(2n)})=0, \\ F_2^{(2n-1)}(\alpha_1,\ldots,\alpha_m,h,x_1,\ldots,x_n,z_1,\ldots,z_{n-1},(t\alpha(t,\vec{v}))',\ldots,(t\alpha(t,\vec{v}))^{(2n)})=0. \end{cases} \tag{2.9}$$

It seems that we can apply Ritt–Wu's method to the above system of polynomial equations, by eliminating $x_1,\ldots,x_n$, $z_1,\ldots,z_{n-1}$, $(t\alpha(t,\vec{v}))'$, …, $(t\alpha(t,\vec{v}))^{(2n)}$, to obtain the symbolic expressions satisfied by $\alpha_1,\ldots,\alpha_m$ and h. However, the involved symbolic computations are so complex that they cannot be finished. In Section 4, an algorithm will be developed for deriving some unknown relations on some objects involving limit processes in a framework of representing and manipulating a class of non-algebraic objects. Applying this algorithm to (2.9), we can obtain symbolic expressions satisfied by $\alpha_1,\ldots,\alpha_m$ and h.

2.2. The LCE of the Lorenz Systems

As an example, in this section we turn to consider the Lorenz systems [10]

$$\begin{cases} \dot{x}_1 = -\sigma x_1 + \sigma x_2, \\ \dot{x}_2 = -x_1 x_3 + r x_1 - x_2, \\ \dot{x}_3 = x_1 x_2 - b x_3. \end{cases} \tag{2.10}$$

The LCE h of (2.10) can be defined as follows:

$$h = \lim_{t\to\infty} \frac{1}{t} \ln \|D_{(x_1^0,x_2^0,x_3^0)}(x_1(t),x_2(t),x_3(t)) \circ \vec{v}\|,$$

where $\vec{v} = (a_1,a_2,a_3) \in S^2$.

From (2.3), we have the following variational systems of (2.10):

$$\begin{cases} \dot{u}_1 = -\sigma u_1 + \sigma u_2, \\ \dot{u}_2 = -x_3 u_1 + r u_1 - u_2 - x_1 u_3, \\ \dot{u}_3 = x_2 u_1 + x_1 u_2 - b u_3. \end{cases} \tag{2.11}$$

From (2.5) and (2.7) we have

$$h + (t\alpha(t,\vec{v}))' = \frac{U^T Df(x) U}{U^T U} = \frac{1}{u_1^2+u_2^2+u_3^2}(u_1,u_2,u_3)\begin{pmatrix} -\sigma & \sigma & 0 \\ r - x_3 & -1 & -x_1 \\ x_2 & x_1 & -b \end{pmatrix}\begin{pmatrix} u_1 \\ u_2 \\ u_3 \end{pmatrix},$$

and

$$\begin{aligned} \dot{z}_1 &= -\sigma z_1^2 + (\sigma - 1)z_1 - x_1 z_2 - x_3 + r, \\ \dot{z}_2 &= -\sigma z_1 z_2 + (\sigma - b) z_2 + x_1 z_1 + x_2. \end{aligned} \tag{2.12}$$

For the Lorenz systems, F_1 and F_2 are as follows:

$$F_1 = \begin{vmatrix} -\sigma - (h + (t\alpha(t,\vec{v}))') & \frac{1}{2}(\sigma + r - x_3) & \frac{1}{2}x_2 \\ \frac{1}{2}(\sigma + r - x_3) & -1 - (h + (t\alpha(t,\vec{v}))') & 0 \\ \frac{1}{2}x_2 & 0 & -b - (h + (t\alpha(t,\vec{v}))') \end{vmatrix},$$

$$F_2 = (h + (t\alpha(t,\vec{v}))')(1 + z_1^2 + z_2^2) + z_1^2 + b z_2^2 - (\sigma + r - x_3)z_1 - x_2 z_2 + \sigma.$$

For $\vec{v} \in \{\vec{v}_1, \vec{v}_2, \vec{v}_3\}$, we have

$$\begin{cases} F_1(\sigma, b, r, h, x_1, x_2, x_3, (t\alpha(t,\vec{v}))') = 0, \\ F_2(\sigma, b, r, h, x_1, x_2, x_3, z_1, z_2, (t\alpha(t,\vec{v}))') = 0. \end{cases} \tag{2.13}$$

By differentiating F_1 and F_2 five times and substituting (2.10), (2.11) and (2.12), we obtain

$$\begin{cases} F_1'(\sigma, b, r, h, x_1, x_2, x_3, (t\alpha(t,\vec{v}))', (t\alpha(t,\vec{v}))'') = 0, \\ F_2'(\sigma, b, r, h, x_1, x_2, x_3, z_1, z_2, (t\alpha(t,\vec{v}))', (t\alpha(t,\vec{v}))'') = 0, \\ F_1''(\sigma, b, r, h, x_1, x_2, x_3, (t\alpha(t,\vec{v}))', \ldots, (t\alpha(t,\vec{v}))''') = 0, \\ F_2''(\sigma, b, r, h, x_1, x_2, x_3, z_1, z_2, (t\alpha(t,\vec{v}))', \ldots, (t\alpha(t,\vec{v}))''') = 0, \\ \vdots \\ F_1^{(5)}(\sigma, b, r, h, x_1, x_2, x_3, (t\alpha(t,\vec{v}))', \ldots, (t\alpha(t,\vec{v}))^{(6)}) = 0, \\ F_2^{(5)}(\sigma, b, r, h, x_1, x_2, x_3, z_1, z_2, (t\alpha(t,\vec{v}))', \ldots, (t\alpha(t,\vec{v}))^{(6)}) = 0. \end{cases} \tag{2.14}$$

3. Representing and Manipulating a Class of Non-Algebraic Objects

The necessity of symbolically dealing with mathematical objects involving limit processes with non-standard analysis [14] has been recognized and some efforts have been made towards doing such tasks (see, e.g., [4, 12]), among which the authors proposed a symbolic manipulation method for these objects [12]. Based on this earlier work and some related methods, here we present a framework for representing and manipulating a class of non-algebraic objects. In this framework, an algorithm can be developed for deriving mechanically some unknown relations on the mathematical objects of our interest.

3.1. The Construction of a Class of Non-Algebraic Objects

Let K be a totally ordered commutative field, and $Q \subset K$, where Q is the field of rational numbers, and $\prec$ is a total ordering on K. We introduce the binary relations $\sim$ and $\ll$ on K. Any a, b and $c \in K$ satisfy the following axioms:

(1) $a \sim a$,
(2) $a \sim b \longmapsto b \sim a$,
(3) $a \sim b, b \sim c \longmapsto a \sim c$,
(4) $a \sim (a + a)$,
(5) $a \sim 1 \longmapsto 0 \prec a$,
(6) $0 \ll 1$,
(7) $a \ll 1, b \ll 1 \longmapsto (a + b) \ll 1$,
(8) $a \ll 1 \longmapsto -a \ll 1$,
(9) $a \ll 1, c \sim 1 \longmapsto c \cdot a \ll 1$.

Intuitively, the meanings of $a \sim a$ and $a \ll b$ are respectively as follows: $a \sim b$ stands for that a has the same order of magnitude as b, i.e., $\lim \frac{a}{b} = M$ and $M \neq 0$, and $a \ll b$ stands for that a is negligible w.r.t. b, particularly, i.e., $\lim \frac{a}{b} = 0$, $a \ll 1$ for that a can be negligible, i.e., $\lim a = 0$.

In addition, for any $a \neq 0 \in K$, we define

$$\text{sign}(a) = + \longleftrightarrow 0 \prec a, \ \text{sign}(a) = - \longleftrightarrow a \prec 0, \text{ and}$$

$$|a| = \begin{cases} a, & \text{if } \text{sign}(a) = +, \\ 0, & \text{if } a = 0, \\ -a, & \text{if } \text{sign}(a) = -. \end{cases}$$

Let $N = \{a \mid a \in K \text{ and } a \ll 1\}$, $C = \{a \mid a \in K \text{ and } |a| \prec 1\}$. It can be proved that there is no model in the field of real numbers, but there is a model in the field $\widetilde{R}$ of real numbers of NSA [14]. Let $I \subset \widetilde{R}$ be the infinitesimals of $\widetilde{R}$, $A = R + I = \{a + i \mid a \in R, \ i \in I\}$, and R be the field of real numbers. Then N can be interpreted as I, and C as A, K as $\widetilde{R}$. For $a \in \widetilde{R}$, we define $\langle a \rangle \in R$ to be the standard part of a, i.e., $\langle\ \rangle : \widetilde{R} \to R$ maps a from $\widetilde{R}$ to R. If $a \in R$, then $a = \langle a \rangle$. Construct a polynomial ring $\widetilde{R}[x_1, \ldots, x_n]$ of indeterminates $x_1, \ldots, x_n$ with coefficients in $\widetilde{R}$.

Definition 3.1. An analytic equality e is of the form $f = \delta$, where $f \in K[x_1, \ldots, x_n]$ and $\delta \ll 1$. Let ϕ denote an empty equality.

Let $x_1 < x_2 < \cdots < x_n$, which induces a total ordering $<$ on polynomials in $K[x_1, \ldots, x_n]$. Let P_1 and P_2 be two finite polynomial sets. Define $P_1 < P_2$, if (1) $P_1 = \Phi$, $P_2 \neq \Phi$, where Φ denotes the empty set, or (2) $f < g$, or (3)

$P_1 - \{f\} < P_2 - \{g\}$, where f and g are the minimal polynomials of P_1 and P_2, respectively.

By this ordering we can define a total ordering on analytic equalities.

Definition 3.2. For two analytic equalities $f = \delta_1$ and $g = \delta_2$ define $f = \delta_1 < g = \delta_2$ if $f < g$.

As usual, we can define the ordering on analytic equality sets in the following way. Let $E_1 = \{f_1 = \delta_1, \dots, f_n = \delta_n\}$ and $E_2 = \{g_1 = \delta'_1, \dots, g_l = \delta'_l\}$ be finite sets of analytic equalities, and define $E_1 < E_2$ if $\{f_1, \dots, f_n\} < \{g_1, \dots, g_l\}$.

3.2. Symbolic Manipulation for Analytic Equalities and an Algorithm

First, we need some definitions and notations, which are the extensions of those from Buchberger [3] and Wu [17].

Definition 3.3. Let $f = \delta_1$ and $g = \delta_2$ be analytic equalities; we define class$(f = \delta) = i$, if i is the largest subscript such that y_i occurs in f, where $y_i \in \{p_1, \dots, p_m, x_1(t), \dots, x_n(t)\}$, and call y_i the *leading variable* of $f = \delta_1$. If $f \in R$, then define class$(f = \delta_1) = 0$. f can be written as

$$f = f_k y_i^k + f_{k-1} y_i^{k-1} + \cdots + f_0;$$

we call k the *degree* of f with respect to y_i, written as $k = \deg(f, y_i)$. If $\deg(g, y_i) < \deg(f, y_i)$, then $g = \delta_2$ is said to be *reduced* with respect to $f = \delta_1$, written as $g = \delta_2$ red / $f = \delta_1$.

In the following, some reasoning methods for analytic equalities are developed, which will lead to an algorithm for constructing necessary conditions for a set of analytic equalities to hold.

Let $f = \delta_1$ and $g = \delta_2$ be analytic equalities, and x be the leading variable of $f = \delta_1$. Write f and g, respectively, as $f = f_k x^k + \cdots$ and $g = g_l x^l + \cdots$. If $g = \delta_2$ is not reduced with respect to $f = \delta_1$, i.e., $k \le l$, then we have

$$f = \delta_1, \ g = \delta_2 \Rightarrow r = \delta, \tag{3.1}$$

where $r = v \cdot g - u \cdot f$, $u = g_l x^{l-k}$, $v = f_k$, $\delta = \delta_1 + \delta_2$.

Note that $r = \delta < g = \delta_2$, and $r = \delta$ is the logic consequence of reasoning from $f = \delta_1$ and $g = \delta_2$, because, by axioms (1)–(9),

$$r = v \cdot g - u \cdot f = v \cdot \delta_2 - u \cdot \delta_1 = \delta_2 + \delta_1 = \delta.$$

(3.1) corresponds to the superposition in term rewriting and S-polynomial in the Gröbner basis method [3, 9].

Let

$$\text{new}(f = \delta_1, g = \delta_2) = \begin{cases} \phi, & \text{if } r \text{ is } 0. \\ r = \delta, & \text{if } \deg(r, x) < k. \\ \text{new}(f = \delta_1, r = \delta), & \text{if } \deg(r, x) \ge k. \end{cases}$$

Obviously, $\text{new}(f = \delta_1, g = \delta_2)$ is the logical consequence of reasoning from $f = \delta_1$ and $g = \delta_2$, and $\text{new}(f = \delta_1, r = \delta) < g = \delta_2$. So, we have

$$f = \delta_1,\ g = \delta_2 \Rightarrow \text{new}(f = \delta_1, g = \delta_2). \tag{3.2}$$

Let E be a finite set of analytic equalities,

$$E \Rightarrow (E - \{g = \delta_2\}) \cup \{\text{new}(f = \delta_1, g = \delta_2)\}. \tag{3.3}$$

Note that $(E - \{g = \delta_2\}) \cup \{\text{new}(f = \delta_1, g = \delta_2)\} < E$. In fact, by (3.3) we describe the reduction of E by the reducer $f = \delta_1$.

We also have

$$E \Rightarrow (E - \{f = \delta\}) \cup \{f_1 = \delta\} \quad \text{or} \quad (E - \{f = \delta\}) \cup \{f_2 = \delta\}, \tag{3.4}$$

if $f = f_1 \cdot f_2$. Note that

$$(E - \{f = \delta\}) \cup \{f_1 = \delta\} < E \quad \text{and} \quad (E - \{f = \delta\}) \cup \{f_2 = \delta\} < E.$$

In the following, we discuss how to construct necessary conditions for a set of analytic equalities to hold, based on the computing methods in the above section, which are of the form $A \Rightarrow B$, where, of course, B is a necessary condition for A to hold. The algorithm can be used for mechanically deriving unknown relations from given relations by constructing necessary conditions for them to hold. Let E be a finite nonempty set of analytic equalities. The idea for developing the algorithm is to use repeated reduction of equalities in E by some "reducers" chosen from E and repeated split of E, if possible, by factorizing the left-hand sides of reducers.

An Algorithm Constructing Necessary Conditions for Analytic Equality Sets

By $\text{mc}(E)$ we denote the set of the equalities with the highest *class* in E, i.e., $\text{mc}(E) = \{e' \mid e' \in E \text{ and } \text{class}(e') \geq \text{class}(e) \text{ for all } e \in E\}$. Define the reducer of E as follows: if $E = \Phi$ then $\text{reducer}(E) = \phi$; otherwise, $\text{reducer}(E) =$ the minimal equality of $\text{mc}(E)$. The reducer of E will be used as the initial equality in the following recursive procedure. Necessary conditions for E to hold can be constructed by calling the procedure $\text{CON}(E, \text{reducer}(E))$, where $\text{CON}(S, e)$ is defined recursively as follows:

(1) If there is $g = \delta_1 \in S$ and $g \neq 0 \in R$, then $\text{CON}(S, e) =$ false.
(2) If $e = \phi$ then $\text{CON}(S, e) = S$.
(3) If $e \neq \phi$, supposing e is of the form $f = \delta$, then if $\delta = 0$ and $f = f_1 \cdot f_2$, then $\text{CON}(S, e) = \text{CON}(S', \text{reducer}(S')) \vee \text{CON}(S'', \text{reducer}(S''))$, where $S' = (S - \{e\}) \cup \{f_1 = \delta\}$, $S'' = (S - \{e\}) \cup \{f_2 = \delta\}$.
(4) If, for every $e' \in S$, e' red $/$ e, then $\text{CON}(S, e) = \text{CON}(S, \text{reducer}(\{\bar{e} \mid \bar{e} \in S \text{ and } \bar{e} \prec e\}))$.
(5) If there is some $e' \in S$ and $\neg\, e'$ red $/e$, then $\text{CON}(S, e) = \text{CON}((S - \{e'\}) \cup \{\text{new}(e, e')\}, e)$.

Remark. In case (1), there is a contradictory equality $g = \delta_1$ in S, so let the condition be the false. In case (2), when e becomes ϕ, a recursive branch of the procedure stops. If all recursive branches stop, then the procedure stops and constructs necessary conditions for S to hold, which are of the form $S_1 \vee \cdots \vee S_r$. In case (3), factorization is used to split S. In case (4), there is no equality in S which is not reduced w.r.t. e. So the procedure recursively calls itself to use new reducer to reduce S. In case (5), there is an equality e' in S, which is not reduced w.r.t. e. So, the procedure reduces S by replacing e' in S with $\text{new}(e, e')$, and then the procedure continues to reduce $(S - \{e'\}) \cup \{\text{new}(e, e')\}$ by e.

It is easy to show that the algorithm terminates.

Proof of Termination. To prove the termination of the algorithm, we only need to note that, in case (3), $S' < S$ and $S'' < S$; in case (4), $\text{reducer}(\{\bar{e} \mid \bar{e} \in S \text{ and } \bar{e} < e\}) < e$; in case (5), $(S - \{e'\}) \cup \{\text{new}(e, e')\} < S$. Thus, the pair of arguments, (S, e), of CON strictly decreases. □

4. Symbolic Computation of the LCE

In the following we will do symbolic computations in $\widetilde{R}$, the field of real numbers of NSA. In this field $a \ll 1$ is interpreted as $a \in I$, where I are infinitesimals of $\widetilde{R}$, and $|a| \sim 1$ is interpreted as a is bounded in $\widetilde{R}$. After we get results in $\widetilde{R}$, we should get back to the field R of real numbers. This can be done by taking the standard parts of the results.

Recall that $\lim_{t \to \infty} \alpha(t, \vec{v}) = 0$ for any $\vec{v} \in S^{n-1}$. Therefore, if we let t_* satisfy $1/t_* \ll 1$, then $\alpha(t_*, \vec{v}) \ll 1$. For simplicity, let $\tau = t_* - t$; then $\frac{d\tau}{dt} = -1$ and $\frac{dx}{dt} = \frac{dx}{d\tau}\frac{d\tau}{dt} = -\frac{dx}{d\tau}$. Thus (2.1) and (2.3) become

$$\begin{cases} \dfrac{dx_1}{d\tau} = -f_1(\alpha_1, \ldots, \alpha_m, x_1, \ldots, x_n), \\ \dfrac{dx_2}{d\tau} = -f_2(\alpha_1, \ldots, \alpha_m, x_1, \ldots, x_n), \\ \quad \vdots \\ \dfrac{dx_n}{d\tau} = -f_n(\alpha_1, \ldots, \alpha_m, x_1, \ldots, x_n), \end{cases} \tag{4.1}$$

and

$$\begin{pmatrix} \dfrac{du_1}{d\tau} \\ \vdots \\ \dfrac{du_n}{d\tau} \end{pmatrix} = - \begin{pmatrix} \dfrac{\partial f_1}{\partial x_1} & \cdots & \dfrac{\partial f_1}{\partial x_n} \\ \vdots & \ddots & \vdots \\ \dfrac{\partial f_n}{\partial x_1} & \cdots & \dfrac{\partial f_n}{\partial x_n} \end{pmatrix} \begin{pmatrix} u_1 \\ \vdots \\ u_n \end{pmatrix}. \tag{4.2}$$

F_1 and F_2 become

$$F_1 = |-(Df(x) + Df(x)^T)/2 - (h + (\tau\bar{\alpha}(\tau, \vec{v}))')I|$$

and

$$F_2 = (h + (\tau\bar{\alpha}(\tau, \vec{v}))')(1 + z_1^2 + \cdots + z_{n-1}^2) + (1, z_1, \ldots, z_{n-1})Df(x)\begin{pmatrix} 1 \\ z_1 \\ \vdots \\ z_{n-1} \end{pmatrix}.$$

Now $\alpha(t_*, \vec{v}) \ll 1$ becomes

$$\bar{\alpha}(\tau, \vec{v})|_{\tau=0} \ll 1.$$

Thus

$$(\tau\bar{\alpha}(\tau, \vec{v}))'|_{\tau=0} = (\bar{\alpha}(\tau, \vec{v}) + \tau\bar{\alpha}'(\tau, \vec{v}))|_{\tau=0} = \bar{\alpha}(\tau, \vec{v})|_{\tau=0}.$$

That is, $\beta = (\tau\bar{\alpha}(\tau, \vec{v}))'|_{\tau=0} \ll 1$.

From the corollary in Section 2.1, we can establish a set of analytic equalities for $\tau = 0$:

$$\begin{cases} F_1(\alpha_1, \ldots, \alpha_m, h, x_1, \ldots, x_n, \beta) = 0, \\ F_2(\alpha_1, \ldots, \alpha_m, h, x_1, \ldots, x_n, z_1, \ldots, z_{n-1}, \beta) = 0, \\ F_1'(\alpha_1, \ldots, \alpha_m, h, x_1, \ldots, x_n, \beta, (\tau\bar{\alpha}(\tau, \vec{v}))'') = 0, \\ F_2'(\alpha_1, \ldots, \alpha_m, h, x_1, \ldots, x_n, z_1, \ldots, z_{n-1}, \beta, (\tau\bar{\alpha}(\tau, \vec{v}))'') = 0, \\ \quad \vdots \\ F_1^{(2n-1)}(\alpha_1, \ldots, \alpha_m, h, x_1, \ldots, x_n, \beta, (\tau\bar{\alpha}(\tau, \vec{v}))'', \ldots, (\tau\bar{\alpha}(\tau, \vec{v}))^{(2n)}) = 0, \\ F_2^{(2n-1)}(\alpha_1, \ldots, \alpha_m, h, x_1, \ldots, x_n, z_1, \ldots, z_{n-1}, \beta, \\ \qquad (\tau\bar{\alpha}(\tau, \vec{v}))''), \ldots, (\tau\bar{\alpha}(\tau, \vec{v}))^{(2n)}) = 0. \end{cases} \tag{4.3}$$

For (4.3) we can take some ordering on $x_1, \ldots, x_n$, $z_1, \ldots, z_{n-1}$, $(\tau\bar{\alpha}(\tau, \vec{v}))''$, $\ldots$, $(\tau\bar{\alpha}(\tau, \vec{v}))^{(2n)}$. Applying the algorithm described to (4.3), by eliminating $x_1, \ldots, x_n$, $z_1, \ldots, z_{n-1}$, $(\tau\bar{\alpha}(\tau, \vec{v}))''$, $\ldots$, $(\tau\bar{\alpha}(\tau, \vec{v}))^{(2n)}$, we can obtain an analytic equality $F(\alpha_1, \ldots, \alpha_m, h, \beta) = 0$. This process is completely automated, because in every step of the algorithm only analytic equalities are generated and the algorithm can terminate. Thus, we have the following result.

Main Theorem 1. *For n-dimensional dynamical systems, there exists an algorithm which can generate an analytic equality $F(\alpha_1, \ldots, \alpha_m, \widehat{h}) = 0$ automatically, where $\widehat{h} = h + \beta$, $\beta \in I$, $I \subset \widetilde{R}$ are infinitesimals, and $\widetilde{R}$ is the field of real numbers of NSA. The maximal LCE h is the standard part of $\widehat{h}$, i.e., $h = \langle \widehat{h} \rangle \in R$, where R is the field of real numbers. Moreover, $\widehat{h}$ is one of the roots of $F(\alpha_1, \ldots, \alpha_m, y) = 0$.*

As an example, we consider the Lorenz systems. In this case, (4.3) becomes

$$\begin{cases} F_1(\sigma, b, r, h, x_1, x_2, x_3, \beta) = 0, \\ F_2(\sigma, b, r, h, x_1, x_2, x_3, z_1, z_2, \beta) = 0, \\ F_1'(\sigma, b, r, h, x_1, x_2, x_3, \beta, (\tau\bar{\alpha}(\tau, \vec{v}))'') = 0, \\ F_2'(\sigma, b, r, h, x_1, x_2, x_3, z_1, z_2, \beta, (\tau\bar{\alpha}(\tau, \vec{v}))'') = 0, \\ F_1''(\sigma, b, r, h, x_1, x_2, x_3, \beta, (\tau\bar{\alpha}(\tau, \vec{v}))'', (\tau\bar{\alpha}(\tau, \vec{v}))''') = 0, \\ F_2''(\sigma, b, r, h, x_1, x_2, x_3, z_1, z_2, \beta, (\tau\bar{\alpha}(\tau, \vec{v}))'', (\tau\bar{\alpha}(\tau, \vec{v}))''') = 0, \\ \qquad \vdots \\ F_1^{(5)}(\sigma, b, r, h, x_1, x_2, x_3, \beta, (\tau\bar{\alpha}(\tau, \vec{v}))'', \ldots, (\tau\bar{\alpha}(\tau, \vec{v}))^{(6)}) = 0, \\ F_2^{(5)}(\sigma, b, r, h, x_1, x_2, x_3, z_1, z_2, \beta, (\tau\bar{\alpha}(\tau, \vec{v}))'', \ldots, (\tau\bar{\alpha}(\tau, \vec{v}))^{(6)}) = 0. \end{cases}$$

Theoretically, we can apply the algorithm to the above system of analytic equalities, by eliminating x_1, x_2, x_3, z_1, z_2, $(\tau\bar{\alpha}(\tau, \vec{v}))'$, …, $(\tau\bar{\alpha}(\tau, \vec{v}))^{(6)}$, to obtain the symbolic expressions satisfied by σ, b, r and h. However, the involved symbolic computations are so complex that they cannot be finished. To overcome this obstruction, we will use our knowledge on some qualitative properties of the Lorenz systems to simplify the computations. We choose the initial values x_1^0, x_2^0, x_3^0, such that $\dot{x}_1(t_*) = 0$ and $\dot{x}_2(t_*) = 0$, that is, $x_2(t_*) = x_1(t_*)$ and $x_3(t_*) = r - 1$. (2.10) and (2.11) become

$$\begin{cases} \dfrac{dx_1}{d\tau} = \sigma x_1 - \sigma x_2, \\ \dfrac{dx_2}{d\tau} = x_1 x_3 - r x_1 + x_2, \\ \dfrac{dx_3}{d\tau} = -x_1 x_2 + b x_3 \end{cases} \tag{4.4}$$

and

$$\begin{cases} \dfrac{du_1}{d\tau} = \sigma u_1 - \sigma u_2, \\ \dfrac{du_2}{d\tau} = x_3 u_1 - r u_1 + u_2 + x_1 u_3, \\ \dfrac{du_3}{d\tau} = -x_2 u_1 - x_1 u_2 + b u_3. \end{cases} \tag{4.5}$$

F_1 and F_2 become

$$F_1 = \begin{vmatrix} \sigma - (h + (\tau\alpha(t, \vec{v}))') & -\frac{1}{2}(\sigma + r - x_3) & -\frac{1}{2}x_2 \\ -\frac{1}{2}(\sigma + r - x_3) & 1 - (h + (\tau\alpha(t, \vec{v}))') & 0 \\ -\frac{1}{2}x_2 & 0 & b - (h + (\tau\alpha(t, \vec{v}))') \end{vmatrix}$$

and

$$F_2 = (h + (\tau\alpha(t,\vec{v}))')(1 + z_1^2 + z_2^2) - z_1^2 - bz_2^2 + (\sigma + r - x_3)z_1 + x_2z_2 - \sigma.$$

Note that $\dot{x}_1(t_*) = 0$ and $\dot{x}_2(t_*) = 0$; thus

$$\frac{dx_1}{d\tau}(0) = 0 \quad \text{and} \quad \frac{dx_2}{d\tau}(0) = 0.$$

Now we can establish a set of analytic equalities for $\tau = 0$:

$$\begin{cases} \sigma x_1 - \sigma x_2 = 0, \\ x_1x_3 - rx_1 + x_2 = 0, \\ F_1(\sigma, b, r, h, x_1, x_2, x_3, \beta) = 0, \\ F_2(\sigma, b, r, h, x_1, x_2, x_3, z_1, z_2, \beta) = 0, \\ F_1'(\sigma, b, r, h, x_1, x_2, x_3, \beta, (\tau\bar{\alpha}(\tau,\vec{v}))'') = 0, \\ F_2'(\sigma, b, r, h, x_1, x_2, x_3, z_1, z_2, \beta, (\tau\bar{\alpha}(\tau,\vec{v}))'') = 0, \\ F_1''(\sigma, b, r, h, x_1, x_2, x_3, \beta, (\tau\bar{\alpha}(\tau,\vec{v}))'', (\tau\bar{\alpha}(\tau,\vec{v}))''') = 0, \\ F_2''(\sigma, b, r, h, x_1, x_2, x_3, z_1, z_2, \beta, (\tau\bar{\alpha}(\tau,\vec{v}))'', (\tau\bar{\alpha}(\tau,\vec{v}))''') = 0. \end{cases} \tag{4.6}$$

For (4.6), we order $z_1 < z_2 < x_1 < (\tau\bar{\alpha}(\tau,\vec{v}))'' < (\tau\bar{\alpha}(\tau,\vec{v}))''' < x_3 < x_2$. Applying the algorithm described in Section 3 to (4.6), by eliminating x_2, x_3, $(\tau\bar{\alpha}(\tau,\vec{v}))'''$, $(\tau\bar{\alpha}(\tau,\vec{v}))''$, x_1, z_1, and z_2, we can obtain an analytic equality $\phi(\sigma_0, b_0, r, h, \beta) = 0$. This process is completely automated. Note that the symbolic computations with three parameters σ, b and r are complex, and we choose any σ_0 and b_0 as the input to the algorithm, i.e., write σ and b as σ_0 and b_0 respectively. Assuming that the readers are not familiar with the algorithm, here we briefly explain this process. As the process is very complicated, we only describe the successful path chosen from all the paths produced in the process. Let the reducer e be $\sigma x_1 - \sigma x_2 = 0$. $x_1x_3 - rx_1 + x_2 = 0$, $F_1 = 0$, $F_2 = 0$, $F_1' = 0$, $F_2' = 0$, $F_1'' = 0$ and $F_2'' = 0$ are reduced by e w.r.t. x_2, and then let the reducer e be $x_1x_3 - rx_1 + x_1 = 0$. The analytic equalities of $F_1 = 0$, $F_2 = 0$, $F_1' = 0$, $F_2' = 0$, $F_1'' = 0$ and $F_2'' = 0$ reduced w.r.t. x_2 are further reduced by the new reducer e w.r.t. x_3, yielding

$$\begin{cases} {}_1F_1(\sigma_0, b_0, r, h, x_1, \beta) = 0, \\ {}_1F_2(\sigma_0, b_0, r, h, x_1, z_1, z_2, \beta) = 0, \\ {}_1F_1'(\sigma_0, b_0, r, h, x_1, \beta, (\tau\bar{\alpha}(\tau,\vec{v}))'') = 0, \\ {}_1F_2'(\sigma_0, b_0, r, h, x_1, z_1, z_2, \beta, (\tau\bar{\alpha}(\tau,\vec{v}))'') = 0, \\ {}_1F_1''(\sigma_0, b_0, r, h, x_1, \beta, (\tau\bar{\alpha}(\tau,\vec{v}))'', (\tau\bar{\alpha}(\tau,\vec{v}))''') = 0, \\ {}_1F_2''(\sigma_0, b_0, r, h, x_1, z_1, z_2, \beta, (\tau\bar{\alpha}(\tau,\vec{v}))'', (\tau\bar{\alpha}(\tau,\vec{v}))''') = 0. \end{cases}$$

Then, let the reducer e be ${}_1F_1'' = 0$. ${}_1F_1 = 0$, ${}_1F_2 = 0$, ${}_1F_1' = 0$, ${}_1F_2' = 0$ and ${}_1F_2'' = 0$ are reduced by e w.r.t. $(\tau\bar{\alpha}(\tau,\vec{v}))'''$. One system of the reduced analytic equalities are as follows:

$$\begin{cases} {}_1F_1(\sigma_0, b_0, r, h, x_1, \beta) = 0, \\ {}_1F_2(\sigma_0, b_0, r, h, x_1, z_1, z_2, \beta) = 0, \\ {}_1F_1'(\sigma_0, b_0, r, h, x_1, \beta, (\tau\bar{\alpha}(\tau,\vec{v}))'') = 0, \\ {}_2F_2'(\sigma_0, b_0, r, h, x_1, z_1, z_2, \beta, (\tau\bar{\alpha}(\tau,\vec{v}))'') = 0, \\ {}_2F_2''(\sigma_0, b_0, r, h, x_1, z_1, z_2, \beta, (\tau\bar{\alpha}(\tau,\vec{v}))'') = 0. \end{cases}$$

Let the new reducer e be ${}_1F_1' = 0$. ${}_1F_1 = 0$, ${}_1F_2 = 0$, ${}_2F_2' = 0$ and ${}_2F_2'' = 0$ are reduced by e w.r.t. $(\tau\bar{\alpha}(\tau,\vec{v}))''$. One system of the reduced analytic equalities is in the following form:

$$\begin{cases} {}_1F_1(\sigma_0, b_0, r, h, x_1, \beta) = 0, \\ {}_1F_2(\sigma_0, b_0, r, h, x_1, z_1, z_2, \beta) = 0, \\ {}_3F_2'(\sigma_0, b_0, r, h, x_1, z_1, z_2, \beta) = 0, \\ {}_3F_2''(\sigma_0, b_0, r, h, x_1, z_1, z_2, \beta) = 0. \end{cases}$$

Furthermore, let the reducer e be ${}_1F_2 = 0$. ${}_1F_1 = 0$, ${}_3F_2' = 0$ and ${}_3F_2'' = 0$ are reduced by e w.r.t. x_1. One system of the corresponding reduced analytic equalities is

$$\begin{cases} {}_1F_1(\sigma_0, b_0, r, h, z_1, z_2, \beta) = 0, \\ {}_4F_2'(\sigma_0, b_0, r, h, z_1, z_2, \beta) = 0, \\ {}_4F_2''(\sigma_0, b_0, r, h, z_1, z_2, \beta) = 0 \end{cases}$$

Let the reducer e be ${}_2F_1 = 0$. ${}_4F_2' = 0$ and ${}_4F_2'' = 0$ are reduced by e w.r.t. z_2. One system of the reduced analytic equalities is

$$\begin{cases} {}_5F_2'(\sigma_0, b_0, r, h, z_1, z_2, \beta) = 0, \\ {}_5F_2''(\sigma_0, b_0, r, h, z_1, z_2, \beta) = 0. \end{cases}$$

Then, let the reducer e be ${}_5F_2' = 0$. ${}_2F_1 = 0$ and ${}_5F_2'' = 0$ are reduced by e w.r.t. z_2, and one system of the corresponding reduced analytic equalities is

$$\begin{cases} {}_4F_1(\sigma_0, b_0, r, h, z_1, \beta) = 0, \\ {}_6F_2''(\sigma_0, b_0, r, h, z_1, \beta) = 0. \end{cases}$$

Let the new reducer e be ${}_4F_1 = 0$. ${}_6F_2'' = 0$ is reduced by e w.r.t. z_1. The reduced analytic equality is

$${}_7F_2''(\sigma_0, b_0, r, h, z_1, \beta) = 0.$$

Finally, let the reducer e be ${}_7F_2'' = 0$. ${}_4F_1 = 0$ is reduced by e w.r.t. z_1. The reduced analytic equality is

$$F(\sigma_0, b_0, r, h, \beta) = 0.$$

F can be factorized automatically into the following form:

$$F(\sigma_0, b_0, r, h, \beta) = \phi_1\phi_2\phi_3\phi_4^2\phi_5^4\phi_6^6,$$

where ϕ_1, ϕ_2, ϕ_3, ϕ_4^2, ϕ_5^4, and ϕ_6^6 contain 2, 29, 147, 19, 28 and 15 terms, respectively. We choose $\phi = \phi_3(\sigma_0, b_0, r, h, \beta)$ from these factors, and the reason will be clear later. ϕ can be written as

$$\phi(\sigma_0, b_0, r, \widehat{h}) = \sum_{i=0}^{26} g_i(\sigma_0, b_0, r)\widehat{h}^i,$$

where $\widehat{h} = h + \beta$ and $g_i(\sigma_0, b_0, r)$, for $0 \le i \le 26$, are polynomials of degree up to 26 (see the appendix for $\sigma_0 = 10, b_0 = 8/3$).

We consider the case for the Lorenz systems with the parameter values $\sigma_0 = 10$, $b_0 = 8/3$ and $r = 28$. By applying Sturm's theorem to $\phi(\sigma_0, b_0, r, \widehat{h}) = 0$, we obtain four real roots σ_1, σ_2, σ_3 and σ_4 of $\phi(\sigma_0, b_0, r, \widehat{h}) = 0$. Among them, σ_4 is in the interval (9008361/1000000, 9008362/1000000). Note that the computations involved in applying Sturm's theorem are exact, and therefore we can compute σ_4 as precisely as desired.

According to the known numerical results with $\sigma_0 = 10$, $b_0 = 8/3$ and $r = 28$, the maximal LCE $h = 0.91 \pm 0.01$ [6, 5], we select σ_4 from σ_1, σ_2, σ_3 and σ_4, and let $\widehat{h} = \sigma_4$. But we have to get back to the standard real number field R. Hence the desired LCE h should be the standard part of $\widehat{h}$, i.e., $h = \langle\widehat{h}\rangle \in R$. Thus, we have already obtained the following result.

Main Theorem 2. *For the Lorenz systems, there exists an algorithm with inputs σ_0 and b_0, which can generate an analytic equality $\phi(\sigma_0, b_0, r, \widehat{h}) = 0$ automatically, where $\widehat{h} \in \widetilde{R}$ and $\widetilde{R}$ is the field of real numbers of NSA, and the maximal LCE h of the Lorenz systems is the standard part of $\widehat{h}$, i.e., $h = \langle\widehat{h}\rangle \in R$, where R is the field of real numbers. Moreover, $\widehat{h}$ is one of the roots of $\phi(\sigma_0, b_0, r, y) = 0$.*

5. Final Remarks

Since the 1980s, with the rapid development of computer science, particularly computer algebra systems and symbolic computation techniques, the new interplays between mathematics and computer science have appeared from some unexpected subjects such as algebraic topology, differential geometry, and dynamical systems. These new developments support the promise of bringing us new insights and powerful mathematical tools to bear on problems in computing. At the same time, such problems have opened new frontiers of exploration and put a challenge for both mathematicians and computer scientists [2]. The authors' work presented in this paper is mainly inspired by the above new developments.

We remark that, compared with numerical methods, our symbolic method has two advantages. (1) It is a continuous approach, more precisely, when computing the LCE by numerical methods we have to fix initial values and parameters of systems and thus obtain discrete results only, i.e., we cannot have the precise relations between the LCE and the corresponding parameters. Using the symbolic method, we can get a definite expression of the LCE with parameters, which is an important basis for discussing the properties of the systems in theory. (2) For some dynamical systems, in fact, it is impossible to numerically compute their LCE. For example, for van der Pol systems, how to compute the LCE of its invariant sets of zero-measure by numerical methods [12, 18]?

The proposed method of symbolic manipulation for some non-algebraic objects is applicable to other problems related to dynamical systems; for example, Hausdorff dimension, entropy and weakened Hilbert 16th problems. Our work on such aspects will appear later.

References

[1] G. Benettin, et al., Lyapunov Characteristic Exponents for Smooth Dynamical Systems and for Hamiltonian Systems: A Method for Computing All of Them, *Meccanica* (March 1980), 9–30.

[2] L. Blum, M. Shub and S. Smale, On a Theory of Computation and Complexity over the Real Numbers: NP-Completeness, Recursive Functions and Universal Machines, *Bull. Amer. Math. Soc.* **20**(1) (1989).

[3] B. Buchberger, Gröbner Bases: An Algorithmic Method in Polynomial Ideal Theory, in *Recent Trends in Multidimensional Systems Theory* (N. K. Bose, ed.), pp. 184–232, D. Reidel Publ. Comp., Dordrecht (1985).

[4] P. Dague, Symbolic Reasoning with Relative Orders of Magnitude, in *Proceedings of IJCAI '93* (Sydney, Australia, 1993), pp. 1509–1514.

[5] V. Franceschini, C. Giberti and Z. Zheng, Characterization of the Lorenz Attractor by Unstable Periodic Orbits, *Nonlinearity* **6** (1993), 251–258.

[6] P. Grassberger and I. Procaccia, Dimensions and Entropies of Strange Attractors from a Fluctuating Dynamics Approach, *Physica* 13D **34** (1984), 34–54.

[7] J. Guckenheimer and P. Holmes, *Nonlinear Oscillations, Dynamical Systems and Bifurcations of Vector Fields*, Springer-Verlag, Berlin Heidelberg (1986).

[8] E. M. James and N. G. Llyod, A Cubic System with Eight Small-Amplitude Limit Cycles, *IMA J. Appl. Math.* **47** (1991), 163–171.

[9] D. E. Knuth and P. B. Bendix, Simple Word Problems in Universal Algebras, in *Computational Algebra* (J. Leach, ed.), pp. 263–297, Pergamon Press, New York (1970).

[10] E. N. Lorenz, Deterministic Non-Periodic Flow, *J. Atmos. Sci.* **20**(1963), 130–141.

[11] S. Ma and S. Ning, Deriving Some New Conditions on the Existence of Eight Limit Cycles for a Cubic System, *Comput. Math. Appl.* **33**(7) (1997), 59–84.

[12] S. Ma, Z. Zheng and S. Ning, Symbolic Manipulation for Some Non-Algebraic Objects and Its Application in Computing the LCE of Van der Pol Equation, *Int. J. Comput. Math.* **57**(3–4) (1995), 219–229.

[13] S. Ning, S. Ma, K. H. Kwet and Z. Zheng, On a Cubic System with Eight Small-Amplitude Limit Cycles, *Appl. Math. Lett.* **7**(4) (1994), 23–27.

[14] A. Robinson, *Non-Standard Analysis*, North-Holland Publ. Co., Amsterdam (1966).

[15] C. Sparrow, *The Lorenz Equations: Bifurcations, Chaos, and Strange Attractors*, Springer-Verlag, New York (1982).

[16] D. Wang, A Class of Cubic Differential Systems with 6-Tuple Focus, *J. Diff. Eqns.* **87** (1990), 305–315.

[17] W.-t. Wu, Basic Principle of Mechanical Theorem Proving in Elementary Geometries, *J. Automat. Reason.* **2**(4) (1986), 221–252.

[18] Z. Zheng, Qualitative Analysis of Van der Pol Type Equation with Periodic Forcing Term, *Acta Math. Sinica* **6**(3) (1990), 243–256.

Appendix

$$F_1 = \begin{vmatrix} -\sigma + (h + (\tau\alpha(t,\vec{v}))') & \frac{1}{2}(\sigma + r - x_3) & \frac{1}{2}x_2 \\ \frac{1}{2}(\sigma + r - x_3) & -1 + (h + (\tau\alpha(t,\vec{v}))') & 0 \\ \frac{1}{2}x_2 & 0 & -b + (h + (\tau\alpha(t,\vec{v}))') \end{vmatrix},$$

$$F_2 = (h + (\tau\alpha(t,\vec{v}))')(1 + z_1^2 + z_2^2) - z_1^2 - bz_2^2 + (\sigma + r - x_3)z_1 + x_2z_2 - \sigma;$$

$f_1 = -h - \tau\bar{\alpha}(\tau,\vec{v})' + b - x_1^2\tau\bar{\alpha}(\tau,\vec{v})' + 2\sigma h + 2\sigma\tau\bar{\alpha}(\tau,\vec{v})' - 4h^2 - 8h\tau\bar{\alpha}(\tau,\vec{v})' - 4\tau\bar{\alpha}(\tau,\vec{v})'^2 - 2\sigma b - 4\sigma h^2 - 4\sigma\tau\bar{\alpha}(\tau,\vec{v})'^2 + 4hb - 4h^2b + 12h^2\tau\bar{\alpha}(\tau,\vec{v})' + 12h\tau\bar{\alpha}(\tau,\vec{v})'^2 + 4\tau\bar{\alpha}(\tau,\vec{v})'b - 4\tau\bar{\alpha}(\tau,\vec{v})'^2b + 4\sigma hb - 8\sigma h\tau\bar{\alpha}(\tau,\vec{v})' + 4\sigma\tau\bar{\alpha}(\tau,\vec{v})'b - 8h\tau\bar{\alpha}(\tau,\vec{v})'b + 4h^3 + 4\tau\bar{\alpha}(\tau,\vec{v})'^3 + \sigma^2b - \sigma^2h - \sigma^2\tau\bar{\alpha}(\tau,\vec{v})' - x_1^2h + x_1^2$,

$f_2 = h + hz_1^2 + hz_2^2 + \tau\bar{\alpha}(\tau,\vec{v})' + \tau\bar{\alpha}(\tau,\vec{v})'z_1^2 + \tau\bar{\alpha}(\tau,\vec{v})'z_2^2 - z_1^2 - bz_2^2 + z_1\sigma + z_1 + x_1z_2 - \sigma$,

$f_1' = -2x_1^2\tau\bar{\alpha}(\tau,\vec{v})' + 2\sigma b^2 - 2hb - 2\tau\bar{\alpha}(\tau,\vec{v})'b - 2\sigma hb - 2\sigma\tau\bar{\alpha}(\tau,\vec{v})'b - \tau\bar{\alpha}(\tau,\vec{v})'' - 2x_1^2h - 8\sigma h\tau\bar{\alpha}(\tau,\vec{v})'' - 8\sigma\tau\bar{\alpha}(\tau,\vec{v})''\tau\bar{\alpha}(\tau,\vec{v})' - 8\tau\bar{\alpha}(\tau,\vec{v})''\tau\bar{\alpha}(\tau,\vec{v})'b + 24h\tau\bar{\alpha}(\tau,\vec{v})''\tau\bar{\alpha}(\tau,\vec{v})' + 4\sigma\tau\bar{\alpha}(\tau,\vec{v})''b - 8h\tau\bar{\alpha}(\tau,\vec{v})''b + 4\tau\bar{\alpha}(\tau,\vec{v})''b + 12h^2\tau\bar{\alpha}(\tau,\vec{v})'' + 2\sigma\tau\bar{\alpha}(\tau,\vec{v})'' - 8h\tau\bar{\alpha}(\tau,\vec{v})'' - 8\tau\bar{\alpha}(\tau,\vec{v})''\tau\bar{\alpha}(\tau,\vec{v})' + 12\tau\bar{\alpha}(\tau,\vec{v})''\tau\bar{\alpha}(\tau,\vec{v})'^2 - \sigma^2\tau\bar{\alpha}(\tau,\vec{v})'' + 2b^2 + 2\sigma bx_1^2 - x_1^2\tau\bar{\alpha}(\tau,\vec{v})'' - 2\sigma b^2r + 2\tau\bar{\alpha}(\tau,\vec{v})'br + 2hbr - 2\sigma\tau\bar{\alpha}(\tau,\vec{v})'x_1^2 - 2\sigma hx_1^2 + 2bx_1^2 - 2b^2r + 2\sigma\tau\bar{\alpha}(\tau,\vec{v})'br + 2\sigma hbr$,

$f_2' = -1 + 2\tau\bar{\alpha}(\tau,\vec{v})'z_2^2b - \sigma - 2\tau\bar{\alpha}(\tau,\vec{v})'z_1^2\sigma + 2hz_1^3\sigma - 2z_1x_1z_2 + \tau\bar{\alpha}(\tau,\vec{v})'' - 2z_1^2 + 2hz_1^2 + 2\tau\bar{\alpha}(\tau,\vec{v})'z_1^2 + 3z_1 - 2\tau\bar{\alpha}(\tau,\vec{v})'z_2^2\sigma + 2z_2^2b\sigma - 2hz_1^2\sigma + 2\tau\bar{\alpha}(\tau,\vec{v})'z_1^3\sigma + 2hz_2^2b - 2hz_2^2\sigma - 2hz_1 + 3\sigma z_1^2 + x_1z_2 + 2hz_2^2\sigma z_1 + 2\tau\bar{\alpha}(\tau,\vec{v})'z_2^2\sigma z_1 - 2z_2^2b\sigma z_1 + 2z_2bx_1z_1 + \sigma^2z_1^2 - z_1\sigma^2 - 2\sigma z_1^3 - 2z_2^2b^2 + 3z_2bx_1 + \tau\bar{\alpha}(\tau,\vec{v})''z_1^2 + \tau\bar{\alpha}(\tau,\vec{v})''z_2^2 + z_1\sigma x_1z_2 - x_1^2 - 2\tau\bar{\alpha}(\tau,\vec{v})'z_1 - z_1br + z_1b - 2hz_2x_1 - 2\tau\bar{\alpha}(\tau,\vec{v})'z_2x_1$,

$f_1'' = -8\tau\bar{\alpha}(\tau,\vec{v})'''\tau\bar{\alpha}(\tau,\vec{v})'b - 8\tau\bar{\alpha}(\tau,\vec{v})''^2b + 24\tau\bar{\alpha}(\tau,\vec{v})''^2\tau\bar{\alpha}(\tau,\vec{v})' - 8\sigma\tau\bar{\alpha}(\tau,\vec{v})'''\tau\bar{\alpha}(\tau,\vec{v})' - 8h\tau\bar{\alpha}(\tau,\vec{v})'''b + 24h\tau\bar{\alpha}(\tau,\vec{v})'''\tau\bar{\alpha}(\tau,\vec{v})' + 4\sigma\tau\bar{\alpha}(\tau,\vec{v})'''b + 24h\tau\bar{\alpha}(\tau,\vec{v})''^2 - 8\sigma\tau\bar{\alpha}(\tau,\vec{v})''^2 + 6b^2x_1^2 - 8\sigma h\tau\bar{\alpha}(\tau,\vec{v})''' - \tau\bar{\alpha}(\tau,\vec{v})''' + 2\sigma b^3 - \sigma^2\tau\bar{\alpha}(\tau,\vec{v})''' - 8\tau\bar{\alpha}(\tau,\vec{v})'''\tau\bar{\alpha}(\tau,\vec{v})' + 4\tau\bar{\alpha}(\tau,\vec{v})'''b + 12\tau\bar{\alpha}(\tau,\vec{v})'''\tau\bar{\alpha}(\tau,\vec{v})'^2 +$

$2\sigma\tau\bar{\alpha}(\tau,\vec{v})''' - 8h\tau\bar{\alpha}(\tau,\vec{v})''' + 12h^2\tau\bar{\alpha}(\tau,\vec{v})''' - 8\tau\bar{\alpha}(\tau,\vec{v})''^2 - 4\sigma\tau\bar{\alpha}(\tau,\vec{v})''b -$
$4\tau\bar{\alpha}(\tau,\vec{v})''b - 4x_1^2\tau\bar{\alpha}(\tau,\vec{v})'' - x_1^2\tau\bar{\alpha}(\tau,\vec{v})''' - 2x_1^4 - 4\sigma\tau\bar{\alpha}(\tau,\vec{v})''x_1^2 - 4\tau\bar{\alpha}(\tau,\vec{v})'bx_1^2 -$
$4hbx_1^2 + 4\sigma\tau\bar{\alpha}(\tau,\vec{v})''br - 2\sigma hb^2 + 2\sigma hb^2r - 2\sigma\tau\bar{\alpha}(\tau,\vec{v})'b^2 + 2\sigma\tau\bar{\alpha}(\tau,\vec{v})'b^2r + 2bx_1^4 +$
$4\tau\bar{\alpha}(\tau,\vec{v})''br - 4\tau\bar{\alpha}(\tau,\vec{v})'b^2 - 2\tau\bar{\alpha}(\tau,\vec{v})'b^2r^2 + 6\tau\bar{\alpha}(\tau,\vec{v})'b^2r - 4hb^2 - 2hb^2r^2 + 6hb^2r -$
$2\sigma b^3r + 2rbx_1^2 + 2rhbx_1^2 + 2r\tau\bar{\alpha}(\tau,\vec{v})'bx_1^2 - 2\sigma hbx_1^2 - 4rb^2x_1^2 + 2\sigma b^2x_1^2 - 2\sigma\tau\bar{\alpha}(\tau,\vec{v})'bx_1^2 -$
$2bx_1^2 + 4b^3 + 2b^3r^2 - 6b^3r,$

$f_2'' = -3 + 2h + 2\tau\bar{\alpha}(\tau,\vec{v})' - 3b + 2x_1^2\tau\bar{\alpha}(\tau,\vec{v})' - 16\tau\bar{\alpha}(\tau,\vec{v})'z_1^2\sigma + 10hz_1^3\sigma - \sigma b + \sigma^2 -$
$6z_1x_1z_2 + 2z_2^2bx_1^2 + 2\sigma bz_1^2 + 4hz_2^2\sigma^2 + \sigma rb + \tau\bar{\alpha}(\tau,\vec{v})''' - 4z_1^2 + 4hz_1^2 + 4\tau\bar{\alpha}(\tau,\vec{v})'z_1^2 -$
$9z_1\sigma + 7z_1 - 2\tau\bar{\alpha}(\tau,\vec{v})'z_2^2\sigma + 2z_2^2b\sigma - 16hz_1^2\sigma + 4\tau\bar{\alpha}(\tau,\vec{v})'z_2^2b^2 + 10\tau\bar{\alpha}(\tau,\vec{v})'z_1^3\sigma +$
$4\tau\bar{\alpha}(\tau,\vec{v})''z_2^2b - 4\tau\bar{\alpha}(\tau,\vec{v})''z_2^2\sigma + 4\tau\bar{\alpha}(\tau,\vec{v})''z_1^3\sigma + 4hz_2^2b^2 - 10\tau\bar{\alpha}(\tau,\vec{v})'z_1^3\sigma^2 - 2hz_2^2\sigma -$
$4\tau\bar{\alpha}(\tau,\vec{v})''z_1^2\sigma - 2\sigma x_1z_2 - 6hz_1 + 2x_1^2h + 19\sigma z_1^2 + 5x_1z_2 + \tau\bar{\alpha}(\tau,\vec{v})'''z_1^2 + \tau\bar{\alpha}(\tau,\vec{v})'''z_2^2 +$
$4\tau\bar{\alpha}(\tau,\vec{v})''z_2^2\sigma z_1 + 2hz_2^2\sigma z_1 + 2\tau\bar{\alpha}(\tau,\vec{v})'z_2^2\sigma z_1 - 2z_2^2b\sigma z_1 - 4\sigma^2z_1^2 - 3z_1\sigma^2 - 10\sigma z_1^3 -$
$2b\sigma z_1 - 7\sigma x_1z_2b - 4z_2^2b^3 + 2x_1^2z_1^2 - 8hz_2^2b\sigma - 4hz_2bx_1z_1 + 6b^2x_1z_1z_2 - 8z_2^2b^2\sigma z_1 +$
$8z_2^2b^2\sigma - 2\sigma x_1^2z_1 + 6hz_1\sigma - 6hz_1\sigma x_1z_2 + 8hz_2^2\sigma z_1b + 8\tau\bar{\alpha}(\tau,\vec{v})'z_2^2\sigma z_1b - 8\tau\bar{\alpha}(\tau,\vec{v})'z_2^2b\sigma -$
$4\tau\bar{\alpha}(\tau,\vec{v})'z_2bx_1z_1 + 6hz_2^2\sigma^2z_1^2 - 10hz_2^2\sigma^2z_1 + 2hz_2^3\sigma x_1 + 6\tau\bar{\alpha}(\tau,\vec{v})'z_1\sigma -$
$6\tau\bar{\alpha}(\tau,\vec{v})'z_1\sigma x_1z_2 + 6\tau\bar{\alpha}(\tau,\vec{v})'z_2^2\sigma^2z_1^2 - 10\tau\bar{\alpha}(\tau,\vec{v})'z_2^2\sigma^2z_1 + 2\tau\bar{\alpha}(\tau,\vec{v})'z_2^3\sigma x_1 +$
$2hz_1^2\sigma x_1z_2 - 6z_2^2b\sigma^2z_1^2 + 10z_2^2b\sigma^2z_1 - 2z_2^3b\sigma x_1 + 8z_2bx_1\sigma z_1^2 + 2\tau\bar{\alpha}(\tau,\vec{v})'z_1^2\sigma x_1z_2 +$
$4\tau\bar{\alpha}(\tau,\vec{v})''z_1^2 + 4hz_1x_1z_2 + \sigma^3z_1 + 4\tau\bar{\alpha}(\tau,\vec{v})'z_1x_1z_2 - 10x_1z_2\sigma z_1^2 - z_2\sigma^2x_1 +$
$4\tau\bar{\alpha}(\tau,\vec{v})'z_1^2\sigma^2 + 6hz_1^4\sigma^2 - 10hz_1^3\sigma^2 + 6\tau\bar{\alpha}(\tau,\vec{v})'z_1^4\sigma^2 + 4hz_1^2\sigma^2 + 4\tau\bar{\alpha}(\tau,\vec{v})'z_2^2\sigma^2 -$
$2bx_1^2z_1^2 - 4z_2^2b\sigma^2 + 2z_1^3\sigma^3 - 3z_1^2\sigma^3 - 2x_1^2z_2^2 - 6\sigma^2z_1^4 + 12z_1\sigma x_1z_2 - z_2brx_1 + 12z_1^3\sigma^2 -$
$3x_1^2 - 6\tau\bar{\alpha}(\tau,\vec{v})'z_1 - 4\tau\bar{\alpha}(\tau,\vec{v})''z_1 - 2hz_1b + 2hz_1br + 2b\sigma z_1r - 2\tau\bar{\alpha}(\tau,\vec{v})'z_1b +$
$2\tau\bar{\alpha}(\tau,\vec{v})'z_1br - 2\sigma bz_1^2r - 4z_1br + z_1b^2 - z_1b^2r + 4z_1b - 2hz_2x_1 - 2\tau\bar{\alpha}(\tau,\vec{v})'z_2x_1 -$
$4\tau\bar{\alpha}(\tau,\vec{v})''z_2x_1 + 4z_1x_1^2 + 7b^2x_1z_2 + 3rb - 3bx_1^2 - \sigma x_1^2 + 6\tau\bar{\alpha}(\tau,\vec{v})'z_2\sigma x_1 -$
$6\tau\bar{\alpha}(\tau,\vec{v})'z_2bx_1 + x_1^2z_2^2\sigma + 6hx_1z_2\sigma - 6hx_1z_2b + 2x_1z_2\sigma^2z_1^2 - 4z_1bx_1^2;$

$$\phi(\sigma_0, b_0, r, \widehat{h}) = \sum_{i=0}^{26} g_i(\sigma_0, b_0, r)\widehat{h}^i, \text{ where}$$

$g_0 = -198484859289600r^6 + 746948224633651\,20r^5 + 863926597237361725\,44r^4 +$
$165807533275340900451\,84r^3 + 1606495263194255209344000r^2 +$
$23931996856322570330282\,2800r + 101559642947516296319052\,78000,$

$g_1 = 963399936245760r^6 - 245805012282556416r^5 - 29190893400266537164\,8r^4 -$
$5476201766351093384755\,2r^3 - 60425922509832654633494\,40r^2 -$
$67256297210738177537897\,7600r - 193921531657129559577484\,29600,$

$g_2 = -1676953233522688r^6 + 23974813726097408r^5 + 22946853331824681779\,2r^4 +$
$437592347500455520962\,56r^3 + 46354630018459144880609\,92r^2 +$
$5101141835042450590490139\,68r + 950308368716025079782417\,0500,$

$g_3 = 878720224854016r^6 + 873657521973428224r^5 + 270842808702290925568r^4 + 33630966637386612987648r^3 + 343645889585162774327888 0r^2 - 4059119378855563007097808 0r - 2289382524087970568018379487,$

$g_4 = 914341883543552r^6 - 1301664575982247936r^5 - 524187030032331436032r^4 - 59465350814684001365760r^3 - 456464590091709254198144 0r^2 + 4240856502833180216166211 2r + 2405220492380890242920452186,$

$g_5 = -1489728728596480r^6 + 3601410875701821 44r^5 + 187460219915061731328r^4 + 13283826021649132574976r^3 + 582087164074662713679776r^2 - 76816786996207910351317384r + 11738425413513964674880 19467,$

$g_6 = 661894339624960r^6 + 755731932755918848r^5 + 144515894798881581056r^4 + 68180376848981471063 04r^3 + 2449566880961511690937 6r^2 - 2549591425395423889149 9528r - 3622603872541189189781357118,$

$g_7 = 5340820668416r^6 - 823891876621582336r^5 - 134654713631522807808r^4 + 76078411778431122728 96r^3 + 380589873036621265963904r^2 + 46350145124567875855822224r + 2587944936282469247752334758,$

$g_8 = -49006006239232r^6 + 325574288035807232r^5 + 35146666009625608192r^4 - 1117900845323087169945 6r^3 + 296808678691923465416704r^2 - 17280170178667987958893632r - 1170496574133006352998283790,$

$g_9 = -30427414790144r^6 - 28691027338723328r^5 - 7787830491802075136r^4 + 408624562030122194534 4r^3 - 610971831894096108327424r^2 + 12087998274491738408216704r + 389033676273323566553647504,$

$g_{10} = 28285054484480r^6 - 22236388532420608r^5 + 877010862255223603 2r^4 - 2593324012156662824 96r^3 + 32211900444808017977958 4r^2 - 12306479503268766459441792r + 695887139191530558276949 28,$

$g_{11} = -8589221560320r^6 + 10865880257789952r^5 - 559122907399572684 8r^4 - 627920141968747397 12r^3 - 6982531052728846255513 6r^2 + 5468480756209576785711104r - 224037561262307076652719392,$

$g_{12} = 1300519452672r^6 - 2763285407465472r^5 + 187558538854400000 0r^4 - 796357337610923212 80r^3 + 32663411659829966725 12r^2 - 1415292361131443858856 96r + 157081601833589610455676000,$

$g_{13} = -96032784384r^6 + 457437446406144r^5 - 394107107698802688r^4 + 537156408256572948 48r^3 - 74673527404884147404 8r^2 - 9239988294845736896501 76r - 60666833682633239295578624,$

$g_{14} = 2717908992r^6 - 48451484123136r^5 + 56886813385555968r^4 - 14445144729166675968r^3 + 15455555219456711393 28r^2 + 471111072279951255328768r + 14154082082187031313831936,$

$g_{15} = 2906803666944r^5 - 5944527084847104r^4 + 2130906321977868288r^3 - 68569192355039664537 6r^2 - 126707051386212335816704r - 188860108319091224296 4992,$

$g_{16} = -73383542784r^5 + 446619545763840r^4 - 176359304464957440r^3 + 159317338058693443584r^2 + 216439358920238981775 36r + 111617605409145632470528,$

$g_{17} = -21527877648384r^4 + 6796795028963328r^3 - 2311577949383059046 4r^2 - 245548567486733333299 2r - 7596397738759288295424,$

$g_{18} = 483108323328r^4 + 31804631285760r^3 + 220696211106850406 4r^2 + 187012227883716476928r + 5962768736200944316416,$

$g_{19} = -11010589065216r^3 - 1391016675917168 64r^2 - 10070806418923585536r - 1885157164832436707328,$

$g_{20} = 238496514048r^3 + 5588570712047616r^2 + 506279865092014080r + 31793594322602328 8832,$

$g_{21} = -130214510198784r^2 - 32762777717440512r - 3392800551043419340 8,$

$g_{22} = 1348422598656r^2 + 1903461317738496r + 242660680210553241 6,$

$g_{23} = -64600450007040r - 117263464381808640,$

$g_{24} = 908121341952r + 3715090011389952,$

$g_{25} = -70585795215360,$

$g_{26} = 619173642240.$

Shucheng Ning
Institute of Mathematics
Academia Sinica
Beijing 100080, P. R. China
e-mail: scning@math.ac.cn

Zhiming Zheng
LMIB – School of Science
Beihang University
Beijing 100083, P. R. China
e-mail: zzheng@pku.edu.cn

Trends in Mathematics: *Differential Equations with Symbolic Computation*, 109–120

Symbolic Computation for Equilibria of Two Dynamic Models

Weinian Zhang and Rui Yan

Abstract. Many practical dynamic models contain complicated nonlinearities that make it difficult to investigate the distribution and qualitative properties of equilibria, which are actually the basic information for further discussion on bifurcations. In this paper effective methods of symbolic computation are introduced for two nonlinear systems.

Mathematics Subject Classification (2000). Primary 68W30; Secondary 34C05.

Keywords. Symbolic computation, equilibrium, qualitative property, polynomial.

1. Introduction

Many practical dynamic models are in the form of ordinary differential equations with complicated nonlinearities. The nonlinearities make it difficult to investigate the qualitative properties of their equilibria or even difficult to give the number of equilibria. For example, the multi-molecular reaction model [2]

$$\begin{cases} \dfrac{dx}{dt} = 1 - ax - x^p y^q, \\ \dfrac{dy}{dt} = b(x^p y^q - y), \end{cases} \tag{1.1}$$

is a polynomial differential system of degree $n = p + q$, where $x, y \geq 0$, integer parameters $p, q \geq 1$, and real parameter $a \geq 0$, $b > 0$. We have to solve the system of polynomial equations

$$\begin{cases} 1 - ax - x^p y^q = 0, \\ b(x^p y^q - y) = 0, \end{cases} \tag{1.2}$$

This work was completed with the support of NSFC (China) Grant 10471101, TRAPOYT, and MOE Doctoral Base Grant.

for equilibria of (1.1). Factoring (1.2) we turn to solve

$$\begin{cases} 1 - ax - x^p y^q = 0, \\ by = 0, \end{cases} \quad \text{or} \quad \begin{cases} 1 - ax - x^p y^q = 0, \\ x^p y^{q-1} - 1 = 0. \end{cases} \tag{1.3}$$

Clearly, from the first system of (1.3) we obtain a unique equilibrium $A : (1/a, 0)$. Moreover, equilibria determined by the second system of (1.3) lie on the line

$$1 - ax - y = 0. \tag{1.4}$$

For $y \geq 0$, we can get $x \leq \frac{1}{a}$. Eliminating y in the second system of (1.3), we have

$$x^p(1 - ax)^{q-1} - 1 = 0. \tag{1.5}$$

However we hardly solve (1.5) for general p and q.

Another example is the nonlinear mathematical model

$$J_s \frac{d^2\theta}{dt^2} + g(\theta) = 0, \tag{1.6}$$

where

$$\begin{aligned} g(\theta) = {} & LK\left(c - \sqrt{c^2 + L^2 \sin\frac{\theta}{2} - cL\sin\theta}\right) \frac{-\frac{L}{2}\sin\theta + c\cos\theta}{\sqrt{c^2 + \frac{L^2}{2} - \frac{L^2}{2}\cos\theta - cL\sin\theta}} \\ & -LK\left(c - \sqrt{c^2 + L^2 \sin\frac{\theta}{2} + cL\sin\theta}\right) \frac{\frac{L}{2}\sin\theta + c\cos\theta}{\sqrt{c^2 + \frac{L^2}{2} - \frac{L^2}{2}\cos\theta + cL\sin\theta}} \\ = {} & -K\sqrt{\Delta}\sin(\theta - \psi)\left(\frac{c}{\sqrt{c^2 + L^2/2 - \sqrt{\Delta}\cos(\theta - \psi)}} - 1\right) \\ & -K\sqrt{\Delta}\sin(\theta + \psi)\left(\frac{c}{\sqrt{c^2 + L^2/2 - \sqrt{\Delta}\cos(\theta + \psi)}} - 1\right), \end{aligned} \tag{1.7}$$

J_s, K, L, c are all positive constants, $\Delta = (L^2/2)^2 + (cL)^2$ and $\psi = \arccos(\frac{L^2/2}{\sqrt{\Delta}})$. It was considered in [5] and [4] to describe the free oscillations of the shaft of a rigid rotor supported symmetrically by four equal bearings at the ends on a low-speed balance platform. For the convenience of computation we take $J_s = 1$. In practice L is always much greater than c (simply denoted by $L \gg c$), for example $L = 4.765m$ and $c = 0.00138m$. Equation (1.6) is a Hamiltonian system and can be reduced equivalently to

$$\frac{d\theta}{dt} = \eta, \quad \frac{d\eta}{dt} = -g(\theta). \tag{1.8}$$

The energy function is $H(\theta,\eta) = \frac{\eta^2}{2} + G(\theta)$, where

$$\begin{aligned} G(\theta) &:= \int_0^\theta g(\zeta)\,d\zeta \\ &= 2Kc(2c - \sqrt{c^2 + L^2/2 - \sqrt{\Delta}\cos(\theta-\psi)} - \sqrt{c^2 + L^2/2 - \sqrt{\Delta}\cos(\theta+\psi)}) \\ &\quad + K\sqrt{\Delta}(2\cos\psi - \cos(\theta-\psi) - \cos(\theta+\psi)). \end{aligned} \tag{1.9}$$

Here $\frac{\eta^2}{2}$ and $G(\theta)$ represent its kinetic energy and potential energy respectively. It is easy to check that $-\pi, 0, \pi$ are zeros of g in the interval $[-\pi,\pi]$, thus the system has at least three equilibria $A_- : (-\pi, 0)$, $O : (0,0)$ and $A_+ : (\pi, 0)$ for $\theta \in [-\pi,\pi]$. However, it is not easy to determine whether the complicated g has other equilibria in the interval $\theta \in [-\pi,\pi]$.

Investigating all equilibria of a system is the first step to study the dynamics of the system. Without knowing qualitative properties of equilibria we cannot discuss global behaviors and bifurcations further. In this paper we show two effective methods of symbolic computation in the investigation of distribution and qualitative properties of equilibria for these nonlinear systems.

2. Equilibria of the Multi-molecular Reaction

As known in Section 1, the other equilibria of system (1.1) are determined by the polynomial in (1.5), denoted by $f(x)$. The difficulties in computation come from both the unspecific degree p, q in the polynomial and the irreducibility over the field $\mathbf{Q}$ of rational numbers. There is not an effective Maple command available to find either the number of its real roots or their locations. Our strategy is to transfer f to another simple polynomial. More concretely, its derivative f' may be reducible over $\mathbf{Q}$ and, by the Rolle's theorem, its zeros provide information about the number and locations of the real roots of f.

With the Maple V7 software, the derivative f' can be factored as

$$f'(x) = x^{p-1}(1-ax)^{q-2}(p - a(p+q-1)x), \tag{2.1}$$

which has three zeros

$$x_1 = 0, \quad x_2 = \frac{p}{a(p+q-1)}, \quad x_3 = \frac{1}{a}, \tag{2.2}$$

where $x_1 < x_2 < x_3$. Moreover, $f(x_1) = f(x_3) = -1$ and

$$f(x_2) = (\frac{p}{a(p+q-1)})^p(1 - \frac{p}{p+q-1})^{q-1} - 1.$$

By the Rolle's theorem, in the first quadrant the second system of (1.3) has no zero when $f(x_2) < 0$, a unique zero $B_0 : (x_2, y_2)$, where

$$y_2 = 1 - ax_2 = \frac{q-1}{p+q-1}, \tag{2.3}$$

when $f(x_2) = 0$, or two zeros $B_- : (x_-, y_-)$ and $B_+ : (x_+, y_+)$, where

$$x_- \in (0, \frac{p}{a(p+q-1)}), \quad x_+ \in (\frac{p}{a(p+q-1)}, \frac{1}{a}), \tag{2.4}$$

when $f(x_2) > 0$. Here the analytic expressions of x_-, x_+, y_- and y_+ cannot be given simply.

Let

$$a_0 = (\frac{p^p(q-1)^{q-1}}{(p+q-1)^{p+q-1}})^{\frac{1}{p}}. \tag{2.5}$$

The above conditions $f(x_2) < 0, f(x_2) = 0$ and $f(x_2) > 0$ are equivalent to $a > a_0, a = a_0$ and $a < a_0$, respectively. Thus we can summarize the above results in the following theorem.

Theorem 2.1. *System* (1.1) *has*

(i) *a unique equilibrium* $A : (\frac{1}{a}, 0)$ *when* $a > a_0$;

(ii) *two equilibria* $A : (\frac{1}{a}, 0)$ *and* $B_0 : (x_2, y_2)$, *where* x_2 *and* y_2 *are given in* (2.2) *and* (2.3), *when* $a = a_0$;

(iii) *three equilibria* $A : (\frac{1}{a}, 0)$, $B_- : (x_-, y_-)$ *and* $B_+ : (x_+, y_+)$, *where* $y_- = 1 - ax_-, y_+ = 1 - ax_+$, *when* $a < a_0$.

Next, we further show how to investigate qualitative properties of equilibria. Our discussion is focused at case (iii), where not all equilibria are determined with their exact coordinates. Let $B : (x, y)$ be one of the equilibria of (1.1) determined by the second system of (1.3). At B the vector field of (1.1) has its linear part

$$\begin{aligned} \begin{pmatrix} P'_x(x,y) & P'_y(x,y) \\ Q'_x(x,y) & Q'_y(x,y) \end{pmatrix} &= \begin{pmatrix} -a - px^{p-1}y^q & -qx^py^{q-1} \\ bpx^{p-1}y^q & bqx^py^{q-1} - b \end{pmatrix} \\ &= \begin{pmatrix} -a - py/x & -q \\ bpy/x & bq - b \end{pmatrix} = \begin{pmatrix} -a - p(1-ax)/x & -q \\ bp(1-ax)/x & bq - b \end{pmatrix}, \end{aligned} \tag{2.6}$$

where the second system of (1.3) and (1.4) are applied. This matrix has its trace

$$T = \frac{(a(p-1) + b(q-1))x - p}{x} \tag{2.7}$$

and determinant

$$D = -\frac{ab(p+q-1)x - bp}{x}. \tag{2.8}$$

Let

$$\begin{aligned} \Delta &:= T^2 - 4D \\ &= \{(a^2 + b^2 - 2ab + 2abp + 2abq - 2a^2p - 2b^2q + a^2p^2 + b^2q^2 + 2abpq)x^2 \\ &\quad + (2ap - 2bp - 2ap^2 - 2bpq)x + p^2\}/x^2. \end{aligned} \tag{2.9}$$

The qualitative properties of the equilibrium B are decided by the relations among T, D and Δ, i.e., B is a saddle if $D < 0$, a node if $D > 0$ and $\Delta \geq 0$, a focus if

$D > 0, \Delta < 0$ and $T \neq 0$, or a center-type equilibrium if $D > 0$ and $T > 0$, as seen in [3] and [6].

Obviously, T has a zero $x^{(1)} = \frac{p}{-a-b+ap+bq}$, D has a zero $x^{(2)} = x_2 = \frac{p}{a(p+q-1)}$, and Δ has two zeros $x^{(3)} = \frac{p}{ap+(\sqrt{bq}+\sqrt{b-a})^2}$, $x^{(4)} = \frac{p}{ap+(\sqrt{bq}-\sqrt{b-a})^2}$, when $b \geq a$ but Δ has no zeros when $b < a$. Those zeros have the following relations of order:

Lemma 2.2. *If $q = 1$ then $x^{(1)} > x^{(2)}$ and, moreover, $x^{(1)} \geq x^{(2)} \geq x^{(4)} \geq x^{(3)}$ when $b \geq a$. If $q > 1$ then $x^{(1)} > x^{(2)}$ when $b < a$; $x^{(1)} \geq x^{(2)} \geq x^{(4)} \geq x^{(3)}$ when $a \leq b \leq \frac{aq}{q-1}$; and $x^{(2)} > x^{(4)} > x^{(1)} > x^{(3)}$ when $b > \frac{aq}{q-1}$.*

Proof. The proof for $q = 1$ is trivial since $x^{(1)} = \frac{p}{a(p-1)}$, $x^{(2)} = \frac{1}{a}$, $x^{(3)} = \frac{p}{ap+(\sqrt{b}+\sqrt{b-a})^2}$ and $x^{(4)} = \frac{p}{ap+(\sqrt{b}-\sqrt{b-a})^2}$. In case $q > 1$,

$$\frac{1}{x^{(2)}} - \frac{1}{x^{(1)}} = \frac{aq + b(1-q)}{p} > 0, \tag{2.10}$$

i.e., $x^{(1)} > x^{(2)}$, when $b < a$. Furthermore, for $b \geq a$, it is obvious that $x^{(3)} \leq x^{(4)}$. The result that $x^{(2)} \geq x^{(4)}$ comes from

$$\frac{1}{x^{(2)}} - \frac{1}{x^{(4)}} = -\frac{(\sqrt{b} - \sqrt{(b-a)q})^2}{p} \leq 0. \tag{2.11}$$

Similarly to show $x^{(1)} \geq x^{(3)}$. From (2.10) we see that $x^{(1)} \geq x^{(2)}$ when $b \leq \frac{aq}{q-1}$ and $x^{(1)} < x^{(2)}$ when $b > \frac{aq}{q-1}$. Similarly, $x^{(4)} \leq x^{(1)}$ when $b \leq \frac{aq}{q-1}$ and $x^{(4)} < x^{(1)}$ when $b > \frac{aq}{q-1}$. □

Theorem 2.3. *A is a stable node of (1.1). Moreover, when $a < a_0 = \left(\frac{p^p(q-1)^{q-1}}{(p+q-1)^{p+q-1}}\right)^{\frac{1}{p}}$, the qualitative properties of the other two equilibria B_+, B_- are described in the following table.*

possibilities of a, b, p, q			B_-	B_+
$q = 1$	$a > b$		stable node	saddle
	$a \leq b$	$f(x^{(4)}) \leq 0$ or $f(x^{(3)}) \geq 0$	stable node	
		$f(x^{(4)}) > 0$ and $f(x^{(3)}) < 0$	stable focus	
$q > 1$	$b > a$		stable node	
	$a \leq b \leq \frac{aq}{q-1}$	$f(x^{(4)}) \leq 0$ or $f(x^{(3)}) \geq 0$	stable node	
		$f(x^{(4)}) > 0$ and $f(x^{(3)}) < 0$	stable focus	
	$b > \frac{aq}{q-1}$	$f(x^{(3)}) > 0$	stable node	
		$f(x^{(3)}) < 0$ and $f(x^{(1)}) > 0$	stable focus	
		$f(x^{(4)}) < 0$	unstable node	
		$f(x^{(1)}) < 0$ and $f(x^{(4)}) > 0$	unstable focus	
		$f(x^{(1)}) = 0$	weak focus	
Remark: $f(x) = x^p(1-ax)^{q-1} - 1$, $x^{(1)} = \frac{p}{-a-b+ap+bq}$, $x^{(2)} = x_2 = \frac{p}{a(p+q-1)}$, $x^{(3)} = \frac{p}{ap+(\sqrt{bq}+\sqrt{b-a})^2}$ and $x^{(4)} = \frac{p}{ap+(\sqrt{bq}-\sqrt{b-a})^2}$.				

Proof. We can give the linear part of the vector field of (1.1) at A easily and see that it has two eigenvalues $\lambda_1 = -a,\ \lambda_2 = -b$. Thus A is a stable node.

As above, let $B : (x, y)$ be B_- or B_+. Clearly, $0 < x < 1/a$. Moreover, from (2.8), $D < 0$ (resp. > 0) if and only if $x > x^{(2)}$ (resp. $< x^{(2)}$), i.e., $x \in (\frac{p}{a(p+q-1)}, \frac{1}{a})$ (resp. $x \in (0, \frac{p}{a(p+q-1)})$). This means that $B = B_+$ (resp. $B = B_-$) by (2.4) . We discuss B_- and B_+ in the following six cases.

Case i: $D < 0$. In this case $B = B_+$ is a saddle.

Case ii: $D > 0, T < 0, \Delta \geq 0$. In this case $B = B_-$ is a stable node. By (2.7)–(2.8) the conditions that $D > 0$ and $T < 0$ are equivalent to $x_- < x^{(2)}$ and $x_- < x^{(1)}$ respectively. Moreover, from (2.9), the polynomial $x^2\Delta$ has its leading coefficient

$$2ab(q-1) + b^2(q-1)^2 + 2ap(bq-a) + b^2 + 2abp + a^2p^2 + a^2 > 0$$

for $q \geq 1$ and $b \geq a$. Thus the condition $\Delta \geq 0$ is equivalent to

$$a > b \quad \text{or} \quad \begin{cases} a \leq b, \\ x_- \geq x^{(4)} \end{cases} \quad \text{or} \quad \begin{cases} a \leq b, \\ 0 < x_- \leq x^{(3)}. \end{cases}$$

Hence the conditions of the case ii can be expressed equivalently as

$$\text{(I)} : \begin{cases} x_- < x^{(2)}, \\ x_- < x^{(1)}, \\ a > b \end{cases} \quad \text{or} \quad \text{(II)} : \begin{cases} x_- < x^{(2)}, \\ x_- < x^{(1)}, \\ a \leq b, \\ x_- \geq x^{(4)} \end{cases} \quad \text{or} \quad \text{(III)} : \begin{cases} x_- < x^{(2)}, \\ x_- < x^{(1)}, \\ a \leq b, \\ 0 < x_- \leq x^{(3)}. \end{cases}$$

When $q = 1$, the relations that $x_- < x^{(2)}$ and that $x_- < x^{(1)}$ hold naturally, so (I) is equivalent to that $a > b$. As the unique zero of $f(x)$ in $(0, x^{(2)})$, x_- must lie in $[x^{(4)}, x^{(2)}]$ since $x_- \geq x^{(4)} > 0$ in (II). By continuity, $f(x^{(4)}) \leq 0$ because $f(0) = -1$ and $f(x^{(2)}) > 0$, so (II) is equivalent to that both $a \leq b$ and $f(x^{(4)}) \leq 0$. In (III), similarly, x_- as the unique zero of $f(x)$ in $(0, x^{(2)})$ must lie in $(0, x^{(3)}]$. Noting that $f(0) = -1$ and $f(x^{(2)}) > 0$, we see that (III) is equivalent to that both $a \leq b$ and $f(x^{(3)}) \geq 0$. When $q > 1$, using the same arguments as above we know that (I) is equivalent to that $a > b$; (II) is equivalent to that both $a \leq b \leq \frac{aq}{q-1}$ and $f(x^{(4)}) \leq 0$; and (III) is equivalent to

$$\begin{cases} a \leq b \leq \frac{aq}{q-1}, \\ f(x^{(3)}) \geq 0 \end{cases} \quad \text{or} \quad \begin{cases} b > \frac{aq}{q-1}, \\ f(x^{(3)}) > 0. \end{cases}$$

Therefore, B_- is a stable node if and only if

$$a > b \quad \text{or} \quad \begin{cases} a \leq b, \\ f(x^{(4)}) \leq 0 \end{cases} \quad \text{or} \quad \begin{cases} a \leq b, \\ f(x^{(3)}) \geq 0 \end{cases}$$

if $q = 1$, or

$$a > b \quad \text{or} \quad \begin{cases} a \leq b \leq \frac{aq}{q-1}, \\ f(x^{(4)}) \leq 0 \end{cases} \quad \text{or} \quad \begin{cases} a \leq b \leq \frac{aq}{q-1}, \\ f(x^{(3)}) \geq 0 \end{cases} \quad \text{or} \quad \begin{cases} b > \frac{aq}{q-1}, \\ f(x^{(3)}) > 0 \end{cases}$$

if $q > 1$.

Case iii: $D > 0, T < 0, \Delta < 0$. In this case $B = B_-$ is a stable focus. Similarly, we can show that the conditions of this case are equivalent to

$$\begin{cases} a < b, \\ f(x^{(3)}) < 0, & \text{if } q = 1, \\ f(x^{(4)}) > 0 \end{cases}$$

or

$$\begin{cases} a \le b \le \frac{aq}{q-1}, \\ f(x^{(3)}) < 0, \\ f(x^{(4)}) > 0 \end{cases} \quad \text{or} \quad \begin{cases} b > \frac{aq}{q-1}, \\ f(x^{(3)}) < 0, & \text{if } q > 1. \\ f(x^{(1)}) > 0 \end{cases}$$

Case iv: $D > 0, T > 0, \Delta \ge 0$. In this case $B = B_-$ is an unstable node and it is similar to show that the conditions of this case are equivalent to

$$\begin{cases} q > 1, \\ b > \frac{aq}{q-1}, \\ f(x^{(4)}) < 0. \end{cases}$$

Case v: $D > 0, T > 0, \Delta < 0$. In this case $B = B_-$ is an unstable focus and the conditions of this case can be proved similarly to be equivalent to

$$\begin{cases} q > 1, \\ b > \frac{aq}{q-1}, \\ f(x^{(1)}) < 0, \\ f(x^{(4)}) > 0. \end{cases}$$

Case vi: $D > 0, T = 0$. In this case $B = B_-$ is of center type. Similarly, the conditions of this case are equivalent to

$$\begin{cases} b > \frac{aq}{q-1}, \\ f(x^{(1)}) = 0. \end{cases}$$

Further computation with the aid of Maple V7 derives the Lyapunov number

$$L_3(a, b, p, q) = \frac{q(b(bq - b - aq))^{1/2}(ap - a + b + 2aq - bq)(ap - a + bq - b)^3}{16b^3p^2(a + b - bq)(aq + b - bq)^3},$$

which vanishes when $b = a((p + 2q - 1)/(q - 1))$. Continue to compute the Lyapunov number

$$L_5(a, b, p, q) = \frac{q(p + q)(q - 1)^{7/2}(p + q - 1)^{5/2}K(p, q)}{18a^4p^4(p + 2q - 1)^{9/2}(p + 2q - 2)^3},$$

where

$$K(p, q) = p^3 + (4q - 3)p^2 + 2(q - 1)p + 4(q - 1)^2 > 0$$

since $p \ge 1, q > 1$. This proves that B_- is a weak focus of multiplicity at most 2.

Up to now, all possibilities of parameters are verified. □

3. Equilibria of the Rotor Oscillation

The equilibria of system (1.6) is determined by zeros of g. Our strategy is to:

(i) Reduce the zero problem of g to the corresponding one of a polynomial by eliminating trigonometric functions and rationalizing;

(ii) Apply the interval isolating computation to find the number and locations of zeros;

(iii) Pick out the extraneous roots by reasoning on polynomial inequalities.

Let $s := L/c$ and rewrite g in (1.7) as

$$g(\theta) = K\sqrt{\Delta}g_0(\theta), \tag{3.1}$$

where

$$\begin{aligned} g_0(\theta) &= -\sin(\theta-\psi)\Big(\frac{\sqrt{2}}{\sqrt{2+2s^2-\sqrt{s^4+4s^2}\cos(\theta-\psi)}}-1\Big) \\ &\quad -\sin(\theta+\psi)\Big(\frac{\sqrt{2}}{\sqrt{2+2s^2-\sqrt{s^4+4s^2}\cos(\theta+\psi)}}-1\Big), \end{aligned} \tag{3.2}$$

and $\psi := \arccos(s/\sqrt{2s^2+4})$. Let $g_{\min}$ denote the minimal of $g_0(\theta)$ on $[0,\pi]$. Clearly $g_{\min}$ is a continuous function of the new parameter s.

Numerically plotting $g_{\min}(s)$ shows that when s is approximately greater than 2.0, i.e., $L \gg 2c$, $g(\theta)$ has no other zeros between 0 and π. The same conclusion can also be made on $(-\pi, 0)$ by symmetry. However, the inevitable error makes the numerical simulation be short of enough persuasion to the existence of zeros of g_0 in a definite interval. The following theorem demonstrates that we are able to prove rigorously the exclusion of other equilibria for any specified parameters L and c when $L \gg c$.

Theorem 3.1. *The system* (1.8) *is* 2π*-periodic and has equilibria* $O : (0,0)$, $A_- : (-\pi,0)$ *and* $A_+ : (\pi,0)$ *for* $\theta \in [-\pi,\pi]$. *For specified* $L = 4.765$ *and* $c = 0.00138$, *system* (1.8) *has exactly these three equilibria in* $[-\pi,\pi]$. *Moreover, when* $L \gg c$ *(for instance* $L = 4.765$ *and* $c = 0.00138$*),* O *is a center and both* A_- *and* A_+ *are saddles. The curve* $H(\theta,\eta) = h_*$, *where* $h_* := 2Kc(2c - 2\sqrt{c^2+L^2}) + 2KL^2$, *has two branches* Γ_+ *and* Γ_-, *which are two heteroclinic orbits both connecting* A_- *and* A_+ *but lie in the upper half-plane and lower half-plane respectively. For each* $h \in (0,h_*)$, *the curve* $H(\theta,\eta) = h$ *is a periodic orbit* Γ_h *around* O *and surrounded by* $\Gamma_\pm$.

Proof. The periodicity of the system is implied by the periodicity of g. Thus it suffices to discuss the system for θ in the interval $[-\pi,\pi]$. It is immediately obvious that the system has at least three equilibria $A_- : (-\pi,0)$, $O : (0,0)$ and $A_+ : (\pi,0)$ for $\theta \in [-\pi,\pi]$. For specified $L = 4.765$ and $c = 0.00138$, we further claim that g has no zeros in the open interval $(0,\pi)$.

Consider g_0 as in (3.2). It can be factorized as $g_0(\theta) = -\frac{U(\theta)}{U_+U_-}$, with $U_\pm = [1 + s^2/2 - \sqrt{s^4/4 + s^2}\cos(\theta \pm \psi)]^{1/2}$ nonsingular and

$$U(\theta) = [\sin(\theta)\cos(\psi) + \cos(\theta)\sin(\psi)]U_- + [\sin(\theta)\cos(\psi) - \cos(\theta)\sin(\psi)]U_+ - 2\sin(\theta)\cos(\psi)U_+U_-. \tag{3.3}$$

Clearly, all zeros of g can be calculated from $U(\theta) = 0$. Take a change of variable $z := \cos\theta$, which in fact defines a homeomorphism from $(0, \pi)$ onto $(-1, 1)$. Let $\tilde{U}(z) := U(\arccos z)$. Rationalizing the equality $\tilde{U}(z) = 0$ leads to an equation

$$(z-1)(z+1)P(z) = 0, \tag{3.4}$$

where

$$\begin{aligned} P(z) := {} & (s^{10} + 8s^8 + 16s^6)z^6 + (-4s^{10} - 22s^8 - 16s^6 + 32s^4)z^5 \\ & + (5s^{10} + 10s^8 - 20s^6 + 28s^4 + 32s^2 - 64)z^4 \\ & + (20s^8 - 120s^4 + 64s^2 + 128)z^3 \\ & + (-5s^{10} - 20s^8 + 40s^6 + 32s^4 - 64s^2)z^2 \\ & + (4s^{10} + 2s^8 - 16s^6 + 24s^4 - 32s^2)z \\ & + (-s^{10} + 2s^8 - 4s^6 + 4s^4) \end{aligned} \tag{3.5}$$

and $s := L/c = 3665.384615$. The polynomial P in the form (3.5) is irreducible over the field $\mathbf{Q}$ of rational numbers, so it is hard to compute real roots of such a polynomial of high degree. In order to find all real roots of P, we employ the command "`realroot`$(P, 1/100000)$" of the Maple V7 software, which immediately gives isolating intervals of width $1/100000$ by its internal symbolic operations

$$\left[\frac{268435411}{268435456}, \frac{67108853}{67108864}\right], \qquad \left[-1, \frac{-131071}{131072}\right],$$

covering all real roots of $P(z)$. These two intervals in the variable z correspond to the intervals

$$[5.789646011 \times 10^{-4}, 5.725382162 \times 10^{-4}], \quad [3.137686400, \pi] \tag{3.6}$$

in θ, respectively.

On the other hand, for $\theta \in (0,\ \psi)$, where $\psi = \arccos(\frac{L^2/2}{\sqrt{\Delta}}) = 5.793099424 \times 10^{-4}$, it can be seen that $\sin(\theta-\psi) < 0$, $\sin(\theta+\psi) > 0$, $\cos(\theta-\psi) > \cos(-\psi) = \frac{L^2/2}{\sqrt{\Delta}}$ and $\cos(\theta + \psi) < \cos(\psi) = \frac{L^2/2}{\sqrt{\Delta}}$. So

$$\sqrt{c^2 + \frac{L^2}{2} - \sqrt{\Delta}\cos(\theta - \psi)} < c, \qquad \sqrt{c^2 + \frac{L^2}{2} - \sqrt{\Delta}\cos(\theta + \psi)} > c. \tag{3.7}$$

Hence in terms of (1.7), $g(\theta) > 0$ on the interval $(0, \psi)$. The first interval in (3.6) is obviously contained in $(0, \psi)$; therefore, the real root of $P(z)$ in the first interval is an additional one arising in the procedure of rationalization of $\tilde{U}(z)$. It is actually not a root of g.

In order to show that the real root of $P(z)$ in the second interval in (3.6) is not a root of g, we notice that the interval is contained in $(2\psi, \pi]$. For $\theta \in (2\psi, \pi - \psi)$, we have that $\sin(\theta-\psi) > 0$, $\sin(\theta+\psi) > 0$ and that

$$\sqrt{c^2 + \frac{L^2}{2} - \sqrt{\Delta}\cos(\theta-\psi)} > c, \qquad \sqrt{c^2 + \frac{L^2}{2} - \sqrt{\Delta}\cos(\theta+\psi)} > c,$$

similar to (3.7). Hence in terms of (1.7), $g(\theta) > 0$ on the interval $(2\psi, \pi-\psi)$. For $\theta \in [\pi-\psi, \pi)$, define

$$g_p(\theta) := -\sqrt{\Delta}\sin(\theta-\psi)\Big(\frac{c}{\sqrt{c^2 + L^2/2 - \sqrt{\Delta}\cos(\theta-\psi)}} - 1\Big),$$

$$g_q(\theta) := \sqrt{\Delta}\sin(\theta+\psi)\Big(\frac{c}{\sqrt{c^2 + L^2/2 - \sqrt{\Delta}\cos(\theta+\psi)}} - 1\Big).$$

For the given $L = 4.765$ and $c = 0.00138$, note the fact that $\frac{\pi}{2} - \psi \le \frac{\theta-\psi}{2} < \frac{\pi}{2} - \frac{\psi}{2}$ and $\sin(\theta-\psi) < |\cos(\theta-\psi)| < 1$ for $\theta \in [\pi-\psi, \pi)$. Employing Maple V7 we have

$$\begin{aligned}\frac{d}{d\theta}g_p(\theta) &= 11.35261440\cos(\theta-\psi)\Big(1 - \frac{0.00138}{\sqrt{22.7052288}\sin(\frac{\theta-\psi}{2})}\Big)\\&\quad + \frac{0.08892847907\sin^2(\theta-\psi)}{(22.7052288)^{3/2}\sin^3(\frac{\theta-\psi}{2})}\\&< 11.35261440\cos(\theta-\psi)\Big(1 - \frac{0.00138}{\sqrt{22.7052288}\sin(\frac{\pi}{2}-\psi)}\Big)\\&\quad + \frac{0.08892847907\sin^2(\theta-\psi)}{(22.7052288)^{3/2}\sin^3(\frac{\pi}{2}-\psi)}\\&= 11.34932655\cos(\theta-\psi) + 0.0008219629813\sin^2(\theta-\psi)\\&< 0.\end{aligned} \tag{3.8}$$

Similarly, noting that $\frac{\pi}{2} \le \frac{\theta+\psi}{2} < \frac{\pi}{2} + \frac{\psi}{2}$ and $\sin(\theta+\psi) < |\cos(\theta+\psi)| < 1$ for $\theta \in [\pi-\psi, \pi)$, we can derive

$$\begin{aligned}\frac{d}{d\theta}g_q(\theta) &= -11.35261440\cos(\theta+\psi)\Big(1 - \frac{0.00138}{\sqrt{22.7052288}\sin(\frac{\theta+\psi}{2})}\Big)\\&\quad - \frac{0.08892847907\sin^2(\theta+\psi)}{(22.7052288)^{3/2}\sin^3(\frac{\theta+\psi}{2})}\\&> -11.35261440\cos(\theta+\psi)\Big(1 - \frac{0.00138}{\sqrt{22.7052288}\sin(\frac{\pi}{2}+\frac{\psi}{2})}\Big)\\&\quad - \frac{0.08892847907\sin^2(\theta+\psi)}{(22.7052288)^{3/2}\sin^3(\frac{\pi}{2}+\frac{\psi}{2})}\\&= -11.34932655\cos(\theta+\psi) - 0.0008219626723\sin^2(\theta+\psi)\\&> 0.\end{aligned} \tag{3.9}$$

Moreover, it is easy to check that $g_p(\pi-\psi) > g_q(\pi-\psi) = 0$ and $g_p(\pi) = g_q(\pi)$. By (3.8) and (3.9) we see that $g(\theta) = g_p(\theta) - g_q(\theta) > 0$ on $[\pi-\psi, \pi)$. Therefore, $g(\theta) > 0$ on $(2\psi, \pi)$ and thus g has no zeros in the second interval in (3.6) except for the point π.

As above, the claimed fact that g has no zeros in the interval $(0, \pi)$ is proved. We can similarly prove that g has no zeros in $(-\pi, 0)$. It follows that system (1.8) has only three equilibria O, A_- and A_+ in $[-\pi, \pi]$.

Now we continue to discuss the qualitative properties of the equilibria and orbits near them. As a basic knowledge in Chapter V of [1], the potential $G(\theta)$ reaches extrema at equilibria, and an equilibrium $(\theta_0, 0)$ is a center (resp. saddle) when $G(\theta_0)$ reaches a minimal (resp. maximal) value. Notice that

$$G''(0) = \frac{K}{2c^2}\Delta[\sin^2(-\psi) + \sin^2\psi] = KL^2 > 0;$$

hence O is a center. Similarly, let us check

$$\begin{aligned} G''(\pm\pi) &= -2K\sqrt{\Delta}\{-\cos\psi\,(\frac{c}{\sqrt{c^2+L^2}} - 1) - \frac{c}{2}\sqrt{\Delta}\frac{\sin^2\psi}{(c^2+L^2)^{3/2}}\} \\ &= -\frac{KL^2D(L,c)}{(c^2+L^2)^{3/2}[(c^2+L^2)^{3/2} + (cL^2+2c^3)]}, \end{aligned} \tag{3.10}$$

where $D(L,c) := L^6 + 2L^4c^2 - c^4L^2 - 3c^6$. Clearly, $D(L,c) \to +\infty$ as $L \to +\infty$. Consequently $D(L,c) > 0$ and $G''(\pm\pi) < 0$ when $L \gg c$, in particular when $L = 4.765m$ and $c = 0.00138m$. That is, both A_- and A_+ are saddles when $L \gg c$. Furthermore, we can verify that $h_* := H(\pm\pi, 0) = G(\pm\pi)$, which is the energy of the system at A_- and A_+ respectively. The orbits $\Gamma_\pm$ connecting A_- and A_+ lie on the curve

$$H(\theta, \eta) = h_*. \tag{3.11}$$

At the center O we have $H(0,0) = G(0) = 0$. Therefore, each $h \in (0, h_*)$ determines a closed curve $H(\theta, \eta) = h$, which is a periodic orbit around O. This completes the proof. □

For any given real system parameters L and c with $L \gg c$, we can derive the same result as given in Theorem 3.1, by following the same procedure as for $L = 4.765$ and $c = 0.00138$.

References

[1] J. K. Hale, *Ordinary Differential Equations*, Wiley-Interscience, New York, 1969.

[2] X. Hou, R. Yan and W. Zhang, Bifurcations of a Polynomial Differential System of Degree n in Biochemical Reactions, *Comput. Math. Appl.* **43**, 1407–1423, 2002.

[3] L. Perko, *Differential Equations and Dynamical Systems*, Springer-Verlag, New York, 1991.

[4] J. Xu, R. Yan and W. Zhang, *An Algorithm for Melnikov Functions and Application to a Chaotic Rotor*, NUS Research Report, Singapore, 2002.

[5] W. D. Zhang and W. Zhang, Modelling and Analysis for Nonlinear Oscillation in a Rigid Rotor (in Chinese), *Turbine Technology* **40**, 84–88, 1998.

[6] Z.-F. Zhang, T.-R. Ding, W.-Z. Huang and Z.-X. Dong, *Qualitative Theory of Differential Equations*, Amer. Math. Soc., Providence, 1992.

Acknowledgment

Many thanks to the referees for their helpful comments and suggestions.

Weinian Zhang
Department of Mathematics
Sichuan University
Chengdu, Sichuan 610064
P. R. China
e-mail: matzwn@126.com

Rui Yan
Department of Electric and Computer Engineering
National University of Singapore
10 Kent Ridge Crescent
Singapore 117576
e-mail: engp1579@nus.edu.sg

Trends in Mathematics: *Differential Equations with Symbolic Computation*, 121–142

Attractive Regions in Power Systems by Singular Perturbation Analysis

Zhujun Jing, Ruiqi Wang, Luonan Chen and Jin Deng

Abstract. This paper aims to investigate attractive regions of operating points for power systems by applying singular perturbation analysis. A time-scale decomposition is performed to illustrate how the critical model can be identified with reduced-order systems and how bifurcation phenomena can be explained with such low order systems. The slow dynamics and fast dynamics including bifurcation conditions and domain of attractor of the stable equilibrium are also analyzed. We show that the attractive region of a stable equilibrium point is composed of the domain enclosed by the stable manifold of a saddle point for the simplified subsystem, and that the size of a stability region is also considerably affected by the voltage magnitude behind the transient reactance. Several numerical examples are used to demonstrate our theoretical results.

Mathematics Subject Classification (2000). Primary 93C70; Secondary 93C95.

Keywords. Power system, singular perturbation, bifurcation, attractive region, stability.

1. Introduction

In recent years, study on stability regions of power systems has attracted considerably attention among power system engineers. Many methods based on dynamical system theory [5, 7, 13, 27], such as bifurcation analysis and singular perturbation, have been used to solve this problem, see the references except [3, 4, 8, 10, 14, 19, 20, 22, 26] in this paper. In particular, because of the existence of various time scales in power systems, singular perturbation theory [6, 12, 15, 16, 17, 18] was applied to study dynamical behavior in several papers [1, 2, 11, 20, 23, 25, 28]. The major advantages of applying singular perturbation theory to voltage stability problems of power systems are that the complicated

This work is supported by the National Key Basic Research Fund (No.G1998020307) and the Chinese Academy of Sciences (KZCX2-SW-118).

original systems can be mathematically simplified without loss of physical meaning. Specifically, the systems can be approximated by two simple subsystems, i.e. slow and fast subsystems, whose dimensions are usually much low. We therefore need only to study the dynamical behavior of the simplified subsystems, and can understand the global properties of the original system with the combination of the behaviors of those two subsystems.

A power system considered in [23, 24] is modelled without automatic voltage regulators, and its dynamics as well as stability were studied by singular perturbation theory. In this paper, we extend the model in [23, 24] to further consider a voltage control device (excitation field control with hard limits) for both single and multimachine systems. Slow and fast dynamics are decomposed respectively by using singular perturbation analysis, and nonlinear properties are also investigated to derive the attractive regions of equilibria. This paper is a continuing work of our previous paper [11], which theoretically gives the stability region of a power system but with only a three-dimensional generator model.

The organization of this paper is as follows: Section 2 briefly describes the necessary techniques of singular perturbation theory. In Section 3, we give the theoretical analysis to multimachine systems using the methods introduced in Section 2. In Section 4, we examine a single machine system from both theoretical analysis and numerical simulation. The domain of attractor is also derived by using results obtained from slow and fast subsystems. Finally we give several general remarks to conclude this paper in Section 5.

2. Singular Perturbed Systems

In this section, we summarize necessary techniques of singular perturbation theory used in this paper. Consider the standard form of a singular perturbed system:

$$\begin{cases} \dot{x_t} = f(x,z,\epsilon), \\ \epsilon \dot{z_t} = g(x,z,\epsilon), \end{cases} \tag{2.1}$$

where $x \in R^n$ and $z \in R^l$, and ϵ is a small parameter. The functions f and g are both assumed to be C^r on a set $U \times I$, where $U \subset R^N$ is open with $N = n + l$ and $r > 1$. Due to this small parameter ϵ, the dynamics of the variable z evolve on a much faster time scale than the dynamics of the variable x. Thus z is commonly referred to as the fast variable and x is called the slow variable.

System (2.1) can be reformulated with a change of time-scale $t = \epsilon\tau$ as

$$\begin{cases} \dot{x_\tau} = \epsilon f(t,x,z,\epsilon), \\ \dot{z_\tau} = g(t,x,z,\epsilon). \end{cases} \tag{2.2}$$

The time scale given by t is said to be slow whereas that for τ is fast. As long as $\epsilon \neq 0$ the dynamics (2.1) and (2.2) are equivalent. Thus (2.1) is called the slow system and (2.2) the fast system. The essential idea in singular perturbation is to deduce the behavior of the solution of the singular perturbation system (2.1) or

(2.2) by studying two limiting cases. In (2.1) letting $\epsilon \to 0$, the limit only makes sense if $g(x, z, 0) = 0$ and is thus given by

$$\begin{cases} \dot{x_t} = f(x, z, 0), \\ 0 = g(x, z, 0), \end{cases} \tag{2.3}$$

which is a differential-algebraic system and is also called the slow subsystem or hybrid system.

Let $\epsilon \to 0$ in (2.2) we obtain the fast subsystem

$$\begin{cases} \dot{x_\tau} = 0, \\ \dot{z_\tau} = g(x, z, 0), \end{cases} \tag{2.4}$$

where the slow variable x can be regarded as the parameter in the vector field of the fast variable. The dynamic properties of the fast subsystem (2.4) depend on the parameter x. The set of equilibria of the fast subsystem is called critical manifold. For sufficiently small ϵ, a normally hyperbolic subset of the critical manifold persists as a locally invariant manifold that is $O(\epsilon)$ close to the critical manifold [5, 12]. This perturbed manifold is called the slow manifold.

By combining the solutions of slow subsystem (2.3) and fast subsystem (2.4) appropriately, the qualitative behavior of the singular perturbed system (2.1) or (2.2) can be approximately determined and the theoretical results as well as proofs are given in [17, 18].

Let $z = H(x, \epsilon)$ be the invariant manifold for (2.1), when H is a sufficiently smooth function of x and ϵ, $z = H(x, \epsilon)$ is slow manifold M_ϵ for (2.1).

According to the implicit function theorem in [13], when $\det(g_z) \neq 0$ (the Jacobian matrix g_z is nonsingular) the system (2.3) becomes

$$\dot{x_t} = f(x, z(x), 0). \tag{2.5}$$

System (2.5) determines the dynamics of the slow subsystem (2.3), where $z = H(x, 0)$ is an order-0 slow manifold M_0.

For $\det(g_z) \neq 0$, by linearizing equation (2.3) at the equilibrium, we have the following linear equation.

$$\dot{x} = Ax, \tag{2.6}$$

where $A = J_1 - J_2(J_4)^{-1}J_3$, $J_1 = \frac{\partial f}{\partial x}$, $J_2 = \frac{\partial f}{\partial z}$, $J_3 = \frac{\partial g}{\partial x}$ and $J_4 = \frac{\partial g}{\partial z} \neq 0$.

The equilibrium point is local stable if all the eigenvalues of matrix A have negative real parts. The stable equilibrium corresponds to the stable operating point in power system.

There are the two limiting cases, which are considered in power systems.

The first limiting case is that $J_4 \neq 0$ and the matrix A becomes singular having simple eigenvalue zero or the matrix A has a simple pair of purely imaginary eigenvalues with the remaining eigenvalues having nonzero real parts. This is a condition of nonhyperbolic equilibrium, which can give rise to a saddle-node bifurcation (SNB) (or transcritical, or pitchfork bifurcation) or Hopf bifurcation.

Obviously, the condition $\det(A) = 0$ defines a voltage stability limit when the critical eigenvalue is related to the voltage response. And a Hopf bifurcation causes the emergence or disappearance of a periodic oscillation from an equilibrium point.

The second limiting case is that fold points [27] correspond to points where Jacobian matrix J_4 is singular, and also corresponds to algebraic singular point in (2.3) or saddle-node bifurcation of the fast subsystem (2.4) or impasse surface [19] that separates in general the state space into disjoint components called causal regions. The fundamental nature of the solutions changes at the algebraic singularity point ($J_4 = 0$) is called a singularity induced bifurcation (SIB) [21].

As is discussed in [21], the singularity of J_4 is a sufficient condition for voltage collapse characterized by an infinitely large eigenvalue of the matrix A near the point of collapse. Certainly the power system will collapse before reaching the singularity point due to having an extremely large positive eigenvalue.

3. Decomposition of Multimachine Dynamics

A power system consisting of m-synchronous machines with the excitation field control can be described in the transient time-scale by the following set of equations:

$$\begin{cases} \dot{\delta} = \omega - \omega_s, \\ M\dot{\omega} = -D(\omega - \omega_s) + P_T - P_G, \\ T'_d\dot{E}' = -E' + E_{fd} - (x_d - x'_d)I_d, \\ T_A\dot{E}_{fdr} = -K_A(V(\delta, E') - V_{ref}) - (E_{fdr} - E_{fd0}), \end{cases} \tag{3.1}$$

where $P_G = E'I_q - (x_q - x_d)I_dI_q$, and

$$\delta = \begin{pmatrix} \delta_1 \\ \vdots \\ \delta_m \end{pmatrix}, \quad \omega = \begin{pmatrix} \omega_1 \\ \vdots \\ \omega_m \end{pmatrix}, \quad E' = \begin{pmatrix} E'_1 \\ \vdots \\ E'_m \end{pmatrix}, \quad E_{fdr} = \begin{pmatrix} E_{fdr1} \\ \vdots \\ E_{fdrm} \end{pmatrix},$$

$$V = \begin{pmatrix} V_1(E', \delta) \\ \vdots \\ V_m(E', \delta) \end{pmatrix}, \quad x_d = \begin{pmatrix} x_{d1} \\ \vdots \\ x_{dm} \end{pmatrix}, \quad x'_d = \begin{pmatrix} x'_{d1} \\ \vdots \\ x'_{dm} \end{pmatrix},$$

$$I_d = \begin{pmatrix} I_{d1}(E', \delta) \\ \vdots \\ I_{dm}(E', \delta) \end{pmatrix}, \quad I_q = \begin{pmatrix} I_{q1}(E', \delta) \\ \vdots \\ I_{qm}(E', \delta) \end{pmatrix}, \quad E_{fd0} = \begin{pmatrix} E_{fd01} \\ \vdots \\ E_{fd0m} \end{pmatrix},$$

$$P_G = \begin{pmatrix} P_{G1}(E', \delta) \\ \vdots \\ P_{Gm}(E', \delta) \end{pmatrix}, \quad P_T = \begin{pmatrix} P_{T1} \\ \vdots \\ P_{Tm} \end{pmatrix}, \quad V_{ref} = \begin{pmatrix} V_{ref1} \\ \vdots \\ V_{refm} \end{pmatrix},$$

M, D, T'_d and T_A are $m \times m$ diagonal matrices containing the mechanical starting times ($M_i = 2H_i$), the damping terms, and the field open circuit time constants respectively [24]. P_T, V_{ref}, ω_s are considered as constant inputs. δ are the rotor angles, ω are the machine frequency deviations, E' are the voltage magnitudes behind the transient reactance, E_{fd} are the field excitation voltages. I_d, I_q and V are the functions of the state variables δ and E', depending on the interconnection between machines and loads, and are written as $I_d(\delta, E')$, $I_q(\delta, E')$ and $V(\delta, E')$. All other parameters are explained in [24]. In order to make a thorough investigation of system stability, ideally all the existing dynamic factors in a power system should be taken into account. However, such an approach, though theoretical preferable, makes the analysis too complicated. It is therefore necessary to simplify the problem without loss of generality. In this paper, we adopt singular perturbation theory to simplify the dynamical models of power systems.

The multimachine model (3.1) can be decomposed into two subsystems, i.e., a slow one consisting of flux-decay model, and a fast one describing electromechanical oscillations and field excitation voltage. To achieve such a decomposition, we mainly follow the approach in [23, 24] to introduce the following parameters:

$$\epsilon = \sqrt{\frac{2H}{\omega_0}}, \quad H_0 = \frac{1}{M_i}\sum_{i=1}^{m} H_i, \quad \omega' = \epsilon\omega, \quad H = \mathrm{diag}[H_i/H_0],$$

and $T_A = \epsilon T$, $\omega_0 = 2\pi f_0 = 120\pi$. Using the above notation, the system (3.1) takes the following singular perturbation form [23, 24]:

$$\begin{cases} \epsilon\dot{\delta} = \omega', \\ \epsilon\dot{\omega}' = P_T - P_G(\delta, E') - \frac{\epsilon}{2H}D\omega', \\ \epsilon\dot{E}_{fdr} = T^{-1}[-K_A(V(\delta, E') - V_{ref}) - (E_{fdr} - E_{fdr0})], \\ \dot{E}' = T_d'^{-1}(E_{fd} - E(\delta, E')). \end{cases} \tag{3.2}$$

In (3.2), clearly δ, ω', and E_{fdr} are fast variables with $3m$ dimensions and E' are slow variables with m dimensions, when ϵ is assumed to be sufficiently small.

Consider the m-dimensional slow manifold in the state space of the system (3.2) defined by the $3m$ equations, which are written as a series expansion in ϵ:

$$\delta = h_1(E') = h_{10} + \epsilon h_{11} + O(\epsilon^2), \tag{3.3}$$

$$\omega' = h_2(E') = h_{20} + \epsilon h_{21} + O(\epsilon^2), \tag{3.4}$$

$$E_{fdr} = h_3(E') = h_{30} + \epsilon h_{31} + O(\epsilon^2). \tag{3.5}$$

The manifold M defined by h_1, h_2 and h_3 is an invariant manifold for the fast dynamics equations are satisfied, which guarantee that a trajectory starting on the manifold will remain on it for all time.

$$\epsilon\dot{\delta} = \epsilon\frac{\partial h_1}{\partial E'}\dot{E}' = \epsilon\frac{\partial h_1}{\partial E'}(T_d')^{-1}(E_{fd} - E(h_1, E')) = \omega' = h_2(E'), \tag{3.6}$$

$$\begin{aligned} \epsilon\dot{\omega}' &= \epsilon\frac{\partial h_2}{\partial E'}\dot{E}' = \epsilon\frac{\partial h_2}{\partial E'}(T_d')^{-1}(E_{fd} - E(h_1, E')) \\ &= P_T - P_G(h_1, E') - \frac{\epsilon}{2H}Dh_2(E'), \end{aligned} \tag{3.7}$$

$$\epsilon \dot{E}_{fdr} = \epsilon \frac{\partial h_3}{\partial E'} \dot{E}' = \epsilon \frac{\partial h_3}{\partial E'} (T'_d)^{-1} (E_{fd} - E(h_1, E'))$$
$$= T^{-1}[-K_A(V(h_1, E') - V_{ref} - (h_3(E') - E_{fdr0})]. \quad (3.8)$$

It is not possible in general to solve the equations (3.6)–(3.8) in order to obtain analytically the slow manifold $M = \{h_1, h_2, h_3\}$ of the system (3.2). However, an approximate slow manifold can be found in the following way:

$$\epsilon(\frac{\partial h_{10}}{\partial E'} + \epsilon \frac{\partial h_{11}}{\partial E'})(T'_d)^{-1}(E_{fd} - E(h_{10}, E') - \frac{\partial E}{\partial \delta} h_{11}\epsilon) = h_{20} + \epsilon h_{21},$$
$$\epsilon(\frac{\partial h_{20}}{\partial E'} + \epsilon \frac{\partial h_{21}}{\partial E'})(T'_d)^{-1}(E_{fd} - E(h_{10}, E') - \frac{\partial E}{\partial \delta} h_{11}\epsilon)$$
$$= P_T - P_G(h_{10}, E') - \epsilon \frac{\partial P_G}{\partial \delta} h_{11} - \frac{\epsilon}{2H} D(h_{20} + \epsilon h_{21}),$$
$$\epsilon(\frac{\partial h_{30}}{\partial E'} + \epsilon \frac{\partial h_{31}}{\partial E'})(T'_d)^{-1}(E_{fd} - E(h_{10}, E') - \frac{\partial E}{\partial \delta} h_{11}\epsilon)$$
$$= T^{-1}[-K_A(V(h_{10}, E') + \frac{\partial V}{\partial \delta} h_{11}\epsilon - V_{ref}) - (h_{30} + \epsilon h_{31} - E_{fdr0})].$$

Equating the coefficients of ϵ^0, we can obtain

$$h_{20} = 0, \quad (3.9)$$
$$P_T - P_G(h_{10}, E') = 0, \quad (3.10)$$
$$h_{30} = E_{fd0} - K_A(V(h_{10}, E') - V_{ref}), \quad (3.11)$$

where h_{10} is the implicit function of E'.

Equating the coefficients of ϵ^1, it follows that

$$\frac{\partial h_{10}}{\partial E'} (T'_d)^{-1}(E_{fd} - E(h_{10}, E')) = h_{21}, \quad (3.12)$$
$$\frac{\partial h_{20}}{\partial E'} (T'_d)^{-1}(E_{fd} - E(h_{10}, E')) = -\frac{\partial P_G}{\partial \delta} h_{11} - \frac{D}{2H} h_{20}, \quad (3.13)$$
$$\frac{\partial h_{30}}{\partial E'} (T'_d)^{-1}(E_{fd} - E(h_{10}, E')) = T^{-1}(-K_A \frac{\partial V}{\partial \delta} h_{11} - h_{31}). \quad (3.14)$$

Suppose that $\frac{\partial P_G}{\partial \delta} \neq 0$; by (3.13) we have $h_{11} = 0$.

From (3.14) and $h_{11} = 0$, we obtain

$$h_{31} = -(T'_d)^{-1} T (\frac{\partial h_{30}}{\partial E'})(E_{fd} - E(h_{10}, E')), \quad (3.15)$$

where

$$\frac{\partial h_{30}}{\partial E'} = -K_A(\frac{\partial V}{\partial \delta} \frac{\partial h_{10}}{\partial E'} + \frac{\partial V}{\partial E'}).$$

Differentiating (3.10) with respect to E', we obtain

$$-\frac{\partial P_G}{\partial \delta} \frac{\partial h_{10}}{\partial E'} - \frac{\partial P_G}{\partial E'} = 0.$$

Thus, we have

$$\frac{\partial h_{10}}{\partial E'} = -(\frac{\partial P_G}{\partial \delta})^{-1}(\frac{\partial P_G}{\partial E'}).$$

We therefore write (3.15) as

$$h_{31} = (T'_d)^{-1} T K_A (M_4 - M_3 M_1^{-1} M_2)(E_{fd} - E(h_{10}, E')), \tag{3.16}$$

where

$$\frac{\partial P_G}{\partial \delta} = M_1, \quad \frac{\partial P_G}{\partial E'} = M_2, \quad \frac{\partial V}{\partial \delta} = M_3, \quad \frac{\partial V}{\partial E'} = M_4,$$

and M_i $(i = 1, 2, 3, 4)$ are matrices.

Finally, the $O(\epsilon^2)$ approximations of the slow manifold M_ϵ can be written as

$$\delta = h_1(E') = h_{10} + O(\epsilon^2), \tag{3.17}$$

$$\omega' = h_2(E') = \epsilon h_{21} + O(\epsilon^2) \tag{3.18}$$

$$= \epsilon \frac{\partial h_{10}}{\partial E'} (T'_d)^{-1} (E_{fd} - E(h_{10}, E')) + O(\epsilon^2), \tag{3.19}$$

$$E_{fdr} = h_3(E') = h_{30} + \epsilon h_{31} + O(\epsilon^2) = E_{fd0} - K_A(V(h_{10}, E') - V_{ref}), \\ + \epsilon (T'_d)^{-1} T K_A (M_4 - M_3 M_1^{-1} M_2)(E_{fd} - E(h_{10}, E')). \tag{3.20}$$

The slow dynamics on the order-one slow manifold M_ϵ, are derived from the slow-subsystem in (3.2) by replacing h_{10} and $h_{30}+\epsilon h_{31}$ into δ and E_{fdr} respectively, i.e.,

$$T'_d \dot{E}' = h_{30} + \epsilon h_{31} - E(h_{10}, E'), \tag{3.21}$$

where h_{10}, h_{30} and h_{31} are given by (3.10), (3.11) and (3.16) respectively. (3.21) has m dimensions, which is much smaller than the $4m$-dimensions original system (3.2). By (3.21) we can examine the dynamical properties of the original system in a relatively easy way.

To analyze the fast dynamics, we investigate (3.2) in the $\tau = t/\epsilon$ time scale. (3.2) can be rewritten as the following form:

$$\begin{cases} \delta_\tau = \omega', \\ \dot{\omega}'_\tau = P_T - P_G(\delta, E') - \frac{\epsilon}{2H} D\omega', \\ \dot{E}_{fdr_\tau} = T^{-1}[-K_A(V(\delta, E') - V_{ref}) - (E_{fdr} - E_{fdr0})], \\ \dot{E}'_\tau = \epsilon T'^{-1}_d (E_{fd} - E(\delta, E')). \end{cases} \tag{3.22}$$

Thus, the fast subsystem is given by

$$\begin{cases} \dot{\delta}_\tau = \omega', \\ \dot{\omega}'_\tau = P_T - P_G - \frac{\epsilon}{2H} D\omega', \\ \dot{E}_{fdr_\tau} = T^{-1}[-K_A(V - V_{ref}) - (E_{fdr} - E_{fdr0})], \end{cases} \tag{3.23}$$

where the slow variables E' act as parameters for the fast-subsystem, which has $3m$ dimensions.

By the slow system (3.21) and fast system (3.23), theoretically we can analyze the nonlinear properties of the original system, such as stability, bifurcation and attractive region for general multimachine power systems in a relatively easy manner due to the reduced dimensions in each subsystem. Next, we use a single machine system to demonstrate the detailed analysis.

4. A Detailed Analysis for Single Machine System

In this section, we investigate the single machine system, i.e., the case $m = 1$ in the systems (3.1) and (3.2), in detail by geometric singular perturbation method and numerical simulation. The system (3.1) is written as

$$\begin{cases} \dot{\delta} = \omega_0 \omega, \\ M\dot{\omega} = -D\omega + P_T - P_G, \\ T_d^{'}\dot{E}^{'} = E_{fd} - E(\delta, E'), \\ T_A \dot{E}_{fdr} = -K_A(V(\delta, E') - V_{ref}) - (E_{fdr} - E_{fd0}), \end{cases} \tag{4.1}$$

where all variables are scale due to $m = 1$.

As the same way as the system (3.2), we have the following singular perturbation form:

$$\begin{cases} \epsilon\dot{\delta} = \omega^{'} = f_1, \\ \epsilon\dot{\omega}^{'} = P_T - P_G(\delta, E') - \frac{\epsilon}{2H} D\omega^{'} = f_2, \\ \epsilon\dot{E}_{fdr} = T_A^{-1}[-K_A(V(\delta, E') - V_{ref}) - (E_{fdr} - E_{fdr0})] = f_3, \\ \dot{E}^{'} = T_d^{'-1}(E_{fd} - E(\delta, E')) = g, \end{cases} \tag{4.2}$$

where δ, ω', and E_{fdr} are fast variables and E' is the slow variable, and D is the damping factor in per unit; T_d' is the direct axis transient time-constant; T_A is the excitation control time-constant; K_A is the control gain which may be varied for special control actions; V_{ref} is the reference bus voltage; E_{fd0} is the reference field voltage. The functions P_G and $E(\delta, E')$ in (4.2) depend on the interaction between machine and load and are represented as [9]

$$P_G = \frac{E'}{x_d' + x} \sin\delta, \tag{4.3}$$

$$E(\delta, E') = \frac{x_d + x}{x_d^{'} + x} E^{'} - \frac{x_d - x_d^{'}}{x_d^{'} + x} \cos\delta. \tag{4.4}$$

The voltage $V(\delta, E^{'})$ refers to the bus voltage at the generator bus terminal and for the SMIB representation; it is easily computed as

$$V(\delta, E^{'}) = \frac{1}{x + x_d^{'}} \sqrt{(x_d' + xE' \cos\delta)^2 + (xE' \sin\delta)^2}. \tag{4.5}$$

E_{fd} is the the field excitation voltage and is considered as output of the wind-up limiter

$$E_{fd} = \begin{cases} E_{fd_{max}} & \text{if} \quad E_{fdr} > E_{fd_{max}}, \\ E_{fdr} & \text{if} \quad E_{fd_{min}} \le E_{fdr} \le E_{fd_{max}}, \\ E_{fd_{min}} & \text{if} \quad E_{fdr} < E_{fd_{min}}. \end{cases} \tag{4.6}$$

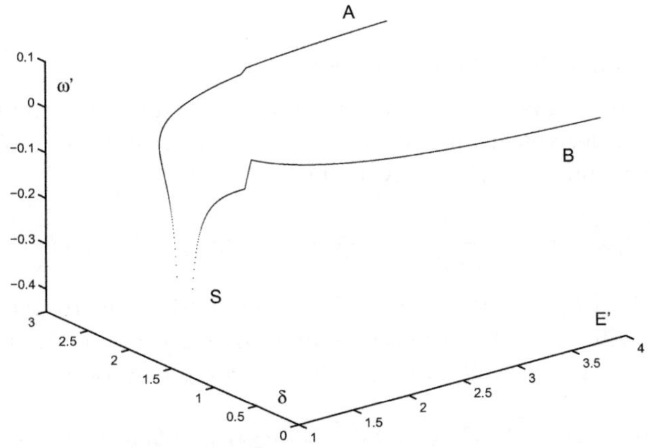

FIGURE 1. Order-one slow manifold M_ϵ (the top half manifold A-S is unstable, and the bottom half manifold B-S is stable, where S is a singular point of the slow manifold).

4.1. Slow Manifold

In this subsection, since δ, ω', and E_{fdr} are scale variables, we consider the one-dimensional slow manifold M_ϵ in the state space of the system (4.1) defined by the three equations in (3.3)–(3.5). The $O(\epsilon^2)$ approximation of slow manifold M_ϵ is also represented by (3.17)–(3.20), and h_{10} is solved by

$$h_{10} = \begin{cases} \sin^{-1}\left(\frac{P_T(x_d'+x)}{E'}\right), & h_{10} \in [0, \frac{\pi}{2}), \\ \pi - \sin^{-1}\left(\frac{P_T(x_d'+x)}{E'}\right), & h_{10} \in [\frac{\pi}{2}, \pi). \end{cases} \tag{4.7}$$

Note that:

1. for $E_{fdr} > E_{fd_{Max}}$ or $E_{fdr} < E_{fd_{Min}}$, we substitute $E_{fd_{Max}}$ or $E_{fd_{Min}}$ for E_{fd} in (3.19) or (3.20);
2. for $E_{fd_{Min}} \leq E_{fdr} \leq E_{fd_{Max}}$, we substitute h_{30} for E_{fd} in (3.19) and (3.20), where h_{30} is still defined by (3.11);
3. from (3.19)–(3.20) and [10] we can show that the frequency deviation during slow transient is of order ϵ, though the angle and field excitation voltage may vary considerably. As the value of the frequency error is small we can assume that the frequency control loop is not excited during the slow flux-voltage transients.

The order-one slow manifold M_ϵ is shown in Figure 1 for E_{fd} ($E_{fd_{min}} = 0$, $E_{fd_{max}} = 5$) and the other parameters are given as follows:

$$H = 5,\quad M = 10,\quad T_{d0}' = 10,\quad T_A = 1,\quad K_A = 190,\quad E_{fd0} = 2,$$
$$x_d' = 0.4,\quad x = 0.5,\quad x_d = 1,\quad V_{ref} = 1.05,\quad P_T = 1.3.$$

Figure 1 shows that the slow manifold M_ϵ is not smooth because M_1^{-1} becomes infinitely large when $\delta = \frac{\pi}{2}$.

4.2. Slow Flux and Voltage Dynamics: SNB

In this subsection, we investigate the slow dynamics on the slow manifold M_ϵ. The slow dynamics is derived from the slow-subsystem in (4.2) by replacing h_{10} and $h_{30} + \epsilon h_{31}$ into δ and E_{fdr} respectively, i.e.,

$$T_d'\dot{E}' = -\frac{x_d + x}{x_d' + x}E' + \frac{x_d - x_d'}{x_d' + x}\cos h_{10} + h_{30} + \epsilon h_{31}, \tag{4.8}$$

where h_{10}, h_{30} and h_{31} are given by (4.7), (3.11) and (3.16) respectively.

Next, we will study the dynamics of (4.8) on the slow manifold M_ϵ. The equilibrium E_o' of system (4.8) satisfies the following equation:

$$q_1(E_o') = -\frac{x_d + x}{x_d' + x}E_o' + \frac{x_d - x_d'}{x_d' + x}\cos h_{10} + h_{30} + \epsilon h_{31} = 0. \tag{4.9}$$

The linearized version of the above equation around an equilibrium point E_o' has the following matrix, by using the definition of the M matrix:

$$A_s = T_d'^{-1}[K_A(M_{3o}M_{1o}^{-1}M_{2o} - M_{4o}) + (K_{3o} + K_{4o}M_{1o}^{-1}M_{2o})] + O(\epsilon), \tag{4.10}$$

where

$$K_3 = -\frac{\partial E}{\partial E'}|_{\delta=h_{10}}, \quad K_4 = \frac{\partial E}{\partial \delta}|_{\delta=h_{10}},$$

and the subscript "o" denotes evaluation at the equilibrium point E_o'. The matrix $M_1 - M_4$ is in (3.16). When ϵ is small, the SNB condition is given by

$$\det[K_A(M_{3o}M_{1o}^{-1}M_{2o} - M_{4o}) + (K_{3o} + K_{4o}M_{1o}^{-1}M_{2o})] = 0. \tag{4.11}$$

However, since it is difficult to compute (4.11), the following simplification is adopted. Using the condition in [9] and $\frac{P_T(x_d'+x)}{E'} \neq 1$, we can assume that

$$\left|\frac{\frac{x_d+x}{x_d'+x}E' - \frac{x_d-x_d'}{x_d'+x}\cos\delta - (E_{fd0} + \epsilon h_{31})}{K_A}\right| \ll V_{ref}. \tag{4.12}$$

Therefore equation (4.9) can be simplified to be

$$(E')^4 - a_0(E')^2 + a_1P_T^2 + a_3 = 0, \tag{4.13}$$

where $a_0 = \frac{2[V_{ref}^2(x+x_d')^2 - x_d'^2]+4}{x^2}$, $a_1 = \frac{4(x_d'+x)^2}{x^2}$, and $a_3 = \frac{[V_{ref}^2(x+x_d')^2 - x_d'^2]^2}{x^4}$. This quadratic equation (4.13) for E'^2 has positive solutions for $0 < P_T < b_1$, where $b_1 = \sqrt{\frac{a_0^2 - 4a_3}{4a_1}}$. Hence, there are two positive solutions for E' in (4.13). By simple analysis, we can obtain the approximation condition for the existence of equilibrium for (4.8).

Lemma 4.1. *If $P_T < b_1$ and condition (4.12) are satisfied, then system (4.8) has two equilibria; when $P_T > b_1$, there is no any equilibrium; when $P_T = b_1$, there is only one equilibrium point S_{sn}, and $P_T = b_1 \equiv P_{T_{SN}}$ is an approximate saddle-node bifurcation point.*

Note that before the SNB bifurcation, there are two equilibrium points close to each other. One is stable and the other is unstable. As these two points approach each other, an annihilation occurs at the saddle-node bifurcation, while at the same time the system's Jacobian matrix becomes singular and the system collapse occurs. A saddle-node bifurcation in the SMIB system (4.1) will result in a slow flux decay felt by generator. This will have a similar drifting effect on the generator terminal and load bus leading the whole system to a voltage collapse. The condition (4.11) is the same as the condition of the SNB for the original system (4.1), but (4.11) is obtained from the reduced-order models, which implies that the singular perturbation method simplifies the difficulties of analysis.

4.3. Fast Dynamics: Electromechanical Oscillations and Dynamics of Field Excitation Voltage

To analyze the fast dynamics, we examine (4.2) in the $\tau = t/\epsilon$ time scale; then (4.2) can be rewritten in the following form:

$$\begin{cases} \dot{\delta} = \omega', \\ \dot{\omega}' = P_T - P_G(\delta, E') - \frac{\epsilon}{2H} D\omega', \\ \dot{E}_{fdr} = T^{-1}[-K_A(V(\delta, E') - V_{ref}) - (E_{fdr} - E_{fdr0})], \\ \dot{E}' = \epsilon T_d'^{-1}(E_{fd} - E(\delta, E')). \end{cases} \tag{4.14}$$

We consider the fast subsystem

$$\begin{cases} \dot{\delta} = \omega', \\ \dot{\omega}' = P_T - P_G - \frac{D}{2H}\epsilon\omega', \\ \dot{E}_{fdr} = T^{-1}[-K_A(V - V_{ref}) - (E_{fdr} - E_{fdr0})], \end{cases} \tag{4.15}$$

where the slow variable E' acts as a parameter for the fast-subsystem and $\dot{} = \frac{d}{d\tau}$.

The equilibrium points of fast subsystem (4.15) satisfy the following conditions:

$$\begin{cases} \omega' = 0, \\ \sin\delta = \dfrac{P_T}{g_1 E'}, \\ \cos\delta = \dfrac{(g_3 - E_{fdr})^2 - g_2(x_d'^2 + x^2E'^2)}{2g_2 x x_d' E'}, \end{cases} \tag{4.16}$$

where the symbols g_1, g_2 and g_3 defined below are introduced for notational convenience:

$$g_1 = \frac{1}{x_d' + x}, \quad g_2 = \frac{K_A^2}{(x_d' + x)^2}, \quad g_3 = K_A V_{ref} + E_{fdr0}.$$

When δ is eliminated from the equilibrium conditions (4.16), we have the following equation in E_{fdr}:

$$P_T^2 = g_1^2 E'^2 - \left(\frac{(g_3 - E_{fdr})^2 - g_2(x_d'^2 + x'^2E'^2)}{2g_2 x x_d'}\right)^2 g_1^2. \tag{4.17}$$

Let

$$a = E'^2 - \frac{P_T^2}{g_1^2}, \quad b = g_2(x_d'^2 + x'^2E'^2), \quad c = 2g_2xx_d'.$$

By analysis of the roots of (4.17) for E_{fdr}, we can derive the condition for the existence of the positive roots of (4.17).

Lemma 4.2. *Assume that the following conditions hold:*

(1) *if* $G_1(E'^2) < P_T^2 < G_0(E'^2)$, *then equation* (4.17) *has two positive roots:*

$$E_{fdr}^{(1)} = g_3 - \sqrt{b + c\sqrt{a}}, \quad E_{fdr}^{(2)} = g_3 - \sqrt{b - c\sqrt{a}}.$$

Thus, the system (4.15) *has exactly two equilibrium points denoted as* $x_l = (\delta_l, 0, E_{fdr_l})$ $(0 < \delta_l < \frac{\pi}{2})$ *and* $x_h = (\delta_h, 0, E_{fdr_h})$ $(\frac{\pi}{2} < \delta_h < \pi)$ *with* $E_{fdr_l} < E_{fdr_h}$;

(2) *if* $P_T^2 = G_0(E'^2)$, *then equation* (4.17) *has one positive root* $E_{fdr}^{SN} = g_3 + \sqrt{b}$. *Thus, the system* (4.15) *has only one equilibrium point* $x_{SN} = (\frac{\pi}{2}, 0, E_{fdr_{SN}})$, *and the system bifurcates at the parameter* $P_{T_{SN}} = \sqrt{G_0(E'^2)}$;

(3) *if* $P_T^2 > G_0(E'^2)$, *then there is no root for equation* (4.17) *and*

$$G_0(E'^2) = g_1^2E'^2, \quad G_1(E'^2) = g_1^2E'^2 - g_1^2\left[\frac{g_2(x_d'^2 + x'^2E'^2) - g_3^2}{2g_2xx_d'}\right]^2,$$

$$E_{fdr_l} = g_3 - h_1, \quad E_{fdr_h} = g_3 - h_2,$$

$$h_{1,2} = \sqrt{g_2(x_d'^2 + x'^2E'^2) \pm 2g_2xx_d'\sqrt{\frac{g_1^2E'^2 - P_T^2}{g_1^2}}}.$$

The parameter chart for the roots of equation (4.17) *is shown in Figure 2.*

For checking the local stability properties of equilibrium points, we use the Jacobian matrix of the system:

$$J = \begin{pmatrix} 0 & 1 & 0 \\ -M_1 & -\frac{\epsilon D}{2H} & 0 \\ -\frac{K_A}{T}M_3 & 0 & -\frac{1}{T} \end{pmatrix},$$

where $M_1 = \frac{\partial P_G}{\partial \delta}$ and $M_3 = \frac{\partial V}{\partial \delta}$. The characteristic polynomial for solving the eigenvalues of J can be derived to be

$$\det(\lambda I - J) = (-\frac{1}{T} - \lambda)(\lambda^2 + \frac{\epsilon D}{2H}\lambda + M_1). \tag{4.18}$$

Using (4.18), we can obtain eigenvalues and stability of equilibrium points for fast subsystem (4.15) as follows.

1. When $D = 0$, the equilibrium points on the branch BC in Figure 3 are unstable due to $M_1 < 0$. The equilibrium points on the branch AB in Figure 3 are nonhyperbolic due to

$$\lambda_{1,2} = \pm i\sqrt{\frac{\sqrt{E'^2 - (x_d' + x)^2P_T^2}}{x_d' + x}}, \quad \lambda_3 = -\frac{1}{T}.$$

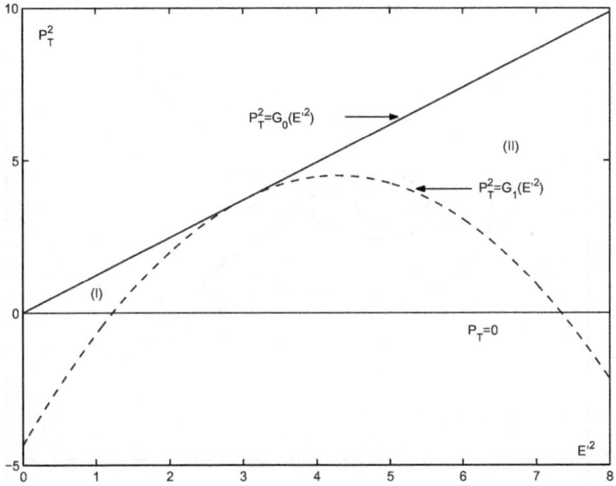

FIGURE 2. Parameter chart for the roots of equation (4.17).

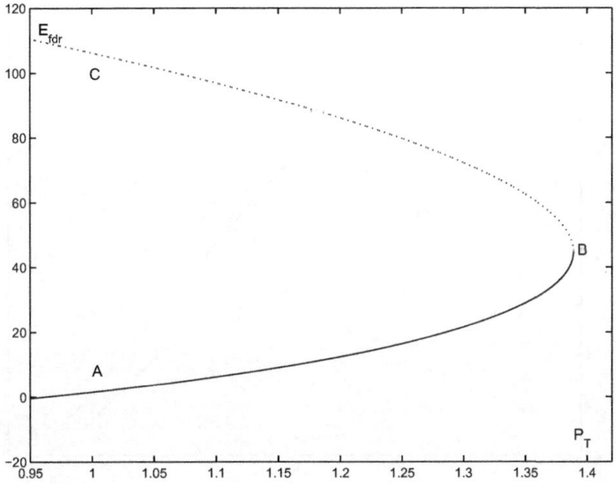

FIGURE 3. Bifurcation diagram of equilibrium point on $(E_{fdr} - P_T)$ plane for fast subsystem (4.15) (the points on the dot line C-B are unstable, and the points on the solid line A-B are stable).

2. When $D > 0$, the equilibrium points on the branch BC in Figure 3 are unstable due to $M_1 < 0$. The equilibrium points on the branch AB are stable due to the eigenvalues

$$\lambda_{1,2} = \frac{1}{2}\left(-\frac{\epsilon D}{M} \pm \sqrt{\frac{\epsilon^2 D^2}{M^2} - \frac{4E'}{x'_d + x}\cos(h_{10})}\right), \quad \lambda_3 = -\frac{1}{T}.$$

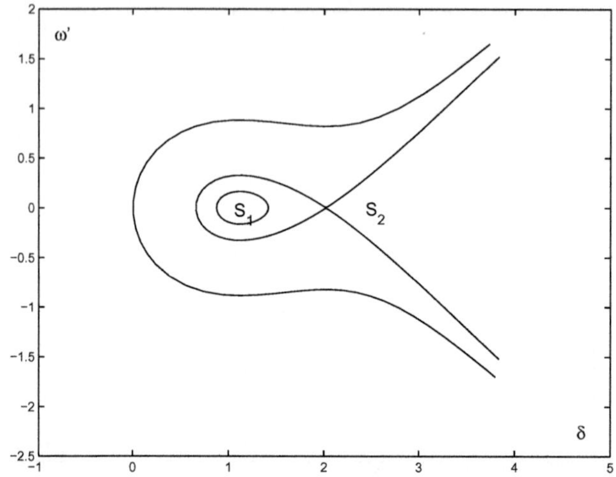

FIGURE 4. Phase portraits for equation (4.15) at $D = 0$ and $P_T = 0.8$. (S_1 is a center point, and S_2 is a saddle point.)

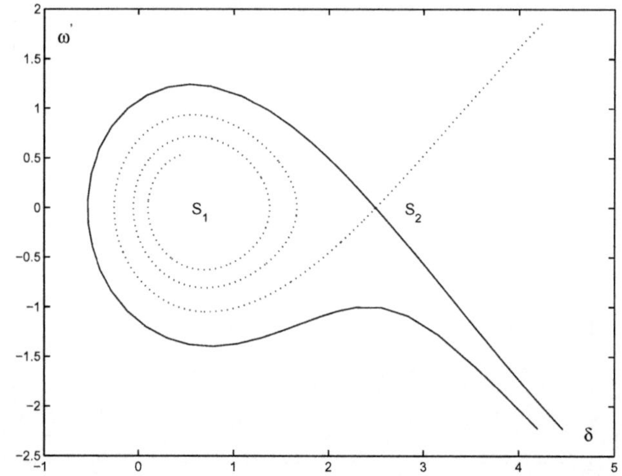

FIGURE 5. Phase portraits for equation (4.15) at $D = 5$ and $P_T = 0.8$. (Real line represents stable manifold, and the dotted line is an unstable manifold. S_1 is a stable point, and S_2 is a saddle point.)

The bifurcation diagram of $E_{fdr} - P_T$ for system (4.15) when $E' = 1.25$ is drawn in Figure 3.

For given E', when $G_1(E'^2) < P_T^2 < G_0(E'^2)$, the projection of the trajectory of (4.15) on $\delta - \omega'$ plane is shown in Figures 4 and 5 for $D = 0$ and $D = 5$ respectively. In fact, the first two equations of (4.15) are independent of the third; we can easily show that the first two equations of the system (4.15) are a Hamiltonian system, where the Hamiltonian function is

$$H(\delta, \omega') = \frac{\omega'^2}{2} - \frac{E'}{x'_d + x} \cos \delta - P_T \delta.$$

By summarizing the above discussions, we have the following theorem for the fast subsystem (4.15).

Theorem 4.3. *When $G_1(E'^2) < P_T^2 < G_0(E'^2)$, the approximate fast motion has two equilibrium points S_1 and S_2, and each of them lying on one branch of the slow manifold. Every point on the slow manifold is an equilibrium point of fast motion. S_2 is unstable, and the stability of S_1 is as follows:*

(1) *if $D = 0$, then S_1 is a center and S_2 is a saddle;*
(2) *if $D > 0$, then S_1 is stable, S_2 is a saddle.*

In Figure 4, we show that the phase portrait and an orbit homoclinic to the saddle S_2 for $D = 0$, $P_T = 0.8$ and $E' = 1.25$. The phase portrait for $D = 5$, $P_T = 0.8$ and $E' = 1.25$ is shown in Figure 5. In fact, the attractive region of S_1 is just the domain enclosed by the stable manifold of saddle S_2. Now we consider the dynamical behavior of the fast-subsystem (4.15). From Figure 5 we obtain that the initial values of state variables $(\delta, \omega', E_{fdr})$ lie in the attractive region of S_1, and the orbit starting from them will eventually converge to the stable equilibrium S_1 with a fast transient.

In the above analysis, slow variable E' actually acts as a parameter. Now we consider the effect of E' on the domain of attractive region. We take $P_T = 0.8$ and $D = 5$. The attractive domain for $E' = 0.8$ is shown in Figure 6 and for $E' = 4$ is shown in Figure 7. By comparing Figure 6 with Figure 7, the attractive domain in Figure 7 is clearly larger than that in Figure 6, which implies that E' has significant effects on the attractive domain.

4.4. Estimation of Attractive Region for the Original System

Until now we just consider the reduced system, i.e., the limiting case of $\epsilon = 0$ to the original system (4.2). For $\epsilon \neq 0$ and sufficiently small, by the geometric singular perturbation theory, we can further construct the dynamical behavior of the system (4.2) through studying the reduced system (4.8) and (4.15). The motion and the order-0 slow manifold are sketched in Figure 8, where S_1 is a stable equilibrium. The trajectory starting from the attractive region of S_1 first approaches AB by spiral lines with fast transient, and then it slowly reaches the equilibrium remaining on the slow manifold. On the other hand, the trajectory starting from point lying on the slow manifold branch AB moves along AB. Figures 8 and 9 indicate the trajectories and solution curves of the original system (4.2) for $P_T = 0.8 < P_{T_{SN}}$ and $D = 5$ with initial value $(\delta, \omega', E', E_{fdr}) = (2.51, 0.001, 1.25, 2.5643)$ lying in the interior of the domain of the attractive region of S_1. We observe that the trajectory starting from the attractive region of S_1, which lies on the slow manifold M_ϵ, converges to the equilibrium S_1 with a fast transient period. Comparing the

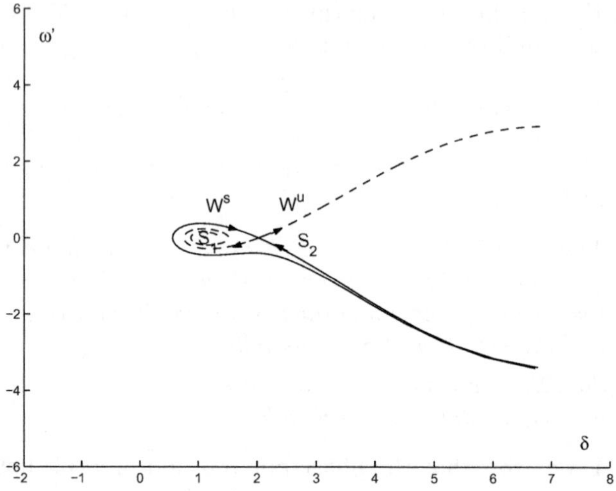

FIGURE 6. Attractive region for $P_T = 0.8$, $D = 5$ and $E' = 0.8$. (W^s is the stable manifold and W^u is the unstable manifold. S_1 is a stable equilibrium, and S_2 is a saddle equilibrium.)

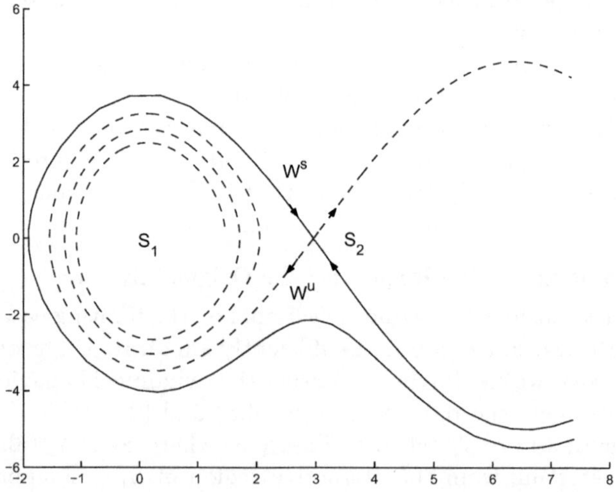

FIGURE 7. Attractive region for $P_T = 0.8$, $D = 5$ and $E' = 4$. (W^s is the stable manifold and W^u is the unstable manifold. S_1 is a stable equilibrium and S_2 is an unstable equilibrium.)

stable and unstable manifolds at saddle S_2 of the fast subsystem (4.15) with the system (4.2) by numerical simulation, Figure 10 shows that the stable and unstable manifolds at S_2 of the system (4.15) coincide well with those of the system (4.2),

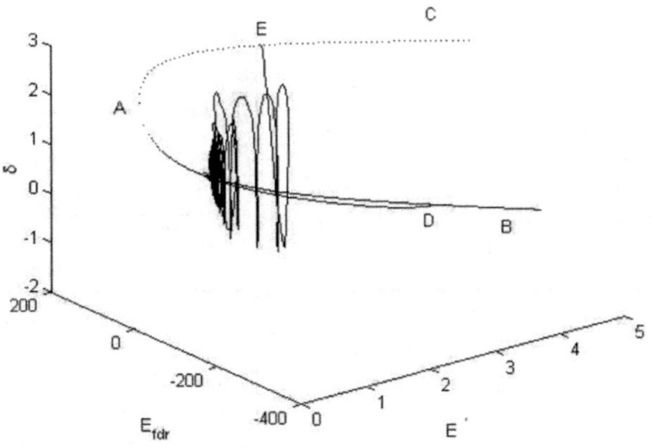

FIGURE 8. Slow manifold and trajectories near the slow manifold.

which implies that we can only use the simple fast system approximately to analyze the stability of the complicated original system. Therefore, the attractive region of the original (4.2) is just the domain enclosed by the stable manifold of the saddle S_2 of system (4.15). Clearly, a stable operating point must be inside the attractive region for a power system.

On the other hand, if a point is out of the region formed by the stable manifold, or flows to the edge SNB of the slow manifold where $P_T = P_{T_{SN}}$, then the variables of the system (4.2) cannot converge to an equilibrium, which is displayed in Figures 11 and 12.

5. Conclusion

In this paper, we mainly examine a dynamical model of power systems, which is modified from [23, 24] by further considering AVRs for multimachine systems. Nonlinear dynamical system is decomposed into two simple subsystems for a relatively general settings of power systems, by including AVRs. The dynamical behavior of the power system was studied using the geometric singular perturbation analysis and numerical simulations. We show that our model does not only consider the control devices, such as AVRs for multimachine systems, but also can be used to derive the attractive regions of equilibria.

We have proven that the attractive region of the system is given by the stable manifold of a saddle point of the reduced fast-subsystem. Therefore, theoretically by observing whether its operating point lies in the attractive region with stability boundary, we can estimate the stability of the power system in the current operating state. By a single-machine system and numerical simulation, we also demonstrate that the complicated original system can be simplified without loss of dynamical behavior and physical meaning. Moreover, we identified the effect of E' on the size of stability region.

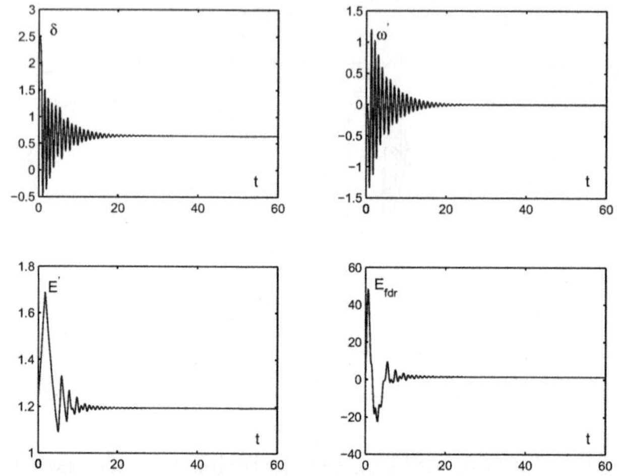

FIGURE 9. Solution curves of (4.15) for $P_T = 0.8$ and $D = 5$ with initial value $(\delta, \omega', E', E_{fdr}) = (2.51, 0.01, 1.25, 2.5643)$ lying in the interior of the stable manifold of Figure 5.

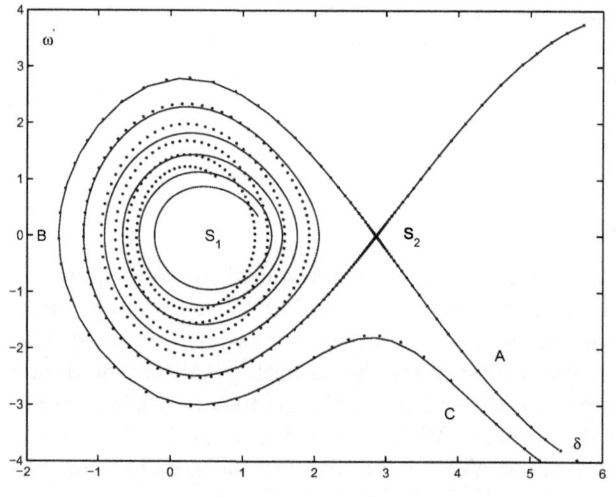

FIGURE 10. Stable manifolds and unstable manifolds for $P_T = 0.8$ and $D = 5$: the dotted line is for fast subsystem at $E' = 2.617636$ and the solid line is for original system.

References

[1] L. Chen and K. Aihara, Stability and Bifurcation Analysis of Differential-Difference-Algebraic Equations. *IEEE Trans. on Circuits & Systems* I **48** (2001), 308–326.

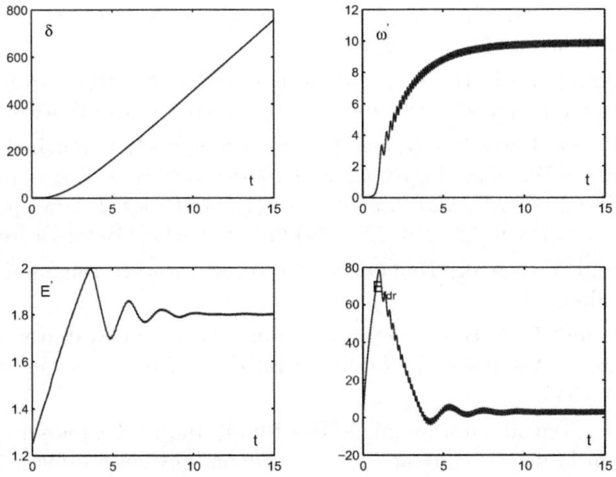

FIGURE 11. Solution curves of (4.2) for $P_T = 0.8$ and $D = 5$ with initial value $(\delta, \omega', E', E_{fdr}) = (2.54, 0.01, 1.25, 2.5643)$ lying outside of the stable manifold of Figure 5.

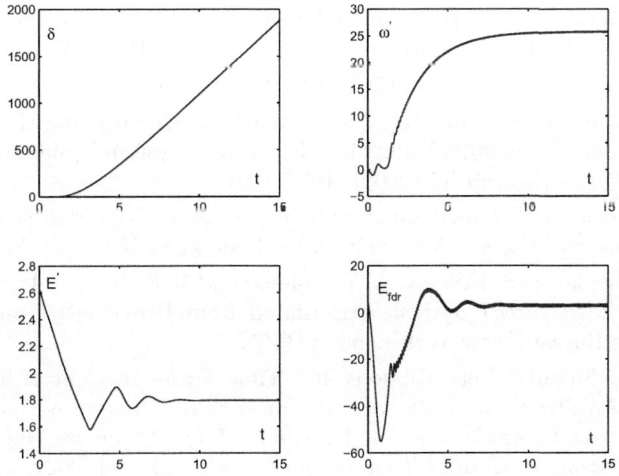

FIGURE 12. Divergence with hard-limits when $P_T > P_{T_{SN}}$ for system (4.2).

[2] J. H. Chow, J. R. Winkelman, M. A. Pai and P. W. Sauer, Singular Perturbation Analysis of Large-scale Power Systems. *Electrical Power & Energy Systems* **12** (1990), 117–126.

[3] C. L. DeMarco and T. J. Overbye, An Energy Based Security Measure for Assessing Vulnerability to Voltage Collapse. *IEEE Trans. on Power Systems* **5** (1990), 419–427.

[4] N. Fenichel, Geometric Singular Perturbation Theory for Ordinary Differential Equation. *J. Diff. Eqns.* **31** (1979), 53–98.

[5] J. Guckenheimer and P. Holmes, *Nonlinear Oscillations, Dynamical System and Bifurcation of Vector Fields.* Springer-Verlag, New York Berlin Heidelberg Tokyo, 1992.

[6] J. Guckenheimer, Towards a Global Theory of Singularly Perturbed Dynamical Systems. *Nonlinear Dynamical Systems and Chaos, Series: Progress in Nonlinear Differential Equations and Their Applications* (H. W. Broer, I. Hoveijn, F. Takens and S. A. V. Gils, eds.), vol. **19**, 213–225. Birkhäuser-Verlag, Basel Boston Berlin, 1996.

[7] J.K. Hale and Y. Kocak, *Dynamics and Bifurcation.* Springer-Verlag, New York Berlin Heidelberg, 1991.

[8] I. A. Hiskens and D. J. Hill, Energy Functions Transient Stability and Voltage Behavior in Power Systems with Nonlinear Loads. *IEEE Trans. on Power Systems* **4** (1989), 1525–1533.

[9] W. Ji and V. Venkatasubramanian, Hard-limit Induced Chaos in a Fundamental Power System Model. *Electrical Power & Energy System* **8** (1996), 279–295.

[10] W. Ji and V. Venkatasubramanian, Dynamics of a Minimal Power System: Invariant Tori and Quasi-periodic Motions. *IEEE Trans. on Circuits & Systems* I **42** (1995), 981–1000.

[11] Z. J. Jing, Z. Y. Jia and Y. H. Gao, Research of the Stability Region in a Power System. *IEEE Trans. on Circuits & Systems* I **50** (2003), 298–303.

[12] C. K. R. T. Jones, Geometric Singular Perturbation Theory. *Lecture Notes in Mathematics* vol. **1609**, 44–118. Springer-Verlag, New York Berlin Heidelberg, 1994.

[13] H. K. Khalil, *Nonlinear System* (2rd edn.), Prentice Hall, Upper Saddle River, 1996.

[14] K. B. Kilani and R. A. Schlueter, An Approach for Determining the Subsystem Experiencing and Producing a Bifurcation in a Power System Dynamical Model. *IEEE Trans. on Power Systems* **15** (2000), 1053–1061.

[15] P. V. Kokotović, H. K. Khalil and J. O'Reilly, *Singular Perturbation Methods in Control: Analysis and Design.* Academic Press, London Sydney New York, 1986.

[16] E. F. Mishchenko, Y. S. Kolesov, A. Y. Kolesov and N. K. Rozov, *Asymptotic Methods in Singularly Perturbed Systems* (translated from Russian by Irene Aleksanova). Consultants Bureau, New York London, 1995.

[17] D. S. Naidu, Singular Perturbations and Time Scales in Control Theory and Applications: An Overview. *Dynamics of Continuous, Discrete & Impulsive Systems (DCDIS) Series B: Applications & Algorithms: Special Issue on Singularly Perturbed Dynamic Systems in Control Technology* (Z. Gajic, ed.) **9** (2002), 233–278.

[18] D. S. Naidu, Singular Perturbations Methodology in Control System. *IEE Control Engineering Series* **34**, Peter Peregrinus Limited, Stevenage Herts, UK, 1988.

[19] K. G. Rajesh and K. R. Padigar, Bifurcation Analysis of a Three Node System with Detailed Models. *Electrical Power & Energy Systems* **5** (1990), 1374–1383.

[20] P. W. Sauer and M. A. Pai, Power System Steady-State Stability and the Load-Flow Jacobian. *IEEE Trans. on Power Systems* **21** (1999), 375–393.

[21] V. Venkatasubramanian, H. Schättler and J. Zaborszky, Local Bifurcation and Feasibility Regions in Differential-Algebraic Systems. *IEEE Trans. on Automatic Control* **40** (1995), 1992–2012.

[22] V. Venkatasubramanian, H. Schättler and J. Zaborszky, Voltage Dynamics: Study of a Generator with Voltage Control, Transmission and Matched MW Load. *IEEE Trans. on Automatic Control* **37** (1992), 1717–1733.

[23] C. D. Vournas, P. W. Sauer and M. A. Pai, Relationship Between Voltage and Angle Stability of Power Systems. *Electrical Power & Energy Systems* **18** (1996), 493–500.

[24] C. D. Vournas, M. A. Pai and P. W. Sauer, The Effect of Automatic Voltage Regulation on the Bifurcation Evolution in Power Systems. *IEEE Trans. on Power System* **11** (1996), 1683–1688.

[25] A. Varghese, P. W. Sauer and M. A. Pai, Synchronous Machine Block Diagram Analysis with Fast Dynamics. *Int. J. of Electrical Power & Energy Systems* **11** (1989), 239–247.

[26] H. O. Wang, E. H. Abed and A. M. A. Hamdan, Bifurcations, Chaos and Crises in Voltage Collapse of a Model Power System. *IEEE Trans. on Circuits & Systems* I **41** (1994), 294–302.

[27] S. Wiggins, *Introduction to Applied Nonlinear Dynamical Systems and Chaos.* World Publishing Corp., New York Berlin Heidelberg, 1990.

[28] N. Yorino, H. Sasaki, Y. Masuda, Y. Tamura, M. Kitagawa and A. Oshimo, On Voltage Stability from the Viewpoint of Singular Perturbation Theory. *Electrical Power & Energy System* **16** (1994), 409–417.

Zhujun Jing
Department of Mathematics
Hunan Normal University
Hunan, Changsha 410081 and
Academy of Mathematics and System Sciences
Chinese Academy of Sciences
Beijing 100080
P. R. China
e-mail: `jingzj@math.ac.cn`

Ruiqi Wang
Department of Electrical Engineering and Electronics
Osaka Sangyo University
Daito, Osaka 574-8530
Japan
e-mail: `rqwang@amss.ac.cn`

Luonan Chen
Department of Electrical Engineering and Electronics
Osaka Sangyo University
Daito, Osaka 574-8530
Japan
e-mail: `chen@elec.osaka-sandai.ac.jp`

Jin Deng
Academy of Mathematics and System Sciences
Chinese Academy of Sciences
Beijing 100080
P. R. China
e-mail: `jindeng@amss.ac.cn`

Trends in Mathematics: *Differential Equations with Symbolic Computation*, 143–157

Algebraic Multiplicity and the Poincaré Problem

Jinzhi Lei and Lijun Yang

Abstract. In this paper we derive an upper bound for the degree of the strict invariant algebraic curve of a polynomial system in the complex project plane under generic condition. The results are obtained through the algebraic multiplicities of the system at the singular points. A method for computing the algebraic multiplicity using Newton polygon is also presented.

Mathematics Subject Classification (2000). 34A05; 34M99.

Keywords. Polynomial system, invariant algebraic curve, Poincaré problem.

1. Introduction

In this paper, we present an approach to establish the upper bound of the degree of the strict invariant algebraic curve of a polynomial system in the complex projective plane $\mathbb{P}^2_{\mathbb{C}}$. A polynomial system in $\mathbb{P}^2_{\mathbb{C}}$ is defined by the vector field

$$\dot{z} = P(z, w), \quad \dot{w} = Q(z, w), \tag{1.1}$$

where P and Q are relatively prime polynomials with complex coefficients.

Definition 1.1. A polynomial $f(z, w)$ is said to be a Darboux polynomial of (1.1) if there exists a polynomial $R_f(z, w)$ such that

$$P(z, w)\frac{\partial f}{\partial z} + Q(z, w)\frac{\partial f}{\partial w} = R_f(z, w)f(z, w). \tag{1.2}$$

We call the zero set $C(f) = \{(z, w) \in \hat{\mathbb{C}}^2 |\, f(z, w) = 0\}$ an invariant algebraic curve, and R_f the cofactor of f. In particular, if $C(f)$ contains no constant irreducible component (i.e., the line $z = z_0$ or $w = w_0$), then f is a strict Darboux polynomial, and $C(f)$ is a strict invariant algebraic curve.

This work was supported by the National Natural Science Foundation of China (10301006).

The study of invariant algebraic curves of a polynomial system goes back to Darboux and Poincaré (see [11]). In general, the Darboux polynomial of the system (1.1) can be found by solving the equation (1.2) for f and R_f. Equation (1.2) is easy to solve if the degree of f is known in advance (for example, see [10, Proposition 1]). However, it is still an open problem, for a given system, to establish the upper bound for the degree of the invariant algebraic curve effectively. This problem is known as the Poincaré problem. It is known that such an upper bound does exist for a given polynomial system, see [11, Corollary 3.1]. However, the uniform upper bound that depends merely on the degree of the system does not exist, for non-trivial examples, see [8]. As a consequence, the practical arithmetic to find the bound from the coefficients is significant for the general solution for finding the invariant algebraic curve of a polynomial system. For more remarks and results on the Poincaré problem, see [2, 4, 11, 12]. The first result to address the Poincaré problem was presented by Carnicer [2] as follows.

Theorem 1.2 (Carnicer's theorem [2]). *Let $\mathcal{F}$ be a foliation of $\mathbb{P}^2_{\mathbb{C}}$ and let C be an algebraic curve in $\mathbb{P}^2_{\mathbb{C}}$. Suppose that C is invariant by $\mathcal{F}$ and there are no dicritical singularities of $\mathcal{F}$ in C. Then*

$$\partial^o C \leq \partial^o \mathcal{F} + 2.$$

In the proof of Carnicer's theorem, the relationship between the sum of the multiplicities of a foliation along the branches of a curve, the degree of the curve, the degree of the foliation and the Euler characteristic of the curve are systematically used. This idea is also used in the present paper. However, no effective method was provided in [2] to determine whether a singular point is dicritical or not. The same inequality had been showed by Cerveau and Lins Neto [3] for those systems of which all the singularities of the invariant algebraic curve are nodal. A more straightforward result was presented by Walcher using elementary method [12]. Walcher's result is restated as follows.

Theorem 1.3 ([12, Theorem 3.4]). *Assume that a vector field X of degree M on $\mathbb{P}^2_{\mathbb{C}}$ admits an irreducible invariant algebraic curve, and if all the stationary points of X at infinity are non-degenerate and non-dicritical, then the degree of the curve cannot exceed $M+1$.*

In Walcher's proof, the Poincaré–Dulac normal forms of the non-degenerate stationary points of a vector field were discussed. In particular, when the stationary point is non-dicritical, the precise information of the number of irreducible semi-invariants of the vector field X was obtained, from which the upper bound of the degree of an invariant algebraic curve is derived. It was also pointed out in [12] that if there are dicritical ones among the non-degenerate stationary points, then the vector field can admit infinitely many (pairwise relatively prime) semi-invariants. Moreover, the condition of non-dicriticality can be verified through the investigation of the linear approximation of the vector field at the stationary points. Thus, Walcher's result provides a practical approach for the Poincaré problem.

In this paper, we present an alternative approach for the Poincaré problem by considering the algebraic multiplicities (see Definition 2.1) of the singular points of the system, and obtain an approximate inequality for the upper bound for the degrees under some generic conditions. The main results of this paper are:

Theorem 1.4. *Consider the differential equation*

$$\frac{\mathrm{d}w}{\mathrm{d}z} = \frac{P(z,w)}{z\,Q(z,w)} \tag{1.3}$$

of degree $M = \max\{\deg P(z,w), \deg z\,Q(z,w)\}$. *Let* (1.3) *admit an irreducible strict Darboux polynomial* $f(z,w)$, $a_1,\ldots,a_k \in \mathbb{C}$ *be all the roots of* $P(0,w)=0$, *and* $a_0 = \infty$, *and* $\mathrm{Mul}(0,a_i)$ *be the algebraic multiplicity of* $(0,a_i)$ *respectively; then*

$$\deg_w f(z,w) \leq \sum_{i=0}^{k} \mathrm{Mul}(0,a_i). \tag{1.4}$$

In particular, if the singularities $(0,a_i)$ *are not algebraic critical, then*

$$\deg_w f(z,w) \leq M\,(k+1). \tag{1.5}$$

Theorem 1.5. *Consider the polynomial system* (1.1) *of degree* $M = \max\{\deg P(z,w), \deg Q(z,w)\}$. *If* (1.1) *has an invariant straight line* L, *and the singular points at* L *are not algebraic critical, and if* (1.1) *admits an irreducible strict Darboux polynomial* $f(z,w)$, *then*

$$\deg f(z,w) \leq M(M+1).$$

Note that, in Theorem 1.5, we do not need the singularities to be non-degenerate, and we will see in the next section that not algebraic critical is weaker than non-dicritical. In Theorem 1.5, we require that (1.1) has an invariant straight line. In fact, it is generic that the line at infinity is invariant. Hence, the condition in Theorem 1.5 is generic.

The rest of this paper is arranged as follows. In Section 2, we will introduce the concept and computing method of algebraic multiplicity. And next, the main theorems are proved. In Section 3, as application the 2D Lotka–Volterra system is studied.

2. Algebraic Multiplicity and the Poincaré Problem

Let $f(z,w)$ be a Darboux polynomial of (1.1). In general, the upper bound of the degree of $f(z,w)$ cannot be determined merely from the equation (1.2). The assumption that $f(z,w)$ is irreducible must be taken into account. If $f(z,w)$ is irreducible, then, without loss of generality, we may perform the transformation $(z,w) \mapsto (z+cw, w)$ $(c \in \mathbb{R})$, if necessary, and assume that $\deg_w f(z,w) = \deg f(z,w)$. Let $m = \deg_w f(z,w)$; then there are m algebraic functions $w_i(z)$ satisfying $f(z,w_i(z)) = 0$ $(i = 1,2,\ldots,m)$. If these m algebraic functions pass through some common singular points, then m can be bounded by the possible number of the algebraic solutions that pass through these singular points. To this end, we will define the algebraic multiplicity as the number of local algebraic solutions as follows.

Definition 2.1. Consider a differential equation

$$\frac{\mathrm{d}w}{\mathrm{d}z} = F(z, w), \tag{2.1}$$

and $(z_0, w_0) \in \mathbb{C}^2$. A formal series

$$w(z) = w_0 + \sum_{i \geq 0} \alpha_i \, (z - z_0)^{\mu_i} \tag{2.2}$$

is said to be a local algebraic solution of (2.1) at (z_0, w_0) if $w(z)$ is a formal series solution of (2.1) with $\alpha_i \neq 0$, $\mu_i \in \mathbb{Q}^+$, and $\mu_i < \mu_{i+1}$ $(\forall i)$. The algebraic multiplicity of (2.1) at (z_0, w_0), denoted by $\mathrm{Mul}(z_0, w_0; F)$ or simply by $\mathrm{Mul}(z_0, w_0)$ while the context is clear, is defined as the number of distinct local non-constant algebraic solutions of (2.1) at (z_0, w_0). If $\mathrm{Mul}(z_0, w_0) = \infty$, then (z_0, w_0) is said to be algebraic critical.

It is evident that algebraic criticality implies dicriticality (i.e., there are infinitely many invariant curves passing through the same point).

When $w_0 = \infty$, let $\bar{w} = 1/w$; then $\bar{w}(z)$ satisfies

$$\frac{\mathrm{d}\bar{w}}{\mathrm{d}z} = -\bar{w}^2 \, F(z, 1/\bar{w}) := \bar{F}(z, \bar{w}),$$

and the algebraic multiplicity $\mathrm{Mul}(z_0, \infty; F)$ is simply defined as $\mathrm{Mul}(z_0, 0; \bar{F})$.

Let $a, b, c \in \mathbb{C}$ with $a, c \neq 0$, and let $W = a\,(w - w_0) + b\,(z - z_0)$, $Z = c\,(z - z_0)$; then $W(Z)$ satisfies an equation of the form

$$\frac{\mathrm{d}W}{\mathrm{d}Z} = \tilde{F}(Z, W). \tag{2.3}$$

It is easy to show that a local algebraic solution of (2.1) at (z_0, w_0) corresponds to a local algebraic solution of (2.3) at $(0, 0)$. Hence we have

$$\mathrm{Mul}(z_0, w_0; F) = \begin{cases} \mathrm{Mul}(0, 0; \tilde{F}) & \text{if } \tilde{F}(Z, 0) \not\equiv 0, \\ \mathrm{Mul}(0, 0; \tilde{F}) + 1 & \text{if } \tilde{F}(Z, 0) \equiv 0. \end{cases}$$

It is evident that, if (z_0, w_0) is a regular point and $F(z, w_0) \not\equiv 0$, then $\mathrm{Mul}(z_0, w_0) = 1$. To estimate the algebraic multiplicity at singular point (z_0, w_0), we can substitute (2.2) into (2.1) to find out all possible formal series solutions. A method for finding the formal series solution of a polynomial system at a singular point is given in [7] using Newton polygon (see [1, 6]). The result and proof are restated below.

Lemma 2.2. *Consider the polynomial system*

$$\frac{\mathrm{d}w}{\mathrm{d}z} = \frac{P(z, w)}{Q(z, w)}, \tag{2.4}$$

where

$$P(z,w) = \sum_{i\geq 0} P_i(z)\, w^i, \quad Q(z,w) = \sum_{i\geq 0} Q_i(z)\, w^i,$$

and

$$P_i(z) = p_{i,0}\, z^{k_i} + p_{i,1}\, z^{k_i+1} + \cdots, \quad Q_i(z) = q_{i,0}\, z^{l_i} + q_{i,1}\, z^{l_i+1} + \cdots, \quad i \geq 0.$$

If $(0,0)$ *is a singular point of* (2.4), *and there exists an index* j, *satisfying:*

(1) $k_j = l_{j-1} - 1$;

(2) *for any* $i \neq j$,

$$\min\{k_i, l_{i-1} - 1\} > k_j + (j-i)\,(p_{j,0}/q_{j-1,0});$$

(3) $p_{j,0}/q_{j-1,0} \in \mathbb{Q}^+$,

then $(0,0)$ *is algebraic critical for the system* (2.4).

Proof. Let $\lambda = p_{j,0}/q_{j-1,0}$, and $u(z) = w(z)\, z^{-\lambda}$; then $u(z)$ satisfies

$$\begin{aligned}\frac{\mathrm{d}u}{\mathrm{d}z} &= \frac{\sum_{i\geq 0}(p_{i,0}\, z^{k_i+i\lambda} - q_{i-1,0}\,\lambda\, z^{l_{i-1}+i\lambda-1} + \text{h.o.t.})\, u^i}{\sum_{i\geq 0}(q_{i,0}\, z^{l_i+(i+1)\lambda} + \text{h.o.t.})u^i} \\ &= \frac{z^{l_{j-1}+j\lambda-1}\sum_{i\geq 0}(p_{i,0}\, z^{k_i-k_j+(i-j)\lambda} - q_{i-1,0}\,\lambda\, z^{l_{i-1}-l_{j-1}+(i-j)\lambda} + \text{h.o.t.})\, u^i}{z^{l_{j-1}+j\lambda}\sum_{i\geq 0}(q_{i,0}\, z^{l_i-l_{j-1}+(i-j)\lambda} + \text{h.o.t.})u^i}.\end{aligned}$$

Taking the conditions of the index j into account, we can rewrite the above equation as

$$\frac{\mathrm{d}u}{\mathrm{d}z} = \frac{z^s\, \hat{P}(z,u)}{z\, \hat{Q}(z,u)},$$

where $\hat{P}(0,u), \hat{Q}(0,u) \not\equiv 0$, and

$$s = \min_{i\geq 0}\{k_i - k_j + (i-j)\,\lambda, l_{i-1} - l_{j-1} + (i-j)\,\lambda\} \in \mathbb{Q}^+.$$

Let $z = \bar{z}^{q_{j-1,0}}$; then

$$\frac{\mathrm{d}u}{\mathrm{d}\bar{z}} = \frac{q_{j-1,0}\, \bar{z}^{s\, q_{j-1,0}-1}\, \hat{P}(\bar{z}^{q_{j-1,0}}, u)}{\hat{Q}(\bar{z}^{q_{j-1,0}}, u)}. \tag{2.5}$$

It is easy to have $s\, q_{j-1,0} \in \mathbb{N}$ and $\hat{P}(\bar{z}^{q_{j-1,0}}, u)$, $\hat{Q}(\bar{z}^{q_{j-1,0}}, u)$ are polynomials of $\bar{z}$ and u. Thus, for any α such that $\hat{Q}(0,\alpha) \neq 0$, (2.5) has a unique solution $u(\bar{z};\alpha)$ which is analytic at $\bar{z} = 0$ and satisfies $u(0;\alpha) = \alpha$. Thus,

$$w(z;\alpha) = z^\lambda\, u(z^{1/q_{j-1,0}};\alpha) = z^\lambda(\alpha + \sum_{i\geq 1}\frac{1}{i!}u_z^{(i)}(0;\alpha)z^{i/q_{j-1,0}})$$

is a solution of (2.4), i.e., $w(z;\alpha)$ is a local algebraic solution of (2.4) for any α such that $\hat{Q}(0;\alpha) \neq 0$. Hence, $(0,0)$ is algebraic critical for (2.4). $\square$

Remark 2.3. (1) Lemma 2.2 is also valid for those equations of which P and Q are Puiseux series of z and w (with slight change in the proof):

$$P(z,w) = \sum_{i,j\geq 0} p_{i,j}\, z^{i/\mu}\, w^{j/\nu}, \quad Q(z,w) = \sum_{i,j\geq 0} q_{i,j}\, z^{i/\mu}\, w^{j/\nu}, \quad \mu,\nu\in\mathbb{N}.$$

(2) From the proof of Lemma 2.2, if the index j satisfies the conditions (1), (2), but $p_{j,0}/q_{j-1,0}\in\mathbb{R}^+\backslash\mathbb{Q}^+$, and let $\lambda = p_{j,0}/q_{j-1,0}$, then (2.4) admits infinity solutions of the form $w(z;\alpha) = z^\lambda\, u(z^{1/s};\alpha)$, where $u(\bar{z};\alpha)$ is the solution of

$$\frac{\mathrm{d}u}{\mathrm{d}\bar{z}} = \frac{\hat{P}(\bar{z}^{1/s},u)}{s\,\hat{Q}(\bar{z}^{1/s},u)}$$

such that $u(0;\alpha) = \alpha$. Thus, (2.4) is dicritical at $(0,0)$, but not necessary algebraic critical.

Lemma 2.4. *Let* $(0,0)$ *be a singular point of* (2.4)*; then either* $(0,0)$ *is algebraic critical, or*

$$\mathrm{Mul}(0,0)\leq\max\{\deg_w P(z,w), \deg_w Q(z,w)+1\}. \tag{2.6}$$

Proof. Let $N = \deg_w P(z,w)$, $M = \deg_w Q(z,w)$, and

$$P(z,w) = \sum_{i=0}^{N} P_i(z)\, w^i, \quad Q(z,w) = \sum_{i=0}^{M} Q_i(z)\, w^i,$$

where

$$P_i(z) = p_{i,0}\, z^{k_i} + p_{i,1}\, z^{k_i+1} + \cdots, \quad Q_i(z) = q_{i,0}\, z^{l_i} + q_{i,1}\, z^{l_i+1} + \cdots.$$

Substitute

$$w(z) = \alpha_0\, z^{\lambda_0} + \text{h.o.t.} \quad (\alpha_0\neq 0, \lambda_0\in\mathbb{Q}^+) \tag{2.7}$$

into (2.4); then

$$\begin{aligned} 0 &= \sum_{i=0}^{M} Q_i(z)(\alpha_0\, z^{\lambda_0} + \text{h.o.t.})^i\, (\alpha_0\lambda_0\, z^{\lambda_0-1} + \text{h.o.t.}) - \sum_{i=0}^{N} P_i(z)\, (\alpha_0\, z^{\lambda_0} + \text{h.o.t.})^i \\ &= \sum_{i=0}^{M} q_{i,0}\, \lambda_0\, \alpha_0^{i+1}\, z^{l_i+(i+1)\,\lambda_0-1} - \sum_{i=0}^{N} p_{i,0}\, \alpha_0^i\, z^{k_i+i\,\lambda_0} + \text{h.o.t.} \end{aligned}$$

Thus, at least two of the exponents

$$l_i + (i+1)\,\lambda_0 - 1, \quad k_j + j\,\lambda_0 \quad (0\leq i\leq M, \quad 0\leq j\leq N)$$

are equal to each other and not larger than any other exponents, and $\alpha_0\neq 0$ that vanishes the coefficient of the lowest degree. If this is the case, (λ_0,α_0) is said to be acceptable to (2.4). Assume that $(0,0)$ is not algebraic critical (i.e., Lemma 2.2 is not satisfied), then the values λ_0 and α_0 can be obtained using Newton polygon [1, 6] as follows. Let Γ be the Newton open polygon of all points (see Figure 1)

$$(i+1, l_i-1), \quad (j,k_j) \quad (0\leq i\leq M, \quad 0\leq j\leq N). \tag{2.8}$$

Let $\Gamma_{i_1}^{i_2}$ be an edge of Γ, with $i_1 < i_2$ to be the horizontal coordinates of the extreme vertices. Let $-\lambda_0$ be the slope of $\Gamma_{i_1}^{i_2}$; then α_0 should satisfy a polynomial of degree $i_2 - i_1$. In particular, (λ_0, α_0) is said to be d-folded if α_0 is a d-folded root of the above polynomial. Thus, for the edge $\Gamma_{i_1}^{i_2}$, there are at most $i_2 - i_1$ pairs of (λ_0, α_0) that are acceptable to (2.4). Thus, there are totally at most $\max\{M+1, N\}$ pairs of (λ_0, α_0) that are acceptable to (2.4).

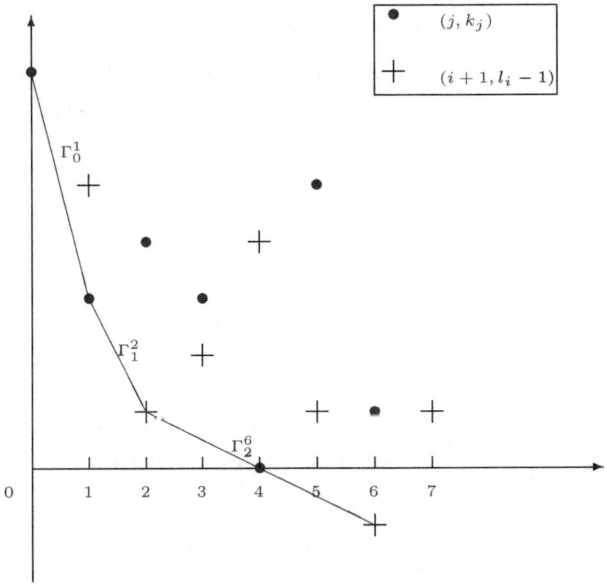

FIGURE 1. Newton polygon

For each (λ_0, α_0) in the first step, let $w(z) = \alpha_0 z^{\lambda_0} + w_1(z)$; then $w_1(z)$ satisfies the equation

$$Q(z, \alpha_0 z^{\lambda_0} + w_1)(\alpha_0 \lambda_0 z^{\lambda_0 - 1} + w_1') - P(z, \alpha_0 z^{\lambda_0} + w_1) = 0. \tag{2.9}$$

Repeat the foregoing argument, if $(0,0)$ is not an algebraic critical point of (2.4), then there are finite solutions of (2.9) of the form

$$w_1(z) = \alpha_1 z^{\lambda_1} + \text{h.o.t.} \quad (\lambda_1 \in \mathbb{Q}^+,\ \lambda_1 > \lambda_0). \tag{2.10}$$

To complete the proof, it is sufficient to show that if (λ_0, α_0) is d-folded, then there are at most d pairs of (λ_1, α_1) with $\lambda_1 > \lambda_0$ which are acceptable to (2.9).

Let

$$\begin{aligned} Q_1(z, w_1) &= Q(z, \alpha_0 z^{\lambda_0} + w_1), \\ P_1(z, w_1) &= P(z, \alpha_0 z^{\lambda_0} + w_1) - \alpha_0 \lambda_0 z^{\lambda_0 - 1} Q(z, \alpha_0 z^{\lambda_0} + w_1); \end{aligned}$$

then $w_1(z)$ satisfies

$$Q_1(z,w_1)\,w_1' - P_1(z,w_1) = 0. \tag{2.11}$$

Write

$$Q_1(z,w_1) = \sum_{i\geq 0} Q_{1,i}(z)\,w_1^i, \quad P_1(z,w_1) = \sum_{i\geq 0} P_{1,i}(z)\,w_1^i$$

and let $l_{1,i}$ and $k_{1,i}$ be the lowest degrees of $Q_{1,i}(z)$ and $P_{1,i}(z)$ respectively, and $r_{1,i} = \min\{k_{1,i}, l_{1,i-1} - 1\}$. We will prove that if (λ_0, α_0) is d-folded, then for any $i > d$,

$$r_{1,d} \leq r_{1,i} + (i-d)\,\lambda_0. \tag{2.12}$$

When (2.12) is satisfied, there are at most d-pairs of (λ_1, α_1) which are acceptable to (2.11) and $\lambda_1 > \lambda_0$. In fact, let (λ_1, α_1) be acceptable to (2.11); then there exist $j_1 < j_2$, such that

$$\lambda_1 = \frac{r_{1,j_1} - r_{1,j_2}}{j_2 - j_1} > \lambda_0$$

and

$$r_{1,d} \geq r_{1,j_1} + (j_1 - d)\,\lambda_1, \quad r_{1,d} \geq r_{1,j_2} + (j_2 - d)\,\lambda_1.$$

If $j_1 > d$ (or $j_2 > d$), then

$$r_{1,d} > r_{1,j_1} + (j_1 - d)\,\lambda_0 \quad (\text{or } \; r_{1,d} > r_{1,j_2} + (j_2 - d)\,\lambda_0)$$

which contradicts (2.12). Hence, $j_1 < j_2 \leq d$, and there are at most d-pairs of (λ_1, α_1) (taking into account that $(0,0)$ is not algebraic critical).

To prove (2.12), let

$$Q(z, \alpha\, z^{\lambda_0}) = \sum_{i\geq 0} \xi_i(\alpha)\, z^{s_i} \qquad (s_0 < s_1 < \cdots),$$

$$P(z, \alpha\, z^{\lambda_0}) = \sum_{i\geq 0} \eta_i(\alpha)\, z^{\tau_i} \qquad (\tau_0 < \tau_1 < \cdots);$$

then

$$Q_{1,i}(z) = \frac{1}{i!}\, z^{-i\,\lambda_0} \sum_{j\geq 0} \xi_j^{(i)}(\alpha_0)\, z^{s_j},$$

$$P_{1,i}(z) = \frac{1}{i!}\, z^{-i\,\lambda_0} \left(\sum_{j\geq 0} \eta_j^{(i)}(\alpha_0)\, z^{\tau_j} - \alpha_0\, \lambda_0\, z^{\lambda_0 - 1} \sum_{j\geq 0} \xi_j^{(i)}(\alpha_0)\, z^{s_j} \right),$$

and hence

$$r_{1,i} \geq \min\{\tau_0, s_0 + \lambda_0 - 1\} - i\,\lambda_0. \tag{2.13}$$

Thus, it is sufficient to show that

$$\min\{k_{1,d}, l_{1,d-1} - 1\} = \min\{\tau_0, s_0 + \lambda_0 - 1\} - d\,\lambda_0. \tag{2.14}$$

To this end, write

$$Q_{1,d-1}(z) = \frac{1}{d!}\xi_0^{(d-1)}(\alpha_0)\, z^{s_0+\lambda_0-d_0\,\lambda_0} + \text{h.o.t.},$$

$$P_{1,d}(z) = \frac{1}{d!}\left(\eta_0^{(d)}(\alpha_0)\, z^{\tau_0} - \alpha_0\,\lambda_0\,\xi_0^{(d)}(\alpha_0)\, z^{s_0+\lambda_0-1}\right)\cdot z^{-d\,\lambda_0} + \text{h.o.t.}$$

and let

$$P(z,\alpha\, z^{\lambda_0}) - \alpha\,\lambda_0\, z^{\lambda_0-1}\, Q(z,\alpha\, z^{\lambda_0}) = \varphi(\alpha)\, z^{v_0} + \text{h.o.t.}$$

Because (λ_0,α_0) is acceptable to (2.4) and d-folded, we have

$$\varphi(\alpha_0) = \cdots = \varphi^{(d-1)}(\alpha_0) = 0,\ \varphi^{(d)}(\alpha_0) \neq 0. \tag{2.15}$$

Therefore, we have the following:

(a) if $\tau_0 < s_0 + \lambda_0 - 1$, then $\varphi(\alpha) = \eta_0(\alpha)$ and $\eta_0^{(d)}(\alpha_0) \neq 0$;
(b) if $s_0 + \lambda_0 - 1 < \tau_0$, then $\varphi(\alpha) = -\lambda_0\,\alpha\,\xi_0(\alpha)$, and hence $\xi_0^{(d)}(\alpha_0) \neq 0$;
(c) if $s_0 + \lambda_0 - 1 = \tau_0$, then $\varphi_0(\alpha) = \eta_0(\alpha) - \alpha\lambda_0\xi_0(\alpha)$, and hence

$$\varphi_0^{(d)}(\alpha_0) = -\lambda_0\xi_0^{(d-1)}(\alpha_0) + (\eta_0^{(d)}(\alpha_0) - \alpha_0\lambda_0\xi_0^{(d)}(\alpha_0)) \neq 0.$$

Thus, we have $\xi_0^{(d-1)}(\alpha_0) \neq 0$ or $\eta_0^{(d)}(\alpha_0) - \alpha_0\lambda_0\xi_0^{(d)}(\alpha_0) \neq 0$.

It is not difficult to verify that (2.14) holds in any one of the above cases, and thus the lemma is concluded. □

From the proof of Lemma 2.4, the local algebraic solutions of (2.4) at (0, 0) can be obtained by repeating the Newton polygon. Moreover, following the procedure, we will either stop by the case that (0, 0) is algebraic critical (Lemma 2.2), or encounter the local algebraic solution of the form

$$w(z) = \sum_{i=0}^{k-1} \alpha_i\, z^{\lambda_i} + u(z),$$

where $(\lambda_{k-1},\alpha_{k-1})$ is 1-folded, and $u(z)$ satisfies an equation

$$\frac{\mathrm{d}u}{\mathrm{d}z} = \frac{\hat{P}(z,u)}{\hat{Q}(z,u)}, \tag{2.16}$$

where $\hat{P},\hat{Q}$ are Puiseux series. Whenever this is the case, we have the following.

Lemma 2.5. *In the equation* (2.16) *that derived from* (2.4) *through the above procedure, let*

$$\hat{P}(z,u) = \hat{p}_{0,0}z^{k_0} + \hat{p}_{1,0}z^{k_1}\, u + h.o.t.,\quad \hat{Q}(z,u) = \hat{q}_{0,0}z^{l_0} + h.o.t.$$

If $(\lambda_{k-1},\alpha_{k-1})$ *is* 1*-folded, and one of the following is satisfied:*

(1) $k_1 \neq l_0 - 1$;
(2) $k_1 = l_0 - 1$, *and* $\hat{p}_{1,0}/\hat{q}_{0,0} \notin (\lambda_{k-1},\infty)\cap\mathbb{Q}^+$,

then (0, 0) *is not algebraic critical of* (2.4).

Proof. Let $u(z)$ be a local algebraic solution of (2.16), expressed as

$$u(z) = \sum_{i\geq k} \alpha_i\, z^{\lambda_i}, \tag{2.17}$$

where $\lambda_i > \lambda_{i-1}$ ($\forall i \geq k$). We will show that (λ_i, α_i) are determined by (2.16) uniquely.

From the proof of Lemma 2.4, we have

$$k_0 - \min\{k_1, l_0 - 1\} > \lambda_{k-1}.$$

Hence, substituting (2.17) into (2.16), and taking into account that $(\lambda_{k-1}, \alpha_{k-1})$ is 1-folded, and either $k_1 \neq l_0 - 1$ or $k_1 = l_0 - 1$, $p_{1,0}/q_{0,0} \notin (\lambda_{k-1}, \infty) \cap \mathbb{Q}^+$, we have $\lambda_k = k_0 - \min\{k_1, l_0 - 1\}$, and α_k is determined uniquely by $p_{0,0}, q_{0,0}, p_{1,0}, k_1, l_0$. Therefore, (λ_k, α_k) is also 1-folded. Let $u(z) = \alpha_k\, z^{\lambda_k} + v(z)$; then $v(z)$ satisfies

$$\frac{\mathrm{d}v}{\mathrm{d}z} = \frac{\hat{p}'_{0,0}\, z^{k'_0} + \hat{p}_{1,0}\, z^{k_1}\, v + \text{h.o.t.}}{\hat{q}_{0,0}\, z^{l_0} + \text{h.o.t.}}, \tag{2.18}$$

where $k'_0 > k_0$. In particular, the conditions in the lemma are also valid for (2.18). Thus, we can repeat the procedure, and hence there is a unique solution $u(z)$ of the form (2.17), and $(0,0)$ is not algebraic critical for (2.4). □

Remark 2.6. In Lemma 2.5, we might also find the solution of the form (2.17) when $k_1 = l_0 - 1$ and $\hat{p}_{1,0}/\hat{q}_{0,0} \in (\lambda_{k-1}, \infty) \cap \mathbb{Q}^+$. However, when this is the case, we can identify two cases:

(1) if $\hat{p}_{1,0}/\hat{q}_{0,0} \in (\lambda_i, \lambda_{i+1}) \cap \mathbb{Q}$ for some $i \geq k-1$, then the condition in Lemma 2.2 is satisfied at the ith step, and $(0,0)$ is algebraic critical;
(2) if $\hat{p}_{1,0}/\hat{q}_{0,0} = \lambda_i$ for some i, then $(0,0)$ is not algebraic critical.

In any case, the procedure can stop in a finite number of steps. Thus, it is effective to find the algebraic multiplicities of (2.4) using the Newton polygon.

Example. Consider the equation

$$(z + w^2)\, w' - (z^2 + \mu w) = 0. \tag{2.19}$$

The Newton polygon of (2.19) is shown in Figure 2. From the Newton polygon, if $\mu \in (1/2, 2) \cap \mathbb{Q}$, then $(0,0)$ is algebraic critical, with local algebraic solutions

$$w(z) = \alpha_0 z^{\mu} + \text{h.o.t.} \quad (\alpha_0 \neq 0).$$

Meanwhile, if $\mu \notin (1/2, 2) \cap \mathbb{Q}$, the possible local algebraic solutions are

$$\begin{aligned} w(z) &= \frac{1}{2-\mu}\, z^2 + \text{h.o.t.} \ \ (\text{if} \ \ \mu \neq 2), \\ w(z) &= \pm\sqrt{2\mu - 1}\, z^{1/2} + \text{h.o.t.} \ \ (\text{if} \ \ \mu \neq 1/2). \end{aligned}$$

When $\mu \neq 2$, let

$$w(z) = \frac{1}{2-\mu}\, z^2 + w_{1,1}(z);$$

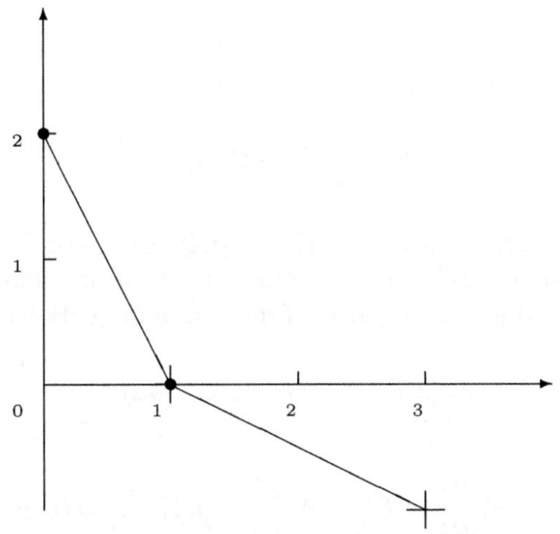

FIGURE 2. Newton polygon of (2.19)

then $w_{1,1}(z)$ satisfies

$$w'_{1,1} = \frac{2\,z^5 - (2-\mu)^3\,\mu w_{1,1} + \text{h.o.t.}}{-(2-\mu)^3\,z + \text{h.o.t.}}.$$

Thus, we conclude the following. If $\mu \in (2,5) \cap \mathbb{Q}$, then $(0,0)$ is algebraic critical, with local algebraic solutions

$$w(z) = \frac{1}{2-\mu}\,z^2 + \alpha_1\,z^\mu + \text{h.o.t.} \quad (\alpha_1 \neq 0).$$

If $\mu \neq 2, 5$, we have the local algebraic solution

$$w(z) = \frac{1}{2-\mu}\,z^2 - \frac{2}{(5-\mu)\,(2-\mu)^3}\,z^5 + \text{h.o.t.}$$

When $\mu \notin [1/2, 2) \cap \mathbb{Q}$, let

$$w(z) = \sqrt{2\mu - 1}\,z^{1/2} + w_{1,2}(z);$$

then $w_{1,2}(z)$ satisfies

$$w'_{1,2} = \frac{2\,z^{5/2} + (2-2\mu)\,z^{1/2}\,w_{1,2} + \text{h.o.t.}}{4\,\mu\,z^{3/2} + \text{h.o.t.}}.$$

Thus, if $\mu \neq 1/5$, we have the local algebraic solution

$$w(z) = \sqrt{2\mu - 1}\,z^{1/2} + \frac{1}{5\,\mu - 1}\,z^2 + \text{h.o.t.}$$

Thus, repeating the above procedure, we can determine, for given μ, the algebraic multiplicity $\text{Mul}(0,0)$ of (2.19). In particular, if $\mu \notin (1/2, \infty) \cap \mathbb{Q}$, then $\text{Mul}(0,0) \leq 3$. □

In the rest of this section, we will prove the main results.

Proof of Theorem 1.4. Let W be the set of all non-constant local algebraic solutions of (1.3) at $(0, a_i)$ for some $0 \le i \le k$. Then

$$|W| = \sum_{i=0}^{k} \mathrm{Mul}(0, a_i).$$

Let $f(z,w)$ be an irreducible strict Darboux polynomial of (1.3), and $m = \deg_w f(z,w)$; then there are m algebraic functions $w_i(z)$ that are defined by $f(z,w) = 0$. It is sufficient to show that any algebraic function $w_i(z) \in W$. To this end, we only need to show that

$$\lim_{z \longrightarrow 0} w_i(z) = \{a_0, a_1, \dots, a_k\}. \tag{2.20}$$

Consider the equation

$$z\, Q(z,w)\, \frac{\partial f}{\partial z} + P(z,w)\, \frac{\partial f}{\partial w} = R_f(z,w)\, f(z,w).$$

Let $z = 0$; then $f(0,w)$ satisfies

$$P(0,w)\, f'_w(0,w) = R_f(0,w)\, f(0,w).$$

Thus $f(0,w)$ is a non-constant multiply of $\prod_{i=1}^{k}(w - w_i)^{l_i}$, $(l_i \ge 0)$. From this it is easy to conclude (2.20).

It is easy to have $\mathrm{Mul}(0,\infty) \le M$. Hence, if the singularities $(0, a_i)$ are not algebraic critical, then

$$\deg_w f(z,w) \le M\,(k+1)$$

from Lemma 2.4. □

Proof of Theorem 1.5. If (1.1) has an invariant straight line L, then we may perform suitable transformation and assume that L is given by

$$a\,z + b\,w + z = 0 \qquad (a \ne 0)$$

and $\deg f(z,w) = \deg_w f((z - b\,w - c)/a, w)$. It is easy to see that the degree of the system does not increase under linear transformation. Let

$$\bar{w} = w, \bar{z} = a\,z + b\,w + c;$$

then $\bar{w}(\bar{z})$ satisfies the equation of the form

$$\frac{d\bar{w}}{d\bar{z}} = \frac{\bar{P}(\bar{z},\bar{w})}{\bar{z}\,\bar{Q}(\bar{z},\bar{w})}, \tag{2.21}$$

where $\bar{P}(\bar{z},\bar{w}), \bar{Q}(\bar{z},\bar{w})$ are polynomials. Moreover, $\bar{f}(\bar{z},\bar{w}) = f((\bar{z} - b\,\bar{w} - c)/a, \bar{w})$ is an irreducible Darboux polynomial of (2.21), and $\deg f(z,w) = \deg_{\bar{w}} \bar{f}(\bar{z},\bar{w})$. Let (a_i, b_i) $(1 \le i \le M)$ be singular points of (1.1) at L; then $(0, b_i)$ are singular points of (2.21) at $\bar{z} = 0$, and not algebraic critical. Hence, applying Theorem 1.4 to (2.21), we have

$$\deg f(z,w) = \deg_{\bar{w}} \bar{f}(\bar{z},\bar{w}) \le M\,(M+1).$$

□

3. Application to 2D Lotka–Volterra System

In this section, we apply Theorem 1.4 to the 2D Lotka–Volterra system

$$\dot{z} = z\,(z + c\,w - 1), \quad \dot{w} = w\,(b\,z + w - a). \tag{3.1}$$

Invariant algebraic curves of the Lotka–Volterra system had been studied by many authors. For recent results on this topic, refer to [9, 5] and the references therein. In [9], the complete list of parameters of which the system has a strict invariant algebraic curve is presented. We will reobtain some of the results through the algebraic multiplicity.

Note that (3.1) is invariant under the following transformations:

$$(z, w, a, b, c) \to (\frac{w}{a}, \frac{z}{a}, \frac{1}{a}, c, b) \quad \text{if } a \neq 0; \tag{3.2}$$

$$(z, w, a, b, c) \to (\frac{1}{z}, (1-c)\frac{w}{z}, 1-b, 1-a, \frac{c}{c-1}) \quad \text{if } c \neq 1. \tag{3.3}$$

Results in this section are also valid under the above transformations.

Since $z = 0$ and $w = 0$ are invariant straight lines of (3.1), Theorem 1.4 is applicable.

Proposition 3.1. *If the 2D L-V system*

$$\frac{\mathrm{d}w}{\mathrm{d}z} = \frac{w\,(b\,z + w - a)}{z\,(z + c\,w - 1)} \tag{3.4}$$

has a strict Darboux polynomial f, then

$$\begin{array}{ll} \deg_w f(z,w) \leq \mathrm{Mul}(0,\infty) + \mathrm{Mul}(0,a) + \mathrm{Mul}(0,0) & \text{if } \ a \neq 0. \\ \deg_w f(z,w) \leq \mathrm{Mul}(0,\infty) + \mathrm{Mul}(0,0) & \text{if } \ a = 0. \end{array}$$

In particular, we have

Proposition 3.2. *If in* (3.4)

$$a \notin \mathbb{Q}^+, \; c \notin \mathbb{Q}^-, \; c - \frac{1}{a} \notin \mathbb{Q}^+\backslash\{1\}, \tag{3.5}$$

then (3.4) *has a strict invariant algebraic curve if and only if*

$$a(1-c) + (1-b) = 0,$$

and the invariant algebraic curve is given by

$$a(z-1) + w = 0.$$

Proof. When (3.5) is satisfied, the singularities $(0,0), (0,a), (0,\infty)$ are not algebraic critical, and

$$\mathrm{Mul}(0,0) = 0, \;\; \mathrm{Mul}(0,a) \leq 1, \;\; \mathrm{Mul}(0,\infty) = 0.$$

Hence, if $f(z,w)$ is a strict irreducible Darboux polynomial, then $\deg_w f = 1$. From this it is easy to conclude the result. □

Proposition 3.2 shows that the algebraic multiplicities may give an exact bound for the degree of the Darboux polynomial in particular cases. However, if there are algebraic critical points among the singularities, (1.4) does not provide the finite value. In this case, as we have seen from Lemma 2.2, there are infinitely many local algebraic solutions. On the other hand, this does not automatically imply that all these local algebraic solutions are algebraic solutions. Hence, we come to the following concrete problem: if a singular point of a system is algebraic critical, how many local algebraic solutions are exactly the algebraic function? It requires additional work to discuss this problem, and one may hope that the solution of this problem should lead to the final resolution of the Poincaré problem.

References

[1] A. D. Bruno, *Local Methods in Nonlinear Differential Equations*, Springer Series in Soviet Mathematics, Springer-Verlag, Berlin, 1989.

[2] M. M. Carnicer, The Poincaré problem in the nondicritical case, *Ann. Math.* **140** (1994), 289–294.

[3] D. Cerveau and A. Lins Neto, Holomorphic foliations in CP(2) having an invariant algebraic curve, *Ann. Inst. Fourier* **41** (1991), 883–903.

[4] A. Campillo and M. M. Carnicer, Proximity inequalities and bounds for the degree of invariant curves by foliations of $\mathbb{P}^2_{\mathbb{C}}$, *Trans. Amer. Math. Soc.* **349**(6) (1997), 2211–2228.

[5] L. Cairó, H. Giacomini and J. Llibre, Liouvillan first integrals for the planar Lotka–Volterra system, *Rendiconti del Circolo Matematico di Palermo* **52** (2003), 389–418.

[6] N. G. Chebotarev, *Theory of Algebraic Functions*, Higher Education Press, Beijing, 1956 (in Chinese, translated from Russian by Dingzhong Xia and Zhizhong Dai).

[7] J. Lei and K. Guan, Analytic expansion of solution passing singular point of second order polynomial system, *Chinese Ann. Math.* Ser. A **22A**(5) (2001), 571–576 (in Chinese).

[8] J. M. Ollagnier, About a conjecture on quadratic vector fields, *J. Pure and Applied Algebra* **165** (2001), 227–234.

[9] J. M. Ollagnier, Liouvillian integration of the Lotka–Volterra system, *Qualitative Theory of Dynamical Systems* **2** (2001), 307–358.

[10] J. V. Pereira, Vector fields, invariant varieties and linear systems, *Ann. Inst. Fourier* (Grenoble) **51**(5) (2001), 1385–1405.

[11] D. Schlomiuk, Algebraic and geometric aspects of the theory of polynomial vector fields, *Bifurcations and Periodic Orbits of Vector Fields* (D. Schlomiuk, ed.), Kluwer Academic, Dordrecht, 1993, pp. 429–467.

[12] S. Walcher, On the Poincaré problem, *J. Diff. Eqns.* **166** (2000), 51–78. doi:10.1006/jdeq.2000.3801

Acknowledgment

The authors would like to thank Professors Zhiming Zheng and Dongming Wang for their work on the organization of the seminar DESC 2004. The authors are also grateful to the referees for their helpful suggestion.

Jinzhi Lei
Zhou Pei-Yuan Center for Applied Mathematics
Tsinghua University
Beijing 100084
P. R. China
e-mail: jzlei@mail.tsinghua.edu.cn, leijinzhi@tsinghua.org.cn

Lijun Yang
Department of Mathematical Science
Tsinghua University
Beijing 100084
P. R. China
e-mail: lyang@math.tsinghua.edu.cn

Trends in Mathematics: *Differential Equations with Symbolic Computation*, 159–171

Formalizing a Reasoning Strategy in Symbolic Approach to Differential Equations

Shilong Ma

Abstract. There is a reasoning strategy, which is an incremental computation, used in symbolic and algebraic approach to differential equations. The center-focus problem can be solved by using this reasoning strategy. In algebraic approach to automated reasoning, the construction of polynomial ideals is at the heart. For polynomials with a known fixed number of variables, the problem of constructing polynomial ideals can be solved by the Gröbner basis method and Wu's method. However, in many cases, the concerned polynomials may contain arbitrarily many variables. Even for the case of polynomials with a fixed number of variables, sometimes we do not know the number in advance, and we only know that there exists such a number. Thus, it is necessary to theoretically study how to construct ideals for polynomial sets with arbitrarily many variables. In this paper, a model for incremental computations, called procedure scheme, is proposed, and based on this model the well limit behavior of incremental computations is studied. It provides an approach to build a new theory by the limit of a sequence of formal theories. A convergent procedure scheme, DISCOVER, is defined in algebraically closed fields. We can formalize the strategy mentioned above using the procedure scheme DISCOVER.

Mathematics Subject Classification (2000). Primary 34C05; Secondary 68W30.

Keywords. Incremental computation, polynomial ideal, limit, well limit behavior, center-focus problem.

1. Introduction

1.1. A Reasoning Schema Used in Symbolic Approach to Differential Equations

There is a reasoning schema used in symbolic approach to differential equations. Given a family of parametrized objects, for example, $\dot{x} = f(x, u)$ with vector $x = (x_1, \ldots, x_n)$ and parameters $u = (u_1, \ldots, u_l)$, and a first order property

---

Research supported by the National 973 Project of China under the grant number G1999032701.

$P(u_1, \ldots, u_l)$ which may be a conjecture, how to automatically derive the conditions on the parameters $u_1, \ldots, u_l$ that characterize the objects having (or not having) the property P in the family?

Formally, that is, how to find a theory Γ such that $\Gamma \vdash P$ or $\Gamma \vdash \neg P$? Let us introduce some notation:

$$\mathrm{Th}(\Gamma) = \{A \mid \Gamma \vdash A\}.$$

Thus, $\Gamma \vdash P$ and $\Gamma \vdash \neg P$ can be denoted $P \in \mathrm{Th}(\Gamma)$ and $\neg P \in \mathrm{Th}(\Gamma)$ respectively.

1.2. Center-focus Problem

The center-focus problem is the key to the second part of Hilbert's 16th problem. Consider differential systems of *center-focus* type

$$\dot{x} = \lambda x + y + P(x, y), \quad \dot{y} = -x + \lambda y + Q(x, y),$$

where $P(x, y)$ and $Q(x, y)$ are homogeneous polynomials of degree $k \geq 2$ and with indeterminate coefficients $u = (u_1, \ldots, u_l)$.

As explained in [14, 8], one can compute a *Liapunov function* $F(x, y)$ and derived polynomials $\eta_2, \eta_4, \ldots, \eta_{2i+2}, \ldots$ in $u_1, \ldots, u_l$ such that

$$\frac{dF(x,y)}{dt} = \eta_2(x^2 + y^2) + \eta_4(x^2 + y^2)^2 + \cdots + \eta_{2i+2}(x^2 + y^2)^{2i} + \cdots,$$

where η_{2i+2} are called *focal values*.

The origin $(0, 0)$ is a *fine focus of order* k if

$$\eta_{2i} = 0 \text{ for } 1 \leq i \leq k, \text{ but } \eta_{2k+2} \neq 0.$$

A fine focus of infinite order is called a *center*. Deciding the origin to be a center or a fine focus of finite order is the key to the construction of limit cycles for the above differential systems.

Wang [14] first applied Ritt–Wu's method [16, 18, 17] and the Gröbner bases method [2] to deal with such systems, and James and Lloyd [3] also did research on this problem with computer algebra systems. Ma and Ning [9] combined Ritt–Wu's method with Budan–Fourier's theorem, a simple and efficient method for locating real roots and proving inequalities, to do such work.

1.3. A Successful Strategy to Solve Center-focus Problem

From the computing experiences of Wang, Ma and Ning [14, 12, 9], a successful strategy is to compute first a few focal values $\eta_2, \ldots, \eta_{2i}$ for $i < j < k$ and guess some simple conditions on the parameters $u_1, \ldots, u_l$, $\Gamma = \{f_1 = 0, \ldots, f_m = 0\}$, make these focal values be zero (i.e., $\eta_2 = 0, \ldots, \eta_{2i} = 0 \in \mathrm{Th}(\Gamma)$), and then simplify the differential system using these conditions.

Then, compute $\eta_{2i+2}, \ldots, \eta_{2j}$ from the simplified differential system, and proceed as follows:

1. if $\eta_{2i+2} = 0 \in \mathrm{Th}(f_1 = 0, \ldots, f_m = 0)$, then check $\eta_{2i+4} = 0, \ldots$;

2. if $\eta_{2i+2} = 0 \notin \text{Th}(f_1 = 0, \ldots, f_m = 0)$ and $\eta_{2i+2} \neq 0 \notin \text{Th}(f_1 = 0, \ldots, f_m = 0)$, then make a guess $h = 0$, consistent with $\{f_1 = 0, \ldots, f_m = 0\}$, such that $\eta_{2i+2} = 0$ and add $h = 0$ into $\{f_1 = 0, \ldots, f_m = 0\}$;

   Remark 1.1. Here, a guess $h = 0$ can be got from a factor h_j of $\eta_{2i+2} = h_1 \cdots h_k$, as $\eta_{2i+2} = 0$ iff $h_1 = 0$, or ..., or $h_k = 0$.

3. if $\eta_{2i+2} \neq 0 \in \text{Th}(f_1 = 0, \ldots, f_m = 0)$, then we have to revise our guess Γ to have a new guess Γ' and simplify the differential system again.

Go on until the goal is proved or disproved.

1.4. Motivation

In algebraic approach to automated reasoning, the construction of polynomial ideals is at the heart. For polynomials with a known fixed number of variables, the problem of constructing polynomial ideals can be solved by using the method of Gröbner bases and Wu's method. However, in many cases, the concerned polynomials may contain arbitrarily many variables. Even for the case of polynomials with a fixed number of variables, sometimes we do not know the number in advance, and we only know that there exists such a number. Thus, it is necessary to theoretically study how to construct ideals for polynomial sets with arbitrarily many variables.

In this paper, a model for incremental computations, called *procedure scheme*, is proposed, and based on this model the well limit behavior of incremental computations is studied. It provides an approach to build a new theory by the limit of a sequence of formal theories. A convergent procedure scheme, DISCOVER, is defined in algebraically closed fields. We can formalize the above strategy using the procedure scheme DISCOVER.

2. Preliminary

Let ALC be the class of algebraically closed fields and $\text{Th}(ALC)$ the first order theory of ALC. $\text{Th}(ALC)$ can be axiomatized in a language L having symbols $0, 1, +, \cdot$, by writing the usual field axioms and adding a sequence of axioms

$$A_n = \forall a_0 \cdots \forall a_{n-1} \exists x (a_0 + a_1 x + \cdots + a_{n-1} x^{n-1} + x^n = 0),$$

$n = 1, 2, \ldots$. The complex number field is a model of ALC. For more details, see [13].

Polynomial and rational equations and inequalities are atomic formulas, which can be written in normal forms as follows.

1. Polynomial equations and inequalities can be written such that their right-hand sides are 0, i.e., in the form $f = 0$ or $f \neq 0$, where f is a polynomial.
2. A rational equation $f/g = 0$ can be reduced to $f = 0 \wedge g \neq 0$; a rational inequality $f/g \neq 0$ can be reduced to $f \neq 0 \wedge g \neq 0$, where f and g are polynomials.

Therefore, we only need consider polynomial equations and inequalities. Noting that the satisfiability of $f \neq 0$ is equivalent to the satisfiability of $fz - 1 = 0$ by introducing a new variable z, we can have the following transformations.

2.1. Transformations of Inequalities and Disjunction of Equations

1. $f \neq 0 \leftrightarrow (\exists z)(fz - 1 = 0)$.
2. $f = 0 \vee g = 0 \leftrightarrow fg = 0$.
3. $\neg(f = 0) \leftrightarrow (\exists z)(fz - 1 = 0)$.

Thus in *ALC*, given a theory Γ containing equations, inequalities and disjunction of equations, Γ can be written as $\{f_1 = 0, \ldots, f_m = 0\}$. We still use Γ to denote this polynomial equation system, i.e., $\Gamma = \{f_1 = 0, \ldots, f_m = 0\}$.

In the following discussions, we only use systems of polynomial equations for the reason explained above.

Let $\Gamma = \{f_1 = 0, \ldots, f_m = 0\}$. Then,

$$\mathrm{Th}(\Gamma) = \{f = 0 \mid f \in \mathrm{Ideal}(f_1, \ldots, f_m)\},$$

that is, $\Gamma \vdash f = 0$ if and only if $f \in \mathrm{Ideal}(f_1, \ldots, f_m)$ [6].

3. The Construction of Ideals for Polynomial Sets with Arbitrarily Many Variables

In algebraic approach to automated reasoning the construction of polynomial ideals and the ideal membership test are at the heart of many problems.

For polynomials with a fixed number of variables, the problems of constructing polynomial ideals and determining ideal or radical ideal membership may be solved by using the method of Gröbner bases [2] and Wu's method [16, 17].

However, in automated differential geometric reasoning, the concerned polynomials may be of the form

$$p = p\left(x_1, \ldots, x_n, \frac{dx_1}{dt}, \ldots, \frac{d^k x_1}{dt^k}, \ldots, \frac{dx_n}{dt}, \ldots, \frac{d^k x_n}{dt^k}, \ldots\right)$$

and thus contain arbitrarily many variables, or even infinitely many variables [15]. In algebraic approach to first order theorem proving in predicate logic, a theorem may also correspond to a set of polynomials with arbitrarily many variables. Thus, it is necessary to theoretically study how to construct ideals for polynomial sets with arbitrarily many variables.

Even for the case of polynomials with a fixed number of variables, we should distinguish between two subcases.

1. Constructing ideals for polynomial sets with a fixed number n of variables. We know the number n in advance.
2. Constructing ideals for polynomial sets with a fixed number of variables. We do not know the number in advance, and we only know that this number exists. For example, when studying the problem of constructing limit cycles

for a differential system, we usually do not know in advance the number of limit cycles.

4. A Model for Incremental Computations

4.1. Long Term Behavior of Incremental Computations

For incremental learning, assume that examples come in sequences, and let E_k be the example set at time k. An incremental knowledge discovery algorithm $\mathcal{A}$ processes E_k and then produces a theory $\Gamma_k = \mathcal{A}(\Gamma_{k-1}, E_k)$ for every k, which is correct with respect to E_k. A reasonable interactive and ongoing computation $\mathcal{A}$ should satisfy the following requirements.

Convergence: For $E_1, \ldots, E_k, \ldots$, the limit (in some sense) of the knowledge sequence $\Gamma_1, \ldots, \Gamma_k, \ldots$ learned by $\mathcal{A}$ should exist.

Correctness: If for any k the discovered knowledge Γ_k is correct with respect to E_k, then the limit of $\Gamma_1, \ldots, \Gamma_k, \ldots$ should be correct with respect to all examples $E_1, \ldots, E_k, \ldots$.

Non-trivialness: For $E_1, \ldots, E_k, \ldots$, the discovered knowledge sequence $\Gamma_1, \ldots,$ $\Gamma_k, \ldots$ learned by $\mathcal{A}$ should be laws abstracted from $E_1, \ldots, E_k, \ldots$, not just copies of them.

4.2. A Definition for Limit Behavior

In the following, we try to capture the idea of the limit of system sequences.

Definition 4.1 ([4]). Let $\Gamma_1, \ldots, \Gamma_k, \ldots$ be a sequence of sets.

$$\overline{\lim}_{k\to\infty}\Gamma_k \equiv \bigcap_{n=1}^{\infty}\bigcup_{m=n}^{\infty}\Gamma_m, \qquad \underline{\lim}_{k\to\infty}\Gamma_k \equiv \bigcup_{n=1}^{\infty}\bigcap_{m=n}^{\infty}\Gamma_m$$

are called the upper limit and lower limit of the above sequence respectively. If $\overline{\lim}_{k\to\infty}\Gamma_k = \underline{\lim}_{k\to\infty}\Gamma_k$, then we say that the (set-theoretic) limit of $\Gamma_1, \ldots, \Gamma_k, \ldots$ exists and denote it by $\lim_{k\to\infty}\Gamma_k$.

Obviously, $\underline{\lim}_{k\to\infty}\Gamma_k \subseteq \overline{\lim}_{k\to\infty}\Gamma_k$. If the limit exists, we also say that $\Gamma_1, \ldots, \Gamma_k, \ldots$ is convergent.

It follows from the definition that given a sequence $\Gamma_1, \ldots, \Gamma_k, \ldots$, an element belongs to the upper limit iff it belongs to infinitely many Γ_k's, and an element belongs to the lower limit iff it belongs to almost every Γ_k.

If the limit exists, then $\Gamma = \lim_{k\to\infty}\Gamma_k$ contains all the elements surviving forever in every Γ_k when k is large enough.

Note that there are two possible cases for the limit Γ. One case is that the limit can be reached at some K, $\Gamma = \Gamma_K = \Gamma_{K+1} = \cdots$. Such limit is called discrete limit. Another case is that the limit cannot be reached and only can be approached infinitely.

Formally introducing limits of sequences of formal systems into logic and computer science and using theory versions as approximations to approach complete formal systems are independent contributions by Li [4].

In the following, we often use $\{\Gamma_k \mid k = 1, 2, \ldots\}$ to stand for the sequence $\Gamma_1, \ldots, \Gamma_k, \ldots$ and do not distinguish them.

4.3. A Model for Incremental Computations

Let Σ be an alphabet, Σ^* be all the sentences over Σ, and Σ^∞ be all of infinite sequences on Σ^*. Φ denotes a Turing computable function on Σ^*; Φ can be extended to Σ^∞ by the following definition.

Definition 4.2 ([5, 6])**.** Let $\bar{\Gamma}$ be a finite set of sentences in Σ^*, and Φ be a given Turing computable function on Σ^*. For any $x = \{x_k \mid k = 1, 2, \ldots\} \in \Sigma^\infty$ and every $k \geq 1$, define

$$\Gamma_0 = \bar{\Gamma},$$
$$\Gamma_k = \Phi(\Gamma_{k-1}, x_k).$$

We extend Φ to Σ^∞ in the following way: Let Φ_{inf} be the function by extending Φ to Σ^∞, and

$$\Phi_{\text{inf}}(\bar{\Gamma}, x) = \begin{cases} \lim_{k\to\infty} \Gamma_k, & \text{if } \lim_{k\to\infty} \Gamma_k \text{ exists,} \\ \text{undefined,} & \text{otherwise.} \end{cases}$$

Φ_{inf} is called a *procedure scheme* on Σ^∞, and $\bar{\Gamma}$ is called an *initial set.*

4.4. Well Limit Behavior of Incremental Computations

Definition 4.3 (Well limit behavior). We say that a convergent sequence $\{\Gamma_k \mid k = 1, 2, \ldots\}$ has *well limit behavior*, if

$$\text{Th}(\lim_{k\to\infty} \Gamma_k) = \lim_{k\to\infty} \text{Th}(\Gamma_k).$$

Remark 4.4. The well limit behavior is similar to some kind of continuous behaviors; however, it is not in Scott's sense. This is because the generated sequence may not be monotonic when a procedure scheme does incremental computation during its execution.

We can give examples to show that not every sequence $\{\Gamma_k \mid k = 1, 2, \ldots\}$ has well limit behavior. Thus, it is desired to find sufficient conditions for a sequence to have well limit behavior.

5. Well Limit Behavior of Term Rewriting Systems

As it is well known, in theory computations can be studied using term rewriting systems [1]. Without loss of generality, we assume that every formal system can be translated into a term rewriting system.

For example, a formal theory Γ can be translated into a term rewriting system R. For convenience, we use $\text{Th}(R)$ to denote the theoretic closure $\text{Th}(\Gamma)$ of Γ, called the *theoretic closure* of R.

Definition 5.1. Let $\preceq$ be a well-founded order on terms and R a term rewriting system. R is called a $\preceq$ well-founded term rewriting system, if for any $l \to r \in R$, we have $l \succeq r$, and for any term $T[x]$ and any substitution θ, $l \succeq r$ implies $T[l^\theta] \succeq T[r^\theta]$.

Theorem 5.2. *If* $\{R_k \mid k = 1, 2, \dots\}$ *is a* $\preceq$ *well-founded term rewriting system sequence and* $\lim_{k\to\infty} R_k$ *exists, then*

$$\mathrm{Th}(\lim_{k\to\infty} R_k) = \lim_{k\to\infty} \mathrm{Th}(R_k).$$

That is, the term rewriting system sequence $\{R_k \mid k = 1, 2, \dots\}$ has well limit behavior. For a proof of this theorem, see [11].

The following corollaries can be derived from the above results. Let R be an infinite term rewriting system, and $\prec$ a well-founded order on terms. If

1. there is a term rewriting system sequence $\{R_k \mid k = 1, 2, \dots\}$ such that $R = \lim_{k\to\infty} R_k$, where every R_k is a finite term rewriting system and lim is set-theoretic limit, and
2. every R_k is a $\prec$ well-founded term rewriting system,

then $\mathrm{Th}(R) = \lim_{k\to\infty} \mathrm{Th}(R_k)$.

That is, for any term e, we have $e \in \mathrm{Th}(G)$ iff there is a K such that $e \in \mathrm{Th}(R_k)$ for any $k \geq K$.

In fact, the above corollary gives a construction method of theoretic closure generated by a class of infinite term rewriting systems in the case of set-theoretic limits.

6. Automated Reasoning and the Construction of Ideals for Polynomial Sets in $\mathbf{K}[x_1, \dots, x_n, \dots]$

In automated geometric reasoning and computer mathematics, the algorithmic solutions of the following deduction problem

$$P_1, \dots, P_m \models Q$$

have been studied, where $P_1, \dots, P_m$ are premises and Q is a conclusion. As it is well known, the deduction problem can be reduced to the construction of ideals of polynomials over $\mathbf{K}[x_1, \dots, x_n]$ and the ideal membership problem, where $\mathbf{K}$ is a field of characteristic zero.

As pointed previously, it is necessary to theoretically study how to construct ideals for polynomial sets with arbitrarily many variables.

We studied the limit behavior of polynomials on $\mathbf{K}[x_1, \dots, x_n, \dots]$. By transforming a polynomial set into a polynomial rewriting system and from the result on term rewriting system sequences, we get the following result on the construction of ideals for polynomial sets in $\mathbf{K}[x_1, \dots, x_n, \dots]$.

Theorem 6.1. *If for every $k \geq 1$, R_k is the polynomial rewriting system corresponding to an irreducible Gröbner basis G_k and $R = \lim_{k\to\infty} R_k$ exists, then*

$$\text{Ideal}(R) = \lim_{k\to\infty} \text{Ideal}(R_k),$$

where $\text{Ideal}(R_k)$ *is the ideal generated by* R_k *on* $\mathbf{K}[x_1, \ldots, x_k]$ *and* $\text{Ideal}(R)$ *is the ideal generated by* R *on* $\mathbf{K}[x_1, \ldots, x_k, \ldots]$.

That is, the polynomial rewriting system sequence $\{R_k \mid k = 1, 2, \ldots\}$ has well limit behavior. For a proof of the theorem, see [11].

The following corollaries can be derived from the above results. Let R be an infinite polynomial rewriting system, and $\prec$ a well-founded order on polynomials. If

1. there is a polynomial rewriting system sequence $\{R_k \mid k = 1, 2, \ldots\}$ such that $R = \lim_{k\to\infty} R_k$, where every R_k is a finite polynomial rewriting system and lim is set-theoretic limit, and
2. every R_k is a $\prec$ well-founded polynomial rewriting system,

then $\text{Ideal}(R) = \lim_{k\to\infty} \text{Ideal}(R_k)$. That is, for any polynomial f, we have $f \in \text{Ideal}(R)$ iff there is a K such that $f \in \text{Ideal}(R_k)$ for any $k \geq K$.

The above corollary gives a construction method of ideals for polynomial sets in $\mathbf{K}[x_1, \ldots, x_n, \ldots]$ in the case of set-theoretic limits.

7. Procedure Scheme DISCOVER

Using Gröbner bases computation, we can define a procedure Cons $(\Gamma,\ h = 0)$ that accepts as input a theory Γ (i.e., a system of equations) and an equation $h = 0$, and produces the decision results: (1) $\Gamma \vdash h = 0$, or (2) $\Gamma \not\vdash h = 0$, and $\Gamma \not\vdash \neg(h = 0)$, or (3) $\Gamma \vdash \neg(h = 0)$.

In what follows, $\Gamma = \{f_1 = 0, \ldots, f_m = 0\}$ and e is an equation of the form $f = 0$.

Procedure Cons(Γ, e):

- **begin**
  - **if** $1 \in \text{GB}(\{f_1, \ldots, f_m, hz - 1\})$ **then return** "$\Gamma \vdash e$";
  - /* z is different from $x_1, \ldots, x_n$ */
  - **if** $1 \in \text{GB}(\{f_1, \ldots, f_m, h\})$ **then return** "$\Gamma \vdash \neg e$";
  - **return** "$\Gamma \not\vdash e$ and $\Gamma \not\vdash \neg e$"
- **end**

Based on the configuration transformation system, we can have a procedure Recons$(\Gamma,\ h = 0)$ that produces a maximal consistent subset of Γ with $h = 0$.

Procedure Recons$(\Gamma, h = 0)$:

- **begin**
  - $\Sigma := \emptyset$;
  - **while** $1 \in \text{GB}(\Gamma \cup \{h\})$ **do**

```
    begin
      Γ := tail(Γ);
      Σ := Σ ∪ {head(Γ)}
    end;
    return Γ
  end
```

Here head(Γ) denotes the first element of Γ, and tail(Γ) denotes the rest of Γ after removing head(Γ).

The following algorithm is based on the the theory reconstruction process. It accepts as input an initial theory E_0, a set of variable-free sentences U_0 and a goal P to be proved or disproved.

Main Algorithm. *Input:* P; /* P is a goal to be discovered, e.g., a conjecture */
$E_0 := \{\}$; $U_0 := \{h_1 = 0, h_2 = 0, \ldots\}$; $E_\infty :=$ DISCOVER(E_0, U_0, P).
The procedure scheme DISCOVER is defined as follows.

```
procedure DISCOVER(E, U, P):
  begin
    e := head(U); /* e is an equation of the form h = 0 */
    if E ⊢ P or E ⊢ ¬P then return E and P (or ¬P) else
    begin
      Cons(E, e);
      if E ⊢ e then begin E := E; U := tail(U) end;
      if E ⊬ e and E ⊬ ¬e then
        begin E := E ∪ {e, f}; U := tail(U) end;
        /* f is an equation of which e is an instance */
      if E ⊢ ¬e then
        begin E := Recons(E, e); U := U end;
      U := U ∪ findnew(E);
      DISCOVER(E, U, P)
    end
  end
```

Here findnew(E) is a procedure for getting new facts with respect to E; head(U) denotes the first element of U, and tail(U) denotes the rest of U after removing head(U), so $U = \text{head}(U) \cup \text{tail}(U)$.

Remark 7.1. To make the above procedure scheme easy to understand, some redundant statements $E := E$ and $U := U$ are included.

8. An Example

Here, as an example, we take a cubic differential system considered in [3, 12, 9]:

$$\begin{aligned} \dot{x} &= \lambda x + y + a_1x^2 - 2b_1xy + (a_3 - a_1)y^2 + a_5x^2y + a_7y^3, \\ \dot{y} &= -x + \lambda y + b_1x^2 + 2a_1xy - b_1y^2 + b_4x^3 + b_5x^2y + (b_6 - a_5)xy^2. \end{aligned} \tag{8.1}$$

To simplify the problem, new coefficients a_8, b_8 and a_9 are introduced in computations by using the relations

$$a_5 = \frac{1}{2}(b_6 + 2a_7 - 2a_9), \quad b_1^2 = \frac{1}{2}(a_8 + b_8), \quad a_1^2 = \tfrac{1}{2}(b_8 - a_8).$$

Now, we are going to show how to use the procedure scheme DISCOVER to find some conditions Γ that ensure the origin to be a fine focus of order 8, i.e.,

$$\eta_{2i} = 0 \ \text{ for } 1 \le i \le 8, \ \text{ but } \eta_{18} \ne 0.$$

This is the goal to be discovered, denoted P. We can have a procedure findnew(D) for computing focal values η's from the differential system D.

1. First, let $\Gamma_0 = \{\}$ and compute $\eta_2 = \lambda$ using findnew(D_1), where D_1 denotes the cubic system (8.1) above. We let $\lambda = 0$, and put it into the current theory, i.e., $\Gamma_1 = \{\lambda = 0\}$; now we have $\eta_2 = 0$. Substituting $\lambda = 0$ into the system (8.1), we get system D_2 simpler than the system (8.1) with λ eliminated.
2. Computing $\eta_4 = b_5 + 4a_3b_1$ using findnew(D_2), for η_4 to be 0 we let $b_5 = 4a_3b_1$ which is logically independent from Γ_1, so we put it into the current theory to obtain $\Gamma_2 = \{\lambda = 0, b_5 = 4a_3b_1\}$. Now we have $\eta_2 = 0, \eta_4 = 0 \in \text{Th}(\Gamma_2)$. Substituting $b_5 = 4a_3b_1$ into the system D_2, we get system D_3 simpler than the system D_2 with b_5 eliminated.
3. Computing $\eta_6 = a_3b_1(2a_9 - 3b_6 - 6b_4 + 10a_3^2 - 4a_1a_3 - 18a_7)$ using the procedure findnew(D_3), for η_6 to be 0 we let $a_3 = 0$ which is logically independent from Γ_2, so we put it into the current theory to obtain

   $$\Gamma_3 = \{\lambda = 0, b_5 = 4a_3b_1, a_3 = 0\}.$$

   Now $\eta_2 = 0, \eta_4 = 0, \eta_6 = 0 \in \text{Th}(\Gamma_3)$. Substituting $a_3 = 0$ into the system D_3, we get system D_4 simpler than the system D_3 with a_3 eliminated.

   Remark 8.1. The choice of $b_1 = 0$ will be rejected in the process, while the choice of $2a_9 - 3b_6 - 6b_4 + 10a_3^2 - 4a_1a_3 - 18a_7 = 0$ will lead to very complex computations. Therefore we choose $a_3 = 0$.

4. Computing

   $$\eta_8 = -a_1b_1(a_7 + b_4)(2a_9 + 7b_4 - 9a_7)$$

   using the procedure findnew(D_4), for η_8 to be 0 we let $b_4 = -a_7$ which is logically independent from Γ_3, so we put it into the current theory to obtain $\Gamma_4 = \{\lambda = 0, b_5 = 4a_3b_1, \, a_3 = 0, b_4 = -a_7\}$. Now $\eta_2 = 0, \ \eta_4 = 0, \eta_6 = 0, \eta_8 = 0 \in \text{Th}(\Gamma_4)$. Substituting $b_4 = -a_7$ into the system D_4, we get system D_5 simpler than the system D_4 with b_4 eliminated.

   Remark 8.2. The choice of $a_1 = 0$ will be rejected in the process, and the choice of

   $$2a_9 - 3b_6 - 6b_4 + 10a_3^2 - 4a_1a_3 - 18a_7 = 0$$

   is right, but will lead to complex computations. The computation with this choice was done in [9]. Here, to make the example easy to be followed by readers, we choose $a_7 + b_4 = 0$.

5. Computing $\eta_{10} = -a_1 b_1 a_9 (20(3b_6 - 4a_8)a_7 - (13b_6 - 20a_8)a_9)$ using the procedure findnew(D_5), for η_{10} to be 0 we let

$$a_7 = \frac{a_9(13b_6 - 20a_8)}{20(3b_6 - 4a_8)}$$

which is logically independent from Γ_4, so we put it into the current theory to obtain

$$\Gamma_5 = \left\{\lambda = 0, b_5 = 4a_3 b_1, a_3 = 0, b_4 = -a_7, a_7 = \frac{a_9(13b_6 - 20a_8)}{20(3b_6 - 4a_8)}\right\}.$$

Now $\eta_2 = 0, \eta_4 = 0, \eta_6 = 0, \eta_8 = 0, \eta_{10} = 0 \in \mathrm{Th}(\Gamma_5)$. Substituting

$$a_7 = \frac{a_9(13b_6 - 20a_8)}{20(3b_6 - 4a_8)}$$

into the system D_5, we get system D_6 simpler than the system D_5 with a_7 eliminated.

Remark 8.3. The choice of $a_9 = 0$ will be rejected in the process.

6. Using the procedure findnew(D_6), we compute

$$\eta_{12} = a_1 b_1 a_9^2((3123b_6^3 - 6270a_8^3 - 12156b_6^2 a_8)a_9 - 24b_6 b_8 (3b_6 - 4a_8)(15b_6 - 28a_8))(3b_6 - 4a_8)^{-2},$$

$$\eta_{14} = -a_1 b_1 b_6^4 b_8^2 G_5 G_7^2 G_9 G_8^{-4}, \quad \text{where}$$

$$G_7 = 15b_6 - 28a_8,$$
$$G_8 = 3213b_6^3 - 6720a_8^3 - 12156b_6^2 a_8 + 15920b_6 a_8^2,$$
$$G_9 = 5G_8^2 G_{10} - 64b_8^2 G_{11},$$
$$G_{10} = 115b_6^2 - 120b_6 a_8 - 144a_8^2,$$
$$G_{11} = 1957968456b_6^6 - 15392836806b_6^5 a_8 + 5123792400a_8^2 b_6^4 - 9370286592a_8^3 b_6^3 + 10052919040a_8^4 b_6^2 - 6038323200a_8^5 b_6 + 1580544000a_8^6,$$

$$\eta_{16} = -a_1 b_1 b_6^5 b_8 G_5 G_7^2 G_{10} G_{12} G_8^{-1} G_{11}^{-2}, \quad \text{where}$$

$$G_{12} = \sum_{k=0}^{10} \alpha_k b_6^k a_8^{10-k}, \quad \text{and } \alpha_k \text{ are large integers,}$$

$$\eta_{18} = a_1 b_1 b_6^5 a_8^6 \, G_5 G_7^2 G_{10} G_{13} G_{11}^{-3}, \quad \text{where}$$

$$G_{13} = \sum_{k=0}^{9} \beta_k b_6^k a_8^{9-k}, \quad \text{and } \beta_k \text{ are large integers.}$$

For the integers α_k and β_k, see [12].

7. For η_{12} to be 0, we let

$$a_9 = \frac{24b_6 b_8 (3b_6 - 4a_8)(15b_6 - 28a_8)}{3123b_6^3 - 6270a_8^3 - 12156b_6^2 a_8}$$

which is logically independent from Γ_5, so we put it into the current theory to obtain

$$\Gamma_6 = \left\{ \lambda = 0,\ b_5 = 4a_3b_1,\ a_3 = 0,\ b_4 = -a_7,\ a_7 = \frac{a_9(13b_6 - 20a_8)}{20(3b_6 - 4a_8)}, \right.$$
$$\left. a_9 = \frac{24b_6b_8(3b_6 - 4a_8)(15b_6 - 28a_8)}{3123b_6^3 - 6270a_8^3 - 12156b_6^2a_8} \right\}.$$

Now $\eta_2 = 0, \eta_4 = 0, \eta_6 = 0, \eta_8 = 0, \eta_{10} = 0, \eta_{12} = 0 \in \mathrm{Th}(\Gamma_6)$.

The following computations are too tedious to be described in detail here, so we just give a very brief description of the computations.

8. For η_{14} to be 0, we let $b_8^2 = 5G_8^2G_{10}/(64G_{11})$ which is logically independent from Γ_6, so we put it into the current theory to obtain
$$\Gamma_7 = \Gamma_6 \cup \left\{ b_8^2 = \frac{5G_8^2G_{10}}{64G_{11}} \right\}.$$
Now $\eta_2 = 0, \eta_4 = 0, \eta_6 = 0, \eta_8 = 0, \eta_{10} = 0, \eta_{12} = 0, \eta_{14} = 0 \in \mathrm{Th}(\Gamma_7)$.
9. For η_{16} to be 0, we let $G_{12} = 0$ which is logically independent from Γ_7, so we put it into the current theory to obtain $\Gamma_8 = \Gamma_7 \cup \{G_{12} = 0\}$. Now $\eta_2 = 0, \eta_4 = 0, \eta_6 = 0, \eta_8 = 0, \eta_{10} = 0, \eta_{12} = 0, \eta_{14} = 0, \eta_{16} = 0 \in \mathrm{Th}(\Gamma_8)$.

   Remark 8.4. For $\eta_{12} = 0$, $\eta_{14} = 0$ and $\eta_{16} = 0$, we can have many choices to make guesses, but they will be rejected in the process. We omit their descriptions here.

10. Finally, we can check $\eta_{18} \neq 0 \in \mathrm{Th}(\Gamma_8)$, thus the goal P is proved.

References

[1] Baader, F. and Nipkow, T., *Term Rewriting and All That.* Cambridge University Press, Cambridge, 1998.

[2] Buchberger, B., Gröbner bases: An algorithmic method in polynomial ideal theory. In: *Recent Trends in Multidimensional Systems Theory* (Bose, N. K., ed.), pp. 184–232. D. Reidel, Dordrecht, Holland, 1985.

[3] James, E. M. and Lloyd, N. G., A cubic system with eight small-amplitude limit cycles. *IMA J. Appl. Math.* **47**: 163–171, 1991.

[4] Li, W., An open logic system (in Chinese). *Science in China* (Sci. Sinica) (ser. A) **10**: 1103–1113, 1992.

[5] Li, W. and Ma, S., A framework for analytic-intelligent agents. In: *Proceedings of the International Conference on Artificial Intelligence* (IC-AI'2000), Las Vegas, USA, 2000.

[6] Li, W. and Ma, S., Limits of theory sequences over algebraically closed fields and applications. *Discrete Appl. Math.* **136**(1): 23–43, 2004.

[7] Ma, S., Sui, Y. and Xu, K., The limits of the Horn logic programs. In: *Proceedings of the 18th International Conference on Logic Programming* (ICLP 2002), Copenhagen, Denmark, LNCS **2401**, p. 467. Springer-Verlag, Berlin Heidelberg, 2002.

[8] Lu, Z. and Ma, S., Centers, foci and limit cycles for polynomial differential systems. In: *Mathematics Mechanization and Applications* (Gao, X.-S. and Wang, D., eds.), pp. 365–387. Academic Press, London, 2000.

[9] Ma, S. and Ning, S., Deriving some new conditions on the existence of eight limit cycles for a cubic system. *Comput. Math. Appl.* **33**: 59–84 1997.

[10] Ma, S. and Ning, S., Practically solving some problems expressed in the first order theory of real closed fields. *Int. J. Comput. Math.* **69**: 265–282, 1998.

[11] Ma, S., A computational study on limit behavior of formal system sequences and its applications (in Chinese). Technical Report, School of Computer Science, Beihang University, December 2003.

[12] Ning, S., Ma, S., Kwek, K. H. and Zheng, Z., A cubic system with eight small-amplitude limit cycles. *Appl. Math. Lett.* **7**: 23–27, 1994.

[13] Rabin, M., Decidable theories. In: *Handbook of Mathematical Logic* (Barwise, ed.), pp. 595–630. North-Holland Publ. Co., 1977.

[14] Wang, D., Mechanical manipulation for a class of differential systems. *J. Symb. Comput.* **12**: 233–254, 1991.

[15] Wang, D., A method for proving theorems in differential geometry and mechanics. *J. Univ. Comput. Sci.* **1**: 658–673, 1995.

[16] Wu, W., On the mechanization of theorem-proving in elementary differential geometry (in Chinese). *Sci. Sinica* Special Issue on Math. (I): 94–102, 1979.

[17] Wu, W., Basic principles of mechanical theorem proving in elementary geometries. *J. Automat. Reason.* **2**: 221–252, 1986.

[18] Wu, W., Mechanical theorem proving in elementary geometry and elementary differential geometry. In: *Proceedings of the 1980 Beijing DD-Symposium*, vol. 2, pp. 1073–1092. Science Press, Beijing, 1982.

Shilong Ma
School of Computer Science and Engineering
Beihang University
Beijing 100083, China
e-mail: `slma@nlsde.buaa.edu.cn`

Trends in Mathematics: *Differential Equations with Symbolic Computation*, 173–200

Looking for Periodic Solutions of ODE Systems by the Normal Form Method

Victor F. Edneral

Abstract. We describe usage of the normal form method and corresponding computer algebra packages for building an approximation of local periodic solutions of nonlinear autonomous ordinary differential equations (ODEs). For illustration a number of systems are treated.

Mathematics Subject Classification (2000). Primary 37M05; Secondary 68T99.

Keywords. Resonant normal form, dynamical system, computer algebra.

1. Introduction

The normal form method is based on a transformation of an ODE system to a simpler set called the normal form. The importance of this method for analyzing of ODEs near stationary point has been recognized for a long time. For the history of this subject see, for instance, [2] or [26]. Definitions of normal form and normalizing transformation can be formulated in different ways for general and special cases. So there are very developed approaches for Hamiltonian systems, see, for example, [12, 31, 5, 6] and [9, Chapters 1 and 2]. For resonant and Belitskii normal forms see [10], [11, Chapter 5 §20] and [3]. There are also many algorithms (and their implementations) for creating normal forms and corresponding transformations. See for the Hamiltonian case an improved algorithm of Deprit and Hori in [35] and its symbolic algebra realization under the REDUCE system in [37]. A matter of numerical creation of normal forms of Hamiltonians is described in [25]. Questions of convergence of the normalizing transformation are discussed in [5, 6, 7, 32, 33]. Concerning algorithms for the creation of normal form in a general case we mention here (in addition to Bruno's books) the papers [39, 40, 38].

In this paper we use the algorithm based on the approach, which was developed by A. D. Bruno [5, 6, 7, 10, 11] for a resonant normal form. The important advantage of this approach is a possibility to investigate a wide class of

This research was supported by the grant of President RF Sc.Sch.-1450.2003.2.

autonomous systems in united, easily algorithmized frame. In particular it provides a constructive way for obtaining the approximations of local families of periodic and conditionally periodic solutions in the form of power/Fourier series for real families and in a form of power series in time dependent exponents for complex ones. For this paper it is especially important that the problem of convergence of used transformations was investigated. This circumstance allows us to hope that approximations of frequencies and corresponding periodic solutions families near stationary points by finite formulas can be done with acceptable precision. Except solutions themselves we can find also approximations of initial conditions, which initiate such periodic solutions. That is, we can produce some elements of a phase analysis.

Another advantage of the used approach is an algorithmic simplicity of the creation of the normal form and the corresponding transformations. We have a direct recurrence formula for this procedure. The usage does not demand keeping of some large intermediate results as it is in other algorithms. The approach is free from a necessity to solve intermediate systems of linear equations and from any restrictions on low resonance cases.

It is also possible to approximate by the proposed method the non-periodic families of solutions ("crude" case). The results are close to the results of the Carleman linearization method. For periodic and conditionally periodic cases the method is a generalization of the Poincaré–Lindstedt approach.

Below we describe the creation of the normal form and the application for building of periodic solutions of well-known second and fourth order equations. Also we will talk briefly about higher order systems.

2. Problem Formulation

Consider the system of autonomous ordinary differential equations

$$\dot{\mathbf{x}} = \mathbf{\Phi}(\mathbf{x}), \tag{2.1}$$

where $\mathbf{x} = (x_1, \ldots, x_n)$ is a vector function of time, $\dot{\mathbf{x}} \stackrel{\text{def}}{=} d\mathbf{x}/dt$ is the time derivative, $\mathbf{\Phi} = (\Phi_1, \ldots, \Phi_n)$ is a vector which is a function of $\mathbf{x}$ and probably of some parameters.

Such a type of equations originates from many scientific and engineering problems where oscillations, vibrations or wave processes take place. The standard way of an investigation of such systems is:

1. Bifurcation analysis, i.e. the investigation of a picture of behavior of system solutions in dependence on parameters. It is especially important if this behavior is changed sharply at some parameters values.
2. Phase portrait, i.e. an investigation of a behavior of system solutions depending on initial conditions.
3. Calculation of system solutions.

If we have solutions of a system in an analytical form, we have a clear picture of the system behavior, but we can have such solutions very rarely. As a rule we have numerical solutions, but a numerical investigation of the above items 1 and 2 is sometimes a complex problem. It is also not simple to obtain numerical solutions in unstable cases. As a consequence of this, we may be interested in some "intermediate" approach, which would be between analytical and numerical methods.

The normal form method is widely used for bifurcation analysis. About methods of such an investigation see, for example, [34, 28, 26]. One can see in these books that the numerical bifurcation analysis is indeed based on the normal form method. The main idea here is in replacing system (2.1) with some "model" system having finite order polynomial right-hand sides and transforming them to the canonical (normal) form. We can make from the lowest no vanishing coefficients of the normal form the qualitative conclusions about the behavior of the original system. It is sufficient to know only lowest orders of the normal form for such an analysis. Sometimes this job can be done by hand, sometimes by computer algebra systems, see, for example, [36].

Here we try to demonstrate that the calculation of high orders of the normal form can be useful also for targets 2 and 3 mentioned above. We restrict our consideration here mainly to the construction of approximations to periodic solutions but we have a material for using this approach in a "crude" case and in the case of conditionally periodic solutions.

The study of systems of type (2.1) in the neighborhood of *stationary* point $\mathbf{x}^0$, where $\mathbf{\Phi}(\mathbf{x}^0) = \mathbf{0}$, typically includes three preliminary steps. Firstly $\mathbf{x}$ is shifted by $-\mathbf{x}^0$ so that $\mathbf{\Phi}(\mathbf{0}) = \mathbf{0}$, i.e. $\mathbf{0}$ is the stationary point to be studied. Each stationary point of the system is considered separately.

The second step is a reduction of the system to a model form where the vector $\mathbf{\Phi}(\mathbf{x})$ is approximated by a vector of polynomials. If in some neighborhood of the stationary point $\mathbf{\Phi}$ is an analytic function of $\mathbf{x}$ then its power series can be used to get a smooth approximation with desired precision. Often this step is made simultaneously with a reduction of the system to its central manifold. In any case, the right-hand sides of the model system will be polynomials without constant terms.

The third step is the transformation of the linear parts matrix to Jordan's form by a complex linear change of $\mathbf{x}$ variables.

After these steps, system (2.1) has the form

$$\dot{y}_i = \lambda_i y_i + \sigma_i y_{i-1} + \tilde{\Phi}_i(\mathbf{y}), \qquad \sigma_1 = 0, \qquad i = 1, \ldots, n, \tag{2.2}$$

where $\mathbf{\Lambda} = (\lambda_1, \ldots, \lambda_n)$ is the vector of eigenvalues of the matrix of the linear part of the system and $\tilde{\mathbf{\Phi}} = (\tilde{\Phi}_1, \ldots, \tilde{\Phi}_n)$ is a vector of polynomials of finite degree without constant and linear terms.

For this paper, we assume that system (2.2) satisfies the following assumptions:

- the system is autonomous and has polynomial nonlinearities;

- $\mathbf{0}$ is a stationary point and the system will be studied near $\mathbf{y} = \mathbf{0}$;
- the linear part of the right-hand side is diagonal and not all eigenvalues are zero, i.e. $\mathbf{\Lambda} \neq \mathbf{0}$.

It is assumed that neither the system is Hamiltonian, nor that it preserves the phase volume nor that it has any internal symmetry.

3. The Normal Form Method

Equations (2.2) can be written in the form

$$\dot{y}_i = \lambda_i y_i + y_i \sum_{\mathbf{q} \in N_i} f_{i,\mathbf{q}} \mathbf{y}^{\mathbf{q}}, \quad i = 1, \ldots, n\,, \tag{3.1}$$

where we use the multi-index notation

$$\mathbf{y}^{\mathbf{q}} = \prod_{j=1}^{n} {y_j}^{q_j},$$

with the power exponent vector $\mathbf{q} = (q_1, \ldots, q_n)$ and the sets

$$N_i = \{\mathbf{q} \in \mathbb{Z}^n : \; q_i \geq -1 \text{ and } q_j \geq 0\,, \text{ if } j \neq i\,, \quad j = 1, \ldots, n\}\,,$$

because the factor y_i has been moved out of the sum in (3.1).

The normalization is done with a near-identity transformation

$$y_i = z_i + z_i \sum_{\mathbf{q} \in N_i} h_{i,\mathbf{q}} \mathbf{z}^{\mathbf{q}}, \quad i = 1, \ldots, n \tag{3.2}$$

and then we will have system (3.1) in the normal form

$$\dot{z}_i = \psi_i(\mathbf{z}) \stackrel{\text{def}}{=} \lambda_i z_i + z_i \sum_{\substack{\langle \mathbf{q}, \mathbf{\Lambda} \rangle = 0 \\ \mathbf{q} \in N_i}} g_{i,\mathbf{q}} \mathbf{z}^{\mathbf{q}}, \quad i = 1, \ldots, n\,. \tag{3.3}$$

The important difference between (3.1) and (3.3) is a restriction on the range of the summation, which is defined by the equation

$$\langle \mathbf{q}, \mathbf{\Lambda} \rangle \stackrel{\text{def}}{=} \sum_{j=1}^{n} q_j \lambda_j = 0\,. \tag{3.4}$$

The h and g coefficients in (3.2) and (3.3) are found by using the recurrence formula

$$g_{i,\mathbf{q}} + \langle \mathbf{q}, \mathbf{\Lambda} \rangle \cdot h_{i,\mathbf{q}} = -\sum_{j=1}^{n} \sum_{\substack{\mathbf{p} + \mathbf{r} = \mathbf{q} \\ \mathbf{p}, \mathbf{r} \in \bigcup_i N_i \\ \mathbf{q} \in N_i}} (p_j + \delta_{ij}) \cdot h_{i,\mathbf{p}} \cdot g_{j,\mathbf{r}} + \tilde{\Phi}_{i,\mathbf{q}}\,, \tag{3.5}$$

where the second summation in the right-hand side is over all integer vectors satisfying the constraint $\mathbf{p} + \mathbf{r} = \mathbf{q}$, and $\tilde{\Phi}_{i,\mathbf{q}}$ is a coefficient of the factor $z_i \mathbf{z}^{\mathbf{q}}$

in the polynomial $\tilde{\Phi}_i$ in (2.2), of which the arguments have been transformed by (3.2). Here $||\mathbf{p}||$ and $||\mathbf{r}|| < ||\mathbf{q}||$, where $||\mathbf{q}|| \stackrel{\text{def}}{=} q_1 + \cdots + q_n$. So (3.5) is a recurrent formulae.

The ambiguity in (3.5) is usually fixed by the conventions

$$\begin{aligned} h_{i,\mathbf{q}} &= 0 \quad \text{if} \quad \langle \mathbf{q}, \mathbf{\Lambda} \rangle = 0, \\ g_{i,\mathbf{q}} &= 0 \quad \text{if} \quad \langle \mathbf{q}, \mathbf{\Lambda} \rangle \neq 0 \end{aligned} \tag{3.6}$$

and then the normalizing transformation is called a "basic" one.

3.1. Local Families of Periodic Solutions

Firstly note that the sums in (3.2) and (3.3) typically include infinitely many terms even though the sum in (3.1) may have only a finite number of terms. The convergence properties of these series were investigated in [5, 6, 7] and, in general, these series diverge. However, using these divergent series, one can find some solutions of the initial system (3.1).

Let

$$\xi_1(\mathbf{z}), \ldots, \xi_s(\mathbf{z}) \tag{3.7}$$

be power series in $\mathbf{z}$ without constant terms. If they converge in some neighborhood of the origin $\mathbf{z} = \mathbf{0}$ then solutions of the system

$$\xi_j(\mathbf{z}) = 0\,, \quad j = 1, \ldots, s, \tag{3.8}$$

make up the *local analytic set*. If series (3.7) are formal (i.e. can diverge in any neighborhood of the origin), then system (3.8) defines the *local formal set*. In the ring of power series there is an ideal generated by series (3.7). If the ideal has a convergent basis then the formal set is analytic.

In [7] it was shown that local analytic sets of periodic solutions of system (3.1) can be found by means of its normal form (3.3). Namely, for the normal form (3.3), define the formal set A to be

$$\begin{aligned} A = \{\mathbf{z} : \quad \psi_i &= \lambda_i z_i \omega \qquad \text{if } \operatorname{Re}\lambda_i = 0; \\ z_i &= 0 \qquad\quad \text{if } \operatorname{Re}\lambda_i \neq 0; \quad i = 1, \ldots, n\}\,, \end{aligned} \tag{3.9}$$

where ω is a free parameter and it is independent of number i. Power series ψ_i are the same as in (3.3).

If *all* imaginary eigenvalues of system (2.2) are rationally dependent numbers (i.e. in the maximal resonant case) and $\mathbf{\Lambda} \neq \mathbf{0}$, the local formal set A is analytic and contains periodic solutions only. Since the ideal generated by this set has a convergent basis, the corresponding system (3.9) has a sense as a system of equations in power series. Each connected branch of set A is a local analytic family of periodic solutions.

From this point of view, in the above resonant case the normal form is adequate to the original system, at least along the set A, and this set includes all local families of periodic solutions. From the practical point of view, such system that has been normalized to some finite order can be used to approximate all families of periodic solutions near the stationary point and the precision of approximation can be increased enough.

If not all eigenvalues are rational dependent we need to split the set A into such subsets that in each of them all coordinates with corresponding rational independent eigenvalues would be zero. Each of these subsets is an analytic set. So in the phase space the set A can have several components. Each such component can have its own frequency ω.

The general case of non-purely imaginary eigenvalues and a definition of analytic sets, which contain local families of conditionally periodic solutions, are given in [7].

3.2. Main Algorithm

The algorithm of the calculation of g and h in (3.2) and (3.3) is based on (3.5) and (3.6). It is convenient to choose the representation of sets of coefficients $g_{i,\mathbf{q}}$ and $h_{i,\mathbf{q}}$ in such a way that they would be combined in homogeneous subgroups where each subgroup has the same order n, i.e. contains only terms with such vector-indexes $\mathbf{q} = \{q_j\}$ that $||\mathbf{q}|| = n$ for each i. One can calculate the sets g and h of the next order by using sets of g and h with smaller order only, i.e. (3.5) is a recurrence formula.

Algorithm. Let n be the dimension of the system. For its normalization till order m we are to do:

(i) For $i = 1, 2, \ldots, n$ do:
Calculate all squared in $\mathbf{y}$ elements in the right-hand side nonlinearity $\tilde{\Phi}_i(\mathbf{y})$ in (2.2), i.e. calculate the subgroup of the first order ($||\mathbf{q}|| = 1$) elements of the set $f_{i,\mathbf{q}}$ in (3.1) and sort it into two subsets depending on the value of the scalar product (3.4). The first set where this product is zero will be the first order subgroup of g_i and the second set after a division by the value of the corresponding scalar product will be the first order subgroup of h_i

(ii) For $k = 2, 3, \ldots, m$ do:
   (a) For $i = 1, 2, \ldots, n$ do:
   Calculate the subgroup of order k of the nonlinear terms $\tilde{\Phi}_i(\mathbf{y})$ in (2.2) for which the substitution $\mathbf{y}$ is evaluated by (3.2) till order $k - 1$ and define coefficients at monomials $z_i\mathbf{z}^{\mathbf{q}}$ as $f_{i,\mathbf{q}}$.
   (b) For $i = 1, 2, \ldots, n$ do:
   Calculate the subgroups of g_i and h_i of order k by a subdivision of set $f_{i,\mathbf{q}}$ into two subsets as in step (i). After that one can supplement the set g_i till full order k and a part of the set h_i without a contribution from the first term of the right-hand side in (3.5).
   (c) For $i = 1, 2, \ldots, n$ do:
   For $j = 1, 2, \ldots, n$ do:
   Supplement the preliminary set of order k of h_i with properly sorted multiplications of *all* elements of such subgroups of $h_{i,\mathbf{p}}$ and $g_{j,\mathbf{r}}$ that their total order, i.e. $||\mathbf{p}+\mathbf{r}|| = k$. Not all these multiplications should be

really calculated because of the factor $(p_j + \delta_{i,j})$ is zero at some values of j index. Before the supplement all elements above are to be divided on the corresponding scalar products also.

The cost of the above algorithm is low in comparison with the cost of evaluation of the right-hand side of the nonlinear system. Under such circumstances it is very important to calculate the right-hand sides very economically, using as much as possible the fact that we need to calculate at each step of (ii) the homogeneous terms of $\tilde{\Phi}_i$ of order k only and all terms of lower orders are not changed during the later operations. The problem of optimization of this evaluation is one of the main limitations for an automatization of generating codes for the right-hand side calculation.

3.3. Computer Algebra Implementation of the Normal Form Method

The calculation of the coefficients of the normal form (3.3) and the corresponding transformation (3.2) with respect to (3.5) and (3.6) was implemented as the NORT package. Earlier attempts of the author to compute sufficiently high orders of the normal form using high level of the REDUCE language [29] were not very successful. Because of this, the NORT package [23, 24] was created. NORT is written in Standard LISP and contains now about 2000 operators. NORT is a package of procedures to treat truncated multivariate power series in arbitrary dimensions. In addition to procedures for arithmetic operations with series, there are special procedures for the creation of normal forms and procedures for substitutions, for calculations of roots (when it is possible), for differentiating, for printing and for inverting multivariate power series, etc. It contains also special procedures for the calculation of Lyapunov's values. NORT can be used as a separate program or as a REDUCE package.

Besides series, expressions in NORT can contain non-negligible variables (parameters). So there is implemented multivariate series-polynomial arithmetic. The complex-valued numerical coefficients of the truncated power series-polynomials may be treated in three different arithmetics: rational, modular, floating point and approximate rational. There are also several options for the output form of these numbers; the output is in a REDUCE readable form. The program uses an internal recurrence representation for its objects. Remark that a garbage collection time for examples below was smaller than 3% of the evaluation time. This can characterize the NORT package as a program with a good enough internal organization.

Unfortunately at this moment the NORT package has no friendly user interface yet. So we create a package for usage with Mathematica [20, 21]. This package works with truncated multivariate power series. The demo version of the *PolynomialSeries* package can be accessed at the `www.mathsource.com` site. Truncation of power series allows one to control correctly the length of truncated series. Therefore, an existing version is enough to support a normal form method. The comparison of the Mathematica package with an earlier version of the normal form package NORT written in LISP demonstrates that the calculations within the Mathematica system are more flexible and convenient but are considerably slower than under LISP.

4. General Scheme of Investigation by the Normal Form Method

The general scheme of the investigation of a nonlinear ODE system by the normal form method near each stationary point is as follows.

1. Recasting the system in a model (free of constant terms polynomial) form.
2. Linear normalization of the system, i.e. the reduction of a linear part of the right-hand sides of the system to Jordan's or to a diagonal form and an investigation of the corresponding eigenvalues.
3. Searching for "resonant" values of parameters, i.e. such values at which groups of rational dependent pure imaginary eigenvalues appear. The system should be investigated in the neighborhood of each such value of parameters.
4. Nonlinear normalization of the system, i.e. the creation of the normal form and the corresponding normalizing transformation (sometimes it is called as recasting the system in a "resonance form").
5. Bifurcation analysis of the system in parameters by observing the lowest no vanishing orders of the corresponding normal form.
6. Building the periodic and conditionally periodic approximations of solution families, which include the stationary point, i.e. *local* solutions.
7. Reducing the order of the normalized system if it is necessary, and a repetition of the investigation above in the neighborhood of each stationary point of the newest system.

5. Second Order Systems as a Transparent Example

All questions of convergence and integrability of normal forms for any second-order systems of type (2.2) near a stationary point have been investigated by Bruno [5, 7] for the case where the stationary point is an elementary singular point, i.e. both eigenvalues of the system are not equal to zero simultaneously. The result of this investigation stated briefly is as follows: the normal forms for such equations are integrable. For each case there are written integrals of the normalized system. The result regarding the convergence is as follows. Let $\lambda_2 \neq 0$ and $\lambda \stackrel{\text{def}}{=} \lambda_1/\lambda_2$. If $\mathrm{Im}(\lambda) \neq 0$ or if $\lambda > 0$ then the transformation is convergent, and we can produce an approximate solution of the original system from known integrals of the normal form by transformation (3.2) with any desirable precision. In other words, the cases of "focus" and "node" can be treated without any additional demands. At a negative irrational λ the convergence will take place if for all nonzero vectors $\mathbf{q}$ with integer elements there exist positive ε, ν such that $|\langle \mathbf{q}, \mathbf{\Lambda} \rangle| > \varepsilon(|q_1| + |q_2|)^{-\nu}$. This condition can be checked before creation of the normal form. This case is a particular case of a saddle point. But at real non-positive rational $\lambda = -m/n \leq 0$ we will have convergence under some additional requirements on the normal form.

This is an interesting case. It includes the cases of a center and a limit circle. Let us look at a couple of examples.

5.1. Duffing's Equation

This is an equation of the second order, which originates from a problem of a mathematical pendulum:

$$\frac{d^2\phi}{dt^2} + \omega_0^2 \sin(\phi) = 0, \tag{5.1}$$

where ϕ is an angle of deviation of the pendulum and it can be approximated by the series (units below have been chosen so that $\omega_0 = 1$) with the change $\phi = \sqrt{6}x$. We get the Duffing's equation

$$\frac{d^2x}{dt^2} = -x + x^3. \tag{5.2}$$

This is a Hamiltonian equation with the full energy

$$H = \frac{1}{2}(\frac{dx}{dt})^2 + \frac{1}{2}x^2 - \frac{1}{4}x^4. \tag{5.3}$$

By linear complex change of variables

$$x = y_1 + y_2, \qquad \frac{dx}{dt} = i\,(y_1 - y_2), \tag{5.4}$$

we can rewrite it in the diagonalized form

$$\frac{dy_1}{dt} = iy_1 - \frac{i}{2}\,(y_1 + y_2)^3, \quad \frac{dy_2}{dt} = -iy_2 + \frac{i}{2}\,(y_1 + y_2)^3. \tag{5.5}$$

Note that the above couple of equations have complex conjugate coefficients at terms with exchange $y_1 \leftrightarrow y_2$. It will take place always if the original equation has real coefficients.

The vector of eigenvalues of a linear part of the system is $\mathbf{\Lambda} = \{i, -i\}$. In accordance with the definition of the normal form (3.3) we will have at sums of this form only terms, where $\langle \mathbf{\Lambda}, \mathbf{p} \rangle = i\,(p_1 - p_2) = 0$, i.e. only terms where $p_1 = p_2$:

$$\begin{aligned} \frac{dz_1}{dt} &= \quad i\, z_1 + z_1\,(g_{1,1,1} \cdot z_1 \cdot z_2 + g_{1,2,2} \cdot z_1^2 \cdot z_2^2 + \cdots), \\ \frac{dz_2}{dt} &= -i\, z_2 + z_2\,(g_{2,1,1} \cdot z_1 \cdot z_2 + g_{2,2,2} \cdot z_1^2 \cdot z_2^2 + \cdots). \end{aligned} \tag{5.6}$$

Condition (3.9) for the analytic set A of the second-order equation with eigenvalues $\lambda_1 = -\lambda_2$ has the form

$$\sum_{k=1,\ldots} g_{1,k,k} \cdot (z_1 \cdot z_2)^k = - \sum_{k=1,\ldots} g_{2,k,k} \cdot (z_1 \cdot z_2)^k.$$

As for any second-order (originally) real equation the equalities $g_{1,i,i} = \bar{g}_{2,i,i}$, $h_{1,i,j} = \bar{h}_{2,j,i}$ take place, this condition has the form

$$\sum_{k=1,\ldots} \operatorname{Re}(g_{1,k,k}) \cdot (z_1 \cdot z_2)^k = 0. \tag{5.7}$$

After the calculation of the normal form for Duffing's equation, it can be revealed that (5.7) is fulfilled identically because all $g_{1,i,i}$ and $g_{2,i,i}$ are purely imaginary here. It means that for Duffing's equation we have the case, which is usually called the "center", when periodic solutions exist for any (small enough) initial conditions.

Indeed, by multiplying the first of equations (5.6) by z_2 and the second one by z_1 we will have $\frac{d(z_1 z_2)}{dt} = 0$ after their addition and (5.6) has a family of solutions:

$$z_1(t) = c_1 e^{+i\omega(c_1 \cdot c_2)t}, \quad z_2(t) = c_2 e^{-i\omega(c_1 \cdot c_2)t},$$

where

$$\omega(z_1 \cdot z_2) \stackrel{\text{def}}{=} 1 + g_{1,1,1}/i + g_{1,2,2}/i + \cdots$$

is a real constant and c_1, c_2 are integration constants.

Now we can obtain the approximation to the original equation (5.2) by substitution of the found z_i into y_i with (3.2) and after that into x. If we choose complex conjugate values for $c_1 = \bar{c}_2$ we will have an approximation of a real family of a periodic solution in the form of truncated Fourier series.

We have obtained these series as series in the $c = c_1$ variable. It is not convenient, and for the final representation we calculated H as a series in c by substituting the found x and $\frac{dx}{dt}$ in (5.3). By inverting this series and substituting c as a series of H into expressions for ω and x we will have the final result as series in H. To save space we show this result here till the fifth order only:

$$\begin{aligned}
\omega &= 1 - \tfrac{3}{4}H - \tfrac{69}{64}H^2 - \tfrac{633}{256}H^3 - \tfrac{110421}{16384}H^4 - \tfrac{1318329}{65536}H^5 + \cdots, \\
x &= \sqrt{2H}\times \\
&\quad [\ \cos(\omega t)(1 + \tfrac{9}{16}H + \tfrac{271}{256}H^2 + \tfrac{10779}{4096}H^3 + \tfrac{243613}{32768}H^4 + \tfrac{2963587}{131072}H^5) \\
&\quad - \cos(3\omega t)H(\tfrac{1}{16} + \tfrac{3}{16}H + \tfrac{1209}{2048}H^2 + \tfrac{127233}{65536}H^3 + \tfrac{6907221}{1048576}H^4) \\
&\quad + \cos(5\omega t)H^2(\tfrac{1}{256} + \tfrac{11}{512}H + \tfrac{3107}{32768}H^2 + \tfrac{25567}{65536}H^3) \\
&\quad - \cos(7\omega t)H^3(\tfrac{1}{4096} + \tfrac{1}{512}H + \tfrac{5805}{524288}H^2) \\
&\quad + \cos(9\omega t)H^4(\tfrac{1}{65536} + \tfrac{21}{131072}H) \\
&\quad - \cos(11\omega t)H^5(\tfrac{1}{1048576}) + \cdots\ ].
\end{aligned} \tag{5.8}$$

The result of this calculation was verified in two ways. The first one was a direct substitution of series (5.8) in the original equation (5.2). After that we had only terms with negligible orders of H. The second way was a comparison of the numerical solutions of Duffing's equation by the Runge–Kutta method (by NAG's d02baf procedure) with the values of the series tabulated at different values of H.

In view of (5.3) we have $H = H(x = \phi/\sqrt{6}, \frac{dx}{dt} = \frac{d\phi}{dt}/\sqrt{6})$. On the other hand, H is physically not small when at a zero velocity a maximum deviation of a pendulum takes place, i.e. when $\phi = \pi/2$ and $\frac{d\phi}{dt} = 0$. In this case we have $H_{\max} \simeq 0.163$. Let us now introduce a function of a maximum relative error during one period $f_{\mathrm{err}}(H)$:

$$f_{\mathrm{err}} = \sup_{t\in[0,2\pi/\omega]} \sqrt{\frac{(x_{\mathrm{series}} - x_{\mathrm{num}})^2 + (dx_{\mathrm{series}}/dt - dx_{\mathrm{num}}/dt)^2}{x_{\mathrm{num}}^2 + (dx_{\mathrm{num}}/dt)^2}}. \tag{5.9}$$

This function indicates a maximum relative deviation in phase space between series (5.8) x_{series} and its symbolically evaluated derivation on the one hand and numerical solutions of (5.2) x_{num}, dx_{num}/dt on the other hand as a function of full energy H. We have

$$\begin{aligned} f_{\mathrm{err}}(H = 0.1) &\simeq 1.8 \times 10^{-8}, \\ f_{\mathrm{err}}(H = 0.125) &\simeq 7.4 \times 10^{-7}, \\ f_{\mathrm{err}}(H_{\max} = 0.163) &\simeq 7.4 \times 10^{-5}. \end{aligned}$$

One can see that the achieved precision can be used for practical goals in a physical range of energies. The use of a ready closed formula can be sometimes more preferable at real-time calculations than the numerical solution of differential equations. Form (5.8) must of course be improved for real usage by standard methods of preparing series for numerical tabulation.

5.2. Van der Pol's Equation

This non-Hamilton equation originated from a problem of vibrations in electronic circuits:

$$\frac{d^2x}{dt^2} = -x + (\varepsilon^2 - x^2)\,\frac{dx}{dt}. \tag{5.10}$$

By linear complex change of variables (5.4) it can be rewritten in the diagonalized form

$$\begin{aligned} \frac{dy_1}{dt} &= \quad i\,y_1 + \frac{1}{2}\,(y_1 - y_2)\,[\varepsilon^2 - (y_1 + y_2)^2], \\ \frac{dy_2}{dt} &= -i\,y_2 + \frac{1}{2}\,(y_2 - y_1)\,[\varepsilon^2 - (y_1 + y_2)^2], \\ \frac{d\varepsilon}{dt} &= 0. \end{aligned}$$

As for Duffing's case these equations have complex conjugate coefficients at terms with exchange $y_1 \leftrightarrow y_2$.

Note that the value $\varepsilon = 0$ is "resonant" here, i.e. only at this value we have a couple of purely imaginary conjugate eigenvalues. In accordance with our common receipt we should consider ε as a small perturbation. We have redefined above the parameter ε as a new variable. Such a trick allows us in practice to free eigenvalues from parameter dependence.

And as for Duffing's equation the sums on the right-hand sides of the normal form will include only terms where $p_1 = p_2$. The third "additional" equation will not be changed. So a difference between (5.6) and the system presented below is only in polynomial dependence of the right sides on ε:

$$\frac{dz_1}{dt} = iz_1 + z_1 \sum_{k=0,1,\dots} [g_{1,1,1,2k} \cdot (z_1 \cdot z_2)\varepsilon^{2k} + g_{1,2,2,2k} \cdot (z_1 \cdot z_2)^2\varepsilon^{2k} + \cdots],$$

$$\frac{dz_2}{dt} = -iz_2 + z_2 \sum_{k=0,1,\dots} [g_{2,1,1,2k} \cdot (z_1 \cdot z_2)\varepsilon^{2k} + g_{2,2,2,2k} \cdot (z_1 \cdot z_2)^2\varepsilon^{2k} + \cdots],$$

$$\frac{d\varepsilon}{dt} = 0. \tag{5.11}$$

For the analytic set (convergent) condition (5.7) we now have the form

$$\mathrm{Re}\,[\sum_{j,k=0,1,\dots} g_{1,j,j,2k}\ \varepsilon^{2k}\ (z_1 \cdot z_2)^j] = 0, \qquad g_{1,0,0,0} \stackrel{\text{def}}{=} 0. \tag{5.12}$$

Contrary to Duffing's equation the above is not satisfied automatically, but because of the implicit function theorem it may be solved in the form

$$z_1 \cdot z_2 = \sum_{k=1,2,\dots} q_k \varepsilon^{2k}.$$

It is easy to see that if the above is satisfied then $z_1 \cdot z_2$ is a constant in time. So we can continue the evaluations as for Duffing's case, but now the constants c_1, c_2 of integration are not free. This is the case of a "limit circle", and this restriction defines the limit circle trajectory

$$z_1(t) = c_1 e^{+i\omega(c_1 \cdot c_2)t}, \quad z_2(t) = c_2 e^{-i\omega(c_1 \cdot c_2)t}, \quad c_1 \cdot c_2 = \sum_{k=1,2,\dots} q_k \varepsilon^{2k}.$$

After that we obtain the approximation to the original equation (5.10) by substituting the above-found z_i into y_i by (3.2) and then into x. If we choose complex conjugate values for $c_1 = \bar{c}_2$ we will have also for Duffing's equation an approximation of the real solution in the form of truncated Fourier series. It is a limit circle trajectory:

$$\omega = 1 - \tfrac{1}{16}\varepsilon^4 + \tfrac{17}{3072}\varepsilon^8 + \tfrac{35}{884736}\varepsilon^{12} - \tfrac{678899}{5096079360}\varepsilon^{16} + \cdots,$$

$$\begin{aligned}
x = \varepsilon \cdot [\ &\cos(\omega t)(2 + \tfrac{1}{64}\varepsilon^4 - \tfrac{23}{49152}\varepsilon^8 - \tfrac{51619}{169869312}\varepsilon^{12} + \tfrac{948555443}{19568944742400}\varepsilon^{16}) \\
&+ \cos(3\omega t)\varepsilon^4(-\tfrac{3}{32} + \tfrac{101}{12288}\varepsilon^4 + \tfrac{24061}{28311552}\varepsilon^8 - \tfrac{279818087}{815372697600}\varepsilon^{12}) \\
&+ \cos(5\omega t)\varepsilon^4(-\tfrac{5}{96} + \tfrac{1865}{110592}\varepsilon^4 - \tfrac{328835}{254803968}\varepsilon^8 - \tfrac{111998015}{293534171136}\varepsilon^{12}) \\
&+ \cos(7\omega t)\varepsilon^8(\tfrac{1379}{110592} - \tfrac{10923199}{3185049600}\varepsilon^4 + \tfrac{21049213549}{183458856960000}\varepsilon^8) \\
&+ \cos(9\omega t)\varepsilon^8(\tfrac{61}{20480} - \tfrac{1769369}{589824000}\varepsilon^4 + \tfrac{161113663733}{237817036800000}\varepsilon^8) \\
&+ \cos(11\omega t)\varepsilon^{12}(-\tfrac{409871}{331776000} + \tfrac{1359229760383}{1872809164800000}\varepsilon^4) \\
&+ \cos(13\omega t)\varepsilon^{12}(-\tfrac{715247}{3715891200} + \tfrac{2076538440769}{5243865661440000}\varepsilon^4)
\end{aligned}$$

$$
\begin{aligned}
&+\cos(15\omega t)\varepsilon^{16}(\tfrac{526426361}{4661213921280}) \\
&+\cos(17\omega t)\varepsilon^{16}(\tfrac{392636471}{29964946636800}) \\
&+\sin(3\omega t)\varepsilon^{2}(-\tfrac{1}{4}+\tfrac{15}{512}\varepsilon^4-\tfrac{779}{1179648}\varepsilon^8-\tfrac{4538017}{6794772480}\varepsilon^{12}) \\
&+\sin(5\omega t)\varepsilon^{6}(\tfrac{85}{2304}-\tfrac{8095}{1327104}\varepsilon^4-\tfrac{1252495}{6115295232}\varepsilon^8) \\
&+\sin(7\omega t)\varepsilon^{6}(\tfrac{7}{576}-\tfrac{99967}{13271040}\varepsilon^4+\tfrac{415949513}{382205952000}\varepsilon^8) \\
&+\sin(9\omega t)\varepsilon^{10}(-\tfrac{9791}{2457600}+\tfrac{117258703}{70778880000}\varepsilon^4) \\
&+\sin(11\omega t)\varepsilon^{10}(-\tfrac{5533}{7372800}+\tfrac{1657839733}{1486356480000}\varepsilon^4) \\
&+\sin(13\omega t)\varepsilon^{14}(\tfrac{21731177}{57802752000}) \\
&+\sin(15\omega t)\varepsilon^{14}(\tfrac{138697}{2774532096})+\cdots].
\end{aligned}
\tag{5.13}
$$

The calculation by the NORT package till the 32nd order in ε took 1.5 minutes on a PentiumPro-200 computer. We had 145 terms for each sum in the normal form (5.11) and 1773 terms for the normalizing transformation from z_i to y_i. The calculated expression for frequency has 9 terms. Note that the power series for the frequency of van der Pol's equation itself has been calculated till the 164th order in ε in [1].

Comparison of this result with a numerical one in terms of (5.9) gives

$$
\begin{aligned}
f_{\text{err}}(\varepsilon^2=0.5) &\simeq 8\times 10^{-10}, \\
f_{\text{err}}(\varepsilon^2=0.75) &\simeq 4\times 10^{-8}, \\
f_{\text{err}}(\varepsilon^2=1.0) &\simeq 1\times 10^{-5}.
\end{aligned}
$$

Besides the solution of equation we can obtain also the expressions for the original conditions, which lie in the limit circle trajectory of (5.10) as series in ε by inversion of series (3.2):

$$
\begin{aligned}
x &= 0, \\
\frac{dx}{dt} &= \varepsilon\cdot(2+\tfrac{17}{96}\varepsilon^4-\tfrac{1577}{552960}\varepsilon^8-\tfrac{102956839}{55738368000}\varepsilon^{12}+\tfrac{48722480822161}{157315969843200000}\varepsilon^{16}+\cdots)
\end{aligned}
$$

and

$$
\begin{aligned}
x &= \varepsilon\cdot(2+\tfrac{1}{96}\varepsilon^4-\tfrac{1033}{552960}\varepsilon^8+\tfrac{1019689}{55738368000}\varepsilon^{12}+\tfrac{9835512276689}{157315969843200000}\varepsilon^{16}+\cdots), \\
\frac{dx}{dt} &= 0.
\end{aligned}
$$

6. Examples of the Fourth Order ODEs

6.1. Henon–Heiles' System

Paper [16] describes an application of the normal form method for building analytic approximations for all (including complex) local families of periodic solutions in the neighborhood of the stationary points of the Henon–Heiles system. The families of solutions are represented as truncated Fourier series in approximated frequencies, and the corresponding trajectories are described by intersections of hypersurfaces, which are defined by pieces of multivariate power series in phase variables of the system. A comparison of the numerical values obtained by a tabulation of the approximate solutions above with results of numerical integration of the Henon–Heiles system displays a good agreement, which is enough for the usage of these approximate solutions for engineering applications.

The Henon–Heiles system was originally appeared in the theory of the particle motion in an axial symmetric gravitation field, more precisely from the problem of a star's motion in the Galactic field [30] as a simple model for numerical experiments. This is a system of two differential equations of the second order:

$$\ddot{x} = -x - 2xy \,, \quad \ddot{y} = -y - x^2 + y^2 \,; \tag{6.1}$$

it can be written in the form (2.1) as a Hamiltonian system with the Hamiltonian function

$$H = \frac{1}{2}[(\dot{x})^2 + (\dot{y})^2 + x^2 + y^2] + x^2 y - \frac{1}{3} y^3 \,. \tag{6.2}$$

The linear change of variables

$$\begin{aligned} x &= y_1 + y_3 \,, \quad & \dot{x} &= -\imath\,(y_1 - y_3)\,, \\ y &= y_2 + y_4 \,, \quad & \dot{y} &= -\imath\,(y_2 - y_4) \end{aligned} \tag{6.3}$$

transforms (6.1) to the form (2.2) required by the method:

$$\begin{aligned} \dot{y}_1 &= -\imath\, y_1 - \imath\,(y_1 + y_3)\,(y_2 + y_4)\,, \\ \dot{y}_2 &= -\imath\, y_2 - \tfrac{\imath}{2}\,[(y_1 + y_3)^2 - (y_2 + y_4)^2]\,, \\ \dot{y}_3 &= \imath\, y_3 + \imath\,(y_1 + y_3)\,(y_2 + y_4)\,, \\ \dot{y}_4 &= \imath\, y_4 + \tfrac{\imath}{2}\,[(y_1 + y_3)^2 - (y_2 + y_4)^2]\,. \end{aligned} \tag{6.4}$$

The eigenvalues of this system are two pairs of complex conjugate imaginary units: $\mathbf{\Lambda} = (-\imath, -\imath, \imath, \imath)$. So this is a deeply resonant problem, the most difficult type of problems to analyse using the perturbation theory.

6.2. A Normal Form of the Henon–Heiles System

For the Henon–Heiles system (6.1), the ratios of all pairs of eigenvalues are ± 1, so this is a pure resonant case, and the set A is analytic. It contains all local families of periodic solutions. The normal form (3.3) for system (6.4) is

$$\dot{z}_i = z_i g_i \stackrel{\text{def}}{=} \lambda_i z_i + z_i \sum_{\substack{q_i \geq -1\,, \\ q_1, \ldots, q_{i-1}, q_{i+1}, \ldots, q_4 \geq 0\,, \\ q_1 + q_2 = q_3 + q_4 > 0}} g_{i,q_1,q_2,q_3,q_4}\, z_1^{q_1} z_2^{q_2} z_3^{q_3} z_4^{q_4}, \quad i = 1,2,3,4\,. \tag{6.5}$$

The set A is defined by the system of equations of the type (3.9):

$$\lambda_i z_i \omega = \lambda_i z_i + z_i \sum g_{i,q_1,q_2,q_3,q_4}\, z_1^{q_1} z_2^{q_2} z_3^{q_3} z_4^{q_4}, \quad i = 1,2,3,4\,, \tag{6.6}$$

where ω is a function in $z_1, \ldots, z_4$ which does not depend on i and time t but it is defined on the set A only.

A search for *all* solutions of a system of type (6.6) is an independent problem of solving systems of equations over the ring of formal power series. It is very important that the formal set A has a basis, which consists of convergent power series. So the families of periodic solutions of (6.6) can be expressed in terms of convergent series. We need to have a reasonable way of solving systems of type (6.6). Here we use a simple factorization.

As the Henon–Heiles system (6.1) is real, in the normal form (6.5) for real x, y the series $g_j(\mathbf{z})$ satisfy the reality conditions

$$\begin{aligned} g_{j+2}(\mathbf{z}) &= \overline{g}_j(\mathbf{z}), \\ z_{j+2} &= \overline{z}_j, \qquad j = 1, 2. \end{aligned}$$

Using our program NORT for the Henon–Heiles system we have obtained

$$\begin{aligned}
g_3 = \imath \cdot [1 &- \frac{7}{3} z_1 z_3^{-1} z_4^2 + \frac{2}{3} z_2 z_4 - \frac{5}{3} z_1 z_3 \\
&+ \frac{413}{54} z_3^{-1} z_1 z_2 z_4^3 + \frac{157}{12} z_2^2 z_4^2 - \frac{59}{4} z_1^2 z_4^2 \\
&- \frac{601}{18} z_1 z_2 z_3 z_4 - \frac{403}{108} z_2^2 z_3^2 + \frac{223}{108} z_1^2 z_3^2 \\
&+ \frac{21203}{432} z_1 z_2^2 z_3^{-1} z_4^4 + \frac{67025}{1296} z_1^3 z_3^{-1} z_4^4 \\
&+ \frac{21206}{405} z_2^3 z_4^3 - \frac{22387}{135} z_1^2 z_2 z_4^3 + \frac{67627}{1080} z_1 z_2^2 z_3 z_4^2 \\
&- \frac{20551}{360} z_1^3 z_3 z_4^2 - \frac{3832}{405} z_2^3 z_3^2 z_4 - \frac{4789}{15} z_1^2 z_2 z_3^2 z_4 \\
&- \frac{46313}{2160} z_1 z_2^2 z_3^3 + \frac{102541}{6480} z_1^3 z_3^3 + O(z^8)], \\
g_4 = \imath \cdot [1 &- \frac{5}{3} z_2 z_4 + \frac{2}{3} z_1 z_3 - \frac{7}{3} z_2 z_3^2 z_4^{-1} \\
&- \frac{785}{108} z_2^2 z_4^2 + \frac{605}{108} z_1^2 z_4^2 + \frac{407}{18} z_1 z_2 z_3 z_4 \\
&+ \frac{53}{4} z_2^2 z_3^2 - \frac{179}{12} z_1^2 z_3^2 - \frac{595}{54} z_1 z_2 z_3^3 z_4^{-1} \\
&- \frac{65495}{1296} z_2^3 z_4^3 + \frac{40139}{432} z_1^2 z_2 z_4^3 + \frac{11291}{135} z_1 z_2^2 z_3 z_4^2 \\
&- \frac{12472}{405} z_1^3 z_3 z_4^2 + \frac{25295}{216} z_2^3 z_3^2 z_4 + \frac{52267}{1080} z_1^2 z_2 z_3^2 z_4 \\
&- \frac{16307}{135} z_1 z_2^2 z_3^3 - \frac{30626}{405} z_1^3 z_3^3 + \frac{77777}{1296} z_2^3 z_3^4 z_4^{-1} \\
&- \frac{130753}{2160} z_1^2 z_2 z_3^4 z_4^{-1} + O(z^8)].
\end{aligned} \tag{6.7}$$

The computation of the normal form and the normalizing transformation for the Henon–Heiles system up to the order 11 by NORT in the rational arithmetic took about 6 seconds on a Pentium-Pro 200 MHz computer and brought 110 terms of the normal form and 1250 terms of the normalizing transformation. There is a significant difference in the speed of the calculations when different types of numerical coefficients are used. Indeed the normal form was computed till the 19-order terms in the rational arithmetic.

6.3. Local Families of Periodic Solutions of the Henon–Heiles System

The equations (6.6) can be recasted (by eliminating ω which is nonzero for nontrivial solutions) in the form

$$\begin{aligned} P_1 &\stackrel{\text{def}}{=} z_1 z_3 \cdot [g_1(\mathbf{z}) + g_3(\mathbf{z})] = 0\,, \\ P_2 &\stackrel{\text{def}}{=} z_2 z_4 \cdot [g_2(\mathbf{z}) + g_4(\mathbf{z})] = 0\,, \\ P_3 &\stackrel{\text{def}}{=} z_1 z_4 \cdot [g_1(\mathbf{z}) + g_4(\mathbf{z})] = 0\,, \\ P_4 &\stackrel{\text{def}}{=} z_2 z_3 \cdot [g_2(\mathbf{z}) + g_3(\mathbf{z})] = 0\,. \end{aligned} \tag{6.8}$$

Finding all local families of periodic solutions of system (6.5) is equivalent to the determination of all solutions of system (6.8). For each solution of (6.8), the corresponding family of solutions of (6.5) has the form

$$z_j = c_j \exp(-\imath\omega t)\,, \quad z_{j+2} = c_{j+2} \exp(\imath\omega t)\,, \quad j = 1, 2\,, \tag{6.9}$$

because of (6.5) and (6.6). The c_j above are integration constants and ω is the parameter from (6.6) which plays a role of a frequency.

It is important that for any reversible system, both the polynomials P_1 and P_2 in (6.8) have the same factor (see [10] and [11, Chapter 5, §10]):

$$\begin{aligned} P_1(\mathbf{z}) &= (z_1^r z_4^s - z_2^r z_3^s) \cdot Q_1(\mathbf{z})\,, \\ P_2(\mathbf{z}) &= (z_1^r z_4^s - z_2^r z_3^s) \cdot Q_2(\mathbf{z})\,, \end{aligned}$$

where r and s are the smallest positive integers that satisfy the equation $\lambda_1 \cdot r - \lambda_2 \cdot s = 0$.

The Henon–Heiles system is reversible and has $s = r = 1$, but due to an additional symmetry we can find by factorizing the preliminary series calculated by NORT program for P_1, P_2 from (6.7) and (6.8):

$$\begin{aligned} P_1 &= \alpha \cdot (z_1^2 z_4^2 - z_2^2 z_3^2) \cdot [1 + \tfrac{85}{18} z_1 z_3 - \tfrac{59}{18} z_2 z_4 + O(z^4)]\,, \\ P_2 &= -\alpha \cdot (z_1^2 z_4^2 - z_2^2 z_3^2) \cdot [1 + O(z^2)]\,, \end{aligned} \tag{6.10}$$

where $\alpha \neq 0$ is a numerical constant and the last factors have constant terms, i.e. they cannot contribute any new local families of solutions (recall that we are interested only in sets of solutions which include the stationary point $\mathbf{z} = \mathbf{0}$). Thus instead of the first pair of equations in (6.8), we have the equation

$$z_1^2 z_4^2 - z_2^2 z_3^2 = 0\,.$$

The equations $z_1z_4 = \pm z_2z_3$ describe a pair of hypersurfaces

$$h_+ = \{\mathbf{z} : z_1z_4 = z_2z_3\} \quad \text{and} \quad h_- = \{\mathbf{z} : z_1z_4 = -z_2z_3\} \tag{6.11}$$

respectively. For the second pair of equations in system (6.8) we have:

if $z_1z_4 = z_2z_3$, then

$$\begin{aligned} P_3 = &\ \beta \cdot z_1z_4 \cdot (3z_1z_3 - z_2z_4) \cdot (z_1z_3 - 3z_2z_4) \cdot \\ &\cdot [1 + \tfrac{977}{180}(z_1z_3 + z_2z_4) + O(z^4)] \,, \\ P_4 = &-\beta \cdot z_2z_3 \cdot (3z_1z_3 - z_2z_4) \cdot (z_1z_3 - 3z_2z_4) \cdot [1 + O(z^2)] \,; \end{aligned}$$

if $z_1z_4 = -z_2z_3$, then

$$\begin{aligned} P_3 = &\ \gamma \cdot z_1z_4 \cdot (z_1z_3 - z_2z_4) \cdot [1 - \tfrac{23}{18}z_1z_3 + \tfrac{49}{18}z_2z_4 + O(z^4)] \,, \\ P_4 = &-\gamma \cdot z_2z_3 \cdot (z_1z_3 - z_2z_4) \cdot [1 + O(z^2)] \,; \end{aligned}$$

where β and γ are some nonzero numerical constants.

Let us define more hypersurfaces

$$\begin{aligned} h_a &= \{\mathbf{z} : 3z_1z_3 = z_2z_4\} \,, \\ h_b &= \{\mathbf{z} : z_1z_3 = 3z_2z_4\} \,, \\ h_c &= \{\mathbf{z} : z_1z_3 = z_2z_4\} \,, \\ h_i &= \{\mathbf{z} : z_i = 0\}, \qquad i = 1, 2, 3, 4 \,. \end{aligned} \tag{6.12}$$

So there are two branches of solutions of (6.8) corresponding to the intersections $h_+ \cap h_a$ and $h_+ \cap h_b$ and one branch corresponding to the intersection $h_- \cap h_c$. Two more branches of solutions correspond to the intersections of hypersurfaces $h_1 \cap h_2$ and $h_3 \cap h_4$. There are also a couple of branches, which correspond to conjugated essentially imaginary families of solutions with zero energy H and a unit frequency ω; the first branch is the $h_1 \cap h_3$ and the second branch is the $h_2 \cap h_4$. All together these branches exhaust all possible local families of periodic solutions of (6.4) by substituting (6.9), binding constants c_j and calculating the corresponding frequency ω as a series in these constants.

These 7 branches of solutions of system (6.8) produce 10 different local families of periodic solutions of system (6.1) with 4 different frequencies (see (6.14) below):

$$\begin{array}{lll} \text{families 1 \& 1}' & (h_+ \cap h_a): & z_1z_4 = z_2z_3\,, \quad 3z_1z_3 = z_2z_4\,; \\ \text{families 2 \& 2}' & (h_+ \cap h_b): & z_1z_4 = z_2z_3\,, \quad z_1z_3 = 3z_2z_4\,; \\ \text{families 3 \& 3}' & (h_- \cap h_c): & z_1z_4 = -z_2z_3\,, \quad z_1z_3 = z_2z_4\,; \\ \text{family 4} & (h_2 \cap h_4): & z_2 = z_4 = 0\,; \\ \text{family 5} & (h_1 \cap h_3): & z_1 = z_3 = 0\,; \\ \text{family 6} & (h_1 \cap h_2): & z_1 = z_2 = 0\,; \\ \text{family 7} & (h_3 \cap h_4): & z_3 = z_4 = 0\,. \end{array} \tag{6.13}$$

Some of these families of solutions were presented in [13]. A duplication of the first three families originates from the symmetry of the original system (6.1) with respect to a time reversion. This reversion changes signs at $\dot{x}$ and $\dot{y}$ there and so doubles the number of families. But time reversion only permutes families 6 and 7 and it is equivalent to a trivial time shift for families 4 and 5.

Using representation (6.9) we can rewrite the table above via integration constants $c_1, \ldots, c_4$ by simple changing $z_i \to c_i$ there. If we are interested in families of real solutions of (6.4), then we should choose in (6.9) $c_1 = \bar{c}_3$ and $c_2 = \bar{c}_4$. However, in the resonant case, more conditions are needed to guarantee real solutions. The first pair of equations in (6.8) are the critical conditions on the phases of the c_i. Because the system (6.1) is autonomous, the solutions contain a common phase shift that can be neglected, and then all constants can be chosen real (or sometimes purely imaginary). After that, each solution of (6.13) (except the essentially complex families 6 and 7 which are determined by a complex constant) determines only 3 of the 4 real constants c_i. So each real family of solutions of (6.9) depends on a single constant.

Using the determined values of c_i via, say, c_1 (for real parametric families 1–5), we substitute them into the system (6.9) and get the corresponding value of the frequency ω as a series of c_1 via (6.6). To find the corresponding families of periodic solutions of (6.1) as truncated series in this constant, we substitute (6.9) into the previously calculated normalizing transformation (3.2) and into the transformation (6.3). Finally the constant c_1 can be fixed by a substitution of the solution (6.1) into the expression for the energy H from (6.2). The energy (which will not depend on time!) is then as a series in one constant (say, c_1), so by inverting this series, the c_1 can be found as series in the energy and then it can be eliminated from all other results. For the real case, the solutions are in the form of a truncated Fourier series.

This yields approximate frequencies for the families of periodic solutions of (6.1) as function of mechanical energy (6.2) which are

$$\begin{aligned}
\omega_{1,4} &= 1 - \tfrac{5}{6}H + \tfrac{17}{48}H^2 + \tfrac{127517}{38880}H^3 + \tfrac{51952319}{3732480}H^4 + \tfrac{1675438657}{111974400}H^5\,, \\
\omega_{2,5} &= 1 - \tfrac{5}{6}H - \tfrac{95}{48}H^2 - \tfrac{54935}{7776}H^3 - \tfrac{22030445}{746496}H^4 - \tfrac{200207485}{1492992}H^5\,, \\
\omega_{3} &= 1 + \tfrac{1}{3}H - \tfrac{2}{3}H^2 + \tfrac{5389}{2430}H^3 - \tfrac{52393}{5832}H^4 + \tfrac{29471957}{729000}H^5\,, \\
\omega_{6,7} &= \pm 1\,,
\end{aligned} \tag{6.14}$$

where the indices on ω correspond to the above numbered branches of solutions. For each frequency the corresponding approximation for a family of periodic solutions is a Fourier series in time and power series in H which are given in [16].

These results were verified in two ways. The first way was a direct substitution of the approximating series into the original equations (6.1). The results contained only terms with negligible orders. The second way was a comparison of the numerical solution ($x_{\rm num}(t)$, $y_{\rm num}(t)$, $\dot{x}_{\rm num}(t)$, $\dot{y}_{\rm num}(t)$) of equations (6.1) computed using the Runge–Kutta method with the tabulated values of the approximate solutions ($x_{\rm app}(t)$, $y_{\rm app}(t)$, $\dot{x}_{\rm app}(t)$, $\dot{y}_{\rm app}(t)$) computed using preliminary calculated formulae. The solutions were computed at the values of energy $H = \frac{1}{24}$, $H = \frac{1}{12}$

and $H = \frac{1}{8}$. These values were chosen from the original paper [30]. The error was computed using the relative mean-square error

$$f_{\text{err}} \stackrel{\text{def}}{=} \max_{t\in[0,2\pi/\omega_i]} \sqrt{\frac{(x_{\text{num}}(t)-x_{\text{app}}(t))^2+(y_{\text{num}}(t)-y_{\text{app}}(t))^2+(\dot{x}_{\text{num}}(t)-\dot{x}_{\text{app}}(t))^2+(\dot{y}_{\text{num}}(t)-\dot{y}_{\text{app}}(t))^2}{x^2_{\text{num}}(t)+y^2_{\text{num}}(t)+\dot{x}^2_{\text{num}}(t)+\dot{y}^2_{\text{num}}(t)}}\,.$$

The results of the numerical comparison (i.e. values of f_{err}) are

	$H = 1/24$:	$H = 1/12$:
for solutions with ω_1 :	1.3×10^{-5}	7.6×10^{-4},
for solutions with ω_2 :	4.7×10^{-5}	4.4×10^{-3},
for solutions with ω_3 :	1.1×10^{-5}	5.0×10^{-4},
for solutions with ω_4 :	1.3×10^{-5}	7.6×10^{-4},
for solutions with ω_5 :	2.9×10^{-5}	1.8×10^{-3}.

Because for the value $H = 1/6$ system (6.1) has a chaotic regime [30], the values $1/24$ and $1/12$ are not physically small values. For the value $H = 1/8$ the maximal corresponding relative mean-square error reaches 10 percents.

We did not numerically check the complex families 6 and 7, which have the special type. Such families exist for any system, which has at least one imaginary eigenvalue [14].

In Figure 1, the intersections of periodic solutions of the Henon–Heiles system (6.1) with Surface Of Section (SOS), which is defined by

$$\text{SOS} = \left\{ x = 0,\ \dot{x} \geq 0,\ H = \frac{1}{12} \right\},$$

are displayed in the coordinates y, $\dot{y}$. Since [30] this surface is often used for the Poincaré mapping. The periodic solution of family 5 lies entirely in the plane $x = \dot{x} = 0$ and bounds one of two domains in the SOS where $\dot{x} \geq 0$. The other local families cross this surface in a single point each, except of the absent at nonzero energy complex families 6 and 7.

It is interesting to compare the Gustavson first integral [27] with the partial integral of motion

$$I \stackrel{\text{def}}{=} z_1 z_3 + z_2 z_4\,,$$

whose derivation is

$$\dot{I} = z_1 z_3 \cdot (g_1 + g_3) + z_2 z_4 \cdot (g_2 + g_4) = P_1 + P_2\,.$$

This derivation is zero along local families of the periodic solutions above.

The power series I is the formal integral till the seventh order terms including. It is also interesting that the derivation $\dot{I}$ computed till the order $n+8, n = 0, \ldots$ can be factorized [13]:

$$\dot{I} = S_n(\mathbf{z}) \cdot (z_1^2 z_4^2 - z_2^2 z_3^2) \cdot [3(z_1 z_4 - z_2 z_3)^2 - (z_1 z_3 + z_2 z_4)^2]\,,$$

where $S_n(\mathbf{z})$ is a polynomial of order l with a nonzero constant term. The existence of the first factor $(z_1^2 z_4^2 - z_2^2 z_3^2)$ follows from (6.10). The factorization is true at

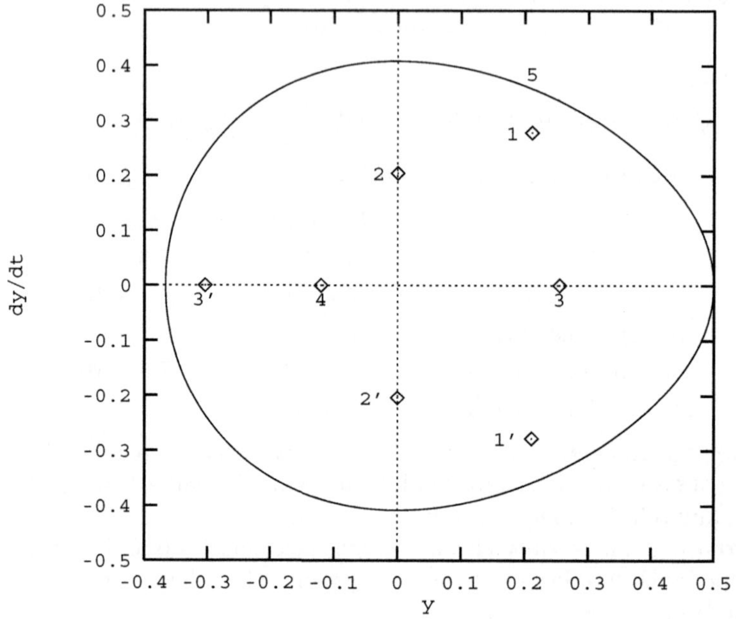

FIGURE 1. Intersections of periodic solutions of the Henon–Heiles system with SOS at the energy level $H = 1/12$.

least till $n = 12$, so we can suppose that the factorized form above takes place in all orders. In this representation it is clear that the derivation above is zero on $h_+ \cup h_-$, $h_3 \cap h_4$, $h_1 \cap h_2$, $h_2 \cap h_4$ and on $h_1 \cap h_3$. So the above I is a constant in time along all periodic families of solutions indeed.

6.4. Visualization of Searching Periodic Solutions

Another very important consequence of the general theory [7] is that if the set A is analytic (i.e. it has a basis which consists of convergent series) then it has such basis not only in the normalized coordinates $\mathbf{z}$, but in coordinates of the original phase space also. So families of solutions of (6.6) can be sought in the original coordinates $x, y, \dot{x}, \dot{y}$.

There are two ways of using this phenomenon, one of which will be discussed in the next section. For the other, rewrite (6.8) into the original coordinates by inverting the transformations (3.2). We will also need to invert the transformations (6.3). This can always be done as this transformation is quasi-identical, but the inversion is computationally lengthy. This will produce functions $z_1(x, y, \dot{x}, \dot{y}), \ldots, z_4(x, y, \dot{x}, \dot{y})$ and then the system (6.6) can be rewritten using the functions z_i. This will produce a system of formal power series equations in the original coordinates, whose solutions describe all local families of periodic solutions

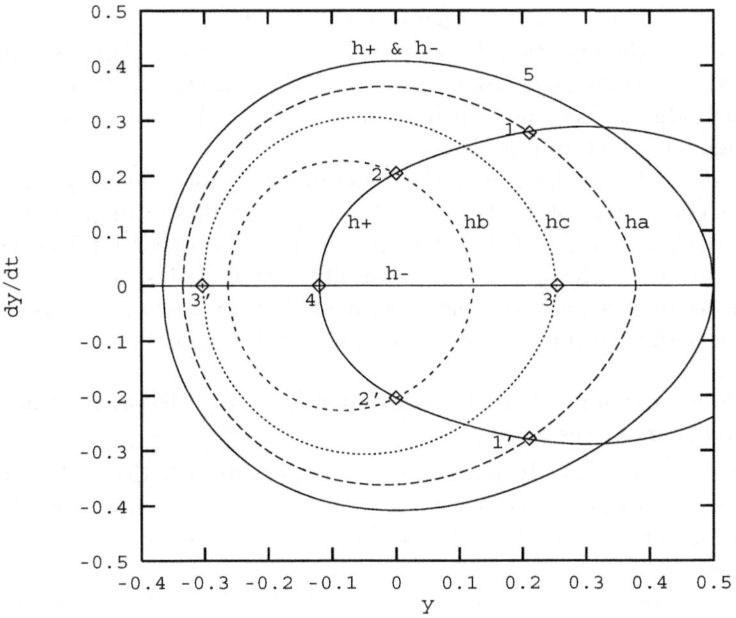

FIGURE 2. Intersections of the SOS with hypersurfaces h_+, h_-, h_a, h_b, h_c and with periodic solutions of the families 1, 1′, 2, 2′, 3, 3′, 4, 5 of the Henon–Heiles system at the energy level $H = 1/12$.

of the system as convergent series. This system (and its solutions), which does not depend explicitly on time, describes the trajectories of periodic solutions.

Rather than evaluating these implicitly defined functions, for a simple analysis it is sufficient to draw each one. For an appropriate visualization, we use the hypersurface SOS, described in the discussion of Figure 1. In Figure 1 we have results of a tabulation of the formulae. In Figure 2 we plot hypersurfaces from (6.11) and (6.12) themselves. In other words if we know the reversed normalizing transformation $\mathbf{z}(x, \dot{x}, y, \dot{y})$ we can rewrite system (6.8) in the original coordinates, and solve it on some surface graphically. For the Henon–Heiles system we have P_i in the factorized form (6.10), so in Figure 2 we plot the intersections of the corresponding surfaces h_+ and h_- with h_a, h_b and h_c. It reproduces points of solutions of families $1, 1', 2, 2', 3$ and $3'$. Note that the intersection $h_+ \cap h_-$ reproduces a point of the solution of family 4 and the whole solution of family 5 from (6.13).

This method can be used if we cannot solve the conditions (3.9) in analytic form. Corresponding hypersurfaces can be tabulated numerically and plotted. In Figure 2, we drew hypersurfaces which are a result of some analysis, i.e. defined by (6.13). But if we draw the hypersurfaces $P_1 = 0$ or $P_2 = 0$ in the original coordinates, then in the SOS we would see curves which look like to $h_+ \cup h_-$, but from the hypersurfaces $P_3 = 0$ and $P_4 = 0$, we would see only distinct points which correspond to the cross sections of periodic orbits with the base surface. This means that all roots of the equation $P_3 = 0$ (or $P_4 = 0$) at other points of this surface are not real. Of course, the cross section of all hypersurfaces $P_1 = 0, \ldots, P_4 = 0$ gives all periodic trajectories. Their plots give a useful graphic representation of some elements of the phase portrait.

We treated in the same way the generalized Henon–Heiles system as a case of a parametric system of fourth order [18]. Some families of its periodic solutions exist only at fixed values of a system parameter, and other families exist in an interval of its value. This is an example of bifurcation analysis by the normal form method. It is remarkable that the system has an additional nontrivial complex family of periodic solutions at one fixed value of the parameter.

6.5. One More Example of the Fourth Order System — Resonant and Non-resonant Cases

In this subsection we study local families of periodic solutions of a Hamilton's system of ordinary differential equations with a cubic nonlinearity [19]. The system appears from the water-wade problem after its reduction to a model system. In a neighborhood of a stationary point we study the system by means of its normal form. We have found the local families of periodic solutions. We demonstrate the importance of a separate investigation of a resonant behavior at corresponding values of parameters.

Let us consider a system with the Hamiltonian

$$H = x_2 y_1 - x_1 y_2 + \frac{1}{2} y_2^2 + \frac{\alpha}{2} x_1^2 + \frac{\beta}{3} x_1^3 - \frac{1}{4} x_1^4 .$$

It is a system of four ordinary differential equations

$$\begin{aligned} \dot{x_1} &= x_2 \,, \\ \dot{x_2} &= y_2 - x_1 \,, \\ \dot{y_1} &= y_2 - \alpha x_1 - \beta x_1^2 + x_1^3 \,, \\ \dot{y_2} &= -y_1 \,. \end{aligned} \tag{6.15}$$

This system is invertible with respect to the involution $(x_1, x_2, y_1, y_2) \to (x_1, -x_2, -y_1, y_2)$. The origin $(0, 0, 0, 0)$ is a stationary point, and the eigenvalues are

$$\{-\sqrt{-1-\sqrt{\alpha}},\ \sqrt{-1-\sqrt{\alpha}},\ -\sqrt{-1+\sqrt{\alpha}},\ \sqrt{-1+\sqrt{\alpha}}\} . \tag{6.16}$$

Below we discuss the cases with positive α. In these cases we have at least one couple of purely imaginary eigenvalues.

Case $\alpha > 1$

In this case only the first couple of eigenvalues is purely imaginary. Let

$$\alpha \to (\omega_0^2 - 1)^2 , \quad \omega_0^2 > 2 ;$$

then the eigenvalues will be

$$\{-i\omega_0,\, i\omega_0,\, -\sqrt{\omega_0^2 - 2},\, \sqrt{\omega_0^2 - 2}\}\,. \tag{6.17}$$

After normalization we have a system in the new coordinates (z_1, z_2, z_3, z_4):

$$\begin{aligned}
\dot{z}_1 &= -\, i\omega_0 z_1 + z_1 P_1(z_1 \cdot z_2, z_3 \cdot z_4) \stackrel{\text{def}}{=} \psi_1\,,\\
\dot{z}_2 &= \quad i\omega_0 z_2 + z_2 P_2(z_1 \cdot z_2, z_3 \cdot z_4) \stackrel{\text{def}}{=} \psi_2\,,\\
\dot{z}_3 &= -\, \sqrt{\omega_0^2 - 2} z_3 + z_3 P_3(z_1 \cdot z_2, z_3 \cdot z_4) \stackrel{\text{def}}{=} \psi_3\,,\\
\dot{z}_4 &= \quad \sqrt{\omega_0^2 - 2} z_4 + z_4 P_4(z_1 \cdot z_2, z_3 \cdot z_4) \stackrel{\text{def}}{=} \psi_4\,.
\end{aligned} \tag{6.18}$$

Here $P_1, \ldots, P_4$ are series calculated with the Mathematica package till the third order in z_i variables. Note that the series P_i in the right-hand side depends on products $z_1 \cdot z_2$ and $z_3 \cdot z_4$ only.

The periodicity condition requires that the local periodic families of solutions should satisfy the condition A. We have nonzero real parts in eigenvalues λ_3, λ_4, so condition (3.9) requires that $z_3 = z_4 = 0$. But the calculated values P_1 and P_2 are such that $P_1(z_1 \cdot z_2, 0) = -P_2(z_1 \cdot z_2, 0)$; therefore one can see that with respect to (6.18) the product $z_1 \cdot z_2$ is constant and we have a one-parametric[1] family of periodic solutions of (6.18):

$$z_1 = \mu \exp(-i \cdot \omega \cdot t)\,, \quad z_2 = \mu \exp(i \cdot \omega \cdot t)\,, \quad z_3 = z_4 = 0\,, \tag{6.19}$$

where the frequency is $\omega = \omega_0 + i \cdot P_1(\mu^2, 0)$ and μ is a real constant. Calculation gives for the first in μ terms

$$\omega = \omega_0 + \mu^2 \frac{\beta^2(20 - 58\omega_0^2) - 9\omega_0^2(\omega_0^2 - 2)(5\omega_0^2 - 2)}{12\omega_0^3(\omega_0^2 - 1)^3(\omega_0^2 - 2)(5\omega_0^2 - 2)} + O(\mu^4)\,. \tag{6.20}$$

The first order[2] solution for (x_1, x_2, y_1, y_2) is

$$(-2\mu \frac{\cos(\omega t)}{\omega_0^2 - 1},\, 2\mu\omega_0 \frac{\sin(\omega t)}{\omega_0^2 - 1},\, 2\mu\omega_0 \sin(\omega t),\, 2\mu \cos(\omega t)) \tag{6.21}$$

(we get it by substituting the z_i from above into the normalizing transformation, as in the previous section [17]).

So we have two external ($\alpha = (\omega_0^2 - 1)^2$ and β), one internal parameter (the constant of integration μ), and a trivial time shift.

Case $0 < \alpha < 1$

In this case all eigenvalues are purely imaginary, i.e., $\omega_0^2 < 2$, and the eigenvalues are

$$\{-i\omega_0,\, i\omega_0,\, -i\sqrt{2 - \omega_0^2},\, i\sqrt{2 - \omega_0^2}\}\,. \tag{6.22}$$

We now have two subcases: the case when eigenvalues are not comparable and the resonant case.

[1] We omit a trivial time shift as a parameter.

[2] Higher order terms in μ are too large for printing.

Case when eigenvalues are not comparable

It means that the fraction $\lambda_1/\lambda_3 = \omega_0/\sqrt{2-\omega_0^2}$ is not a rational number. The normalized equation will have the same form as (6.18), but with all purely imaginary eigenvalues. We omit here a conditionally periodic two frequencies case. But periodic families can be here only if we suppose that $z_3 = z_4 = 0$ or $z_1 = z_2 = 0$. If $z_3 = z_4 = 0$ as in the case above, we will have the same frequency (6.20) and the same solution (6.21). So this family exists for $\alpha > 0$.

If $z_1 = z_2 = 0$, we have the other frequency

$$\omega = \sqrt{2-\omega_0^2} + \mu^2 \frac{\beta^2(96-58\omega_0^2) - 9\omega_0^2(\omega_0^2-2)(5\omega_0^2-8)}{12\omega_0^2(\omega_0^2-1)^3(2-\omega_0^2)^{3/2}(5\omega_0^2-8)} + O(\mu^4)\,, \tag{6.23}$$

and the first approximation to solutions

$$(2\mu\frac{\cos(\omega t)}{\omega_0^2-1},\ -2\mu\sqrt{2-\omega_0^2}\frac{\sin(\omega t)}{\omega_0^2-1},\ 2\mu\sqrt{2-\omega_0^2}\sin(\omega t),\ 2\mu\cos(\omega t))\,. \tag{6.24}$$

But one can see poles in expressions for a frequency above. There are poles at $\omega_0^2 = 0, 2/5, 1, 2, 8/5$, or correspondingly at $\alpha = 1, 9/25, 0, 1, 9/25$. Let us see which eigenvalues correspond to these values:

$$\begin{array}{ll} \omega_0^2 = 0, 1, 2 & \text{— Jordan's matrix in the linear part}\,, \\ \omega_0^2 = 2/5, 8/5 & \text{— Resonant case of the linear part;} \\ \{\lambda_i\} = \{-i\sqrt{\frac{2}{5}},\ i\sqrt{\frac{2}{5}},\ -2i\sqrt{\frac{2}{5}},\ 2i\sqrt{\frac{2}{5}}\}\,. & \end{array} \tag{6.25}$$

So there are domains in ω_0 (or α) where series are slowly convergent and can loose the sense. It is related to the characteristics of a linear part. We omit here the case of Jordan's form of a linear part and study the resonant case.

Resonant case

At $\omega_0^2 = 2/5, 8/5$ all eigenvalues are comparable. In our case we have the third order resonance $(1:2)$, i.e. $\lambda_1/\lambda_3 = 1/2$. The order of resonance is defined as the sum of a numerator and a denominator of this fraction. After normalization in this case we have the system

$$\begin{aligned} \dot{z_1} &= -\ i\sqrt{\tfrac{2}{5}}z_1 + z_1 P_1(z_1, z_2, z_3, z_4)\,, \\ \dot{z_2} &= \quad i\sqrt{\tfrac{2}{5}}z_2 + z_2 P_2(z_1, z_2, z_3, z_4)\,, \\ \dot{z_3} &= -\ 2i\sqrt{\tfrac{2}{5}}z_3 + z_3 P_3(z_1, z_2, z_3, z_4)\,, \\ \dot{z_4} &= \quad 2i\sqrt{\tfrac{2}{5}}z_4) + z_4 P_4(z_1, z_2, z_3, z_4)\,. \end{aligned} \tag{6.26}$$

We can choose its solution in the form

$$\begin{array}{ll} z_1 = a\exp(-i\cdot\omega\cdot t)\,, & z_2 = a\exp(i\cdot\omega\cdot t)\,, \\ z_3 = (\mu - ic)\exp(-2i\cdot\omega\cdot t)\,, & z_4 = (\mu + ic)\exp(2i\cdot\omega\cdot t) \end{array} \tag{6.27}$$

(we fix below a trivial time shift by choosing constant a in z_1, z_2 as a purely real).

After substitution of these variables into condition (3.9) (4 equations here, the same as in [16], and Henon–Heiles' system above) we get firstly the condition $a \cdot c = 0$. It has two solutions.

Subcase $c = 0$

The rest of the periodicity condition (3.9) gives connection between values a and μ. Resolving this connection we have for the frequency

$$\omega = \sqrt{\frac{2}{5}} + \sqrt{\frac{5}{2}}(\frac{25}{18}\beta\mu - \frac{125}{6}\mu^2 + \frac{30625}{864}\beta^2\mu^2) + O(\mu^4)\,. \tag{6.28}$$

Subcase $a = 0$

The frequency is

$$\omega = \sqrt{\frac{2}{5}} + \sqrt{\frac{5}{2}}\frac{125}{31104}\mu^2(455\beta^2 - 216) + O(\mu^4)\,. \tag{6.29}$$

One can see that the values of frequency have no singularities at resonant studying. The solutions have no poles too. The general recommendation is: one should separately study the resonances of orders, smaller than an order of normal form used for an analysis.

So for the local families of periodic solutions of the cubic system above with two external parameters we can conclude that there are two domains in one of the parameters. In the first domain we found one family of periodic solutions with one internal parameter, and for the second domain there are two families with one internal parameter. It is shown that low-level resonances should be studied separately.

7. Conclusions

Here we can conclude that the obtaining of high order normal forms enables us to produce closed formulas for a quantitative approximation of periodic solutions of autonomous nonlinear ODEs.

The normal form method can also be successfully applied to bifurcation analysis and to a phase portrait investigation.

We also used this method for investigating conditionally periodic solutions in the double pendulum system [22] and for the evaluation of cyclicity in planar cubic systems [15] in connection with Hilbert's sixteenth problem.

References

[1] Andersen, G. M. and Geer, J. F., Power series expansions for the frequency and period of the limit cycle of the van der Pol equation. *SIAM J. Appl. Math.* **42** (1983), 678–693.

[2] Arnold, V. I. and Anosov, D. V. (eds.), *Dynamical Systems I. Encyclopaedia of Mathematical Sciences*. Springer-Verlag, New York, 1988.

[3] Bibikov, Yu. N., *Local Theory of Nonlinear Analytic Ordinary Differential Equations*. LNM **702**, Springer-Verlag, New York, 1979.

[4] Boege, W., Gebauer, R. and Kredel, H., Some examples for solving systems of algebraic equations by calculating Groebner bases. *J. Symb. Comput.* **1** (1986), 83–98.

[5] Bruno (Brjuno), A. D., Analytical form of differential equations I. *Trans. Mosc. Mat. Soc.* **25** (1971), 131–288.

[6] Bruno (Brjuno), A. D., Analytical form of differential equations II. *Trans. Mosc. Mat. Soc.* **26** (1972), 199–239.

[7] Bruno, A. D., *Local Method in Nonlinear Differential Equations.* Part I – The Local Method of Nonlinear Analyses of Differential Equations, Part II – The Sets of Analyticity of a Normalizing Transformation. Springer Series in Soviet Mathematics, Springer-Verlag, Berlin New York, 1989.

[8] Bruno, A. D., Bifurcation of the periodic solutions in the case of a multiple pair of imaginary eigenvalues. *Selecta Mathematica* (formerly *Sovietica*) **12**(1) (1993), 1–12.

[9] Bruno, A. D., *The Restricted 3-Body Problem: Plane Periodic Orbits*. De Gruyter Expositions in Mathematics, v. **17**, De Gruyter, Berlin New York, 1994.

[10] Bruno, A. D., Normal forms. *J. Math. Comput. Simul.* **45** (1998), 413–427.

[11] Bruno, A. D., *The Power Geometry in Algebraic and Differential Equations*. Elsevier, North-Holland, 2000.

[12] Deprit, A., Canonical transformation depending on a small parameter. *Celestial Mechanics* **1**(1) (1969), 12–30.

[13] Edneral, V. F., Computer generation of normalizing transformation for systems of nonlinear ODE. *Proceedings of the 1993 International Symposium on Symbolic and Algebraic Computation* (Kiev, Ukraine, July 1993) (M. Bronstein, ed.), pp. 14–19. ACM Press, New York, 1993.

[14] Edneral, V. F., Complex periodic solutions of autonomous ODE systems with analytical right sides near an equilibrium point. *Fundamentalnaya i Prikladnaya Matematika* **1**(2) (1995), 393–398 (in Russian).

[15] Edneral, V. F., Computer evaluation of cyclicity in planar cubic system. *Proceedings of the 1997 International Symposium on Symbolic and Algebraic Computation* (Hawaii, USA, July 1997) (W. Küchlin, ed.), pp. 305–309. ACM Press, New York, 1997.

[16] Edneral, V. F., A symbolic approximation of periodic solutions of the Henon–Heiles system by the normal form method. *J. Math. Comput. Simul.* **45** (1998), 445–463.

[17] Edneral, V. F., About normal form method. *Proceedings of the Second Workshop on Computer Algebra in Scientific Computing* (CASC '99, Munich, Germany, 1999) (Ganzha et al., eds.), pp. 51–66. Springer-Verlag, Berlin Heidelberg, 1999.

[18] Edneral, V. F., Bifurcation analysis of low resonant case of the generalized Henon–Heiles system. *Proceedengs of the Fourth Workshop on Computer Algebra in Scientific Computing* (CASC 2001, Konstanz, Germany, 2001) (Ganzha et al., eds.), pp. 167–176. Springer-Verlag, Berlin Heidelberg, 2001.

[19] Edneral, V. F., Periodic solutions of a cubic ODE system. *Proceedings of the Fifth Workshop on Computer Algebra in Scientific Computing* (CASC 2003, Passau, Germany, September 20–26, 2003) (Ganzha et al., eds.), pp. 77–80. Tech. Univ. München, Munich, 2003.

[20] Edneral, V. F. and Khanin, R., Multivariate power series and normal form calculation in Mathematica. *Proceedings of the Fifth Workshop on Computer Algebra in Scientific Computing* (CASC 2002, Big Yalta, Ukraine, September 2002) (Ganzha et al., eds.), pp. 63–70. Tech. Univ. München, Munich, 2002.

[21] Edneral, V. F. and Khanin, R., Application of the resonant normal form to high order nonlinear ODEs using MATHEMATICA. *Nuclear Inst. and Methods in Physics Research* A **502**(2–3) (2003), 643–645.

[22] Edneral, V. F. and Khanin, R., Investigation of the double pendulum system by the normal form method in MATHEMATICA. *Programming and Computer Software* **30**(2) (2004), 115–117. Translated from *Programmirovanie* **30**(2).

[23] Edneral, V. F. and Khrustalev, O. A., The normalizing transformation for nonlinear systems of ODEs. The realization of the algorithm. *Proceedings of International Conference on Computer Algebra and its Application in Theoretical Physics* (USSR, Dubna, September 1985), pp. 219–224. JINR Publ., Dubna, 1985 (in Russian).

[24] Edneral, V. F. and Khrustalev, O. A., Program for recasting ODE systems in normal form. *Sov. J. Programmirovanie* 5 (1992), 73–80 (in Russian).

[25] Godziewski, K. and Maciejewski, A. J., *Celest. Mech. Dyn. Astron.* **49** (1990), 1.

[26] Guckenheimer, J. and Holmes P., *Nonlinear Oscillations, Dynamical Systems and Bifurcations of Vector Fields*. Springer-Verlag, New York, 1986.

[27] Gustavson, A. G., On constructing formal integrals of a Hamiltonian system near an equilibrium point. *Astronomical J.* **71** (1966), 670–686.

[28] Hassard, B. D., Kazarinoff, N. D. and Wan, Y. H., *Theory and Applications of Hopf Bifurcation*. Cambridge Univ. Press, Cambridge, 1981.

[29] Hearn, A. C., *REDUCE User's Manual.* Rand Publication, CP78, 1987.

[30] Henon, M. and Heiles, C., The applicability of the third integral of motion: Some numerical experiments. *Astronomical J.* **69** (1964), 73–79.

[31] Hori, G. I., Theory of general perturbations with unspecified canonical variables. *J. Japan Astron. Soc.* **18**(4) (1966), 287–296.

[32] Ito, H., Convergence of Birkhoff normal forms for integrable systems. *Comment. Math. Helv.* **64** (1989), 412–461.

[33] Ito, H., Integrability of Hamiltonian systems and Birkhoff normal forms in the simple resonance case. *Math. Ann.* **292** (1992), 411–444.

[34] Marsden, J. E., McCracken, M., *The Hopf Bifurcation and Its Applications.* Springer Applied Math. Series, v. **19**, Springer-Verlag, New York, 1976.

[35] Mersman, W. A., A new algorithm for Lie transformation. *Celestial Mechanics* **3**(1) (1970), 81–89.

[36] Rand, R. and Armbruster, D., *Perturbation Methods, Bifurcation Theory and Computer Algebra*. Springer-Verlag, New York, 1987.

[37] Shevchenko, I. I. and Sokolsky, A. G., Algorithms for normalization of Hamiltonian systems by means of computer algebra. *Comp. Phys. Comm.* **77** (1993), 11–18.

[38] Vallier, L., An algorithm for the computation of normal forms and invariant manifolds. *Proceedings of the 1993 International Symposium on Symbolic and Algebraic Computation* (Kiev, Ukraine, July 1993) (M. Bronstein, ed.), pp. 225–233. ACM Press, New York, 1993.

[39] Walcher, S., On differential equations in normal form. *Math. Ann.* **291** (1991), 293–314.

[40] Walcher, S., On transformations into normal form. *J. Math. Anal. Appl.* **180** (1993), 617–632.

Acknowledgments

The author is very grateful to Prof. Alexandr D. Bruno for important discussions and essential advises.

Victor F. Edneral
Skobeltsyn Institute of Nuclear Physics
Moscow State University
Leninskie Gory, Moscow 119992
Russia
e-mail: edneral@theory.sinp.msu.ru

Trends in Mathematics: *Differential Equations with Symbolic Computation*, 201–212

Algorithmic Reduction and Rational General Solutions of First Order Algebraic Differential Equations

Guoting Chen and Yujie Ma

Abstract. First order algebraic differential equations are considered. A necessary condition for a first order algebraic differential equation to have a rational general solution is given: the algebraic genus of the equation should be zero. Combining with Fuchs' conditions for algebraic differential equations without movable critical point, an algorithm is given for the computation of rational general solutions of these equations if they exist under the assumption that a rational parametrization is provided. It is based on an algorithmic reduction of first order algebraic differential equations with algebraic genus zero and without movable critical point to classical Riccati equations.

Mathematics Subject Classification (2000). Primary 34A09, 68W30; Secondary 14Q05, 34M15.

Keywords. Algebraic differential equation, rational general solution, rational parametrization.

1. Introduction

The study of first order algebraic differential equations can be dated back to C. Briot and T. Bouquet [2], L. Fuchs [11] and H. Poincaré [20]. M. Matsuda [17] gave a modern interpretation of the results using the theory of differential algebraic function field of one variable, and Eremenko [9] studied the bound of the degrees of the rational solutions of a first order algebraic differential equation by using the approach of [17].

From an algorithmic point of view, many authors have been interested in the constructions of closed form solutions for differential equations (this problem

The second author is partially supported by the NKBRSF of China (no. 2004CB318000), the NNSF (no. 10301032) and by a CNRS–K. C. WONG fellowship during his visit to the Laboratoire P. Painlevé, Université de Lille 1, France.

can be traced back to the work of Liouville). In [21], Risch gave an algorithm for finding closed forms for integration. In [14], Kovacic presented a method for solving second order linear homogeneous differential equations. In [26], Singer proposed a method for finding Liouvillian solutions of general linear differential equations. In [15], Li and Schwarz gave a method to find rational solutions for a class of partial differential equations. All these works are limited to linear cases.

For algebraic (nonlinear) differential equations there are some studies in this direction. For Riccati equations, polynomial solutions are considered in [4] and algorithms for the computation of rational solutions are given in [3, 14]. In [5], the algebraic solutions of a general class of first order and first degree algebraic differential equations were studied and the degree bound of algebraic solutions in the nondicritical case was given. In [10, 16], algorithms for the computation of rational general solutions or polynomial solutions are given for some kinds of algebraic differential equations.

Another motivation of our work is differential algebraic geometry. In the series of papers [29, 30, 31], Wu studied algebraic differential geometry from several different points of view. In [31], the author presents an algorithmic method of solving arbitrary systems of algebraic differential equations by extending the characteristic set method to the differential case. The Devil's problem of Pommaret is solved in detail as an illustration.

In this paper, we consider the computation of rational general solutions of first order algebraic differential equations by using methods from algebraic geometry. We give a necessary condition for a first order algebraic differential equation to have a rational general solution: that is, the algebraic genus of the equation should be zero. Combining with Fuchs' conditions for first order algebraic differential equations without movable critical points, we obtain an algorithm for the computation of rational general solutions under the assumption that a rational parametrization is provided. It is based on an algorithmic reduction of first order algebraic differential equation of algebraic genus zero and without movable critical point to classical Riccati equations.

2. Rational General Solutions of First Order Algebraic Differential Equations

We first present some results from algebraic geometry, which is used in the following.

Let $f(x, y)$ be an irreducible polynomial over $\mathbb{C}$. We say that $f(x, y) = 0$ is a rational curve if there exist two rational functions $\phi(t)$, $\psi(t) \in \mathbb{C}(t)$ such that

(i) for all but a finite set of $t_0 \in \mathbb{C}$, $(\phi(t), \psi(t))$ is a point of f;
(ii) with a finite number of exceptions, for every point (x_0, y_0) of f there is a unique $t_0 \in \mathbb{C}$ such that $x_0 = \phi(t_0)$, $y_0 = \psi(t_0)$.

It is impossible to avoid having a finite number of exceptions in the above conditions. They arise from two sources. One is the fact that a rational function is not defined for some values of the variable, and the other is the presence of singular points on the curve.

The following results are well known in algebraic geometry [28].

Theorem 2.1. *An algebraic curve is rational if and only if its genus is zero.*

Theorem 2.2. (a) *Every rational transform of a rational curve is a rational curve.*

(b) *If λ is transcendental over $\mathbb{C}$ and if $\mathbb{C} \subset F \subset \mathbb{C}(\lambda)$, $F \neq \mathbb{C}$, then there is an element $\mu \in F$, transcendental over $\mathbb{C}$, such that $F = \mathbb{C}(\mu)$.*

(c) *If a curve $f(x,y) = 0$ satisfies* (i) *for rational functions $\phi(\lambda), \psi(\lambda)$ which are not both constants, then there exist rational functions $\tilde{\phi}(\lambda), \tilde{\psi}(\lambda)$ for which both* (i) *and* (ii) *are satisfied, and the curve is rational.*

The three statements in Theorem 2.2 are all equivalent. It is called Lüroth's theorem.

Now we consider a first order algebraic differential equation in the form

$$F(z, w, w') = 0, \tag{2.1}$$

where F is a polynomial in w, w' with rational coefficients in $\mathbb{C}(z)$.

Definition 2.3. A general solution $w(z, \lambda)$ of (2.1) is called a rational general solution if it is rational in z and λ.

If $F(z, w, w') = 0$ admits a rational general solution, then it is free from movable critical point for poles are the only singularities of the solution which change their position if one varies the initial data $c \in \mathbb{C}$.

We now prove the following theorem on rational general solution of a first order algebraic differential equation.

Theorem 2.4. *If a first order irreducible algebraic differential equation*

$$F(z, w, w') = 0$$

admits a non-constant rational general solution, then the genus of $F(z, w, w') = 0$ with respect to w, w' is zero for any z, except for finitely many exceptions.

Proof. Let $w = r(z, \lambda)$ be the rational general solution of $F = 0$ with the arbitrary constant λ. Then $w(z) = r(z, \lambda)$ and $w'(z) = \frac{\partial r}{\partial z}$ are rational functions and they satisfy the equation $F(z, w, w') = 0$.

Let z be fixed and consider the curve $f_z(x, y) = F(z, x, y) = 0$. Denote $\phi_z(\lambda) = w(z)$ and $\psi_z(\lambda) = w'(z)$. If $\psi_z(\lambda)$ is a constant, then $w(z) = zw'(z) + \lambda$ is of genus zero and hence the genus of $f_z(x, y) = 0$ is also zero. If $\psi_z(\lambda)$ is not a constant, consider the point $(\phi_z(\lambda), \psi_z(\lambda))$ of $f_z(x, y) = 0$ for the parameter λ in the transcendental extension field $\mathbb{C}(\lambda)$. It is clear that (i) is satisfied for all but finitely many λ. Hence $f_z(x, y) = 0$ is a rational curve and its genus is zero by Theorems 2.1 and 2.2. □

Motivated by this theorem, we present the following definition.

Definition 2.5. The algebraic genus of a first order algebraic differential equation $F(z, w, w') = 0$ is defined to be the genus of $F(z, w, w') = 0$ with respect to w and w'.

3. Reduction of First Order Algebraic Differential Equations

For a first order algebraic differential equation $F(z, w, w') = 0$, Fuchs' theorem presents necessary conditions for the equation to be free from movable critical point. By the Painlevé theorem, we know that Fuchs' conditions are sufficient (see [11, 13, 18, 20]).

3.1. Fuchs Theorem

Let $D(z, w)$ be the p–discriminant of the equation $F(z, w, w') = 0$; it is a polynomial in w, whose coefficients are analytic functions of z [13].

The conditions, necessary to secure that the first order differential equation

$$F(z, w, w') = 0$$

of degree m shall have no movable critical point, are:

1. The coefficient $A_0(z, w)$ is independent of w and therefore reduces to a function of z alone or to a constant. The equation may then be divided throughout by A_0 and takes the form

$$w'^m + \psi_1(z, w)w'^{m-1} + \cdots + \psi_{m-1}(z, w)w' + \psi_m(z, w) = 0$$

in which the coefficients ψ are polynomials in w, and analytic, except for isolated singular points, in z.

2. If $w = \eta(z)$ is a root of $D(z, w) = 0$, and $p = \omega(z)$ is a multiple root of $F(z, \eta, \eta') = 0$, such that the corresponding root of $F(z, w, w') = 0$ regarded as a function of $w - \eta(z)$ is branched, then

$$\omega(z) = \frac{d\eta}{dz}.$$

3. If the order of any branch is α, so that the equation is effectively of the form

$$\frac{d}{dz}\{w - \eta(z)\} = c_k\{w - \eta(z)\}^{\frac{k}{\alpha}},$$

then $k \geq \alpha - 1$.

3.2. Reduction to Classical Riccati Equation

Consider now a first order algebraic differential equation $F(z, w, w') = 0$ of genus zero and without movable critical point. One can find a parametrization of the rational curve $F(z, x, y) = 0$ in the form $x = r_1(t, z)$ and $y = r_2(t, z)$ with $r_1(t, z)$ and $r_2(t, z)$ rational functions in t and z. By the inversion of rational curves we know that t is rational function in z, x and y. For algorithms on parametrization and inversion of rational curves we refer to [1, 22, 23, 24, 25, 27]. One has

$$\frac{dt}{dz} = \left(r_2(t, z) - \frac{\partial r_1}{\partial z}\right) / \frac{\partial r_1}{\partial t} = \frac{P(t, z)}{Q(t, z)}, \tag{3.1}$$

where P and Q are polynomials in t and z. Since $F(z, w, w') = 0$ has no movable critical point, one knows that (3.1) also has no movable critical point as t is rational function in z, x and y. By Fuchs' theorem, we have that equation (3.1) is a Riccati equation [12, Chapter II, §7], that is

$$\frac{dt}{dz} = A(z)t^2 + B(z)t + C(z), \tag{3.2}$$

where A, B, C are rational functions in z. We distinguish two cases according to $A(z)$.

Case 1. If $A(z) \not\equiv 0$, we consider the change of variables $t(z) = -u(z)/A(z)$. One has

$$u'(z) + u^2 = (B(z) + A'(z)/A(z))u - C(z)A(z),$$

in which the coefficient $A(z)$ is reduced to -1. Next we make the change $u = v + \beta(z)$ to reduce the coefficient of u to zero by choosing an appropriate β. We obtain finally a classical Riccati equation in the form

$$v' + v^2 = r(z) \in \mathbb{C}(z). \tag{3.3}$$

Algorithms for the computation of rational solutions of classical Riccati equations are available in the literature (see for example [3, 14]).

If $r(z) \not\equiv 0$, then a rational solution of equation (3.3) is equivalent to an exponential solution $e^{\int v(z)dz}$ of the linear differential equation

$$y'' = r(z)y. \tag{3.4}$$

Proposition 3.1. *If the Riccati equation* (3.3) *with* $r(z) \not\equiv 0$ *has a general rational solution, then* $r(z)$ *has the form*

$$r(z) = \sum_{i=1}^{m} \left(\frac{\beta_i}{(z - z_i)^2} + \frac{\gamma_i}{(z - z_i)} \right),$$

in which $4\beta_i = n_i^2 - 1$ *where* n_i *is an integer* ≥ 2.

Proof. Suppose that $v(z)$ is a rational solution of equation (3.3). Let $z_1, \ldots, z_m$ be the poles of r. According to Kovacic's algorithm (see [14]), $v(z)$ should be in the form

$$v(z) = \sum_{i=1}^{m} \sum_{j=1}^{\nu_i} \frac{a_{ij}}{(z - z_i)^j} + \sum_{k=1}^{d} \frac{1}{z - c_k} + f(z),$$

where the ν_i are known, a_{ij} are known up to two choices each, d is known, and $f \in \mathbb{C}[z]$ is known up to two choices. Hence there may be an arbitrary parameter only in the determination of the c_k.

Let

$$P(z) = \prod_{k=1}^{d} (z - c_k),$$

and

$$\omega(z) = \sum_{i=1}^{m}\sum_{j=1}^{\nu_i} \frac{a_{ij}}{(z-z_i)^j} + f(z).$$

Then $v = P'/P + \omega$ and $y = e^{\int v} = Pe^{\int \omega}$ is a solution of the linear differential equation (3.4). Hence P is a polynomial solution of degree d of the following linear equation:

$$P'' + 2\omega P' + (\omega' + \omega^2 - r)P = 0. \tag{3.5}$$

One can determine whether it has a general polynomial solution or not.

Furthermore if (3.3) admits a rational general solution, then writing $r(z) = p(z)/q(z)$, according to [32], one has

(a) $\deg(p) - \deg(q) \le -2$;
(b) $r(z)$ has only double poles, and hence

$$r(z) = \sum_{i=1}^{m} \left(\frac{\beta_i}{(z-z_i)^2} + \frac{\gamma_i}{(z-z_i)} \right),$$

in which $4\beta_i = n_i^2 - 1$ where n_i is an integer ≥ 2. □

Therefore a possible rational solution of equation (3.3) with $r(z)$ as in the proposition should be

$$v(z) = \sum_{i=1}^{m} \frac{a_i}{z-z_i} + \sum_{k=1}^{d} \frac{1}{z-c_k},$$

where $a_i^2 - a_i - \beta_i = 0$. Hence to determine a rational general solution one needs to compute polynomial solutions of equation (3.5) in order to determine the c_k.

Case 2. If $A(z) \equiv 0$, then one can integrate easily the linear equation

$$t' = B(z)t + C(z)$$

to get the general solution

$$t(z) = \left(\int C(z)dz + \lambda \right) e^{\int B(z)dz},$$

where λ is an arbitrary constant. An effective algorithm is given in [21] for integration in closed forms. One may find rational solutions in this way. It is clear that in this case one may get rational general solutions only if both $\int C(z)dz$ and $e^{\int B(z)dz}$ are rational functions.

3.3. First Order Algebraic Differential Equations with Constant Coefficients

As an application, we consider a first order algebraic differential equation with constant coefficients:

$$F(w, w') = 0. \tag{3.6}$$

This kind of equations was systematically studied in [2] and an algorithm is given in [10] for the determination of the rational general solution of the equation if it exists.

When using the above reduction for equation (3.6), it is clear that z is not involved in the equation, and we henceforth get a Riccati equation with constant coefficients.

As above there are two cases to be considered. In case 1, if equation (3.6) has a non-constant rational solution $w(z)$ then $w(z+\lambda)$ is a general rational solution. Since the equation $u' + u^2 = c$ for a constant $c \neq 0$ does not have a rational solution, then one can reduce equation (3.6) to the equation $u' + u^2 = 0$; in this case we have the general solution

$$u = \frac{1}{z + \lambda}.$$

In case 2, the equation can be converted to $u' = bu + c$ with constant b, c. Then $u = \lambda e^{bz} - \frac{c}{b}$ if $b \neq 0$, and $u = cz + \lambda$ if $b = 0$, where λ is an arbitrary constant.

Summarizing the above, we then have the following

Corollary 3.2. *Let $F(w, w') = 0$ be a first order irreducible algebraic differential equation with constant coefficients. Then it has a non-constant rational general solution if and only if it can be reduced either to a linear equation $u' = c$ for some constant c or to a Riccati equation of the form $u' + u^2 = 0$.*

Example 1. Consider

$$F(y, y') = y'^4 - 8\,y'^3 + (6 + 24\,y)\,y'^2 + 257 + 528\,y^2 - 256\,y^3 - 552\,y.$$

This example comes from [10]. One finds by computation in Maple that its algebraic genus is zero and it has the following rational parametrization:

$$y = \frac{17}{16} - 27t + \frac{2187}{2}t^2 + 531441t^4, \quad y' = 78732t^3 + 81t - 1.$$

And the corresponding Riccati equation is $t'(z) = \frac{1}{27}$. Hence $t = \frac{1}{27}z + \lambda$ and the general solution of the differential equation $F(y, y') = 0$ is

$$y(z) = \frac{17}{16} - 27(\frac{1}{27}z + \lambda) + \frac{2187}{2}(\frac{1}{27}z + \lambda)^2 + 531441(\frac{1}{27}z + \lambda)^4,$$

where λ is an arbitrary constant.

4. Algorithm and Example

We can now give the following algorithm on seeking for a rational general solution of a first order algebraic differential equation.

Algorithm. *Input:* A first order algebraic differential equation $F(z,w,w')=0$.

Output: A rational general solution of $F(z,w,w')=0$ if it exists.

1. Determine the irreducibility of the equation. If $F(z,w,w')=0$ is reducible, then factorize it and go to step (2) for each branch curve of $F(z,w,w')=0$, else go to step 2 directly.
2. Compute the algebraic genus g of $F(z,w,w')=0$. If $g\neq 0$, then the equation does not admit any rational general solution by Theorem 2.4, else go to step 3.
3. Determine the Fuchs conditions of $F(z,w,w')=0$. If the conditions are not satisfied, then the algorithm terminates, else go to step 4.
4. If a rational parametrization of $F(z,w,w')=0$ in the form $w=r_1(t,z)$ and $w'=r_2(t,z)$ with $r_1(t,z)$ and $r_2(t,z)$ rational functions in t and z is provided, then go to step 5.
5. Compute the derivative
$$\frac{dt}{dz}=(r_2(t,z)-\frac{\partial r_1}{\partial z})/\frac{\partial r_1}{\partial t}$$
which is a Riccati equation of the form (3.2) by the Fuchs theorem.
6. Reduce the above Riccati equation to a classical Riccati equation (3.3) and compute a rational solution using the algorithm in [3, 14].

Remark 4.1. Assume that $F(z,w,w')=0$ has algebraic genus 0 for all z (or for all z with some finite exceptions); in general, the rational parametrization that we obtain using the the algorithm in [1, 24, 25, 27] are not rational functions in z (it could appear that algebraic elements are over $\mathbb{C}(z)$ as $\sqrt{z}$, see [25]). Our algorithm works only with those equations $F(z,w,w')=0$ (for instance, first order algebraic differential equations with constant coefficients) for which a rational parametrization may be provided.

Remark 4.2. Our algorithm for the case of constant coefficients is the same as the classical one, which is described for instance in [19, Chapter IV, Section I, pages 62–65].

Example 2. Consider the following equation

$$F(z,w,w')=w'^2+\frac{2w}{z}w'-4\,zw^3+\frac{\left(1+12\,z^2\right)w^2}{z^2}-12\,\frac{w}{z}+\frac{4}{z^2}.$$

Its algebraic genus is zero. One gets the following rational parametrization by Maple:

$$w = r_1 = \frac{t^2z^2 + 4t^2 - 6tz + 1 + 4z^2}{4z(-z+t)^2},$$

$$w' = r_2 = -\frac{-4z^3 + 13tz^2 + t + 2t^2z^3 - 10t^2z + t^3z^4 + t^3z^2 + 4t^3}{4z^2(-z+t)^3}.$$

One obtains the following Riccati equation

$$t' = \frac{(z^2+2)}{2(z^2+1)}t^2 + \frac{z}{z^2+1}t + \frac{3}{2(z^2+1)}.$$

Continuing the reduction procedure of Section 3.2, one obtains

$$u' + u^2 = Bu + C,$$

where

$$B = -\frac{z^3}{(z^2+1)(z^2+2)} \quad \text{and} \quad C = \frac{3(z^2+2)}{4(z^2+1)^2}.$$

And finally one obtains the following classical Riccati equation

$$v' + v^2 = -6(z^2+2)^{-2},$$

which has a rational general solution as follows:

$$v(z) = -\frac{z}{z^2+2} + \frac{1}{z-\lambda} + \frac{1}{z+2/\lambda}$$

with λ an arbitrary constant. By substitutions one finally has the following solution of the equation $F(z, w, w') = 0$:

$$w(z) = \frac{z^2\lambda^2 - 2\,z\lambda^3 + 4\,z\lambda + 4 + \lambda^4 - 3\,\lambda^2}{(z\lambda + 2 - \lambda^2)^2\, z}.$$

5. Conclusion

In this paper, we present an algebraic geometry approach to the study of first order algebraic differential equations. The necessary and sufficient conditions of the rational general solutions of the Riccati equations are studied in [6]. In [7], we prove that the degree of a rational general solution of a first order algebraic differential equation of degree d is less than or equal to d. The algebraic geometry approach is also used to obtain bound for the number of the rational solutions of a first order algebraic differential equation with algebraic genus greater than one. In [8], the algebraic general solutions of first order algebraic differential equations were studied by using the birational transformations of algebraic curves, and an algorithm was presented to get an algebraic general solution of first order algebraic differential equations without movable critical point if the algebraic general solution exists. It is interesting to present an effective algorithm to find the rational general solutions or the rational solutions of a first order algebraic differential equation.

References

[1] S. S. Abhyankar and C. L. Bajaj, Automatic parameterization of rational curves and surfaces III. Algebraic plane curves. *Comput. Aided Geom. Design* **5** (1988), 309–321.

[2] C. Briot and T. Bouquet, Intégrations des équations différentielles au moyen des fonctions elliptiques. *Journal de l'Ecole Polytechnique* t. XXI.

[3] M. Bronstein, Linear ordinary differential equations: Breaking through the order 2 barrier. In: *Proceedings of ISSAC 1992*, pp. 42–48. ACM Press, New York (1992).

[4] J. G. Campbell, A criterion for the polynomial solutions of a certain Riccati equation. *American Math. Monthly* **61** (1952), 388–389.

[5] M. M. Carnicer, The Poincaré problem in the nondicritical case. *Ann. Math.* **140**(2) (1994), 289–294.

[6] G. Chen and Y. Ma, A computer algebra approach to first order algebraic differential equations. In: *Constructive and Invariant Methods in Algebraic and Differential Equations — Proceedings of the Sixth International Workshop on Mathematics Mechanization* (Shanghai, China, May 19–21, 2004).

[7] G. Chen and Y. Ma, Rational solutions of algebraic differential equations. In: *Proceedings of the Third International Congress of Chinese Mathematicians* (Hong Kong, December 17–22, 2004).

[8] G. Chen and Y. Ma, Algebraic solutions of a first order algebraic differential equation. Preprint, Laboratory of P. Painlevé, France (2004).

[9] A. Eh. Eremenko, Rational solutions of first-order differential equaitons. *Ann. Acad. Sci. Fenn.* **23** (1998), 181–190.

[10] R. Feng and X. Gao, Rational general solutions of algebraic ordinary differential equations. In: *Proceedings of ISSAC 2004*, pp. 155–162. ACM Press, New York (2004).

[11] L. Fuchs, Ueber Differentialgleichungen deren integrale feste verzweigungspunkte besitzen. *Sitzungsberichte Acad. Berlin* (1884).

[12] V. V. Golubev, *Lectures on the Analytic Theory of Differential Equations* (2nd edn.). Gos. Izd. Tekh. Teor. Lit., Moscow Leningrad (1950) (in Russian; Chinese translation and German translation are available).

[13] E. L. Ince, *Ordinary Differential Equations.* Longmans, Green and Co. Ltd., London (1927).

[14] J. J. Kovacic, An algorithm for solving second order linear homogeneous differential equations. *J. Symb. Comput.* **2** (1986), 3–43.

[15] Z. Li and F. Schwarz, Rational solutions of Riccati-like partial differential equations. *J. Symb. Comput.* **31**(6) (2001), 691–716.

[16] Y. Ma and X. Gao, Polynomial solutions of algebraic differential equations. In: *Computer Mathematics*, Lecture Notes Ser. Comput. **9**, pp. 92–101. World Scientific, Singapore River Edge (2001).

[17] M. Matsuda, *First Order Algebraic Differential Equations — A Differential Algebraic Approach*, Lecture Notes in Math. **804**, Springer-Verlag, Berlin (1980).

[18] P. Painlevé, Leçons sur la théorie analytique des équations différentielles (1896).

[19] É. Picard, *Traité D'Analyse*, Tome III (1908).

[20] H. Poincaré, Sur un théorème de M. Fuchs. *Acta Math.* **7** (1885), 1–32. Also in *Oeuvres de Henri Poincaré*, vol. **1**, Gauthiers-Villars et Cie, Paris (1928).

[21] R. H. Risch, The problem of integration in finite terms, *Trans. Amer. Math. Soc.* **139** (1969), 167–189.

[22] T. W. Sederberg, D. C. Anderson and R. N. Goldman, Implicit representation of parametric curves and surfaces, *Computer Vision, Graphics, and Image Processing* **28** (1984), 72–84.

[23] T. W. Sederberg, J. Zheng, Algebraic methods for computer aided geometric design. In: *Handbook of Computer Aided Geometric Design*, pp. 363–387. North-Holland, Amsterdam (2002).

[24] J. R. Sendra and F. Winkler, Symbolic parametrization of curves. *J. Symb. Comput.* **12** (1991), 607–631.

[25] J. R. Sendra and F. Winkler, Parametrization of algebraic curves over optimal field extension. *J. Symb. Comput.* **23** (1997), 191–207.

[26] M. F. Singer, Liouvillian solutions of linear differential equations with Liouvillian coefficients. *J. Symb. Comput.* **11** (1991), 251–273.

[27] M. van Hoeij, Rational parametrizations of algebraic curves using a canonical divisor. *J. Symb. Comput.* **23** (1997), 209–227.

[28] R. J. Walker, *Algebraic Curves*. Princeton University Press, Princeton (1950).

[29] W.-t. Wu, Mechanical derivation of Newton's gravitational laws from Kepler's laws. MM Research Preprints, no. **1** (1987), 53–61.

[30] W.-t. Wu, On the foundation of algebraic differential geometry. MM Research Preprints, no. **3** (1989), 1–26.

[31] W.-t. Wu, On algebrico-differential equations-solving. MM Research Preprints, no. **22** (2003), 1–13.

[32] W. Yuan, A note on the Riccati differential equation. *J. Math. Anal. Appl.* **277** (2003), 367–374.

Acknowledgment

The authors are grateful to the anonymous referees for their valuable comments and suggestions, which improved the manuscript, and for pointing out some inaccuracies.

Guoting Chen
Laboratoire P. Painlevé
UMR CNRS 8524
UFR de Mathématiques
Université de Lille 1
59655 Villeneuve d'Ascq
France
e-mail: `Guoting.Chen@math.univ-lille1.fr`

Yujie Ma
Key Laboratory of Mathematics Mechanization
Chinese Academy of Sciences
Beijing 100080
P. R. China
e-mail: `yjma@mmrc.iss.ac.cn`

Trends in Mathematics: *Differential Equations with Symbolic Computation*, 213–238

Factoring Partial Differential Systems in Positive Characteristic

Moulay A. Barkatou, Thomas Cluzeau and Jacques-Arthur Weil

with an appendix by M. van der Put:
Classification of Partial Differential Modules in Positive Characteristic

Abstract. An algorithm for factoring differential systems in characteristic p has been given by Cluzeau in [7]. It is based on both the reduction of a matrix called p-curvature and eigenring techniques. In this paper, we generalize this algorithm to factor partial differential systems in characteristic p. We show that this factorization problem reduces effectively to the problem of simultaneous reduction of commuting matrices.

In the appendix, van der Put shows how to extend his classification of differential modules, used in the work of Cluzeau, to partial differential systems in positive characteristic.

Mathematics Subject Classification (2000). 68W30, 16S32, 15A21, 16S50, 35G05.

Keywords. Computer algebra, linear differential equations, partial differential equations, D-finite system, modular algorithm, p-curvature, factorization, simultaneous reduction of commuting matrices.

Introduction

The problem of factoring D-finite partial differential systems in characteristic zero has been recently studied by Li, Schwarz and Tsarëv in [21, 22] (see also [28]). In these articles, the authors show how to adapt Beke's algorithm (which factors ordinary differential systems, see [9] or [26, 4.2.1] and references therein) to the partial differential case. The topic of the present paper is an algorithm that factors D-finite partial differential systems in characteristic p. Aside from its theoretical value, the interest of such an algorithm is its potential use as a first step in the construction of a modular factorization algorithm; in addition, it provides useful modular filters, e.g., for detecting the irreducibility of partial differential systems.

T. Cluzeau initiated this work while being a member of Laboratoire STIX, École polytechnique, 91128 Palaiseau Cedex, France.

Concerning the ordinary differential case in characteristic p, factorization algorithms have been given by van der Put in [23, 24] (see also [26, Ch. 13]), Giesbrecht and Zhang in [11] and Cluzeau in [7, 8]. In this paper, we study the generalization of the one given in [7]. Cluzeau's method combines the use of van der Put's classification of differential modules in characteristic p based on the p-curvature (see [23] or [26, Ch. 13]) and the approach of the eigenring factorization method (see [27, 2, 26]) as set by Barkatou in [2].

In the partial differential case, we also have notions of p-curvatures and eigenrings at our disposal, but van der Put's initial classification of differential modules in characteristic p cannot be applied directly, so we propose an alternative algorithmic approach. To develop a factorization algorithm (and a partial generalization of van der Put's classification) of D-finite partial differential systems, we rebuild the elementary parts from [7, 8] (where most proofs are algorithmic and independent of the classification) and generalize them to the partial differential context.

In the appendix, van der Put develops a classification of "partial" differential modules in positive characteristic which sheds light on our developments, and comes as a good complement to the algorithmic material elaborated in this paper.

We follow the approach of [7], that is, we first compute a maximal decomposition of our system before reducing the indecomposable blocks. The decomposition phase is separated into two distinct parts: we first use the p-curvature to compute a *simultaneous decomposition* (using a kind of "isotypical decomposition" method), and then, we propose several methods to refine this decomposition into a maximal one.

The generalization to the partial differential case amounts to applying simultaneously the ordinary differential techniques to several differential systems. Consequently, since in the ordinary differential case we are almost always reduced to performing linear algebra on the p-curvature matrix, our generalization of the algorithm of [7] relies on a way to reduce simultaneously commuting matrices (the p-curvatures).

A solution to the latter problem has been sketched in [8]; similar ideas can be found in papers dealing with numerical solutions of zero-dimensional polynomial systems such as [10]. The essential results are recalled (and proved) here for self-containedness.

The paper is organized as follows. In the first part, we recall some definitions about (partial) differential systems and their factorizations. We then show how to generalize to the partial differential case some useful results concerning p-curvatures, factorizations and rational solutions of the system: we generalize the proofs given in [7, 8]. After a section on simultaneous reduction of commuting matrices, the fourth part contains the factorization algorithms. Finally, in Section 5, we show how the algorithm in [7] can be directly generalized (with fewer efforts than for the partial differential case) to other situations: the case of "local" differential systems and that of difference systems.

1. Preliminaries

In this section, we recall some classical definitions concerning differential systems in several derivations. When there is only one derivation ($m = 1$ in what follows), we recover the ordinary definitions of differential field, ring of differential operators, ... We refer to [26, Ch. 2 and Ap. D] for more details on all these notions.

1.1. *D*-Finite Partial Differential Systems

Let $m \in \mathbb{N}^*$ and let $\mathscr{F} = k(x_1, \ldots, x_m)$ be the field of rational functions in the m variables $x_1, \ldots, x_m$ with coefficients in a field k.

For i in $\{1, \ldots, m\}$, let $\partial_i := \frac{d}{dx_i}$ be the operator "derivation with respect to the ith variable" and let $\Theta := \{\partial_1, \ldots, \partial_m\}$ be the commutative monoïd generated by the ∂_i. Following the terminology of [26, Ap. D], we say that $(\mathscr{F}, \Theta)$ is a *partial differential field* or Θ*-field.* The *field of constants* of $(\mathscr{F}, \Theta)$ is

$$\mathscr{C} := \{f \in \mathscr{F}\,;\, \forall\, \delta \in \Theta,\, \delta(f) = 0\}.$$

Definition 1.1. Let $(\mathscr{F}, \Theta)$ be a partial differential field. The *ring of partial differential operators* with coefficients in $\mathscr{F}$ denoted $\mathscr{F}[\Theta]$ is the non-commutative polynomial ring over $\mathscr{F}$ in the variables ∂_i, where the ∂_i satisfy $\partial_i\, \partial_j = \partial_j\, \partial_i$, for all i, j and $\partial_i\, f = f\, \partial_i + \partial_i(f)$, for all $f \in \mathscr{F}$.

Definition 1.2. A *system of partial (linear) differential equations* or *(linear) partial differential system* is given by a finite set of elements of the ring $\mathscr{F}[\Theta]$. To every partial differential system S, we associate the (left) ideal (S) generated by the elements of S.

Definition 1.3. A partial differential system S is said to be *D-finite* if the $\mathscr{F}$-vector space $\mathscr{F}[\Theta]/(S)$ has finite dimension.

D-finite partial differential systems correspond with $\mathscr{F}[\Theta]$-modules, i.e., with vector spaces of finite dimension over $\mathscr{F}$ that are left modules for the ring $\mathscr{F}[\Theta]$ (see [26, Ap. D], and the next section in positive characteristic). In other words, a *D*-finite partial differential system is a partial differential system whose solutions only depend on a finite number of constants.

Throughout this paper, the partial differential systems that we consider are *D*-finite partial differential systems written in the form

$$\begin{cases} \Delta_1(y) = 0 & \text{with} \quad \Delta_1 := \partial_1 - A_1, \\ \quad\vdots & \\ \Delta_m(y) = 0 & \text{with} \quad \Delta_m := \partial_m - A_m, \end{cases} \tag{1.1}$$

where the $A_i \in \mathbb{M}_n(\mathscr{F})$ are square matrices of size $n \in \mathbb{N}^*$ with coefficients in $\mathscr{F}$ and the Δ_i commute. This implies the following relations, called *integrability conditions*, on the matrices A_i (see [26, Ap. D] for example):

$$\partial_i(A_j) - \partial_j(A_i) - A_i\, A_j + A_j\, A_i = 0, \text{ for all } i,\, j. \tag{1.2}$$

The (D-finite) partial differential system given by (1.1) will sometimes be noted $[A_1, \ldots, A_m]$; this is convenient when one wants to refer to the matrices A_i or to the operators Δ_i.

There exist algorithms to test whether a given partial differential system S is D-finite and if so, to write it into the form (1.1). For example, this can be achieved by computing a *Janet basis* (also called *involutive basis* in the literature) of S (see [16, 17, 12, 4]). These bases can be viewed as some kind of (non-reduced) Gröbner bases. A Janet basis of the system yields a basis of the quotient $\mathscr{F}[\Theta]/(S)$. And, the fact that this basis is finite is then equivalent to the fact that the system is D-finite. The matrices A_i can be obtained by computing the action of the ∂_i on the basis of the quotient.

Let $\mathscr{M}$ be an $\mathscr{F}[\Theta]$-module of dimension n over $\mathscr{F}$. Let $(e_1, \ldots, e_n)$ and $(f_1, \ldots, f_n)$ be two bases of $\mathscr{M}$ related by

$$(f_1, \ldots, f_n) = (e_1, \ldots, e_n)\, P$$

where $P \in \mathrm{GL}_n(\mathscr{F})$ is an invertible element of $\mathbb{M}_n(\mathscr{F})$. If $[A_1, \ldots, A_m]$ and $[B_1, \ldots, B_m]$ are respectively the partial differential systems associated with $\mathscr{M}$ with respect to the bases $(e_1, \ldots, e_n)$ and $(f_1, \ldots, f_n)$, then, for all $i \in \{1, \ldots, m\}$, $B_i = P^{-1}\,(A_i\, P - \partial_i(P))$.

In the sequel, to simplify the notations, we will note

$$P[A_i] := P^{-1}\,(A_i\, P - \partial_i(P)).$$

1.2. Factorization and Eigenrings

In this subsection, we define some notions about factorization of partial differential systems that are used in the sequel. We have seen in the last subsection, that a partial differential system over $(\mathscr{F}, \Theta)$ can be thought of as a left module over $\mathscr{F}[\Theta]$. This classical approach has the advantage of enabling one to apply directly the general theorems on modules [15] (like the Jordan–Hölder theorem, Schur's lemma, the Krull–Schmidt theorem) to partial differential systems. This allows a better understanding of the problems arising in the study of partial differential systems.

Let $(\mathscr{F}, \Theta)$ be a partial differential field. Two partial differential systems $S_1 = [A_1, \ldots, A_m]$ and $S_2 = [B_1, \ldots, B_m]$ over $(\mathscr{F}, \Theta)$ are called *equivalent differential systems* (or *similar*) if the associated $\mathscr{F}[\Theta]$-modules are isomorphic. A simple computation shows that S_1 and S_2 are equivalent if, and only if, there exists a matrix $P \in \mathrm{GL}_n(\mathscr{F})$ such that, $B_i = P[A_i]$, for all i.

Let $S = [A_1, \ldots, A_m]$ be a partial differential system over $(\mathscr{F}, \Theta)$ and denote by $\mathscr{M}$ the associated $\mathscr{F}[\Theta]$-module. A subspace $\mathscr{W} \subset \mathscr{M}$ is said to be *invariant* if $\Delta_i \mathscr{W} \subset \mathscr{W}$, for all i. One can see easily that $\mathscr{W} \subset \mathscr{M}$ is invariant if, and only if, $\mathscr{W}$ is a submodule of $\mathscr{M}$.

The partial differential system S is called a *reducible partial differential system* if the $\mathscr{F}[\Theta]$-module $\mathscr{M}$ is reducible, i.e., if there exists a submodule $\mathscr{W}$ of $\mathscr{M}$ such that $0 \neq \mathscr{W} \neq \mathscr{M}$. Otherwise, S is said to be *irreducible*.

The partial differential system S is called a *decomposable partial differential system* if $\mathscr{M}$ is decomposable, i.e., if $\mathscr{M} = \mathscr{W}_1 \oplus \mathscr{W}_2$ where $\mathscr{W}_i \neq 0$. Otherwise, S is *indecomposable.*

The partial differential system S is called a *completely reducible partial differential system* if $\mathscr{M}$ is completely reducible, i.e., if it is a direct sum of irreducible submodules.

In matrix terms, S is reducible, or resp. decomposable, if there exists a system $[B_1, \ldots, B_m]$ equivalent to S over $\mathscr{F}$ such that, for all i, B_i has the following *reduced form*

$$B_i = \begin{pmatrix} B_{1,1} & B_{1,2} & \ldots & B_{1,r} \\ 0 & B_{2,2} & \ddots & \vdots \\ \vdots & \ddots & \ddots & B_{r-1,r} \\ 0 & \ldots & 0 & B_{r,r} \end{pmatrix},$$

or resp. *decomposed form*

$$B_i = \begin{pmatrix} B_{1,1} & & 0 \\ & \ddots & \\ 0 & & B_{r,r} \end{pmatrix}.$$

Definition 1.4. Let $S = [A_1, \ldots, A_m]$ be a partial differential system. *Factoring S* means deciding whether it is reducible or irreducible, decomposable or indecomposable, and, in the reducible (resp. decomposable) case, find an invertible matrix P such that $P[A_i]$ has a reduced (resp. decomposed) form, for all i.

Thus, factoring a partial differential system means factoring *simultaneously* the systems $\partial_i(Y) = A_i Y$. Particularly, we already see that if one of these systems is irreducible over $\mathscr{F}$, then the system $[A_1, \ldots, A_m]$ is irreducible over $\mathscr{F}$ as well.

In the ordinary differential case, when one wants to factor a reducible differential system, a very useful object is the eigenring associated with the differential system; indeed, non-trivial elements of this ring provide factorizations of the differential system (see [27, 2, 26] for example).

Definition 1.5. The *eigenring* $\mathscr{E}(S)$ of a partial differential system $S = [A_1, \ldots, A_m]$ is the set of all $P \in \mathbb{M}_n(\mathscr{F})$ satisfying $\partial_i(P) = P\,A_i - A_i\,P$, for all i.

The eigenring of a partial differential system S is isomorphic to the ring of endomorphisms $\mathrm{End}(S)$ of the associated $\mathscr{F}[\Theta]$-module $\mathscr{M}$. Indeed, it is not difficult to see that a map $u : \mathscr{M} \longrightarrow \mathscr{M}$ belongs to $\mathscr{E}(S)$ if, and only if, u is an $\mathscr{F}$-linear map satisfying $u \circ \Delta_i = \Delta_i \circ u$, for all i.

In the sequel, we will also use the *partial eigenrings* $\mathscr{E}_i(S)$ consisting of all $P \in \mathbb{M}_n(\mathscr{F})$ satisfying $P\,\Delta_i = \Delta_i\,P$. We clearly have $\mathscr{E}(S) = \bigcap_{i=1}^m \mathscr{E}_i(S)$.

Remark 1.1. The following facts are standard (e.g., [2, 26]) for usual differential equations and generalize easily to the case of D-finite partial differential equations.

$\mathscr{E}(S)$ is a finite dimensional $\mathscr{C}$-subalgebra of $\mathbb{M}_n(\mathscr{F})$ which contains $\mathscr{C}\, I_n$. As a consequence, any element of $\mathscr{E}(S)$ has a minimal (and characteristic) polynomial with coefficients in $\mathscr{C}$.

The eigenrings of two equivalent partial differential systems are isomorphic as $\mathscr{C}$-algebras.

If $\mathscr{E}(S)$ is a division ring, then S is indecomposable.

If S is irreducible, then $\mathscr{E}(S)$ is a division ring (Schur's lemma). The converse is false. However, if S is completely reducible and if $\mathscr{E}(S)$ is a division ring, then S is irreducible.

2. Partial Differential Systems in Positive Characteristic

Let p be a prime number and $r \in \mathbb{N}^*$. Consider the partial differential field $(\mathbb{K}, \Theta)$ where $\mathbb{K} := k(x_1, \ldots, x_m)$ with $k = \mathbb{F}_q$ for $q = p^r$. The *partial constant field* of $\mathbb{K}$ with respect to, say, ∂_1 is $\mathscr{C}_1 := \ker_{\mathbb{K}}(\partial_1) = k(x_1^p, x_2, \ldots, x_m)$. The constant field of $(\mathbb{K}, \Theta)$ is

$$\mathscr{C} := \bigcap_{i=1}^{m} \mathscr{C}_i = k(x_1^p, x_2^p, \ldots, x_m^p).$$

Note that $\mathbb{K}$ is a $\mathscr{C}$-vector space of dimension p^m and a $\mathscr{C}_i$-vector space of dimension p.

In the following, we consider partial differential systems $[A_1, \ldots, A_m]$ with coefficients in $(\mathbb{K}, \Theta)$ and, to avoid pathologies, we assume that the prime number p is strictly greater than the size n of the A_i.

Following the theory of differential equations in characteristic p, we now introduce *partial p-curvatures.*

Definition 2.1. Let $[A_1, \ldots, A_m]$ be a partial differential system over $(\mathbb{K}, \Theta)$. The *partial p-curvatures of* $[A_1, \ldots, A_m]$ are the $\mathbb{K}$-linear operators $\Delta_i^p = (\partial_i - A_i)^p$, for $i \in \{1, \ldots, m\}$, acting on $\mathbb{K}^n$.

The proof of the following lemma is then immediate.

Lemma 2.1. *Let* $S = [A_1, \ldots, A_m]$ *be a partial differential system over* $(\mathbb{K}, \Theta)$*. All the partial p-curvatures* Δ_i^p *commute and belong to the eigenring* $\mathscr{E}(S)$*. In particular, the minimal (and characteristic) polynomial of each* Δ_i^p *has its coefficients in* $\mathscr{C} = k(x_1^p, \ldots, x_m^p)$*.*

Note that in [18, 5, p. 189] (see also [19, VII, p. 222]), Katz defines a notion of p-curvature in the case of several derivations and remarks the links between this p-curvature and the eigenring of the system (refined in Lemma 2.1). In [19], he gives a method for computing the partial p-curvatures (see also [26, Ch. 13] or [7]). For all i in $\{1, \ldots, m\}$, it consists in computing the index p element in the Lie sequence $(A_{i,(j)})_{j \in \mathbb{N}}$ associated with $[A_i]$ which is defined by:

$$A_{i,(0)} := I_n \text{ and } \forall j \geq 0,\ A_{i,(j+1)} := \Delta_i(A_{i,(j)}) = \partial_i(A_{i,(j)}) - A_i\, A_{i,(j)}.$$

In [23] (see also [26, Ch. 13]), van der Put gives a classification of differential modules in characteristic p. A consequence of this classification for the factorization problem is that the Jordan form of the p-curvature leads to all the factorizations of the system. In [7] (see also [24, 8]), this is made algorithmic and combining this to the approach of the eigenring factorization method proposed by Barkatou in [2], the author develops an algorithm for factoring differential systems in characteristic p and provides elementary effective proofs of the key results (that can also be viewed from van der Put's classification).

In the sequel, we build upon the approach of [7] to generalize the main steps of the van der Put classification that are needed for the algorithm; in the appendix, van der Put shows how to completely generalize his classification to partial differential modules.

2.1. Rational Solutions

Let $S = [A_1, \ldots, A_m]$ be a partial differential system over $(\mathbb{K}, \Theta)$. The space of rational solutions (or solutions in $\mathbb{K}^n$) of the system S is the set

$$\mathrm{Sol}_{\mathbb{K}}(S) = \{Y \in \mathbb{K}^n;\ \forall i,\ \Delta_i(Y) = 0\}.$$

One can show that $\mathrm{Sol}_{\mathbb{K}}(S)$ is a vector space over the field of constants $\mathscr{C}$ of dimension $\leq n$.

The first algorithmic use of the p-curvature stems from Cartier's lemma ([18, Theorem 5.1]).

Lemma 2.2 (Cartier). *Let $S = [A_1, \ldots, A_m]$ be a partial differential system over $(\mathbb{K}, \Theta)$. The partial p-curvatures Δ_i^p are all zero if, and only if, S admits a basis of rational solutions, i.e. solutions in $\mathbb{K}^n$.*

Note that S admits a basis of rational solutions if, and only if, S has a fundamental matrix of rational solutions, i.e., a matrix $P \in \mathrm{GL}_n(\mathbb{K})$ satisfying $\Delta_i(P) = \partial_i(P) - A_i\,P = 0$, for all i. In other words, S admits a basis of rational solutions if, and only if, there exists $P \in \mathrm{GL}_n(\mathbb{K})$ such that $P[A_i] = 0$, for all i.

Although a proof of the above lemma can be found in [18, Theorem 5.1], we propose a new constructive proof for further algorithmic use.

Proof. The implication "$\Leftarrow$" is trivial so we only need to prove "$\Rightarrow$". Consider first the differential field $(k(x_2, \ldots, x_m)(x_1), \partial_1)$ which has $\mathscr{C}_1$ as constant field, and view Δ_1 as a differential operator acting on $k(x_2, \ldots, x_m)(x_1)^n$; as it satisfies $\Delta_1^p = 0$, Cartier's lemma in the ordinary differential case (e.g., [7, Lemma 3.3]) implies the existence of some $P_1 \in \mathrm{GL}_n(k(x_2, \ldots, x_m)(x_1))$ such that $P_1^{-1}\,\Delta_1\,P_1 = \partial_1$. For all i in $\{1, \ldots, m\}$, let $\tilde{\Delta}_i = P_1^{-1}\,\Delta_i\,P_1 := \partial_i - B_i$ for some matrices B_i having coefficients in $k(x_1, \ldots, x_m)$. The integrability conditions imply that $\partial_1(B_i) = 0$ so that the B_i have their coefficients in $\mathscr{C}_1$, for all i. Now, we use the hypothesis $\tilde{\Delta}_2^p = 0$ and we apply Cartier's lemma in the ordinary differential case to $\tilde{\Delta}_2$: there exists $P_2 \in \mathrm{GL}_n(\mathscr{C}_1)$ such that $P_2^{-1}\,\tilde{\Delta}_2\,P_2 = \partial_2$. Moreover $P_2 \in \mathrm{GL}_n(\mathscr{C}_1)$ implies that ∂_1 commutes with P_2 and thus

$$P_2^{-1}\,P_1^{-1}\,\Delta_1\,P_1\,P_2 = P_2^{-1}\,\tilde{\Delta}_1\,P_2 = P_2^{-1}\,\partial_1\,P_2 = \partial_1.$$

Applying this process recursively, we finally find an invertible matrix $P = P_1 \cdots P_m$ with coefficients in $k(x_1, \ldots, x_m)$ such that $P^{-1}\,\Delta_i\,P = \partial_i$, for all i; the result follows. □

This proof exhibits an algorithm to compute a fundamental matrix of rational solutions of a partial differential system whose partial p-curvatures vanish.

Algorithm SimRatSols.
Input: A partial differential system $S = [A_1, \ldots, A_m]$ with the $A_i \in \mathbb{M}_n(\mathbb{K})$ and whose partial p-curvatures vanish.
Output: A fundamental matrix of rational solutions of $[A_1, \ldots, A_m]$.
1. For i from 1 to m, set $A_i^{[1]} := A_i$.
2. For i from 1 to m do:
   2a. Compute a fundamental matrix P_i of rational solutions of the differential system (viewed as a system in one variable) $\partial_i(Y) = A_i^{[i]}\,Y$.
   2b. For j from 1 to m, compute $A_j^{[i+1]} := P_i^{-1}\,(A_j^{[i]}\,P_i - \partial_j(P_i))$.
3. Return $P_1 \cdots P_m$.

Remark 2.1. When only one of the partial p-curvatures is zero, then, after a change of basis, the system (1.1) can be written as

$$\begin{cases} \Delta_1(y) = 0 \quad \text{with} \quad \Delta_1 := \partial_1, \\ \quad\vdots \\ \Delta_m(y) = 0 \quad \text{with} \quad \Delta_m := \partial_m - A_m, \end{cases} \tag{2.1}$$

so that the integrability conditions (1.2) imply $\partial_1(A_j) = 0$ for all $j \in \{2, \ldots, m\}$. We can thus deduce that the partial differential system no longer depends on the variable x_1 but rather on x_1^p.

An alternative to Algorithm SimRatSols is to use the "Katz' projector formula"; this will be studied (and used) at the end of the next subsection.

In general (when the partial p-curvatures do not vanish), in characteristic p, computing rational solutions is an ordinary linear algebra problem which can be set (and solved) in two ways.

- An iterative method: since for all i,

$$\mathbb{K} \cong \bigoplus_{j=0}^{p-1} \mathscr{C}_i\, x_i^j,$$

  any element Y of $\mathbb{K}^n$ can be written as $Y = \sum_{i=0}^{p-1} C_i\, x_1^i$ with $C_i \in \mathscr{C}_1^n$. The equation $\Delta_1(Y) = 0$ is then seen as an $np \times np$ linear system for the entries of the C_i. Let $Y_{1,1}, \ldots, Y_{1,r_1}$ denote a basis (over $\mathscr{C}_1$) of solutions in $\mathbb{K}^n$ of

$\Delta_1(Y) = 0$ obtained from this linear system. As the Δ_i commute, the space generated over $\mathscr{C}_1 \cap \mathscr{C}_2$ by this basis is stable under Δ_2. Set

$$Y_2 := \sum_{i=1}^{r_1} \sum_{j=0}^{p-1} c_{i,j}\, Y_{1,i}\, x_2^j.$$

The equation $\Delta_2(Y_2) = 0$ translates into an $r_1 p \times r_1 p$ linear system for the $c_{i,j} \in \mathscr{C}_1 \cap \mathscr{C}_2$. Solving this system yields a basis $Y_{2,1}, \ldots, Y_{2,r_2}$ (over $\mathscr{C}_1 \cap \mathscr{C}_2$) of solutions in $\mathbb{K}^n$ of $\{\Delta_1(Y) = 0,\ \Delta_2(Y) = 0\}$. Iterating this process, we finally find a basis over $\mathscr{C}$ of rational solutions of $[A_1, \ldots, A_m]$.

- A direct (less interesting) method proceeds as follows: as $\mathbb{K}$ is a $\mathscr{C}$ vector space (of dimension p^m over $\mathscr{C}$), the system

$$\{\Delta_1(Y) = 0, \ldots, \Delta_m(Y) = 0\}$$

translates into m linear systems of size np^m over $\mathscr{C}$, from which a basis (over $\mathscr{C}$) of rational solutions is obtained.

As observed in [7, 3.2.1] (see also [8]), this leads to an immediate algorithm for computing the eigenring (by computing rational solutions of a partial differential system of dimension n^2).

2.2. Scalar Partial p-Curvatures

We consider the case when all the partial p-curvatures Δ_i^p are scalar, that is, for all i, $\Delta_i^p - \lambda_i I_n$ with $\lambda_i \in \mathscr{C} = k(x_1^p, \ldots, x_m^p)$ (see Lemma 2.1).

First consider individually the system $\partial_1(Y) = A_1 Y$ (also noted $[A_1]$) with coefficients in the differential field $\mathbb{K} = k(x_2, \ldots, x_m)(x_1)$ endowed with the derivation ∂_1 and having $\mathscr{C}_1$ as constant field. In [8] (see also [24, 26]), partial fraction decomposition shows that if $\Delta_1^p = \lambda_1 I_n$ with $\lambda_1 \in \mathscr{C}_1$, then there exists

$$\nu_1 \in \overline{k(x_2, \ldots, x_m)}(x_1)$$

such that $[A_1]$ is equivalent (over $\overline{k(x_2, \ldots, x_m)}(x_1)$) to $[\nu_1 I_n]$. Now Theorem 3.7 of [7] applies and its proof shows that in fact $\mu_1 = \mathrm{Tr}(A_1)/n \in \mathbb{K}$ satisfies $\partial_1^{p-1}(\mu_1) + \mu_1^p = \lambda_1$ and the system $[A_1]$ is thus equivalent over $\mathbb{K}$ to $[\mu_1 I_n]$.

Proposition 2.1. *Let $S = [A_1, \ldots, A_m]$ be a partial differential system over $(\mathbb{K}, \Theta)$. All the partial p-curvatures Δ_i^p are scalar if, and only if, the system S is equivalent over $\mathbb{K}$ to a "diagonal system". In other words, for all i, $\Delta_i^p = \lambda_i I_n$ with $\lambda_i \in \mathscr{C}$ if, and only if, there exists $P \in \mathrm{GL}_n(\mathbb{K})$ such that $P[A_i] = \mu_i I_n$ with $\mu_i \in \mathbb{K}$, for all i.*

Proof. Suppose, without loss of generality, that $m = 2$. Consider a partial differential system $[A_1, A_2]$ satisfying $\Delta_1^p = \lambda_1 I_n$ and $\Delta_2^p = \lambda_2 I_n$ with $\lambda_1, \lambda_2 \in \mathscr{C} = k(x_1^p, x_2^p)$. Set

$$(\mu_1, \mu_2) = (\mathrm{Tr}(A_1)/n, \mathrm{Tr}(A_2)/n)$$

and consider the partial differential system $[A_1 - \mu_1 I_n, A_2 - \mu_2 I_n]$. By construction, its partial p-curvatures vanish. Moreover the integrability condition for this new partial differential system is satisfied: indeed, after some simplifications, this condition can be written as $\partial_1(\mu_2) = \partial_2(\mu_1)$ which is equivalent to $\mathrm{Tr}(\partial_1(A_2)) = \mathrm{Tr}(\partial_2(A_1))$ and, from (1.2), to $\mathrm{Tr}(A_2\, A_1) = \mathrm{Tr}(A_1\, A_2)$. Then, Lemma 2.2 shows the existence of an invertible matrix P with coefficients in $\mathbb{K}$ such that $P[A_1 - \mu_1\, I_n] = P[A_2 - \mu_2\, I_n] = 0$, that is,

$$P^{-1}\left((A_1 - \mu_1\, I_n)\, P - \partial_1(P)\right) = P^{-1}\left((A_2 - \mu_2\, I_n)\, P - \partial_2(P)\right) = 0$$

and the result follows. □

The proof of the next lemma, from [18], exhibits a "Katz' formula" (see [18], Formulas 5.1.2 and 5.1.7, p. 191) to compute a fundamental matrix of rational solutions when all the partial p-curvatures are zero.

Let $S = [A_1, \ldots, A_m]$ be a partial differential system over $(\mathbb{K}, \Theta)$. It is clear that the space of rational solutions $\mathrm{Sol}_{\mathbb{K}}(S)$ of S is included in $\bigcap_{i=1}^{m} \ker(\Delta_i^p)$ (the common kernel of the partial p-curvatures Δ_i^p).

Lemma 2.3 (Katz). *Let $S = [A_1, \ldots, A_m]$ be a partial differential system over $(\mathbb{K}, \Theta)$. Then*

$$\bigcap_{i=1}^{m} \ker(\Delta_i^p) = \mathrm{Sol}_{\mathbb{K}}(S) \otimes_{\mathscr{C}} \mathbb{K}.$$

Proof (adapted from [18]). Assume, for simplicity, that the denominators of A_i do not vanish at $x_i = 0$. For all $i \in \{1, \ldots, m\}$, we define

$$\mathrm{Pr}_i : \mathbb{K}^n \to \mathbb{K}^n,\ v \mapsto \sum_{k=0}^{p-1} \frac{(-x_i)^k}{k!}\, \Delta_i^k(v),$$

and we verify that:

- for all $v \in \mathbb{K}^n$,
$$\Delta_i(\mathrm{Pr}_i(v)) = -\,(-x_i)^{p-1}\, \Delta_i^p(v)$$
so that Pr_i sends $\ker(\Delta_i^p)$ into $\ker(\Delta_i)$;
- for all $i,\, j \in \{1, \ldots, m\}$ such that $i \neq j$,
$$\Delta_j(\mathrm{Pr}_i(v)) = \mathrm{Pr}_i(\Delta_j(v))$$
so that the Pr_i commute.

Now set $\mathrm{Pr} := \prod_{i=1}^{m} \mathrm{Pr}_i$. This operator from $\mathbb{K}^n$ to $\mathbb{K}^n$ satisfies the following property: for all $i \in \{1, \ldots, m\}$, if $\Delta_i^p(v) = 0$, then $\Delta_i(\mathrm{Pr}(v)) = 0$. From [18, Formula 5.1.2, p. 191], the formula for $\mathrm{Pr}(v)$ can be expanded to obtain

$$\mathrm{Pr}(v) = \sum_{\omega} \prod_{i=1}^{m} \frac{(-x_i)^{\omega_i}}{\omega_i!} \prod_{i=1}^{m} \Delta_i^{\omega_i}(v),$$

where the sum is taken over all r-uples $\omega = (\omega_1, \ldots, \omega_r)$ of integers such that $0 \leq \omega_i \leq p-1$. This projector sends $\bigcap_{i=1}^{m} \ker(\Delta_i^p)$ to $\mathrm{Sol}_{\mathbb{K}}(S)$ and the (Taylor) formula [18, Formula 5.1.7, p. 191] induces the identity on $\bigcap_{i=1}^{m} \ker(\Delta_i^p)$ and proves the lemma (compare to the proof of [7, Theorem 3.8]). □

This thus yields an explicit formula for the calculation of a fundamental matrix of rational solutions of a partial differential system whose partial p-curvatures vanish and which satisfies further that $(0,\ldots,0)$ does not cancel the denominator of the A_i.

From this, we obtain the following algorithm that diagonalizes partial differential systems having scalar partial p-curvatures.

Algorithm ScalpCurv.
Input: A partial differential system $S = [A_1,\ldots,A_m]$ satisfying $\Delta_i^p = \lambda_i I_n$, for all i.
Output: A matrix P and the $P[A_i] = \mu_i I_n$.
1. For all i, compute $\mu_i := \mathrm{Tr}(A_i)/n$.
2. For all i, set $B_i := A_i - \mu_i I_n$ and compute the Lie sequences $B_{i,(j)}$.
3. Let $P := \prod_{i=1}^{m} \sum_{j=0}^{p-1} \frac{(-x_i)^j}{j!} B_{i,(j)}$.
4. Return P and the $P[A_i] = \mu_i I_n$.

The correctness of this algorithm follows directly from Proposition 2.1, Lemma 2.3 and their proofs. In Step 3, to apply "Katz' formula", we have to make sure that $(0,\ldots,0)$ does not vanish the denominator of the B_i; if it is the case, then we shift with respect to the corresponding variable. The calculation in Step 2, can be accelerated using [7, Lemma 3.4] and the fact that the Lie sequences of the $[A_i]$ have already been computed to obtain the Δ_i^p.

2.3. Nilpotent Partial p-Curvatures

In the sequel, the characteristic (resp. minimal) polynomial of a matrix M will be noted $\chi(M)$ (resp. $\chi_{\min}(M)$).

We now treat the case when all the partial p-curvatures are nilpotent. Here, we use a method adapted from [24, 26] to handle the partial differential case.

Assume that all partial p-curvatures are nilpotent so, for all $i \in \{1,\ldots,m\}$, $\chi_{\min}(\Delta_i^p) = X^{d_i}$ with $d_i \in \mathbb{N}^*$. The case $d_i = 1$ for all i has already been addressed in Subsection 2.1, so we assume that there exists an i such that $d_i > 1$.

The reasoning is the same as in the iterative method for computing rational solutions given at the end of Subsection 2.1. We have $\chi_{\min}(\Delta_1^p) = X^{d_1}$ so, as shown in [24, 26], one can find a basis of solutions of $\Delta_1(Y) = 0$ in

$$\mathbb{K}^n + \mathbb{K}^n\, l_1 + \cdots + \mathbb{K}^n\, l_1^{d_1-1}$$

where l_1 satisfies $\partial_1(l_1) = 1/x_1$ (note that a general natural algorithm to perform this task — in characteristic zero — is given in [3] and is easily adapted to our setting). So, the solutions of $\Delta_1(Y) = 0$ are of the form

$$Y_{1,i} = \sum_{j=0}^{d_i-1} Y_{1,i,j}\, l_1^j$$

with $Y_{1,i,j} \in \mathbb{K}^n$. We now search for solutions of $\Delta_2(Y) = 0$ of the form $\sum_i c_i\, Y_{1,i}$ where the c_i are constant with respect to ∂_1. Viewing the c_i as functions in the variable x_2, the relation $\Delta_2 \left(\sum_i c_i(x_2)\, Y_{1,i}\right) = 0$ yields a linear differential system (S_{Δ_2}) for the $c_i(x_2)$. Now $\chi_{\min}(\Delta_2^p) = X^{d_2}$, so we know that we can find a basis of solutions of (S_{Δ_2}) in

$$\mathbb{K}^n + \mathbb{K}^n\, l_2 + \cdots + \mathbb{K}^n\, l_2^{d_2-1}$$

with $\partial_2(l_2) = 1/x_2$ (using again a method like in [3]). Iterating this process yields a basis of solutions in $\mathbb{K}^n[l_1, \ldots, l_m]$. Let P denote the invertible matrix whose columns are (generated by) the components in $\mathbb{K}^n$ of these solutions; then, for all i, $P[A_i]$ has a reduced form with zeros as diagonal blocks.

The case when, for all i, $\chi_{\min}(\Delta_i^p) = (X - a_i)^{d_i}$ with $a_i \in \mathbb{K}$ can then be handled since it reduces to the nilpotent case by using the tools from the previous subsection: indeed, letting $\mu_i := \mathrm{Tr}(A_i)/n$ and $B_i := A_i - \mu_i\, I_n$, we factor the partial differential system $[B_1, \ldots, B_m]$ having nilpotent partial p-curvatures, and then we shift back to deduce the factorization of $[A_1, \ldots, A_m]$. Note that this particular case appears naturally when we want to adapt van der Put's method for the computation of the maximal decomposition of a partial differential system $[A_1, \ldots, A_m]$ satisfying $\chi(\Delta_i^p) = F_i^{m_i}$, for all i (see Subsection 4.2, [24, 26] or [7, 8]).

We now have the building blocks for factoring at our disposal. The key will be to reduce the problem to the simultaneous reduction of the (commuting) partial p-curvature matrices, so we address this problem first before proceeding to factorization.

3. Simultaneous Reduction of Commuting Matrices

Let K be a field and V be a vector space of finite dimension n over K. Let $\mathscr{L} = \{\phi_1, \ldots, \phi_s\}$ be a set of s commuting linear endomorphisms of V; V can be viewed as a left $K[X_1, \ldots, X_s]$-module by defining $X_j.v = \phi_j(v)$ for all $v \in V$, $j \in \{1, \ldots, s\}$. We shall denote this module $(V, \mathscr{L})$.

We say that $\mathscr{L}$ is *reducible*, *decomposable* or *completely reducible* over K if the $K[X_1, \ldots, X_s]$-module $(V, \mathscr{L})$ is reducible, decomposable or completely reducible.

In all of this section, $M_1, \ldots, M_s$ are s square matrices of size n with coefficients in K. We further assume that the M_i commute, i.e.,

$$\forall\, i,\, j,\; [M_i, M_j] := M_i\, M_j - M_j\, M_i = 0.$$

We set $\Omega := \{M_1, \ldots, M_s\}$. Viewing the M_i as commuting linear transformations written in the standard basis of K^n, we naturally define the terms Ω reducible, decomposable and completely reducible.

3.1. Simultaneous Decomposition

Recall first (see [14, Ch. 4, 9]) that if Ω is indecomposable, then the minimal polynomial of any $N \in \Omega$ is a power of an irreducible polynomial over K.

Suppose now that Ω is decomposable and let $\mathscr{M} = (K^n, \Omega)$ be the corresponding $K[X_1, \ldots, X_s]$-module. We decompose $\mathscr{M}$ as

$$\mathscr{M} = \mathscr{W}_1 \oplus \cdots \oplus \mathscr{W}_d,$$

where the $\mathscr{W}_i$ are indecomposable. Now from [14, Ch. 4, 9], we know that with respect to a basis of K^n adapted with this decomposition, each element of Ω has a decomposed form. Moreover, the minimal polynomial of each diagonal block is a power of an irreducible polynomial. In other words, there exists a $P \in \mathrm{GL}_n(K)$ such that for all N in Ω, $P^{-1}\, N\, P$ has the form

$$P^{-1}\, N\, P = \begin{pmatrix} N_1 & & 0 \\ & \ddots & \\ 0 & & N_d \end{pmatrix}, \tag{3.1}$$

where, for all j, $\chi_{\min}(N_j) = F_j^{m_j}$ with F_j irreducible over K.

Definition 3.1. A *simultaneous decomposition of* Ω is the given of $P \in \mathrm{GL}_n(K)$ such that, for all $N \in \Omega$, $P^{-1}\, N\, P$ has the form (3.1).

In the following, we shall show how to compute a simultaneous decomposition of Ω. The key to this computation is the (obvious) lemma:

Lemma 3.1. *Assume that there exists an N in Ω such that $\chi(N) = F_1 \cdots F_h$ with $h \geq 2$ and the F_j pairwise coprime. Then, we can compute $P \in \mathrm{GL}_n(K)$ such that, for all N' in Ω, $P^{-1}\, N'\, P$ has a decomposed form.*

Proof. We know from the kernel decomposition theorem that if $\chi(N) = F_1 \cdots F_h$ with the F_j pairwise coprime, then $K^n = \bigoplus_{j=1}^{h} \ker(F_j(N))$. Now, as the matrices N and N' commute, $\ker(F_j(N))$ is stable under N' and the result follows. □

Following this lemma, one can easily construct a recursive rational algorithm to compute a simultaneous decomposition of Ω (see [8]).

We now propose to use another approach to compute a simultaneous decomposition. The idea underlying this method can be found in [10, 8].

Consider the matrix

$$M := t_1\, M_1 + \cdots + t_s\, M_s, \tag{3.2}$$

with coefficients in $K[t_1, \ldots, t_s]$. Here $t_1, \ldots, t_s$ are indeterminates over K. Note that, in practice (see [10, 8]), the calculations are performed after having specialized the t_i to random values.

For all $i \in \{1, \ldots, s\}$, there exists a unique couple of matrices (S_i, N_i) with S_i semi-simple (that is diagonalizable over $\overline{K}$) and N_i nilpotent such that $M_i = S_i + N_i$ and $[S_i, N_i] = 0$. Such a decomposition $M_i = S_i + N_i$ is called the SN *decomposition of* M_i.

Remark 3.1. The eigenvalues of M_i in $\overline{K}$ coincide with the eigenvalues of S_i in $\overline{K}$. In other words, M_i and S_i have the same characteristic polynomial.

Lemma 3.2. *With the above notations, let*

$$S = t_1\, S_1 + \cdots + t_s\, S_s \quad \text{and} \quad N = t_1\, N_1 + \cdots + t_s\, N_s.$$

Then $M = S + N$ is the SN *decomposition of the matrix $M = \sum_{i=1}^{s} t_i\, M_i$.*

Proof. We have to show that S is semi-simple, N is nilpotent and $[S, N] = 0$. We know (see [5, Théorème 19.6, p. 294] or [20]) that for all i, S_i and N_i are polynomials in M_i. Consequently, as the M_i are pairwise commuting matrices, we have $[S_i, S_j] = [N_i, N_j] = [N_i, S_j] = 0$, for all i, j. The matrices $S_1, \ldots, S_s$ are thus pairwise commuting and semi-simple matrices. Thus they are simultaneously diagonalizable over $\overline{K}$, that is, there exists an invertible P with coefficients in $\overline{K}$ such that $P^{-1}\, S_i\, P$ is diagonal, for all i. The fact that S is semi-simple follows immediately. If we note u_i the nilpotence index of N_i: $N_i^{u_i} = 0$ and $N_i^{l} \neq 0$ for all $l < u_i$, then a direct calculation shows that N is nilpotent with nilpotence index at most $u_1 + \cdots + u_s$. Finally, the equality $[S, N] = 0$ is clear since $[S_i, N_j] = 0$ for all i, j. □

Corollary 3.1. *With the previous notations, let $(v_1, \ldots, v_n) \in \overline{K}^n$ be a basis of common eigenvectors of $S_1, \ldots, S_s$. Let $\lambda_{i,j}$ be the eigenvalue of S_i associated with v_j, i.e., $S_i\, v_j = \lambda_{i,j}\, v_j$. Then*

$$S\, v_j = \Big(\sum_{i=1}^{s} t_i\, \lambda_{i,j}\Big)\, v_j$$

and, in particular, S has all its eigenvalues in $\sum_{i=1}^{s} t_i\, \overline{K} \subset \overline{K}[t_1, \ldots, t_s]$.

An interesting consequence of this corollary is that the eigenvalues of M can be computed without computing first those of the M_i. To proceed, it suffices to factor into products of linear forms over $\overline{K}[t_1, \ldots, t_s]$ the determinant of M (for example, we can use the algorithm given in [13, Ap.]). Indeed, we know that $\det(M)$ equals $(-1)^n$ times the product of the eigenvalues of M. Now, from Corollary 3.1, these eigenvalues are linear forms in the t_i with coefficients in $\overline{K}$ and thus $\det(M)$ necessarily factors into linear forms over $\overline{K}[t_1, \ldots, t_s]$.

We obtain the following algorithm that computes a simultaneous decomposition of $\{M_1, \ldots, M_s\}$ (see [10, 8]).

Algorithm SimDec (Simultaneous Decomposition).
Input: $\Omega = \{M_1, \ldots, M_s\}$ (with $M_i \in \mathbb{M}_n(K)$ pairwise commuting matrices).
Output: $P \in \mathbb{M}_n(K)$ giving a simultaneous decomposition of Ω.

1. Let $M := t_1\, M_1 + \cdots + t_s\, M_s$.
2. Compute $\chi(M)$ and factor it over $K(t_1, \ldots, t_s)$: let $\chi(M) = F_1^{m_1} \cdots F_d^{m_d}$ with F_i coprime irreducibles over $K(t_1, \ldots, t_s)$.
3. For $i \in \{1, \ldots, d\}$, do:
   3a. Compute a basis $e_i = (e_{i,1}, \ldots, e_{i,n_i})$ of $\ker(F_i^{m_i}(M))$ (choose the $e_{i,j}$ independent of the t_i).
4. Return the invertible P having the e_{ij} as columns.

Remark 3.2. This algorithm does not necessarily provide a maximal decomposition of Ω. However, if the associated module $\mathscr{M}$ is semi-simple, then the result of this algorithm corresponds to the *isotypical decomposition* of $\mathscr{M}$.

Note that the fact that factoring a partial differential system leads to reducing a linear combination with indeterminate coefficients of matrices already appears in a natural way when we consider integrable systems with constant coefficients. Indeed, let $M_1, \ldots, M_s$ be s commuting matrices with coefficients in $\mathbb{C}$. The D-finite partial differential system $\frac{d}{dt_i} Y = M_i Y$, $1 \leq i \leq s$ admits $\exp(M_1 t_1 + \cdots + M_s t_s)$ as a fundamental matrix of solutions. Thus, if we want to calculate this exponential of matrix, we have first to reduce the matrix $M_1 t_1 + \cdots + M_s t_s$ to a diagonal form (when possible) or a triangular form.

3.2. Reduction of Indecomposable Blocks

Suppose now that Ω is indecomposable. This implies (see [14, Ch. 4, 9]) that there exists a $P \in \mathrm{GL}_n(K)$ such that for all $N \in \Omega$, the matrix $P^{-1} N P$ has the reduced form

$$P^{-1} N P = \begin{pmatrix} N_{1,1} & N_{1,2} & \cdots & N_{1,r} \\ 0 & N_{2,2} & \ddots & \vdots \\ \vdots & \ddots & \ddots & N_{r-1,r} \\ 0 & \cdots & 0 & N_{r,r} \end{pmatrix}, \tag{3.3}$$

where, for all j, $\chi_{\min}(N_{j,j}) = F$ with F is irreducible over K.

Definition 3.2. Assume that Ω is indecomposable. A *(maximal) simultaneous reduction of* Ω is the given of $P \in \mathrm{GL}_n(K)$ such that, for all $N \in \Omega$, $P^{-1} N P$ has the form (3.3).

To compute a simultaneous reduction of Ω, we can once again use the matrix

$$M = t_1 M_1 + \cdots + t_s M_s.$$

We know that $\chi_{\min}(M) = F^m$ with F irreducible over $K(t_1, \ldots, t_s)$. Reducing this single matrix M over $K(t_1, \ldots, t_s)$, we obtain a simultaneous reduction of Ω (for details, see [8]). This leads to the following algorithm.

Algorithm SimRed (Simultaneous Reduction).
Input: $\Omega = \{M_1, \ldots, M_s\}$ (with $M_i \in \mathbb{M}_n(K)$) indecomposable.
Output: $P \in \mathbb{M}_n(K)$ giving a simultaneous reduction of Ω.

1. Let $M := t_1 M_1 + \cdots + t_s M_s$.
2. Compute the polynomial $\chi_{\min}(M) = F^m$.
3. For $i \in \{1, \ldots, m\}$, set $\mu_i := F^i(M)$ and $E_i := \ker(\mu_i)$.
4. Compute a basis of V adapted with the flag $(E_i)_i$ (choose one that does not depend on the t_i, see proof below).
5. Return the matrix P having the elements of this basis as columns.

Proposition 3.1. *The algorithm* SimRed *above computes a simultaneous reduction of* Ω.

Proof. Let $\mu := F(M)$. We have $\mu^m = 0$ and $\mu^i \neq 0$, for $i \in \{1, \ldots, m-1\}$. Let $E_i := \ker(\mu^i)$. It is clear that $E_m = K^n$ and $E_i \subset E_{i+1}$ such that $(E_i)_i$ is a flag of K^n. Let $\mathscr{B}$ be a basis of K^n adapted with this flag (i.e., a basis computed from a basis of E_1 extended into a basis of E_2, ...) and that does not depend on the t_i; this is always possible because simultaneous reduction exists. The matrix of μ with respect to $\mathscr{B}$ has a reduced form with zeros as diagonal blocks and the matrix of M with respect to $\mathscr{B}$ has a reduced form. Calling P the matrix formed by the vectors of $\mathscr{B}$, the reduced forms of the M_i can be retrieved by conjugating by P (or by specializing $(t_1, \ldots, t_s)$ respectively into $(1, 0, \ldots, 0), \ldots, (0, \ldots, 0, 1)$ in the reduced form of M). □

4. Factoring Partial Differential Systems in Positive Characteristic

Let $[A_1, \ldots, A_m]$ be a partial differential system with coefficients in $(\mathbb{K}, \Theta)$ with $\mathbb{K} = k(x_1, \ldots, x_m)$ and $k = \mathbb{F}_q$ for $q = p^r$. We already know that factoring individually the system $\partial_i(Y) = A_i Y$ can be done by applying the algorithm developed in [7]. To achieve this, we use the partial p-curvature Δ_i^p as well as the partial eigenring $\mathscr{E}_i(S)$. This can be done since during the algorithm of [7], we are always reduced to performing linear algebra either on the p-curvature or on an element of the eigenring. Now, if we want to factor the system $[A_1, \ldots, A_m]$, then we have to factor simultaneously the systems $\partial_i(Y) = A_i Y$; we are thus naturally led to reduce simultaneously the partial p-curvatures Δ_i^p which commute from Lemma 2.1.

As in the ordinary differential case, we first give a method to decompose the system and then, we show how to reduce indecomposable blocks.

4.1. Simultaneous Decomposition

The first step to decompose a partial differential system consists in computing a *simultaneous decomposition* of the system.

Definition 4.1. Let $[A_1, \ldots, A_m]$ be a partial differential system with coefficients in $(\mathbb{K}, \Theta)$. A *simultaneous decomposition of* $[A_1, \ldots, A_m]$ is given by $P \in \mathrm{GL}_n(\mathbb{K})$ such that:

1. for all i, $P[A_i] = \begin{pmatrix} B_i^{[1]} & & 0 \\ & \ddots & \\ 0 & & B_i^{[d]} \end{pmatrix}$;
2. for all i, the partial p-curvature of each system $\partial_i(Y) = B_i^{[j]} Y$ has a characteristic polynomial of the form $F_{i,j}^{m_{i,j}}$ with $F_{i,j}$ irreducible.

Proposition 4.1. *Let* $[A_1, \ldots, A_m]$ *be a partial differential system with coefficients in* $(\mathbb{K}, \Theta)$. *The matrix* $P \in \mathrm{GL}_n(\mathbb{K})$ *obtained by applying Algorithm* SimDec *to* $\{\Delta_1^p, \ldots, \Delta_m^p\}$ *provides a simultaneous decomposition of* $[A_1, \ldots, A_m]$.

Proof. For any polynomial Q, the spaces $\ker(Q(\Delta_i^p))$ are stable under the Δ_j since for all i, j, $[\Delta_i^p, \Delta_j^p] = 0$. So P obviously achieves conditions (i) and (ii) of Definition 4.1. □

This induces an algorithm for computing a simultaneous decomposition of a partial differential system $[A_1, \ldots, A_m]$:

- Compute the partial p-curvatures Δ_i^p of $[A_1, \ldots, A_m]$;
- Return $P := \mathrm{SimDec}(\{\Delta_1^p, \ldots, \Delta_m^p\})$.

Example 4.1. Let $\mathbb{K} := \mathbb{F}_p(x_1, x_2)$ with $p = 3$ and consider the D-finite partial differential system $[A_1, A_2]$ where A_1 and A_2 are the following matrices:

$$A_1 = \begin{pmatrix} 1 & x_1 x_2 \\ 0 & 1 \end{pmatrix},$$

$$A_2 = \begin{pmatrix} a_{1,1}^{(2)} & \frac{1}{2} f_2(x_2) x_2 x_1^4 + \frac{1}{2} f_3(x_2) x_1^2 + f_4(x_2) \\ -\dfrac{2 f_2(x_2)}{x_2} & f_1(x_2) + f_2(x_2) x_1^2 \end{pmatrix},$$

where

$$a_{1,1}^{(2)} = \frac{x_1 - 2x_1^3 x_2 f_2(x_2) - f_3(x_2) x_1 + x_1 x_2 (f_1(x_2) + f_2(x_2) x_1^2)}{x_1 x_2}$$

and f_1, f_2, f_3 and f_4 are functions in the variable x_2.

Case 1. First let us have a look at the case

$$\{f_1(x_2) = x_2^4,\ f_2(x_2) = x_2,\ f_3(x_2) = x_2^6,\ f_4(x_2) = 2\, x_2^6 + 2\, x_2^4\}.$$

Following the algorithm given above, we compute the partial p-curvatures Δ_1^p and Δ_2^p, and then, we apply SimDec to $\{\Delta_1^p, \Delta_2^p\}$: to proceed with the second step, we form the matrix $M = t_1\, \Delta_1^p + t_2\, \Delta_2^p$ and we compute and factor its characteristic polynomial $\chi(M)$. We find

- $\chi(M)(X) = 2\, t_2\, x_2^{15}\, X + 2\, t_1\, t_2\, x_2^{12} + 2\, X\, t_2\, x_2^{12} + X\, t_2\, x_2^3 + 2\, t_2\, x_2^{15} t_1 + t_1\, t_2\, x_2^3 + X^2 + 2\, X\, t_1 + {t_1}^2 + {t_2}^2\, x_2^{18} + {t_2}^2 x_2^{15} + {t_2}^2 x_2^{24} + 2\, {t_2}^2 x_2^{27}.$

The fact that $\chi(M)$ is irreducible over $\mathbb{K}(t_1, t_2)$ implies that the partial p-curvatures cannot be simultaneously reduced (nor decomposed) and consequently, the partial differential system $[A_1, A_2]$ is irreducible over $\mathbb{K}$.

Case 2. Now, if

$$\{f_1(x_2) = 2\, x_2,\ f_2(x_2) = 0,\ f_3(x_2) = 2x_2^6 + x_2,\ f_4(x_2) = x_2 + x_2^2\},$$

then, applying the same process, we find

- $\chi(M)(X) = (X + t_1 + 2\, t_2 + t_2\, x_2^3 + t_2\, x_2^{15})\,(X + t_1 + 2\, t_2\, x_2^3),$

so that the system is decomposable.

Applying the method of Algorithm SimDec, we find

$$P := \begin{pmatrix} 1 & 2\,\dfrac{x_2\left(x_2 + x_2^4 + x_2^{10} + 2\,x_2^3 x_1^2 + x_2^{11} + x_2^6 + 2\,x_1^2 + x_1^2 x_2^{15}\right)}{2\,x_2^3 + 2 + x_2^{15}} \\ 0 & 1 \end{pmatrix}.$$

We can then verify that this matrix decomposes simultaneously the differential systems $[A_1]$ and $[A_2]$ (and thus the partial differential system):

$$P[A_1] = \begin{pmatrix} 1 & 0 \\ 0 & 1 \end{pmatrix}$$

and

$$P[A_2] = \begin{pmatrix} \dfrac{x_2^6 + 2\,x_2^2 + 2\,x_2 + 1}{x_2} & 0 \\ 0 & 2\,x_2 \end{pmatrix}.$$

More generally, we can see that the factorization of the system is the following:

- If $f_2(x_2) \neq 0$, then the system is irreducible;
- If $f_2(x_2) = 0$, then the system is decomposable.

4.2. Maximal Decomposition

Once a simultaneous decomposition has been computed, we may restrict the study to each block separately. We are now confronted to the case when the partial differential system $[A_1, \ldots, A_m]$ has partial p-curvatures satisfying $\chi(\Delta_i^p) = F_i^{m_i}$ with F_i irreducible and $m_i \geq 1$. If some $m_i = 1$, then $[A_1, \ldots, A_m]$ is irreducible and the factorization stops.

Let $\mathscr{M}$ denote the (partial) differential module associated with the system $[A_1, \ldots, A_m]$. We want to find a *maximal decomposition* of $\mathscr{M}$, i.e., a decomposition

$$\mathscr{M} = \mathscr{W}_1 \oplus \cdots \oplus \mathscr{W}_d,$$

where the $\mathscr{W}_i$ are indecomposable. As a result, we will write the differential system $[A_1, \ldots, A_m]$ in block diagonal form where the modules corresponding to the diagonal blocks are indecomposable. Here, the techniques from the previous section do not apply because a matrix $P \in \mathrm{GL}_n(\mathbb{K})$ that decomposes simultaneously the Δ_i^p does not necessarily decompose the differential systems $\partial_i(Y) = A_i\,Y$.

To handle this case, we can use the eigenring. In [7, Proposition 4.7], it is shown that there exists a "separating" element in the eigenring. This is a matrix T with characteristic polynomial $\chi(T) = F_1 \cdots F_d$ such that $\gcd(F_i, F_j) = 1$ and $\chi(T_{|\mathscr{W}_i}) = F_i$. Applying a classical result of the eigenring factorization method (see [2, Theorem 2] or [8, Proposition 6]) to this element T yields a maximal decomposition of $\mathscr{M}$.

In practice, such a separating element can be found by taking random elements in the eigenring. In case of failure, one can use the idempotent decomposition of the eigenring from [11] to obtain a maximal decomposition.

As noted in [7], one can also adapt here the method proposed by van der Put in [24, 26]. Let a_i denote a root of F_i, i.e., the image of X in $\mathbb{K}[X]/(F)$. Let $\mathbb{K}^+ := \mathbb{K}(a_1, \ldots, a_m)$. Applying the algorithm of Subsection 4.1 over $\mathbb{K}^+$, we are reduced to studying a differential module $\mathscr{M}^+$ over $\mathbb{K}^+$ having p-curvatures with characteristic polynomial of the form $(X - a_i)^{m_i}$. The latter can be reduced (over $\mathbb{K}^+$) using Subsection 2.3 and, thus, we obtain a differential module $\mathscr{M}^+$ (over $\mathbb{K}^+$) with a maximal decomposition (and a complete reduction of the indecomposable blocks). Now, $\mathbb{K}^+$ has a structure of differential module over $\mathbb{K}$ and we have $\mathscr{M} = \mathscr{M}^+ \otimes_{\mathbb{K}} \mathbb{K}^+$: from this, we recover a basis of $\mathscr{M}$ over $\mathbb{K}$ and, then, a maximal decomposition of $\mathscr{M}$ (and the indecomposable blocks are fully reduced).

This last method can turn out to be costly because it may require to work in an unnecessary algebraic extension. In the next section, we give a simple rational alternative to handle the reduction of indecomposable partial differential systems.

4.3. Reducing Indecomposable Blocks

Definition 4.2. Let $[A_1, \ldots, A_m]$ be an indecomposable partial differential system with coefficients in $(\mathbb{K}, \Theta)$. A *(maximal) simultaneous reduction* of $[A_1, \ldots, A_m]$ is given by an invertible matrix P such that:

1. for all i, $P[A_i] = \begin{pmatrix} B_i^{[1]} & * & \cdots & * \\ 0 & B_i^{[2]} & \ddots & \vdots \\ \vdots & \ddots & \ddots & * \\ 0 & \cdots & 0 & B_i^{[r]} \end{pmatrix}$;
2. for all i, the partial p-curvature of each system $\partial_i(Y) = B_i^{[j]}\, Y$ has a minimal polynomial of the form F_i with F_i irreducible.

Proposition 4.2. *Let $S = [A_1, \ldots, A_m]$ be an indecomposable partial differential system with coefficients in $(\mathbb{K}, \Theta)$. The matrix $P \in \mathrm{GL}_n(\mathbb{K})$ obtained by applying Algorithm* SimRed *to $\{\Delta_1^p, \ldots, \Delta_m^p\}$ provides a simultaneous reduction of $[A_1, \ldots, A_m]$.*

Proof. In the proof of Proposition 3.1, we have constructed an element μ and an invertible matrix P such that $P^{-1}\,\mu\, P = S$ where S is block triangular with zeros as diagonal blocks. Now, we remark that, after turning $(t_1, \ldots, t_s)$ into some $(0, \ldots, 0, 1, 0, \ldots, 0)$ (the 1 is in the ith position) the element $\mu^{[i]}$ obtained is a non-zero and non-invertible element in the partial eigenring $\mathscr{E}_i(S)$. Then, for the same reasons as in the proof of [2, Theorem 1], a direct calculation shows that, $B_i := P[A_i]$ has a reduced form (compare to the proof of [7, Proposition 5.1]). □

We obtain thus the following method to compute a simultaneous reduction of indecomposable partial differential systems.

- Compute the partial p-curvatures Δ_i^p of $[A_1, \ldots, A_m]$;
- Return $P := \mathrm{SimRed}(\{\Delta_1^p, \ldots, \Delta_m^p\})$.

5. Two Other Generalizations

We have shown in the previous sections 3 and 4 how to generalize the algorithm of [7] to factor partial differential systems in characteristic p. We will now see that this algorithm can be directly adapted to other situations as well. We will sketch the algorithms corresponding to [7] in the case of one variable but, following the approach of Sections 3 and 4 to generalize [7] to the multivariate case, one would obtain algorithms for factoring (integrable) partial local systems and (integrable) partial difference systems.

5.1. "Local" Factorizations

In this subsection, we give the elements needed to generalize the algorithm factoring differential systems with coefficients in $\mathbb{K} = k(x)$ with $k = \mathbb{F}_q$ for some $q = p^r$ to the case where the coefficients belong to $\mathbb{K}((x))$.

Let $[A]$ be a differential system with $A \in \mathbb{M}_n(\mathbb{K}((x)))$. The notions of p-curvature and eigenring can be defined as in the ordinary differential case. Noting that $\mathbb{K}((x))$ is a C_1-field ([15, Definition 11.5, p. 649]), we deduce that the classification of differential modules in characteristic p given by van der Put in [23] (see also [26, Ch. 13]) can be applied in this case. Consequently, to construct an algorithm as the one given in [7] when $A \in \mathbb{M}_n(\mathbb{K}(x))$, it only remains to specify how to factor a polynomial with coefficients in $\mathbb{F}_q((x))$ where $q = p^r$: this can be done by using the standard Newton/Puiseux theorem (see [1, Lecture 12] for example):

Lemma 5.1. *Let $F(Y) = Y^n + a_{n-1}(x)\,Y^{n-1} + \cdots + a_0(x)$ be a monic polynomial with coefficients in $k((x))$ with $k = \overline{\mathbb{F}}_p$ and $p > n$. There exists an $r \in \mathbb{N}^*$ such that p does not divide r, and $F(Y) = \prod_{i=1}^{n}(Y - \nu_i)$ with $\nu_i \in k((x^{1/r}))$.*

All the ingredients needed have thus been given and by applying this theorem to the characteristic polynomial of the p-curvature, we obtain an immediate generalization of the algorithm given in [7] to the case where the system has coefficients in $\mathbb{K}((x))$.

Remark 5.1. In Lemma 5.1, if F denotes the characteristic polynomial of the p-curvature, then the ν_i are related to what we call the *exponential parts* of the system. More precisely, we can define a notion of exponential parts in characteristic p in the same way as in characteristic zero and show that they are exactly the reduction modulo p of the exponential parts in characteristic zero: this is detailed in [9] (see also [8, Ch. 2]).

5.2. Factorizations of Difference Systems

The algorithm developed to factor differential systems $Y(x)' = A(x)\,Y(x)$ in characteristic p can be generalized to the case of difference systems

$$Y(x+1) = A(x)\,Y(x).$$

The differential field $(k(x), ')$ where $k = \mathbb{F}_q$ with $q = p^r$ is replaced by the difference field $(k(x), \sigma)$ where σ is defined by $\sigma(x) = x + 1$ and $\sigma(f) = f$ for all $f \in k$. The constant field $\{a \in k\,;\, \sigma(a) = a\}$ is then $k(x^p - x)$ (see [25, Ch. 5] or [11, Theorem 3.1]).

As in the differential case, there exists a natural notion of p-curvature.

Definition 5.1. Let $\sigma(Y) = AY$ with $A \in \mathrm{GL}_n(k(x))$ ($k = \mathbb{F}_q$, $q = p^r$) be a difference system. Its *p-curvature* is the product of matrices

$$A(x+p-1)\cdots A(x+1)\,A(x).$$

A classification of difference modules in characteristic p (similar to that of [23] or [26, Ch. 13] in the differential case) is given in [25, Ch. 5]. It implies that the Jordan form of the p-curvature gives all the factorizations of the difference system. The equation $y^{(p-1)} + y^p = \lambda$ with $\lambda \in k(x^p)$ is replaced by

$$u(x+p-1)\cdots u(x+1)\,u(x) = \lambda$$

with $\lambda \in k(x^p - x)$. When the p-curvature is scalar, then the method used to improve [7, Lemma 3.6] and to obtain [7, Theorem 3.7] cannot be imitated; indeed, if we suppose $p > n$ and try to adapt the proof, the solution $\mathrm{Tr}(A)/n$ of $y^{(p-1)} + y^p = \lambda$ is replaced by the solution $\det(A)^{1/n}$ of $u(x+p-1)\cdots u(x+1)\,u(x) = \lambda$ and we lose the rationality of this solution.

One can define an eigenring as well (see for example [2, 11]): let $\sigma(Y) = AY$ be a difference system with $A \in \mathbb{M}_n(k(x))$. *The eigenring* $\mathcal{E}(A)$ *of* $[A]$ is the set defined by

$$\mathcal{E}(A) = \{P \in \mathbb{M}_n(k(x)) \,|\, \sigma(P)\,A = A\,P\}.$$

All the elements needed to develop an algorithm similar to that of [7] are collected and the algorithm follows naturally. Note further that:

- The results of [2] stay true in the difference case ([2] is written in the general setting of pseudo linear equations);
- The algorithm of Giesbrecht and Zhang [11] can be used to factor Ore polynomials thus, in particular, difference operators.

References

[1] S. S. Abhyankar. *Algebraic geometry for scientists and engineers.* Mathematical Surveys and Monographs, vol. **35**. AMS, 1990.

[2] M. A. Barkatou. On the reduction of matrix pseudo-linear equations. Technical Report RR 1040, Rapport de Recherche de l'Institut IMAG, 2001.

[3] M. A. Barkatou and E. Pflügel. An algorithm computing the regular singular formal solutions of a linear differential system. *J. Symb. Comput.* **28**(4–5), 1999.

[4] Y. A. Blinkov, C. F. Cid, V. T. Gerdt, W. Plesken and D. Robertz. The Maple package "Janet" II: Polynomial systems. In: *Proceedings of the Workshop on Computer Algebra and Scientific Computing* (CASC), Passau, 2003. `http://wwwmayr.informatik.tu-muenchen.de/CASC2003/`

[5] L. Chambadal and J. L. Ovaert. *Algèbre linéaire et algèbre tensorielle.* Dunot Université, Paris, 1968.

[6] A. Chambert-Loir. Théorèmes d'algébricité en géométrie diophantienne. *Séminaire Bourbaki*, exposé No. **886**, Mars 2001.

[7] T. Cluzeau. Factorization of differential systems in characteristic p. In: *Proceedings of the 2003 International Symposium on Symbolic and Algebraic Computation*, pp. 58–65. ACM Press, New York, 2003.

[8] T. Cluzeau. Algorithmique modulaire des équations différentielles linéaires. Thèse de l'Université de Limoges, Septembre 2004.

[9] T. Cluzeau and M. van Hoeij. A modular algorithm for computing the exponential solutions of a linear differential operator. *J. Symb. Comput.* **38**(3): 1043–1076, 2004.

[10] R. Corless,P. Gianni and B. Trager. A reordered Schur factorization method for zero-dimensional polynomial systems with multiple roots. In: *Proceedings of the 1997 International Symposium on Symbolic and Algebraic Computation*, pp. 133–140. ACM Press, New York, 1997.

[11] M. Giesbrecht and Y. Zhang. Factoring and decomposing Ore polynomials over $\mathbb{F}_p(t)$. In: *Proceedings of the 2003 International Symposium on Symbolic and Algebraic Computation*, pp. 127–134. ACM Press, New York, 2003.

[12] M. Hausdorf and M. Seiler. Involutives basis in MuPAD — Part I: Involutive divisions. *MathPad* **11**(1): 51–56, 2002.

[13] M. van Hoeij, J.-F. Ragot, F. Ulmer and J.-A. Weil. Liouvillian solutions of linear differential equations of order three and higher. *J. Symb. Comput.* **11**: 1–17, 1998.

[14] N. Jacobson. *Lectures in abstract algebra II: Linear algebra.* Graduate Texts in Mathematics, vol. **31**. Springer, New York, 1953.

[15] N. Jacobson. *Basic algebra II.* W. H. Freeman and Compagny, San Francisco, 1980.

[16] M. Janet. Sur les systèmes aux dérivées partielles. *Journal de Math.*, 8-ème série, III, 65–151, 1924.

[17] M. Janet. *Leçons sur les systèmes d'équations aux dérivées partielles.* Cahiers Scientifiques, **IV**, Gauthiers-Villars, 1929.

[18] N. Katz. Nilpotent connections and the monodromy theorem: Applications of a result of Turritin. *Publ. Math. I. H. E. S.* **39**: 355–412, 1970.

[19] N. Katz. A conjecture in the arithmetic theory of differential equations. *Bull. Soc. Math. France* **110**: 203–239, 1982.

[20] A. H. M. Levelt. The semi-simple part of a matrix. *Algorithmen in de Algebra*, 85–88, 1994. `http://www.math.ru.nl/medewerkers/ahml/other.htm`

[21] Z. Li, F. Schwarz and S. P. Tsarev. Factoring zero-dimensional ideals of linear partial differential operators. In: *Proceedings of the 2002 International Symposium on Symbolic and Algebraic Computation*, pp. 168–175. ACM Press, New York, 2002.

[22] Z. Li, F. Schwarz and S. P. Tsarev. Factoring systems of linear PDEs with finite-dimensional solution spaces. *J. Symb. Comput.* **36**(3–4): 443–471, 2003.

[23] M. van der Put. Differential equations in characteristic p. *Compositio Math.* **97**: 227–251, 1995.

[24] M. van der Put. Modular methods for factoring differential operators. Manuscript, 1997.

[25] M. van der Put and M. F. Singer. *Galois theory of difference equations.* Lecture Notes in Mathematics, vol. **1666**. Springer, Berlin, 1997.

[26] M. van der Put and M. F. Singer. *Galois theory of linear differential equations.* Grundlehren der mathematischen Wissenschaften, vol. **328**. Springer, Berlin, 2003.

[27] M. F. Singer. Testing reducibility of linear differential operators: A group theoretic perspective. *Appl. Alg. Eng. Comm. Comp.* **7**(2): 77–104, 1996.

[28] M. Wu. On the factorization of differential modules. In this volume.

Acknowledgments

The authors would like to thank Alban Quadrat for helpful explanations and references concerning D-finite partial differential systems and Janet bases, and Marius van der Put for his comments and the appendix that follows.

Appendix: Classification of Partial Differential Modules in Positive Characteristic

By Marius van der Put
University of Groningen, Department of Mathematics
P.O. Box 800, 9700 AV Groningen, The Netherlands
M.van.der.Put@math.rug.nl

(1) *Introduction.* In [23, 26] a classification of differential (resp. difference in [25]) modules over a differential field K of characteristic $p > 0$ with $[K : K^p] = p$ is given. The differential modules in question can be seen as ordinary matrix differential equations. Here we show how to extend this to, say, the case $[K : K^p] = p^m$ with $m > 1$ (compare [1], 6.6 Remarks (1)). The (partial) differential modules are the partial differential equations considered in this paper. In order to simplify the situation, we will, as in the paper, avoid the skew field that may arise in the classification. The algorithmic results of the paper are made more transparent from the classification that we will work out.

(2) *Assumptions and notation.* Let K be a field of characteristic $p > 0$ and let K_0 be a subfield such that the universal differential module Ω_{K/K_0} has dimension $m \geq 1$ over K. There are elements $x_1, \ldots, x_m \in K$ such that $\{dx_1, \ldots, dx_m\}$ is a basis of Ω_{K/K_0}. Then $x_1, \ldots, x_m$ form a p-basis of K/K_0 which means that the set of monomials $\{x_1^{a_1} \cdots x_m^{a_m} \mid 0 \leq a_i < p \text{ for all } i\}$ is a basis of K over K^pK_0. We will write $C := K^pK_0$. For $i \in \{1, \ldots, m\}$, the derivation ∂_i of K/K_0 is given by $\partial_i x_j = \delta_{i,j}$. Clearly, the $\{\partial_i\}$ is a set of commuting operators. Put $\mathcal{D} := K[\partial_1, \ldots, \partial_m]$. This is a ring of differential operators and the partial differential equations that one considers (in the paper and here) are left $\mathcal{D}$-modules M of finite dimension over K. We note that M is a cyclic module (and thus $M \cong \mathcal{D}/J$ for some left ideal J of finite codimension) if $\dim_K M \leq p$. If $\dim_K M > p$, then in general M is not cyclic (compare [2], Exercise 13.3, p. 319). For notational convenience we will write $\mathcal{D}$-module for left $\mathcal{D}$-module of finite dimension over K.

(3) *Classification of the irreducible $\mathcal{D}$-modules.* Similarly to [23], one can prove the following statements. The center Z of $\mathcal{D}$ is $C[t_1, \ldots, t_m]$. The latter is a (free)

polynomial ring in the variables $\{t_i := \partial_i^p\}_{i=1}^m$. Consider any maximal ideal $\underline{m} \subset Z$ and put $L = Z/\underline{m}$. Then $L \otimes_Z \mathcal{D}$ is a central simple algebra over L of dimension p^{2m}. The well-known classification implies that this algebra is isomorphic to a matrix algebra $\mathrm{Matr}(p^{m_1}, D)$ where D is a (skew) field having dimension p^{2m_2} over its center L. Clearly $m_1 + m_2 = m$. The unique simple left module M of this algebra is $D^{p^{m_1}}$ and has dimension $p^{-m}[L : C]p^{m_1+2m_2} = p^{m_2}[L : C]$ over K. In particular, if the dimension of M over K is $< p$, then $L \otimes_Z \mathcal{D}$ is isomorphic to $\mathrm{Matr}(p^m, L)$.

Let M be an irreducible $\mathcal{D}$-module. Then M is also a Z-module of finite dimension over C. The irreducibility of M implies that $\underline{m}M = 0$ for some maximal ideal $\underline{m}$ of Z (write again $L = Z/\underline{m}$). Hence M is a simple left module over $L \otimes_Z \mathcal{D}$. If one knows the structure of the algebras $L \otimes_Z \mathcal{D}$, then the classification of irreducible $\mathcal{D}$-modules is complete.

(4) *Isotypical decomposition.* Let M be any $\mathcal{D}$-module. Put $I := \{a \in Z \mid aM = 0\}$, the annulator of M. Then $I \subset Z$ is an ideal of finite codimension. Thus M is also a Z/I-module of finite dimension. Let $\{\underline{m}_1, \ldots, \underline{m}_s\}$ denote the set of maximal ideals containing I. This is the support of M. Then the Artin ring Z/I is the direct product of the local Artin rings $Z_{\underline{m}_i}/(I)$. One writes $1 = e_1 + \cdots + e_m$, where e_i is the unit element of the ring $Z_{\underline{m}_i}/(I)$. Then $M = \oplus M_i$, with $M_i = e_i M$. This is a module over $Z_{\underline{m}_i}/(I)$. Since Z is the center of $\mathcal{D}$ each M_i is again a $\mathcal{D}$-module. Moreover, the annulator of M_i is an ideal with radical $\underline{m}_i$. The above decomposition will be called the isotypical decomposition of M. The classification of $\mathcal{D}$-modules is in this way reduced to the classification of $\mathcal{D}$-modules which are annihilated by a power of some maximal ideal $\underline{m}$ of Z. The latter depends on the structure of $Z/\underline{m} \otimes_Z \mathcal{D}$.

(5) *Restricting the class of* $\mathcal{D}$*-modules.* Let S denote the set of maximal ideals $s = \underline{m}$ in Z such that the algebra $Z/\underline{m} \otimes_Z \mathcal{D}$ is isomorphic to $\mathrm{Matr}(p^m, Z/\underline{m})$. In the sequel we will only consider $\mathcal{D}$-modules with support in S. The differential modules M, considered in this paper, satisfy $\dim_K M < p$. According to (3), their support is in S. We note that S depends on the fields $K_0 \subset K$. There are examples where S is the set of all maximal ideals of Z.

(6) *Classification of the* $\mathcal{D}$*-modules with support in* $\{s\}$*, where* $s \in S$. We fix a maximal ideal $s = \underline{m} \in S$. The above Tannakian category will be denoted by $(\mathcal{D}, s)$. We note that the tensor product $M_1 \otimes M_2$ of two objects in this category is defined as $M_1 \otimes_K M_2$, provided with the action of ∂_i (for $i = 1, \ldots, m$) given by

$$\partial_i(m_1 \otimes m_2) = (\partial_i m_1) \otimes m_2 + m_1 \otimes (\partial_i m_2).$$

Consider the category (Z, s) of the finitely generated Z-modules N, with support in $\{s\}$. The Tannakian structure of this category is determined by the definition of a tensor product. The tensor product of two modules N_1, N_2 in (Z, s) is $N_1 \otimes_C N_2$ equipped with the operations t_i given by

$$t_i(n_1 \otimes n_2) = (t_i n_1) \otimes n_2 + n_1 \otimes (t_i n_2).$$

The aim is to produce an equivalence $\mathcal{F}_s : (Z, s) \to (\mathcal{D}, s)$ of Tannakian categories. Once this is established, the required classification is reduced to classifying the objects of (Z, s). The functor $\mathcal{F}_s$ is defined as $\mathcal{F}_s(N) = M := K \otimes_C N$. The right-hand side is clearly a (left) $K[t_1, \dots, t_m]$-module. It suffices to extend this to a left $\mathcal{D}$-module by defining the operation of the ∂_i on M.

Let $\widehat{Z}_s$ denote the completion of the local ring $Z_{\underline{m}}$. Let $(\widehat{Z}_s, s)$ denote the category of the $\widehat{Z}_s$-modules of finite dimension over C. The categories (Z, s) and $(\widehat{Z}_s, s)$ are clearly the "same". Put $\widehat{\mathcal{D}}_s = \widehat{Z}_s \otimes_Z \mathcal{D}$ and let $(\widehat{\mathcal{D}}_s, s)$ denote the category of the left $\widehat{\mathcal{D}}_s$-modules which have finite dimension over K. Then the categories $(\mathcal{D}, s)$ and $(\widehat{\mathcal{D}}_s, s)$ are the "same". Therefore it suffices to construct an equivalence $\mathcal{F}_s : (\widehat{Z}, s) \to (\widehat{\mathcal{D}}_s, s)$.

For this purpose we need a free, rank one, $\widehat{Z}_s \otimes_C K$-module $\mathcal{Q}_s = \widehat{Z}_s \otimes_C Ke$, such that its structure of $\widehat{Z}_s \otimes_C K$-module extends to that of a left $\widehat{\mathcal{D}}_s$-module. Given $\mathcal{Q}_s$, the functor $\mathcal{F}_s$ is defined by $N \mapsto M := N \otimes_{\widehat{Z}_s} \mathcal{Q}_s$. Then M is a left $\widehat{\mathcal{D}}_s$-module by $\lambda(n \otimes \mu e) = n \otimes (\lambda\mu)e$. It is easily verified that $\mathcal{F}_s$ is indeed an equivalence of Tannakian categories. We note that M is equal to $N \otimes_C K$ as $\widehat{Z}_s \otimes_C K$-module, and our construction extends this to a left $\widehat{\mathcal{D}}_s$-module structure.

(7) *The construction of* $\mathcal{Q}_s$. By assumption $A_0 := Z/\underline{m} \otimes_Z \mathcal{D}$ is isomorphic to $\mathrm{Matr}(p^m, Z/\underline{m})$. Let I be the (unique) simple left module of A_0. Then the morphism $A_0 \to \mathrm{End}_{Z/\underline{m}}(I)$ is a bijection. In particular, the commutative subalgebra $Z/\underline{m} \otimes_C K$ of A_0 acts faithfully on I. By counting dimensions over C, one sees that I is in fact a free $Z/\underline{m} \otimes_C K$-module with generator, say, e. Thus we have found a left A_0-module structure on $Z/\underline{m} \otimes_C Ke$. Now $\mathcal{Q}_s$ is constructed by lifting this structure, step by step, to a left $\widehat{\mathcal{D}}_s$-module structure on $\widehat{Z}_s \otimes_C Ke$. This is in fact equivalent to lifting a given isomorphism $A_0 \to \mathrm{Matr}(p^m, Z/\underline{m})$ to an isomorphism $\widehat{\mathcal{Q}}_s \to \mathrm{Matr}(p^m, \widehat{Z}_s)$. The method of [1] for the case $m = 1$ can be extended here. For notational convenience we present here a proof for the case $p = 2$ and $m = 2$.

We note that $\widehat{Z}_s \otimes_C K = \widehat{Z}_s[x_1, x_2]$ has a free basis $\{1, x_1, x_2, x_1x_2\}$ over $\widehat{Z}_s$. Consider the free module $\widehat{Z}_s[x_1, x_2]e$. We have to construct operators ∂_1 and ∂_2 on this module such that $\partial_1\partial_2 - \partial_2\partial_1 = 0$ and $\partial_i^2 = t_i$ for $i = 1, 2$. Put $\partial_i e = \ell_i e$ for $i = 1, 2$. Then the conditions are $\partial_i(\ell_i) + \ell_i^2 - t_i = 0$ for $i = 1, 2$ and $\partial_1(\ell_2) - \partial_2(\ell_1) = 0$. Suppose that we have found ℓ_1, ℓ_2 such that these equalities hold modulo $\underline{m}^s$. Then we want to change the ℓ_i in $\ell_i + r_i$ with $r_1, r_2 = 0 \bmod (\underline{m}^s)$ such that the required equalities hold module $\underline{m}^{s+1}$. This step suffices for the proof of the statement. It amounts to solving

$$\partial_i(r_i) = -\partial(\ell_i) - \ell_i^2 + t_i \bmod \underline{m}^{s+1} \quad \text{and}$$

$$\partial_1(r_2) - \partial_2(r_1) = \partial_2(\ell_1) - \partial_1(\ell_2) \bmod \underline{m}^{s+1}.$$

The right-hand sides of the equalities are already 0 mod $\underline{m}^s$. Write $r_1 = r_1(0,0) + r_1(1,0)x_1 + r_1(0,1)x_2 + r_1(1,1)x_1x_2$ and similarly for r_2. The right-hand side of the first equation with $i = 1$ is killed by the operator ∂_1 and therefore contains only the terms $1, x_2$. This leads to a unique determination of $r_1(1,0)$ and $r_1(1,1)$

and the $r_1(0,0), r_1(0,1)$ can be chosen freely. Similarly, the terms $r_2(0,1), r_2(1,1)$ are determined and the terms $r_2(0,0), r_2(1,0)$ can be chosen freely. The second equation reads

$$r_2(1,0) - r_1(0,1) + r_2(1,1)x_2 - r_1(1,1)x_1 = \partial_2(\ell_1) - \partial_1(\ell_2) \bmod \underline{m}^{s+1}.$$

The right-hand side R uses only the terms $1, x_1, x_2$. Moreover, $\partial_1(R) = \partial_1\partial_2(\ell_1) = \partial_2\partial_1(\ell_1)$ and this is equal to $\partial_2(\partial_1(\ell_1) + \ell_1^2 - t_1)$. Hence the coefficients of x_1 of the two sides are equal. The same holds for the coefficients of x_2. The coefficient of 1 on the two sides can be made equal for a suitable choice of $r_1(0,1)$ and/or $r_2(1,0)$.

(8) *Final remarks.* Let (Z, S) denote the Tannakian category of the Z-modules, having finite dimension over C and with support in S. Let $(\mathcal{D}, S)$ denote the category of the left $\mathcal{D}$-modules having finite dimension over K and with support in S (as Z-module). One can "add" the equivalences $\mathcal{F}_s$ in an obvious way to an equivalence $\mathcal{F} : (Z, S) \to (\mathcal{D}, S)$. For an object M of $(\mathcal{D}, S)$, there is an object N of (Z, S) such that $\mathcal{F}(N) = M$. Then M, as module over $K[t_1, \ldots, t_m]$, describes in fact the p-curvature of M. Since $N \otimes_C K \cong M$, one can say that N represents already the p-curvature of M. In particular, the characteristic (and minimal) polynomials for the t_i have their coefficients in $C = K_0K^p$.

As observed before, classifying the left $\mathcal{D}$-modules of finite dimension over K and with support in S is equivalent to classifying the Z-modules of finite dimension over C and with support in S. The latter is done by decomposing an object into isotypical components. Hence we may restrict our attention to a single maximal ideal $s = \underline{m} \in S$. The modules N that we want to classify are in fact the finitely generated modules over the complete regular local ring $\widehat{Z}_s \cong L[[d_1, \ldots, d_m]]$ which are annihilated by a power of the maximal ideal $\underline{m}$. Unlike the case $m = 1$, no reasonable classification (or moduli spaces) seems possible. One observation can still be made. The module N has a sequence of submodules $0 = N_0 \subset N_1 \subset \cdots \subset N_t = N$ such that each quotient N_{i+1}/N_i is isomorphic to the module $L = Z/\underline{m}$. In other words, N is a multiple extension of the module L.

Moulay A. Barkatou, Thomas Cluzeau and Jacques-Arthur Weil
LACO, Université de Limoges
123 avenue Albert Thomas
87060 Limoges, France
e-mail: {moulay.barkatou,thomas.cluzeau,jacques-arthur.weil}@unilim.fr

Trends in Mathematics: *Differential Equations with Symbolic Computation*, 239–254

On the Factorization of Differential Modules

Min Wu

Abstract. Differential modules are modules over rings of linear (partial) differential operators which are finite-dimensional vector spaces. We present a generalization of the Beke–Schlesinger algorithm that factors differential modules. The method requires solving only one set of associated equations for each degree d of a potential factor.

Mathematics Subject Classification (2000). 13N05; 13N10; 13N15; 16S32; 16S36; 32C38; 33F10; 68W30.

Keywords. Differential (sub)module, ring of linear differential operators, integrable system, hyperexponential solution.

1. Introduction

The problem of factoring linear ordinary differential equations or operators has been studied for a long time since Beke [1] and Schlesinger [12] introduced the idea of associated equations and developed a factorization algorithm for linear ordinary differential equations (operators). This idea together with some improvements [2, 13] on the Beke–Schlesinger algorithm inspired the study of reducing the factorization problem to that of finding hyperexponential solutions of associated systems. Making use of the main algorithm in [8] for computing hyperexponential solutions of systems of linear partial differential equations (PDEs), Li and others [9, 10] generalize the Beke–Schlesinger factorization algorithm to systems of linear PDEs (in one or several unknowns) with finite-dimensional solution spaces. However, in their algorithm, there is a combinatorial explosion caused by trying out all the possible sets of leading derivatives for a potential factor. To avoid this combinatorial explosion, we formulate the factorization problem in terms of differential modules and then reduce the factorization problem to that of finding one-dimensional differential submodules, and further to that of computing hyperexponential solutions of associated systems.

Throughout this paper, fields are always assumed to be commutative, modules and vector spaces are left modules and left vector spaces. The notation $(\cdot)^\tau$

denotes the transpose of vectors or matrices. We write $\mathrm{M}_n(k)$ (resp. $\mathrm{GL}_n(k)$) for the set of all $n \times n$ (resp. invertible) matrices with entries in a field k.

The paper is organized as follows. Section 2 introduces the notion of differential modules and gives some straightforward generalizations of results in Chapters 1 and 2 of [11]. Section 3 outlines the idea of the generalized Beke–Schlesinger factorization method, by reducing the problem of finding d-dimensional submodules of a differential module to that of finding 1-dimensional submodules of its exterior power and by linking 1-dimensional submodules and hyperexponential solutions of the associated integrable systems. Section 4 describes a factorization algorithm for differential modules.

2. Preliminaries

2.1. Differential Modules

Let k be a field (of characteristic zero).

Definition 2.1. A *derivation* on the field k is a map $\delta : k \mapsto k$ satisfying

$$\delta(a+b) = \delta(a) + \delta(b), \quad \delta(ab) = \delta(a)b + a\delta(b),$$

for all $a, b \in k$.

The field k equipped with a set $\{\delta_1, \ldots, \delta_m\}$ of pairwise commuting derivations is called a (*partial*) *differential field*. An element c of k is called a *constant w.r.t.* δ_i if $\delta_i(c) = 0$. An element c of k is called a *constant* if c is a constant w.r.t. all the δ_i. All constants of k form a subfield of k, which we denote by C.

Let k be a differential field equipped with the pairwise commuting derivations $\delta_1, \ldots, \delta_m$. The *ring of linear differential operators* over k is the ring $k[\partial_1, \ldots, \partial_m]$ of noncommutative polynomials in $\partial_1, \ldots, \partial_m$ with the following multiplicative rule:

$$\partial_i \partial_j = \partial_j \partial_i \quad \text{and} \quad \partial_i a = a\partial_i + \delta_i(a),$$

for $a \in k$ and $i, j \in \{1, \ldots, m\}$. Denote $\mathbb{D} = k[\partial_1, \ldots, \partial_m]$. Definition D.3 in [11] states that M is a *differential module* if M is a $\mathbb{D}$-module that is also a finite-dimensional vector space over k. Hence the dimension of the differential module M is understood as the dimension of M as a vector space over k.

2.2. Constructions on Differential Modules

We observe that the notion of ordinary differential modules introduced in Chapter 1 of [11] is a special case of differential modules where $m = 1$. As a straightforward generalization, the constructions on differential modules can be carried on as follows.

A (*differential*) *submodule* N of M is a k-vector subspace $N \subseteq M$ such that $\partial_i(N) \subseteq N$ for $i = 1, \ldots, m$.

Let N be a submodule of M. Then M/N endowed with the map ∂_i given by

$$\partial_i(w + N) = \partial_i(w) + N,$$

for $w \in M$ and $i = 1, \dots, m$, is the *quotient differential module.*

The *direct sum* of two differential modules M_1 and M_2 is $M_1 \oplus M_2$ endowed with the map ∂_i given by

$$\partial_i(w_1 + w_2) = \partial_i(w_1) + \partial_i(w_2),$$

for $w_1 \in M_1$, $w_2 \in M_2$ and each i.

The *tensor product* $M_1 \otimes M_2$ of two differential modules is $M_1 \otimes_k M_2$ endowed with the map ∂_i given by

$$\partial_i(w_1 \otimes w_2) = \partial_i(w_1) \otimes w_2 + w_1 \otimes \partial_i(w_2),$$

for $w_1 \in M_1$, $w_2 \in M_2$ and each i.

The dth *exterior power* $\wedge^d M$ of a differential module M is the k-vector space $\wedge_k^d M$ endowed with the map ∂_i given by

$$\partial_i(w_1 \wedge \cdots \wedge w_d) = \sum_{j=1}^{d} w_1 \wedge \cdots \wedge \partial_i(w_j) \wedge \cdots \wedge w_d,$$

for $w_1, \dots, w_d \in M$ and each i.

A *morphism* $\phi : M_1 \to M_2$ is a k-linear map ϕ such that $\phi \circ \partial_i = \partial_i \circ \phi$ for $i = 1, \dots, m$, namely, ϕ is a $\mathbb{D}$-linear map.

Two differential modules are said to be *isomorphic* if there exists a bijective morphism between them.

The *internal* Hom, $\mathrm{Hom}(M_1, M_2)$ of two differential modules is the k-vector space $\mathrm{Hom}_k(M_1, M_2)$ of all k-linear maps from M_1 to M_2 endowed with the map ∂_i on $\mathrm{Hom}_k(M_1, M_2)$ given by

$$\partial_i(\ell)(w_1) = -\ell(\partial_i(w_1)) + \partial_i(\ell(w_1)),$$

for $w_1 \in M_1$, $\ell \in \mathrm{Hom}_k(M_1, M_2)$ and $i = 1, \dots, m$. A special case of the internal Hom is the *dual module* of a differential module M, which is defined to be M^*=$\mathrm{Hom}_k(M, \mathbf{1_k})$ where $\mathbf{1_k}$ denotes the differential module kw with $\partial_i(w) = 0$ for each i.

3. Reduction of the Factorization Problem

By *factoring* a differential module, we mean finding its proper submodules. In the sequel, we will outline the idea of the generalized Beke–Schlesinger algorithm for factoring differential modules.

3.1. From M to $\wedge^d M$

Let M be a differential module. An element w of $\wedge^d M$ is said to be *decomposable* if w is a wedge product of a finite number of elements of M, i.e., $w = w_1 \wedge \cdots \wedge w_d$ with $w_i \in M$.

We have the following generalization of Lemma 10 in [3] or the corresponding statement in Section 4.2.1 of [11].

Proposition 3.1. *A differential module M has a d-dimensional submodule if and only if $\wedge^d M$ has a 1-dimensional submodule generated by a decomposable element.*

Proof. Let N be a d-dimensional submodule of M with a basis $w_1, \ldots, w_d$ over k. Suppose that

$$\partial_i (w_1, \ldots, w_d)^\tau = A_i (w_1, \ldots, w_d)^\tau \quad \text{for } i = 1, \ldots, m,$$

where $A_i = (a_{ist})_{1 \leq s, t \leq d} \in \mathrm{M}_d(k)$. Then $\wedge_k^d N$ is a k-vector subspace of $\wedge_k^d M$ generated by $w_1 \wedge \cdots \wedge w_d$. Moreover,

$$\begin{aligned} \partial_i (w_1 \wedge \cdots \wedge w_d) &= \sum_{s=1}^{d} w_1 \wedge \cdots \wedge \partial_i(w_s) \wedge \cdots \wedge w_d \\ &= \sum_{s=1}^{d} w_1 \wedge \cdots \wedge \left(\sum_{t=1}^{d} a_{ist}\, w_t \right) \wedge \cdots \wedge w_d = \mathrm{tr}(A_i)\,(w_1 \wedge \cdots \wedge w_d) \in \wedge^d N, \end{aligned}$$

where $\mathrm{tr}(A_i)$ denotes the trace of the matrix A_i, for $i = 1, \ldots, m$. So $\wedge^d N$ is a 1-dimensional submodule of $\wedge^d M$.

Conversely, let $u \in \wedge^d M$ be a decomposable element which generates a 1-dimensional submodule of $\wedge^d M$. Suppose that $u = w_1 \wedge \cdots \wedge w_d$ with $w_i \in M$. Since $u \neq 0$, $w_1, \ldots, w_d$ are linearly independent over k and there exists a basis B of M containing $w_1, \ldots, w_d$. Pick arbitrarily a finite number of distinct $b_1, \ldots, b_s$ in $B \setminus \{w_1, \ldots, w_d\}$. Since $w_1, \ldots, w_d, b_1, \ldots, b_s$ are linearly independent over k, so are $b_1 \wedge u, \ldots, b_s \wedge u$. In particular, $b \wedge u \neq 0$ for any $b \in B \setminus \{w_1, \ldots, w_d\}$.

Consider a map $\phi_u : M \to \wedge^{d+1} M$ defined by $v \mapsto v \wedge u$. One can verify that $\ker(\phi_u)$ is a k-vector space. Let $v \in \ker(\phi_u)$. Then $v \wedge u = 0$ and

$$0 = \partial_i(v \wedge u) = \partial_i(v) \wedge u + v \wedge \partial_i(u) = \partial_i(v) \wedge u + v \wedge (au) \quad \text{for some } a \in k,$$

which implies that $\partial_i(v) \wedge u = 0$ and $\partial_i(v) \in \ker(\phi_u)$ for each i. So $\ker(\phi_u)$ is a $\mathbb{D}$-module.

Clearly, $\oplus_{i=1}^d k w_i \subseteq \ker(\phi_u)$. Suppose that $w \in \ker(\phi_u) \subset M$. Then there exist $b_1, \ldots, b_s \in B \setminus \{w_1, \ldots, w_d\}$ such that

$$w = \sum_{i=1}^{d} \lambda_i w_i + \sum_{j=1}^{s} \xi_j b_j \quad \text{with } \lambda_i,\ \xi_j \in k.$$

Then $0 = w \wedge u = \sum_{j=1}^{s} \xi_j \left(b_j \wedge u \right)$. The linear independence of $b_1 \wedge u, \ldots, b_s \wedge u$ therefore implies that $\xi_j = 0$ for $j = 1, \ldots, s$. So $\ker(\phi_u) = \oplus_{i=1}^d k w_i$ is a d-dimensional submodule of M. □

Proposition 3.1 converts the problem of finding d-dimensional submodules of a differential module M into that of finding all those 1-dimensional submodules of $\wedge^d M$ whose generator is decomposable, and thus reduces the factorization problem to its "subproblem"of finding 1-dimensional submodules.

Remark 3.2. Proposition 3.1 and the proof remain valid if M is infinite-dimensional over k.

3.2. One-Dimensional Submodules of Differential Modules

In this section, we study how to find 1-dimensional submodules of differential modules.

We recall the definition of integrable systems given in Appendix D.1 of [11]: let $A_1, \dots, A_m$ be $n \times n$ matrices with entries in a differential field k; the system

$$\{ \delta_1(Z) = A_1 Z, \ \dots, \ \delta_m(Z) = A_m Z \} \tag{3.1}$$

is called an *integrable system* over k if $A_1, \dots, A_m$ satisfy the following integrability condition:

$$[A_i, A_j] = \delta_i(A_j) - \delta_j(A_i), \qquad \text{for any } i, j, \tag{3.2}$$

where $[A_i, A_j] := A_i A_j - A_j A_i$ is the commutator of the matrices A_i and A_j.

Let M be a differential module of dimension n and $e_1, \dots, e_n$ be a basis of M over F. Suppose that

$$\partial_i(e_1, \dots, e_n)^\tau = B_i(e_1, \dots, e_n)^\tau, \quad \text{with } B_i \in \mathrm{M}_n(k), \ i = 1, \dots, m.$$

Let $w = \sum_{i=1}^n a_i e_i \in M$ with $a_i \in k$. Then

$$\partial_i(w) = \partial_i((a_1, \dots, a_n)(e_1, \dots, e_n)^\tau) = (\delta_i(a_1, \dots, a_n) + (a_1, \dots, a_n)B_i)\,(e_1, \dots, e_n)^\tau,$$

for $i = 1, \dots, m$; hence the condition

$$\partial_1(w) = 0, \ \dots, \ \partial_m(w) = 0 \tag{3.3}$$

translates to

$$\delta_i(a_1, \dots, a_n)^\tau = -B_i^\tau(a_1, \dots, a_n)^\tau, \quad i = 1, \dots, m. \tag{3.4}$$

In other words, the coefficient vector $Z := (a_1, \dots, a_n)^\tau$ of w satisfies the system (3.1) where $A_i = -B_i^\tau$. It follows from (3.4) that the A_i satisfy the integrability condition (3.2) and that (3.1) is an integrable system over k, which is called the *integrable system associated with* M w.r.t. a basis $e_1, \dots, e_n$. If we choose another basis $f_1, \dots, f_n$ of M over k with

$$(f_1, \dots, f_n) = (e_1, \dots, e_n)T \quad \text{for some } T \in \mathrm{GL}_n(k),$$

then by replacing Z with TZ^* in (3.1) we obtain the integrable system for the new basis:

$$\{\delta_1(Z^*) = A_1^* Z^*, \ \dots, \ \delta_m(Z^*) = A_m^* Z^*\},$$

where $A_i^* = T^{-1}A_i T - T^{-1}\delta_i(T)$ for each i. Two integrable systems $\{\delta_i(Z) = A_i Z\}_{1 \le i \le m}$ and $\{\delta_i(Z^*) = A_i^* Z^*\}_{1 \le i \le m}$ over k of dimension n are said to be *equivalent* if there exists a $T \in \mathrm{GL}_n(k)$ such that $A_i^* = T^{-1}A_i T - T^{-1}\delta_i(T)$ for each i. From the above statement, the integrable systems associated with the same differential module w.r.t. different bases are equivalent to each other.

It is clear that any integrable system of form (3.1) comes from a differential module $M := k^n$ with the canonical basis $\{e_1, \dots, e_n\}$ and the ∂_i given by the formulas

$$\partial_i(e_1, \dots, e_n)^\tau = -A_i^\tau(e_1, \dots, e_n)^\tau \quad \text{for } i = 1, \dots, m.$$

To investigate "solutions" of integrable systems, we introduce

Definition 3.3. A field $K \supseteq k$ is called a *differential extension field over* k if all derivations $\delta_1, \ldots, \delta_m$ on k can be extended to K and the extended maps pairwise commute.

Let K be a differential extension field over k. A vector $Z^* \in K^n$ is called a *solution* of the integrable system (3.1) if $\delta_i(Z^*) = A_i Z^*$ for each i.

First, we recall some definitions introduced in [10]. A nonzero element h of K is said to be *hyperexponential over* k *w.r.t.* δ_i if $\frac{\delta_i(h)}{h}$ belongs to k. The element $h \in K$ is said to be *hyperexponential over* k if h is hyperexponential over k w.r.t. all δ_i. Two hyperexponential elements h_1 and h_2 of K are said to be *equivalent* over k, denoted by $h_1 \sim h_2$, if their ratio is the product of a constant of K and an element of k.

A vector $H \in K^n$ is said to be *hyperexponential over* k *w.r.t.* δ_i if $H = hV$ where $V \in k^n$ and $h \in K$ is hyperexponential over k w.r.t. δ_i. The vector $H \in K^n$ is said to be *hyperexponential over* k if H is hyperexponential over k w.r.t. all δ_i. Observe that an equivalent condition for $H \in K^n$ being hyperexponential over k is that $H = hV$ with $V \in k^n$ and $h \in K$ hyperexponential over k. Indeed, if H is hyperexponential over k then $H = h_i V_i$ with $V_i \in k^n$ and $\frac{\delta_i(h_i)}{h_i} \in k$ for each i. From $h_i V_i = h_j V_j$, it follows that $\frac{h_i}{h_j} \in k$ and hence

$$k \ni \delta_i\left(\frac{h_i}{h_j}\right) = \frac{\delta_i(h_i)h_j - h_i\delta_i(h_j)}{h_j^2} = \frac{\delta_i(h_i)}{h_i}\frac{h_i}{h_j} - \frac{h_i}{h_j}\frac{\delta_i(h_j)}{h_j},$$

which implies that $\frac{\delta_i(h_j)}{h_j} \in k$ for each i. So h_j is hyperexponential over k and $H = h_j V_j$ is of the desired form.

Remark 3.4. The definition of hyperexponential vectors in Section 4 of [10] is slightly different from ours. In [10], a nonzero vector $H^* = (h_1^*, \ldots, h_n^*)^\tau \in K^n$ is defined to be hyperexponential over k if each component h_i^* is either zero or hyperexponential over k. However, we can show that if such H^* is a hyperexponential solution of a submodule L in the terminology of [10], then L must have a solution of form hV where V is a rational vector and h is hyperexponential over k. Indeed, by re-arrangement of indices we may assume that

$$h_1^* = r_{11}u_1, \ \ldots, \ h_{j_1}^* = r_{1,j_1}u_1, \ \ldots, \ h_{j_{s-1}+1}^* = r_{s1}u_s, \ \ldots, \ h_{j_s}^* = r_{s,j_s}u_s,$$
$$h_{j_s+1}^* = \cdots = h_n^* = 0,$$

where the r_{ij} are rational elements and the u_i are hyperexponential elements that are pairwise inequivalent. In other words, we group the components of H^* by equivalence classes and move all the zero components to the end. Hence for any $f \in L$,

$$0 = f(H^*) = R_1 u_1 + \cdots + R_s u_s,$$

where the R_i are rational elements. Since the u_i are inequivalent to each other, we have $R_1 = \cdots = R_k = 0$ by Proposition 3.3 in [10]. It follows that

$$f(u_1(r_{11}, \ldots, r_{1,j_1}, \underbrace{0, \ldots, 0}_{n-j_1})^\tau) = R_1 u_1 = 0,$$

for any $f \in L$ and therefore $u_1(r_{11}, \ldots, r_{1,j_1}, 0, \ldots, 0)^\tau$ is a solution of L, which is hyperexponential by our definition. So these two definitions of hyperexponential vectors are compatible.

For convenience of later discussion, we give the following proposition, which is an analogue to Proposition 5.1 in [10]. It describes a correspondence between 1-dimensional submodules of a differential module and hyperexponential solutions of the associated integrable systems. Although this proposition is obvious in the ordinary case, we give a detailed proof here because the integrability condition should be taken into account in the partial case.

Proposition 3.5. *Let M be a differential module of dimension n and* (3.1) *be the integrable system associated with M w.r.t. a basis $e_1, \ldots, e_n$. Then M has a 1-dimensional submodule if and only if* (3.1) *has a hyperexponential solution.*

Proof. Let $H = hV$ be a hyperexponential solution of (3.1) where $V \in k^n$ and h is a hyperexponential element of some differential extension field K. Set

$$u = H^\tau(e_1, \ldots, e_n)^\tau \quad \text{and} \quad w = \frac{u}{h} = V^\tau(e_1, \ldots, e_n)^\tau \in M.$$

Recall that (3.1) is a translation of the condition (3.3). Since H is a solution of (3.1), we have $\partial_i(u) = 0$ and therefore

$$\partial_i(w) = \partial_i\left(\frac{u}{h}\right) = \frac{\partial_i(u)h - \delta_i(h)u}{h^2} = -\frac{\delta_i(h)}{h}\frac{u}{h} = -\frac{\delta_i(h)}{h} w \in kw,$$

for $i = 1, \ldots, m$. So kw is a 1-dimensional submodule of M. Suppose that we also have $H = h_2 V_2$ where V_2 is another vector in k^n and $h_2 \in K$ is hyperexponential over k. Set $w_2 = V_2^\tau(e_1, \ldots, e_n)^\tau \in M$. In the same way, $\partial_i(w_2) = -\frac{\delta_i(h_2)}{h_2} w_2$ for each i. Since $hV = h_2 V_2$, $h = a h_2$ for some nonzero $a \in k$ and thus $V = a^{-1} V_2$. So $kw = k a^{-1} w_2 = k w_2$. This means that the hyperexponential solution H of (3.1) induces uniquely a 1-dimensional submodule kw of M, which is called the *1-dimensional submodule associated with H*.

Conversely, let N be a 1-dimensional submodule of M generated by w and suppose that $\partial_i(w) = a_i w$ with $a_i \in k$ for $i = 1, \ldots, m$. It follows that

$$\partial_j(\partial_i(w)) - a_i a_j w = \delta_j(a_i) w,$$

which has a left hand-side invariant under the permutation of (i, j). This implies that $\delta_j(-a_i) = \delta_i(-a_j)$ for any i, j. By [10], there is a well-defined hyperexponential element h over k such that $\delta_i(h) = -a_i h$ for $i = 1, \ldots, m$. Therefore $\partial_i(hw) = 0$ for each i. This again implies, from the translation of (3.3), that $h(f_1, \ldots, f_n)^\tau$ is a solution of the system (3.1) where $f_1, \ldots, f_n$ are coordinates of w under the basis $e_1, \ldots, e_n$. In addition, $h(f_1, \ldots, f_n)^\tau$ is hyperexponential over k. □

Proposition 3.5 gives a method of constructing 1-dimensional submodules of a differential module via hyperexponential solutions of the associated integrable systems. Note that this construction is independent from the choice of the associated integrable systems. Indeed, let

$$S: \quad \{\delta_1(Z) = A_1 Z, \ \ldots, \ \delta_m(Z) = A_m Z\}$$

and

$$S^*: \quad \{\delta_1(Z^*) = A_1^* Z^*, \ \ldots, \ \delta_m(Z^*) = A_m^* Z^*\}$$

be integrable systems associated with M w.r.t. the bases $e_1, \ldots, e_n$ and $f_1, \ldots, f_n$, respectively. Then S and S^* are equivalent, and there is a $T \in \mathrm{GL}_n(k)$ such that

$$(f_1, \ldots, f_n) = (e_1, \ldots, e_n)T \tag{3.5}$$

and $A_i^* = T^{-1}A_iT - T^{-1}\delta_i(T)$ for each i. One can verify that T^{-1} viewed as a linear transformation from the set of solutions of S to the set of solutions of S^* is a bijection. If $H_1 := hV$ is a hyperexponential solution of S with $V \in k^n$ and h hyperexponential over k, then $H_2 := T^{-1}H_1$ is a hyperexponential solution of S^*. Therefore the 1-dimensional submodules of M associated with H_1 and H_2 are generated by $(e_1, \ldots, e_n)V$ and $(f_1, \ldots, f_n)T^{-1}V$, respectively. From (3.5), these two generators are equal, so are the 1-dimensional submodules they generate.

We now study the structure of all 1-dimensional submodules of differential modules. Let M be a differential module of dimension n. Suppose that

$$S: \quad \{\delta_1(Z) = A_1 Z, \ \ldots, \ \delta_m(Z) = A_m Z\},$$

with $A_i = (a_{ist}) \in \mathrm{M}_n(k)$, is the integrable system associated with M w.r.t. a basis $e_1, \ldots, e_n$, and M^* is the dual module of M with the basis $\{e_1^*, \ldots, e_n^*\}$ such that $e_i^*(e_j)$ equals 1 if $i = j$ and 0 otherwise. Since $\partial_i(e_1, \ldots, e_n)^\tau = -A_i^\tau(e_1, \ldots, e_n)^\tau$ for each i,

$$\partial_i\left(e_j^*\right)(e_t) = -e_j^*(\partial_i(e_t)) + \delta_i\left(e_j^*(e_t)\right) = -e_j^*\left(-\sum_{s=1}^{n} a_{ist}e_s\right) = a_{ijt},$$

for $t = 1, \ldots, n$; hence $\partial_i\left(e_j^*\right) = \sum_{s=1}^{n} a_{ijs}e_s^*$ and

$$\partial_i(e_1^*, \ldots, e_n^*)^\tau = A_i(e_1^*, \ldots, e_n^*)^\tau, \quad i = 1, \ldots, m.$$

Thus $(e_1^*, \ldots, e_n^*)^\tau$ is a "solution" of the system S.

Assume that k contains a nonconstant a. Then $\delta_\ell(a) \neq 0$ for some $\ell \in \{1, \ldots, m\}$. By Proposition 2.9 in [11], M^* as a $k[\partial_\ell]$-module has a cyclic vector w such that M^* is generated by $\{w, \partial_\ell(w), \partial_\ell^2(w), \ldots\}$ over k; therefore $M^* = k[\partial_\ell]w$. As $\dim_k M^* = n$, the vectors $w, \partial_i(w), \partial_i^2(w), \ldots, \partial_i^n(w)$ are linearly dependent over k for each $i \in \{1, \ldots, m\}$. By linear algebra, we can find linear ordinary operators $L_i \in k[\partial_i]$ of minimal order such that $L_i(w) = 0$. Clearly, $\mathrm{ord}(L_i) \leq n$ for each i and in particular, $\mathrm{ord}(L_\ell) = n$ since $w, \partial_\ell(w), \ldots, \partial_\ell^{n-1}(w)$ form a basis of M^*. Suppose that

$$(e_1^*, \ldots, e_n^*)^\tau = P(w, \partial_\ell(w), \ldots, \partial_\ell^{n-1}(w))^\tau \quad \text{with } P \in \mathrm{GL}_n(k). \tag{3.6}$$

Denote by L the system $\{L_1(z) = 0, \ldots, L_m(z) = 0\}$ of linear PDEs. From (3.6), every hyperexponential solution of S has the form

$$H = P\left(h, \delta_\ell(h), \ldots, \delta_\ell^{n-1}(h)\right)^\tau$$

where h is a hyperexponential solution of L. Applying the algorithm HyperexponentialSolutions in [10] to the system L, we obtain

$$\begin{array}{ll} h_1, & \{r_{11}, \ \ldots, \ r_{1,t_1}\}, \\ \vdots & \qquad \vdots \\ h_s, & \{r_{s1}, \ \ldots, \ r_{s,t_s}\}, \end{array}$$

where $h_1, \ldots, h_s$ are hyperexponential elements of a differential extension field K, $h_i \not\sim h_j$ whenever $i \neq j$, and $r_{i1}, \ldots, r_{i,t_i}$ are elements of k that are linearly independent over C. According to Proposition 3.4 in [10], every hyperexponential solution h of L has the form

$$h = h_i(c_1 r_{i1} + \cdots + c_{t_i} r_{i,t_i}),$$

where $i \in \{1, \ldots, s\}$ and $c_1, \ldots, c_{t_i} \in C$, not all zero. For $1 \le i \le s$ and $1 \le j \le t_i$, we compute

$$P\left(h_i r_{ij}, \delta_\ell(h_i r_{ij}), \ldots, \delta_\ell^{n-1}(h_i r_{ij})\right)^\tau = h_i V_{ij},$$

where $V_{ij} \in k^n$ and the first entry of V_{ij} is r_{ij}. Since $r_{i1}, \ldots, r_{i,t_i}$ are linearly independent over C_F, so are the vectors $V_{i1}, \ldots, V_{i,t_i}$ for each i. Therefore, we obtain

$$\begin{array}{ll} h_1, & \{V_{11}, \ \ldots, \ V_{1,t_1}\}, \\ \vdots & \qquad \vdots \\ h_s, & \{V_{s1}, \ \ldots, \ V_{s,t_s}\}, \end{array}$$

such that every hyperexponential solution H of S has the form

$$H = h_i(c_1 V_{i1} + \cdots + c_{t_i} V_{i,t_i}),$$

where $i \in \{1, \ldots, s\}$ and $c_1, \ldots, c_{t_i} \in C$, not all zero. Set

$$N_i = k(e_1, \ldots, e_n)(c_1 V_{i1} + \cdots + c_{t_i} V_{i,t_i}), \quad i = 1, \ldots, s,$$

where the c_j are arbitrarily chosen elements of C, not all zero.

From the proof of Proposition 3.5, $N_1, \ldots, N_s$ are 1-dimensional submodules of M. We show in the sequel that

$$I_i = \{N_i \mid c_1, \ldots, c_{t_i} \in C, \text{ not all zero}\}, \quad i = 1, \ldots, s,$$

constitute a partition of the set of all 1-dimensional submodules of M by the equivalence relation "$\simeq$", the isomorphism between differential modules. First, we see that each I_i is a well-defined equivalence class w.r.t. "$\simeq$". Suppose that N_i and N_i' are two 1-dimensional submodules of M belonging to I_i. There exist two sets of elements $c_1, \ldots, c_{t_i}$ and $c_1', \ldots, c_{t_i}'$ of C, not all zero, such that

$$N_i = k\,\underbrace{(e_1, \ldots, e_n)(c_1 V_{i1} + \cdots + c_{t_i} V_{i,t_i})}_{w_i},$$

and

$$N_i' = k\,\underbrace{(e_1,\ldots,e_n)(c_1'V_{i1}+\cdots+c_{t_i}'V_{i,t_i})}_{w_i'}.$$

From the proof of Proposition 3.5, we have

$$\partial_l(w_i) = -\frac{\delta_l(h_i)}{h_i}w_i \quad \text{and} \quad \partial_l(w_i') = -\frac{\delta_l(h_i)}{h_i}w_i', \quad l=1,\ldots,m.$$

Then the correspondence $w_i \mapsto w_i'$ from N_i to N_i' is an isomorphism of differential modules. So $N_i \simeq N_i'$ and I_i is a well-defined equivalence class. Let $N = kw$ be a 1-dimensional submodule of M where $w = (e_1,\ldots,e_n)V \in M$. From the proof of Proposition 3.5, there exists a hyperexponential element h such that hV is a hyperexponential solution of S. There exist $i \in \{1,\ldots,s\}$ and $c_1,\ldots,c_{t_i} \in C$, not all zero, such that $hV = h_i(c_1V_{i1}+\cdots+c_{t_i}V_{i,t_i})$. It follows that $h = ah_i$ for some $a \in k$, $V = a^{-1}(c_1V_{i1}+\cdots+c_{t_i}V_{i,t_i})$ and

$$kw = k(e_1,\ldots,e_n)V = k(e_1,\ldots,e_n)\,(c_1V_{i1}+\cdots+c_{t_i}V_{i,t_i}).$$

So N belongs to the equivalence class I_i. Now let N_i and N_j be two 1-dimensional submodules in the equivalence classes I_i and I_j, respectively. There are two sets of elements $c_1,\ldots,c_{t_i}$ and $c_1',\ldots,c_{t_j}'$ of C, not all zero, such that

$$N_i = k\,\underbrace{(e_1,\ldots,e_n)(c_1V_{i1}+\cdots+c_{t_i}V_{i,t_i})}_{w_i},$$

and

$$N_j = k\,\underbrace{(e_1,\ldots,e_n)(c_1'V_{j1}+\cdots+c_{t_j}'V_{j,t_j})}_{w_j}.$$

Suppose that $N_i \simeq N_j$, given by the correspondence $\phi : w_i \mapsto aw_j$ with $a \in k$. Note that $\partial_l(w_i) = -\frac{\delta_l(h_i)}{h_i}w_i$ and $\partial_l(w_j) = -\frac{\delta_l(h_j)}{h_j}w_j$ for $l = 1,\ldots,m$. On one hand,

$$\phi \circ \partial_l(w_i) = \phi\left(-\frac{\delta_l(h_i)}{h_i}w_i\right) = -a\frac{\delta_l(h_i)}{h_i}w_j,$$

and on the other hand,

$$\partial_l \circ \phi(w_i) = \partial_l(aw_j) = \left(\delta_l(a) - a\frac{\delta_l(h_j)}{h_j}\right)w_j,$$

for each l. Hence

$$\frac{\delta_l\left(\frac{h_j}{ah_i}\right)}{\frac{h_j}{ah_i}} = \frac{\delta_l(h_j)}{h_j} - \frac{\delta_l(h_i)}{h_i} - \frac{\delta_l(a)}{a} = 0, \quad \text{for each } l,$$

which implies that $\frac{h_j}{ah_i}$ is a constant of K, a contradiction with the fact $h_i \not\sim h_j$. So $N_i \not\simeq N_j$ whenever $i \neq j$ and $\{I_1,\ldots,I_s\}$ is a partition of all 1-dimensional submodules of M given the equivalence relation "$\simeq$".

We now describe an algorithm for finding 1-dimensional submodules of differential modules.

Algorithm OneDimSubMods (for finding 1-dimensional submodules of a differential module).

Input: A differential module M with a basis $e_1, \ldots, e_n$ and the actions of ∂_i on this basis:

$$\partial_i(e_1, \ldots, e_n)^\tau = B_i(e_1, \ldots, e_n)^\tau, \quad i = 1, \ldots, m,$$

where $B_1, \ldots, B_m$ are $n \times n$ matrices with entries in k.

Output: All 1-dimensional submodules of M.

1. [Find the cyclic vector.] Construct the integrable system associated with M

$$S: \quad \{\delta_1(Z) = A_1 Z, \ldots, \delta_m(Z) = A_m Z\}$$

where $A_i = -B_i^\tau$. Let M^* be the dual module of M with the basis $\{e_1^*, \ldots, e_n^*\}$ such that

$$\partial_i(e_1^*, \ldots, e_n^*)^\tau = A_i(e_1^*, \ldots, e_n^*)^\tau, \quad i = 1, \ldots, m.$$

Find a cyclic vector w of M^* such that M^* is generated by $w, \partial_\ell(w), \ldots, \partial_\ell^{n-1}(w)$ over k for some $\ell \in \{1, \ldots, m\}$. Then there is a $P \in \mathrm{GL}_n(k)$ such that

$$(e_1^*, \ldots, e_n^*)^\tau = P(w, \partial_\ell(w), \ldots, \partial_\ell^{n-1}(w))^\tau.$$

By linear algebra, find linear ordinary operators $L_i \in k[\partial_i]$ of minimal order such that $L_i(w) = 0$ for $i = 1, \ldots, m$. We then obtain a system of linear PDEs:

$$L: \quad \{L_1(z) = 0, \ldots, L_m(z) = 0\}.$$

2. [Compute hyperexponential solutions.] Apply the algorithm Hyperexponential-Solutions in [10] to compute all hyperexponential solutions of L. If the output is NULL then exit [M has no 1-dimensional submodules]. Otherwise, suppose that the output is

$$\begin{array}{ll} h_1, & \{r_{11}, \ldots, r_{1,t_1}\}, \\ \vdots & \quad\vdots \\ h_s, & \{r_{s1}, \ldots, r_{s,t_s}\}, \end{array}$$

where $h_1, \ldots, h_s$ are hyperexponential elements of a differential extension field, $h_i \not\sim h_j$ whenever $i \neq j$, and $r_{i1}, \ldots, r_{i,t_i}$ are elements of k that are linearly independent over C. Let

$$V_{ij} = h_i^{-1} P\left(h_i r_{ij}, \delta_\ell(h_i r_{ij}), \ldots, \delta_\ell^{n-1}(h_i r_{ij})\right)^\tau \in k^n,$$

for $i = 1, \ldots, s$ and $j = 1, \ldots, t_i$. Then $V_{i1}, \ldots, V_{i,t_i}$ are linearly independent over C and

$$\begin{array}{ll} h_1, & \{V_{11}, \ldots, V_{1,t_1}\}, \\ \vdots & \quad\vdots \\ h_s, & \{V_{s1}, \ldots, V_{s,t_s}\}, \end{array}$$

describe the structure of hyperexponential solutions of S: every hyperexponential solution of S has the form

$$H = h_i(c_1 V_{i1} + \cdots + c_{t_i} V_{i,t_i}),$$

with $i \in \{1, \dots, s\}$ and $c_1, \dots, c_{t_i} \in C$, not all zero.

3. [Retrieve 1-dimensional submodules.] Set

$$I_i = \{k(e_1, \dots, e_n)(c_1 V_{i1} + \cdots + c_{t_i} V_{i,t_i}) \mid c_1, \dots, c_{t_i} \in C, \text{ not all zero}\},$$

for $i = 1, \dots, s$. Then $\{I_1, \dots, I_s\}$ is a partition of all 1-dimensional submodules of M by the equivalence relation "$\simeq$".

3.3. Decomposability of Elements of $\wedge^d M$

Let M be a differential module. In order to find all d-dimensional submodules of M, by Proposition 3.1 it suffices to find all those 1-dimensional submodules of $\wedge^d M$ whose generators are decomposable.

Apply the algorithm OneDimSubMods to $\wedge^d M$ to obtain its 1-dimensional submodules kw where $w \in \wedge^d M$ may contain some unspecified constants $c_1, \dots, c_t$. To test the decomposability of w, consider the map

$$\phi_w : M \to \wedge^{d+1} M, \quad v \mapsto v \wedge w.$$

From the proof of Proposition 3.1, w is decomposable if and only if $\ker(\phi_w)$ is of dimension d. The latter is equivalent to the condition that the matrix P of the map ϕ_w has rank $n - d$. Hence identifying the decomposability of w amounts to a rank computation of P, i.e., identifying the constants $c_1, \dots, c_t$ in w such that all $(n-d+1) \times (n-d+1)$ minors of P are zero and P has at least a nonzero $(n-d) \times (n-d)$ minor, which amounts to solving a nonlinear system in $c_1, \dots, c_t$. We observe that this is the Plücker relations described in [11, 15] (for more details, see [5, 7, 6]). If this nonlinear system has no solutions in $\overline{C}$, the algebraic closure of C, then M has no d-dimensional submodules. Otherwise, $\ker(\phi_w)$ is a d-dimensional submodule of M.

Remark 3.6. There are alternative ways to compute ranks of parameterized matrices, for example, the Gauss method with branching, a Gröbner basis method using the linear structure [4] or the algorithm described in [14] for computing ranks of parameterized linear systems. These methods may be more efficient than computing minors.

4. Factorization Algorithm

Based on the results in previous sections, we now describe an algorithm for factoring differential modules.

Algorithm FactorDiffMod (for factoring differential modules).

Input: A differential module M with a basis $\{e_1, \dots, e_n\}$ and the actions of ∂_i on this basis:

$$\partial_i(e_1, \dots, e_n)^\tau = B_i(e_1, \dots, e_n)^\tau, \quad \text{for } i = 1, \dots, m, \tag{4.1}$$

where $B_1, \dots, B_m$ are $n \times n$ matrices with entries in k.

Output: All d-dimensional submodules of M given by their bases and the actions of ∂_i on the bases, for $0 < d < n$.

1. [Construct the exterior power.] From (4.1), construct a basis $\{f_1, \ldots, f_{\tilde{N}}\}$ of $\wedge^d M$ with $\tilde{N} = \binom{n}{d}$ and the matrices $\tilde{B}_i \in \mathrm{M}_{\tilde{N}}(k)$ such that

$$\partial_i(f_1, \ldots, f_{\tilde{N}})^\tau = \tilde{B}_i(f_1, \ldots, f_{\tilde{N}})^\tau, \quad i = 1, \ldots, m.$$

2. [Compute 1-dimensional submodules.] Apply the algorithm OneDimSubMods to compute all 1-dimensional submodules of $\wedge^d M$. If $\wedge^d M$ has no 1-dimensional submodules, then exit [M has no d-dimensional submodules]. Otherwise, suppose that kw is a 1-dimensional submodule of $\wedge^d M$ where w may contain some unspecified constants $c_1, \ldots, c_t$.

3. [Test the decomposability.] For each w obtained in step 2, consider the map

$$\phi_w : M \to \wedge^{d+1} M, \quad v \mapsto v \wedge w.$$

Construct the matrix P of ϕ_w, which is an $\binom{n}{d+1} \times n$ matrix with entries in $k(c_1, \ldots, c_t)$. The condition that all $(n-d+1) \times (n-d+1)$ minors of P are zero and that P has at least one nonzero $(n-d)$ minor yields a nonlinear system in $c_1, \ldots, c_t$. If this nonlinear system has no solutions in $\overline{C}$ then exit [M has no d-dimensional submodules]. Otherwise, substitute the values of $c_1, \ldots, c_t$ into P and compute a basis $\{\alpha_1, \ldots, \alpha_d\}$ of the rational kernel of P with $\alpha_j \in k^n$.

4. [Retrieve d-dimensional submodules.] Set $v_j = (e_1, \ldots, e_n)\alpha_j$ for $j = 1, \ldots, d$. Then $\oplus_{j=1}^d k v_j$ is a d-dimensional submodule of M.

We now apply the algorithm FactorDiffMod to redo Example 1 in [9].

Example. Let $\mathbb{D} = \overline{\mathbb{Q}}(x, y)[\partial_x, \partial_y]$ where $\partial_x = \frac{\partial}{\partial x}$ and $\partial_y = \frac{\partial}{\partial y}$ are the usual differential operators w.r.t. x and y, respectively. Let M be a differential module with a basis $\{e_1, e_2, e_3\}$ such that

$$\partial_x(e_1, e_2, e_3)^\tau = B_x(e_1, e_2, e_3)^\tau, \quad \partial_y(e_1, e_2, e_3)^\tau = B_y(e_1, e_2, e_3)^\tau$$

where

$$B_x = \begin{pmatrix} 0 & -\dfrac{y}{4x} & -\dfrac{1}{4} \\ -1 & \dfrac{2-xy}{4x} & -\dfrac{2+xy}{4y} \\ 0 & \dfrac{y^2}{4x} & \dfrac{y}{4} \end{pmatrix}, \quad B_y = \begin{pmatrix} 0 & -\dfrac{1}{4} & -\dfrac{x}{4y} \\ 0 & -\dfrac{2+xy}{4y} & \dfrac{2x-x^2y}{4y^2} \\ -1 & \dfrac{y}{4} & \dfrac{x}{4} \end{pmatrix}.$$

To compute two-dimensional submodules of M, construct the second exterior power $\wedge^2 M$ of M with a basis $\{f_1 := e_1 \wedge e_2,\ f_2 := e_1 \wedge e_3,\ f_3 := e_2 \wedge e_3\}$ and compute the integrable system associated with $\wedge^2 M$:

$$S: \quad \{\partial_x(Z) = A_{xx} Z, \quad \partial_y(Z) = A_{yy} Z\}$$

where $Z = (z_1, z_2, z_3)^\tau$ is a vector of unknowns and

$$A_{xx} = \begin{pmatrix} \frac{xy-2}{4x} & -\frac{y^2}{4x} & 0 \\ \frac{2+xy}{4y} & -\frac{y}{4} & 1 \\ -\frac{1}{4} & \frac{y}{4x} & -\frac{1}{2x} \end{pmatrix}, \quad A_{yy} = \begin{pmatrix} \frac{2+xy}{4y} & -\frac{y}{4} & -1 \\ \frac{x^2y-2x}{4y^2} & -\frac{x}{4} & 0 \\ -\frac{x}{4y} & \frac{1}{4} & \frac{1}{2y} \end{pmatrix}.$$

Let M_2^* be the dual module of $\wedge^2 M$ with the basis $\{f_1^*, f_2^*, f_3^*\}$ such that

$$\partial_x(f_1^*, f_2^*, f_3^*)^\tau = A_{xx}(f_1^*, f_2^*, f_3^*)^\tau \quad \text{and} \quad \partial_y(f_1^*, f_2^*, f_3^*)^\tau = A_{yy}(f_1^*, f_2^*, f_3^*)^\tau.$$

We find that f_1^* is a cyclic vector of M_2^* and moreover,

$$(f_1^*, f_2^*, f_3^*)^\tau = P\left(f_1^*, \partial_x(f_1^*), \partial_x^2(f_1^*)\right)^\tau$$

with

$$P = \begin{pmatrix} 1 & 0 & 0 \\ \frac{xy-2}{4x} & -\frac{y^2}{4x} & 0 \\ -\frac{3(xy-2)}{8x^2} & \frac{3y^2}{8x^2} & -\frac{y^2}{4x} \end{pmatrix}.$$

By linear algebra, we find two linear ordinary differential operators, both of minimal order,

$$L_x = \partial_x^3 + \frac{3}{x}\partial_x^2 + \frac{3-xy}{4x^2}\partial_x - \frac{y}{8x^2} \in \overline{\mathbb{Q}}(x,y)[\partial_x]$$

and

$$L_y = \partial_y^3 + \frac{6-2xy}{y(xy-6)}\partial_y^2 + \frac{23xy-x^2y^2-42}{4y^2(xy-6)}\partial_y + \frac{x^2y^2-30xy+72}{8y^3(xy-6)} \in \overline{\mathbb{Q}}(x,y)[\partial_y]$$

such that L_x and L_y both annihilate f_1^*.

Applying the algorithm HyperexponentialSolutions in [10] to the system

$$L: \quad \{L_x(z_1) = 0, \quad L_y(z_1) = 0\},$$

we find that every hyperexponential solution h of the system L has the form

$$h = ce^{\int(-\frac{1}{2x}dx + \frac{1}{2y}dy)} = c\sqrt{\frac{y}{x}}, \qquad \text{for any } c \in \overline{\mathbb{Q}},$$

where $e^{\int(-\frac{1}{2x}dx+\frac{1}{2y}dy)}$ denotes a hyperexponential function h_0 in x, y such that

$$\partial_x(h_0) = -\frac{1}{2x}h_0 \quad \text{and} \quad \partial_y(h_0) = \frac{1}{2y}h_0.$$

Hence every hyperexponential solution of S has the form

$$Z = P(h,\, \partial_x(h),\, \partial_x^2(h))^\tau = \sqrt{\frac{y}{x}}\left(c,\, c\cdot\frac{x}{y},\, 0\right)^\tau, \qquad \text{for any } c \in \bar{\mathbb{Q}},$$

and all 1-dimensional submodules of $\wedge^2 M$ are of the form $\bar{\mathbb{Q}}(x,y)w$ where

$$w = (f_1, f_2, f_3)\left(1, \frac{x}{y}, 0\right)^\tau = e_1 \wedge e_2 + \frac{x}{y} e_1 \wedge e_3 \in \wedge^2 M. \tag{4.2}$$

To test the decomposability of w, consider the map

$$\phi_w : M \to \wedge^3 M, \quad v \mapsto v \wedge w.$$

The matrix of ϕ_w is $P = \left(0, -\frac{x}{y}, 1\right)$ and then has rank one. So w is decomposable. (In fact, from (4.2) one can see directly that $w = e_1 \wedge \left(e_2 + \frac{x}{y} e_3\right)$ is a decomposable element of $\wedge^2 M$.) A basis for the rational kernel of P is

$$\left\{(1,0,0)^\tau, \left(0, 1, \frac{x}{y}\right)^\tau\right\}.$$

Set $v_1 = (e_1, e_2, e_3)(1,0,0)^\tau = e_1$ and $v_2 = (e_1, e_2, e_3)\left(0, 1, \frac{x}{y}\right)^\tau = e_2 + \frac{x}{y} e_3$. Then $\ker(\phi_u) = kv_1 \oplus kv_2$ is a two-dimensional submodule of M and the actions of ∂_x and ∂_y on the basis $\{v_1, v_2\}$ are given by

$$\partial_x (v_1, v_2)^\tau = F_x (v_1, v_2)^\tau, \quad \partial_y (v_1, v_2)^\tau = F_y (v_1, v_2)^\tau$$

where

$$F_x = \begin{pmatrix} 0 & -\dfrac{y}{4x} \\ -1 & \dfrac{1}{2x} \end{pmatrix}, \qquad F_y = \begin{pmatrix} 0 & -\dfrac{1}{4} \\ -\dfrac{x}{y} & -\dfrac{1}{2y} \end{pmatrix}.$$

5. Conclusion and Future Work

In this paper, we present an algorithm for factoring differential modules. By factoring the differential modules associated with systems of linear PDEs with finite-dimensional solution spaces, the algorithm FactorDiffMod improves the factorization algorithm in [9]. Further work will include the refinement of the step in the algorithm OneDimSubMods which deals with computing hyperexponential solutions of integrable systems and the improvement for computing ranks of parameterized matrices. The generalization of the factorization algorithm to difference modules and differential-difference modules will also be studied.

References

[1] E. Beke. Die Irreduzibilität der homogenen linearen Differentialgleichungen. *Mathematische Annalen* **45**: 278–294, 1894.

[2] M. Bronstein. An improved algorithm for factoring linear ordinary differential operators. In: *Proc. Int. Symp. on Symbolic Algebraic Comput.* (ISSAC '92) (M. Giesbrecht, ed.), pp. 336–340. ACM Press, New York, 1992.

[3] E. Compoint and J. A. Weil. Absolute reducibility of differential operators and Galois groups. *J. Algebra* **275**(1): 77–105, 2004.

[4] D. Cox, J. Little and D. O'Shea. *Ideals, Varieties, and Algorithms.* Springer, New York, 1996.

[5] P. Griffiths and J. Harris. *Principles of Algebraic Geometry.* John Wiley and Sons, New York, 1978.

[6] J. Harris. *Algebraic Geometry: A First Course.* Springer, New York, 1994.

[7] W. V. D. Hodege and D. Pedoe. *Methods of Algebraic Geometry*, vol. **1**. Cambridge University Press, Cambridge, 1947.

[8] Z. Li and F. Schwarz. Rational solutions of Riccati-like partial differential equations. *J. Symb. Comput.* **31**: 691–716, 2001.

[9] Z. Li, F. Schwarz and S. Tsarev. Factoring zero-dimensional ideals of linear partial differential operators. In: *Proc. Int. Symp. on Symbolic Algebraic Comput.* (ISSAC '02) (T. Mora, ed.), pp. 168–175. ACM Press, New York, 2002.

[10] Z. Li, F. Schwarz and S. Tsarev. Factoring systems of linear PDE's with finite-dimensional solution spaces. *J. Symb. Comput.* **36**: 443–471, 2003.

[11] M. van der Put and M. F. Singer. *Galois Theory of Linear Differential Equations.* Grundlehren der Mathematischen Wissenschaften, vol. **328**. Springer, Heidelberg, 2003.

[12] L. Schlesinger. *Handbuch der Theorie der Linearen Differentialgleichungen.* Teubner, Leipzig, 1895.

[13] F. Schwarz. A factorization algorithm for linear ordinary differential equations. In: *Proc. Int. Symp. on Symbolic Algebraic Comput.* (ISSAC '89) (G. Gonnet, ed.), pp. 17–25. ACM Press, New York, 1989.

[14] W. Y. Sit. A theory for parametric linear systems. In: *Proc. Int. Symp. on Symbolic Algebraic Comput.* (ISSAC '91) (S. M. Watt, ed.), pp. 112–121. ACM Press, New York, 1991.

[15] S. P. Tsarev. Problems that appear during factorization of ordinary linear differential operators. *Program. Comput. Softw.* **20**(1): 27–29, 1994.

Acknowledgments

I am indebted to my two co-advisors Prof. Manuel Bronstein and Prof. Ziming Li for their invaluable instructions and advices on my writing and revising the paper. I would like also to thank the referees for their helpful suggestions on the algorithmic consideration of my work.

Min Wu
Key Laboratory of Mathematics Mechanization
Academy of Mathematics and System Sciences
Beijing 100080, China and
Graduate School of the Chinese Academy of Sciences
Chinese Academy of Sciences
Beijing 100039, China
e-mail: `mwu@mmrc.iss.ac.cn`

Trends in Mathematics: *Differential Equations with Symbolic Computation*, 255–290

Continuous and Discrete Homotopy Operators and the Computation of Conservation Laws

Willy Hereman, Michael Colagrosso, Ryan Sayers, Adam Ringler, Bernard Deconinck, Michael Nivala and Mark Hickman

In memory of Ryan Sayers (1982–2003)

Abstract. We introduce calculus-based formulas for the continuous Euler and homotopy operators. The 1D continuous homotopy operator automates integration by parts on the jet space. Its 3D generalization allows one to invert the total divergence operator. As a practical application, we show how the operators can be used to symbolically compute local conservation laws of nonlinear systems of partial differential equations in multi-dimensions.

Analogous to the continuous case, we also present concrete formulas for the discrete Euler and homotopy operators. Essentially, the discrete homotopy operator carries out summation by parts. We use it to algorithmically invert the forward difference operator. We apply the discrete operator to compute fluxes of differential-difference equations in $(1+1)$ dimensions.

Our calculus-based approach allows for a straightforward implementation of the operators in major computer algebra systems, such as Mathematica and Maple. The symbolic algorithms for integration and summation by parts are illustrated with elementary examples. The algorithms to compute conservation laws are illustrated with nonlinear PDEs and their discretizations arising in fluid dynamics and mathematical physics.

Mathematics Subject Classification (2000). Primary: 37K05, 37J35, 35Q35; Secondary: 37K10, 35Q58, 37K60.

Keywords. Homotopy operator, conservation law, integrability testing.

Research supported in part by the National Science Foundation (NSF) under grants nos. DMS-9732069, DMS-9912293, CCR-9901929, FRG-DMS-0139093, and DMS-9810726.

1. Introduction

This chapter focuses on symbolic methods to compute polynomial conservation laws of partial differential equations (PDEs) in multi-dimensions and differential-difference equations (DDEs, semi-discrete lattices). We only treat (1+1)-dimensional DDEs where time is continuous and the spacial variable has been discretized.

There are several strategies to compute conservation laws of PDEs. Some methods use a generating function [2], which requires the knowledge of key pieces of the Inverse Scattering Transform [1]. Other methods use Noether's theorem to get conservation laws from variational symmetries. More algorithmic methods, some of which circumvent the existence of a variational principle [6, 7, 26, 36], require the solution of a determining system of ODEs or PDEs. Despite their power, only a few of these methods have been implemented in computer algebra systems (CAS), such as Mathematica, Maple, and REDUCE. See [17, 36] for reviews.

The most modern techniques [9] rely on tools from the calculus of variations, including basic operations on vector fields and differential forms. Such tools are available in *Vessiot*, a general purpose suite of Maple packages designed by Anderson [10] for computations on jet spaces. The *Vessiot* package `DE_APPLS` offers commands (but no fully automated code) for constructing conservation laws from symmetries by Noether's theorem.

We advocate a direct approach for the computation of conservation laws without recourse to generalized or adjoint symmetries. We use the following procedure: build candidate densities as linear combinations (with undetermined coefficients) of terms that are homogeneous under the scaling symmetry of the PDE. If no such symmetry exists, one is constructed by introducing weighted parameters. Subsequently, use the variational derivative to compute the coefficients, and, finally, use the homotopy operator to compute the flux. Furthermore, our approach can be adapted to nonlinear DDEs [18, 21, 22].

Our method works for evolution equations with polynomial and transcendental terms and does not require a Lagrangian formulation. Built on tools from (variational) calculus and linear algebra, our method is entirely algorithmic and can be implemented in leading CAS. Implementations [17, 18] in Mathematica and Maple can be downloaded from [13, 20].

Our earlier algorithm [17, 18] worked only for nonlinear PDEs in one spacial variable. In this chapter we present an algorithm that works for systems of PDEs in multi-dimensions that appear in fluid mechanics, elasticity, gas dynamics, general relativity, (magneto-) hydro-dynamics, etc. The new algorithm produces densities in which all divergences and divergence-equivalent terms have been removed.

During the development of our methods we came across tools from the calculus of variations and differential geometry that deserve attention in their own right. These tools are the variational derivative, the higher Euler operators, and the homotopy operator.

To set the stage, we address a few issues arising in multivariate calculus:

(i) To determine whether or not a vector field $\mathbf{F}$ is *conservative*, i.e. $\mathbf{F} = \boldsymbol{\nabla} f$ for some scalar field f, one must verify that $\mathbf{F}$ is *irrotational* or *curl free*, that is $\boldsymbol{\nabla} \times \mathbf{F} = \mathbf{0}$. The field f can be computed via standard integrations [32, pp. 518, 522].

(ii) To test if $\mathbf{F}$ is the curl of some vector field $\mathbf{G}$, one must check that $\mathbf{F}$ is *incompressible* or *divergence free*, i.e. $\boldsymbol{\nabla} \cdot \mathbf{F} = 0$. The components of $\mathbf{G}$ result from solving a coupled system of first-order PDEs [32, p. 526].

(iii) To verify whether or not a scalar field f is the divergence of some vector function $\mathbf{F}$, no theorem from vector calculus comes to the rescue. Furthermore, the computation of $\mathbf{F}$ such that $f = \boldsymbol{\nabla} \cdot \mathbf{F}$ is a nontrivial matter. In single variable calculus, it amounts to computing the primitive $F = \int f \, dx$.

In multivariate calculus, all scalar fields f, including the components F_i of vector fields $\mathbf{F} = (F_1, F_2, F_3)$, are functions of the independent variables (x, y, z). In differential geometry one addresses the above issues in much greater generality. The functions f and F_i can now depend on arbitrary functions $u(x, y, z), v(x, y, z)$, etc. and their mixed derivatives (up to a fixed order) with respect to the independent variables (x, y, z). Such functions are called *differential functions* [33]. As one might expect, carrying out the gradient-, curl-, or divergence-test requires advanced algebraic machinery. For example, to test whether or not $f = \boldsymbol{\nabla} \cdot \mathbf{F}$ requires the use of the variational derivative (Euler operator) in 3D. The actual computation of $\mathbf{F}$ requires integration by parts. That is where the homotopy operator comes into play.

In 1D problems the continuous total homotopy operator[1] reduces the problem of symbolic integration by parts to an integration with respect to a single auxiliary variable. In 2D and 3D, the homotopy operator allows one to invert the total divergence operator and, again, reduce the problem to a single integration. At the moment, no major CAS have reliable routines for integrating expressions involving *unknown* functions and their derivatives. As far as we know, no CAS offer a function to test if a differential function is a divergence. Routines to symbolically invert the total divergence are certainly lacking.

The continuous homotopy operator is a universal, yet little known, tool that can be applied to many problems in which integration by parts (of arbitrary functions) in multi-variables plays a role. The reader is referred to [33, p. 374] for a history of the homotopy operator. One of the first uses of the homotopy operator in the context of conservation laws can be found in [6, 7] and references therein. A clever argument why the homotopy operator actually works is given in [7, p. 582]. In [5], Anco gives a simple algebraic formula to generate conservation laws of scaling invariant PDEs based on the computation of adjoint symmetries. Like ours, Anco's approach is algorithmic and heavily relies on scaling homogeneity. His approach does not require the use of the homotopy integral formula.

A major motivation for writing this chapter is to demystify the homotopy operators. Therefore, we purposely avoid differential forms and abstract concepts

[1] Henceforth, homotopy operator instead of total homotopy operator.

such as the variational bicomplex. Instead, we present calculus formulas for the homotopy operators which makes them readily implementable in major CAS.

By analogy with the continuous case, we also present formulas for the discrete versions of the Euler and homotopy operators. The discrete homotopy operator is a powerful tool to invert the forward difference operator, whatever the application is. It circumvents the necessary summation (by parts) by applying a set of variational derivatives followed by a one-dimensional integration with respect to an auxiliary variable. We use the homotopy operator to compute conserved fluxes of DDEs. Numerous examples of such DDEs are given in [34]. Beyond DDEs, the discrete homotopy operator has proven to be useful in the study of difference equations [24, 30]. To our knowledge, CAS offer no tools to invert the forward difference operator. Our discrete homotopy operator overcomes these shortcomings.

As shown in [24, 30], the parallelism between the continuous and discrete cases can be made rigorous and both theories can be formulated in terms of variational bicomplexes. To make our work accessible to as wide an audience as possible, we do not explicitly use the abstract framework. Aficionados of *de Rham* complexes may consult [8, 9, 11, 27] and [24, 30, 31]. The latter papers cover the discrete variational bicomplexes.

2. Examples of Nonlinear PDEs

We consider nonlinear systems of evolution equations in $(3+1)$ dimensions,

$$\mathbf{u}_t = \mathbf{G}(\mathbf{u}, \mathbf{u}_x, \mathbf{u}_y, \mathbf{u}_z, \mathbf{u}_{2x}, \mathbf{u}_{2y}, \mathbf{u}_{2z}, \mathbf{u}_{xy}, \mathbf{u}_{xz}, \mathbf{u}_{yz}, \ldots), \tag{2.1}$$

where $\mathbf{x} = (x, y, z)$ are space variables and t is time. The vector $\mathbf{u}(x, y, z, t)$ has N components u_i. In the examples we denote the components of $\mathbf{u}$ by u, v, w, etc. Subscripts refer to partial derivatives. For brevity, we use $\mathbf{u}_{2x}$ instead of $\mathbf{u}_{xx}$, etc. and write $\mathbf{G}(\mathbf{u}^{(n)})$ to indicate that the differential function $\mathbf{G}$ depends on derivatives up to order n of $\mathbf{u}$ with respect to x, y, and z. We assume that $\mathbf{G}$ does not explicitly depend on $\mathbf{x}$ and t. There are no restrictions on the number of components, order, and degree of nonlinearity of the variables in $\mathbf{G}$.

We will predominantly work with polynomial systems, although systems involving one transcendental nonlinearity can also be handled. If parameters are present in (2.1), they will be denoted by lower-case Greek letters.

Example 2.1. The coupled Korteweg–de Vries (cKdV) equations [1],

$$u_t - 6\beta u u_x + 6 v v_x - \beta u_{3x} = 0, \qquad v_t + 3 u v_x + v_{3x} = 0, \tag{2.2}$$

where β is a nonzero parameter, describes interactions of two waves with different dispersion relations. System (2.2) is known in the literature as the Hirota–Satsuma system. It is completely integrable [1, 23] when $\beta = \frac{1}{2}$.

Example 2.2. The sine-Gordon (sG) equation [12, 29], $u_{2t} - u_{2x} = \sin u$, can be written as a system of evolution equations,

$$u_t = v, \qquad v_t = u_{2x} + \sin u. \tag{2.3}$$

This system occurs in numerous problems of mathematics and physics, ranging from surfaces with constant mean curvature to superconductivity.

Example 2.3. The (2+1)-dimensional shallow-water wave (SWW) equations [14],

$$\mathbf{u}_t + (\mathbf{u}\cdot\boldsymbol{\nabla})\mathbf{u} + 2\boldsymbol{\Omega} \times \mathbf{u} = -\boldsymbol{\nabla}(h\theta) + \tfrac{1}{2}h\boldsymbol{\nabla}\theta,$$
$$\theta_t + \mathbf{u}\cdot(\boldsymbol{\nabla}\theta) = 0, \quad h_t + \boldsymbol{\nabla}\cdot(h\mathbf{u}) = 0, \tag{2.4}$$

describe waves in the ocean using layered models. Vectors $\mathbf{u} = u(x,y,t)\mathbf{i}+v(x,y,t)\mathbf{j}$ and $\boldsymbol{\Omega} = \Omega\mathbf{k}$ are the fluid and angular velocities, respectively. $\mathbf{i}$, $\mathbf{j}$, and $\mathbf{k}$ are unit vectors along the x, y, and z-axes. $\theta(x,y,t)$ is the horizontally varying potential temperature field and $h(x,y,t)$ is the layer depth. The dot $(\cdot)$ stands for Euclidean inner product and $\boldsymbol{\nabla} = \frac{\partial}{\partial x}\mathbf{i} + \frac{\partial}{\partial y}\mathbf{j}$ is the gradient operator. System (2.4) is written in components as

$$u_t + uu_x + vu_y - 2\Omega v + \tfrac{1}{2}h\theta_x + \theta h_x = 0, \quad v_t + uv_x + vv_y + 2\Omega u + \tfrac{1}{2}h\theta_y + \theta h_y = 0,$$
$$\theta_t + u\theta_x + v\theta_y = 0, \quad h_t + hu_x + uh_x + hv_y + vh_y = 0. \tag{2.5}$$

3. Key Definitions — Continuous Case

Definition 3.1. System (2.1) is said to be *dilation invariant* if it is invariant under a scaling (dilation) symmetry.

Example 3.2. The cKdV system (2.2) is invariant under the scaling symmetry

$$(x,t,u,v) \rightarrow (\lambda^{-1}x, \lambda^{-3}t, \lambda^2 u, \lambda^2 v), \tag{3.1}$$

where λ is an arbitrary scaling parameter.

Definition 3.3. We define the *weight*,[2] W, of a variable as the exponent p in λ^p which multiplies the variable.

Example 3.4. We will replace x by $\lambda^{-1}x$. Thus, $W(x) = -1$ or $W(\partial/\partial x) = 1$. From (3.1), we have $W(\partial/\partial t) = 3$ and $W(u) = W(v) = 2$ for the cKdV equations.

Definition 3.5. The *rank* of a monomial is defined as the total weight of the monomial. An expression is *uniform in rank* if its monomial terms have equal rank.

Example 3.6. All monomials in *both* equations of (2.2) have rank 5. Thus, (2.2) is uniform in rank.

Weights of dependent variables and weights of $\partial/\partial x, \partial/\partial y$, etc. are assumed to be non-negative and rational. Ranks must be positive natural or rational numbers. The ranks of the equations in (2.1) may differ from each other. Conversely, requiring uniformity in rank for each equation in (2.1) allows one to compute the weights of the variables (and thus the scaling symmetry) with linear algebra.

[2] For PDEs modeling physical phenomena, the weights are the remnants of physical units after non-dimensionalization.

Example 3.7. For the cKdV equations (2.2), one has

$$W(u) + W(\partial/\partial t) = 2W(u) + 1 = 2W(v) + 1 = W(u) + 3,$$
$$W(v) + W(\partial/\partial t) = W(u) + W(v) + 1 = W(v) + 3,$$

which yields $W(u) = W(v) = 2$, $W(\partial/\partial t) = 3$, leading to (3.1).

Dilation symmetries, which are special Lie-point symmetries, are common to many nonlinear PDEs. However, non-uniform PDEs can be made uniform by extending the set of dependent variables with auxiliary parameters with appropriate weights. Upon completion of the computations one equates these parameters to 1.

Example 3.8. The sG equation (2.3) is not uniform in rank unless we replace it by

$$u_t = v, \quad v_t = u_{2x} + \alpha \sin u, \quad \alpha \in \mathbb{R}. \tag{3.2}$$

Using the Maclaurin series for the sin function, uniformity in rank requires

$$W(u) + W(\partial/\partial t) = W(v),$$
$$W(v) + W(\partial/\partial t) = W(u) + 2 = W(\alpha) + W(u) = W(\alpha) + 3W(u) = \cdots .$$

This forces us to set $W(u) = 0$. Then, $W(\alpha) = 2$. By allowing the parameter α to scale, (3.2) becomes scaling invariant under the symmetry

$$(x, t, u, v, \alpha) \to (\lambda^{-1}x, \lambda^{-1}t, \lambda^0 u, \lambda^1 v, \lambda^2 \alpha),$$

corresponding to $W(\partial/\partial x) = W(\partial/\partial t) = 1, W(u) = 0, W(v) = 1, W(\alpha) = 2$. The first and second equations in (3.2) are uniform of ranks 1 and 2, respectively.

As in (3.2), the weight of an argument of a transcendental function is always 0.

Definition 3.9. System (2.1) is called *multi-uniform* in rank if it admits more than one dilation symmetry (not the result of adding parameters with weights).

Example 3.10. Uniformity in rank for the SWW equations (2.5) requires, after some algebra, that

$$W(\partial/\partial t) = W(\Omega),\ W(\partial/\partial y) = W(\partial/\partial x) = 1,\ W(u) = W(v) = W(\Omega) - 1,$$
$$W(\theta) = 2W(\Omega) - W(h) - 2,$$

with $W(h)$ and $W(\Omega)$ free. The SWW system is thus multi-uniform. The symmetry

$$(x, y, t, u, v, \theta, h, \Omega) \to (\lambda^{-1}x, \lambda^{-1}y, \lambda^{-2}t, \lambda u, \lambda v, \lambda\theta, \lambda h, \lambda^2\Omega), \tag{3.3}$$

which is most useful for our computations later on, corresponds to $W(\partial/\partial x) = W(\partial/\partial y) = 1, W(\partial/\partial t) = 2, W(u) = W(v) = 1, W(\theta) = 1, W(h) = 1$, and $W(\Omega) = 2$. A second symmetry,

$$(x, y, t, u, v, \theta, h, \Omega) \to (\lambda^{-1}x, \lambda^{-1}y, \lambda^{-2}t, \lambda u, \lambda v, \lambda^2\theta, \lambda^0 h, \lambda^2\Omega), \tag{3.4}$$

matches $W(\partial/\partial x) = W(\partial/\partial y) = 1$, $W(\partial/\partial t) = 2$, $W(u) = W(v) = 1$, $W(\theta) = 2$, $W(h) = 0$, $W(\Omega) = 2$.

4. Conserved Densities and Fluxes of Nonlinear PDEs

Definition 4.1. A scalar differential function $\rho(\mathbf{u}^{(n)})$ is a conserved *density* if there exists a vector differential function $\mathbf{J}(\mathbf{u}^{(m)})$, called the associated *flux*, such that

$$\mathrm{D}_t\,\rho + \mathrm{Div}\,\mathbf{J} = 0 \tag{4.1}$$

is satisfied on solutions of (2.1).

Equation (4.1) is called a local[3] *conservation law*[4] [33], and Div is called the total divergence.[5] Clearly, $\mathrm{Div}\,\mathbf{J} = (\mathrm{D}_x, \mathrm{D}_y, \mathrm{D}_z)\cdot(J_1, J_2, J_3) = \mathrm{D}_x J_1 + \mathrm{D}_y J_2 + \mathrm{D}_z J_3$. In the case of one spacial variable (x), (4.1) reduces to

$$\mathrm{D}_t\,\rho + \mathrm{D}_x J = 0, \tag{4.2}$$

where both density ρ and flux J are scalar differential functions.

The flux $\mathbf{J}$ in (4.1) is not uniquely defined. In 3D, the flux can only be determined up to a curl. Indeed, if $(\rho, \mathbf{J})$ is a valid density-flux pair, so is $(\rho, \mathbf{J} + \nabla \times \mathbf{K})$ for an arbitrary vector differential function $\mathbf{K} = (K_1, K_2, K_3)$. Recall that $\nabla \times \mathbf{K} = (\mathrm{D}_y K_3 - \mathrm{D}_z K_2, \mathrm{D}_z K_1 - \mathrm{D}_x K_3, \mathrm{D}_x K_2 - \mathrm{D}_y K_1)$. In 2D, the flux is only determined up to a divergence-free vector $\mathbf{K} = (K_1, K_2) = (\mathrm{D}_y\theta), -\mathrm{D}_x\theta)$, where θ is an arbitrary scalar differential function. In (4.2) the flux is only determined up to an arbitrary constant.

In the 1D case,

$$\mathrm{D}_t \rho(u^{(n)}) = \frac{\partial \rho}{\partial t} + \sum_{k=0}^{n} \frac{\partial \rho}{\partial u_{kx}} \mathrm{D}_x^k u_t. \tag{4.3}$$

where $u^{(n)}$ is the highest-order term present in ρ. Upon replacement of u_t, u_{tx}, etc. from $u_t = G$, one gets

$$\mathrm{D}_t \rho = \frac{\partial \rho}{\partial t} + \rho(u)'[G],$$

where $\rho(u)'[G]$ is the Fréchet derivative of ρ in the direction of G. Similarly,

$$\mathrm{D}_x J(u^{(m)}) = \frac{\partial J}{\partial x} + \sum_{k=0}^{m} \frac{\partial J}{\partial u_{kx}} u_{(k+1)x}. \tag{4.4}$$

Generalization of (4.3) and (4.4) to multiple dependent variables is straightforward.

Example 4.2. Taking $\mathbf{u} = (u, v)$,

$$\begin{aligned}
\mathrm{D}_t \rho(u^{(n_1)}, v^{(n_2)}) &= \frac{\partial \rho}{\partial t} + \sum_{k=0}^{n_1} \frac{\partial \rho}{\partial u_{kx}} \mathrm{D}_x^k u_t + \sum_{k=0}^{n_2} \frac{\partial \rho}{\partial v_{kx}} \mathrm{D}_x^k v_t, \\
\mathrm{D}_x J(u^{(m_1)}, v^{(m_2)}) &= \frac{\partial J}{\partial x} + \sum_{k=0}^{m_1} \frac{\partial J}{\partial u_{kx}} u_{(k+1)x} + \sum_{k=0}^{m_2} \frac{\partial J}{\partial v_{kx}} v_{(k+1)x}.
\end{aligned}$$

---

[3] We only compute densities and fluxes free of integral terms.

[4] In electromagnetism, this is the continuity equation relating charge density ρ to current $\mathbf{J}$.

[5] Gradient, curl, and divergence are in rectangular coordinates.

We will ignore densities and fluxes that explicitly depend on $\mathbf{x}$ and t. If $\mathbf{G}$ is polynomial then most, but not all, densities and fluxes are also polynomial.

Example 4.3. The first four density-flux pairs for the cKdV equations (2.2) are

$$\begin{aligned}
\rho^{(1)} &= u, \qquad J^{(1)} = -3\beta u^2 + 3v^2 - \beta u_{2x} \ (\text{any}\ \beta),\\
\rho^{(2)} &= u^2 - 2v^2, \quad J^{(2)} = -4\beta u^3 + \beta u_x^2 - 2\beta u u_{2x} + 2v_x^2 - 4vv_{2x} \ (\text{any}\ \beta),\\
\rho^{(3)} &= uv, \qquad J^{(3)} = 3u^2 v + 2u^3 - u_x v_x + u_{2x} v + u v_{2x} \ (\beta = -1),\\
\rho^{(4)} &= (1+\beta)u^3 - 3uv^2 - \tfrac{1}{2}(1+\beta)u_x^2 + 3v_x^2,\\
J^{(4)} &= -\tfrac{9}{2}\beta(1+\beta)u^4 + 9\beta u^2 v^2 - \tfrac{9}{2}v^4 + 6\beta(1+\beta)u u_x^2 - 3\beta(1+\beta)u^2 u_{2x}\\
&\quad + 3\beta v^2 u_{2x} - \tfrac{1}{2}\beta(1+\beta)u_{2x}^2 + \beta(1+\beta)u_x u_{3x} - 6\beta v u_x v_x + 12 u v_x^2\\
&\quad - 6uvv_{2x} - 3v_{2x}^2 + 6v_x v_{3x} \quad (\beta \neq -1).
\end{aligned} \tag{4.5}$$

The above densities are uniform in ranks 2, 4 and 6. Both $\rho^{(2)}$ and $\rho^{(3)}$ are of rank 4. The corresponding fluxes are also uniform in rank with ranks 4, 6, and 8. In [17], a few densities of rank ≥ 8 are listed, which only exist when $\beta = \frac{1}{2}$.

In general, if in (4.2) $\operatorname{rank}\rho = R$ then $\operatorname{rank} J = R + W(\partial/\partial t) - 1$. All the terms in (4.1) are also uniform in rank. This comes as no surprise since (4.1) vanishes on solutions of (2.1); hence it "inherits" the dilation symmetry of (2.1).

Example 4.4. The first few densities [3, 15] for the sG equation (3.2) are

$$\begin{aligned}
\rho^{(1)} &= 2\alpha\cos u + v^2 + u_x^2, \quad J^{(1)} = -2u_x v,\\
\rho^{(2)} &= 2u_x v, \qquad J^{(2)} = 2\alpha\cos u - v^2 - u_x^2,\\
\rho^{(3)} &= 6\alpha v u_x \cos u + v^3 u_x + v u_x^3 - 8 v_x u_{2x},\\
\rho^{(4)} &= 2\alpha^2\cos^2 u - 2\alpha^2 \sin^2 u + 4\alpha v^2 \cos u + 20\alpha u_x^2 \cos u + v^4\\
&\quad + 6v^2 u_x^2 + u_x^4 - 16 v_x^2 - 16 u_{2x}^2.
\end{aligned} \tag{4.6}$$

$J^{(3)}$ and $J^{(4)}$ are not shown due to length. Again, all densities and fluxes are uniform in rank (before α is equated to 1).

Example 4.5. The first few conserved densities and fluxes for (2.5) are

$$\rho^{(1)} = h, \quad \mathbf{J}^{(1)} = \begin{pmatrix} uh \\ vh \end{pmatrix}, \qquad \rho^{(2)} = h\theta, \quad \mathbf{J}^{(2)} = \begin{pmatrix} uh\theta \\ vh\theta \end{pmatrix},$$

$$\rho^{(3)} = h\theta^2, \quad \mathbf{J}^{(3)} = \begin{pmatrix} uh\theta^2 \\ vh\theta^2 \end{pmatrix},$$

$$\rho^{(4)} = (u^2+v^2)h + h^2\theta, \qquad \mathbf{J}^{(4)} = \begin{pmatrix} u^3 h + uv^2 h + 2uh^2\theta \\ v^3 h + u^2 v h + 2vh^2\theta \end{pmatrix},$$

$$\rho^{(5)} = v_x\theta - u_y\theta + 2\Omega\theta,$$

$$\mathbf{J}^{(5)} = \frac{1}{6}\begin{pmatrix} 12\Omega u\theta - 4uu_y\theta + 6uv_x\theta + 2vv_y\theta + u^2\theta_y + v^2\theta_y - h\theta\theta_y + h_y\theta^2 \\ 12\Omega v\theta + 4vv_x\theta - 6vu_y\theta - 2uu_x\theta - u^2\theta_x - v^2\theta_x + h\theta\theta_x - h_x\theta^2 \end{pmatrix}. \tag{4.7}$$

All densities and fluxes are multi-uniform in rank, which will substantially simplify the computation of the densities. Under either of the two scaling symmetries, (3.3) or (3.4), $\mathrm{rank}(\mathbf{J}) = \mathrm{rank}(\rho) + 1$. With the exception of $\rho^{(2)}$ and $\mathbf{J}^{(2)}$, the ranks of the densities under (3.3) and (3.4) differ by one. The same holds for the fluxes.

5. Tools from the Calculus of Variations

In this section we introduce the variational derivative (Euler operator), the higher Euler operators (also called Lie–Euler operators) from the calculus of variations, and the homotopy operator from homological algebra. These tools will be applied to the computation of densities and fluxes in Section 7.

5.1. Continuous Variational Derivative (Euler Operator)

Definition 5.1. A scalar differential function f is a *divergence* if and only if there exists a vector differential function $\mathbf{F}$ such that $f = \mathrm{Div}\,\mathbf{F}$. In 1D, we say that a differential function f is *exact*[6] if and only if there exists a scalar differential function F such that $f = \mathrm{D}_x F$. Obviously, $F = \mathrm{D}_x^{-1}(f) = \int f\,dx$ is then the primitive (or integral) of f.

Example 5.2. Consider

$$f = 3\,u_x\,v^2\,\sin u - u_x^3\,\sin u - 6\,v\,v_x\,\cos u + 2\,u_x\,u_{2x}\,\cos u + 8\,v_x v_{2x}, \tag{5.1}$$

which we encountered [3] while computing conservation laws for (3.2). The function f is exact. Indeed, upon integration by parts (by hand), one gets

$$F = 4\,v_x^2 + u_x^2\,\cos u - 3\,v^2\,\cos u. \tag{5.2}$$

Currently, CAS like Mathematica, Maple[7] and REDUCE fail this integration!

Example 5.3. Consider

$$f = u_x v_y - u_{2x} v_y - u_y v_x + u_{xy} v_x. \tag{5.3}$$

It is easy to verify that $f = \mathrm{Div}\,\mathbf{F}$ with

$$\mathbf{F} = (u v_y - u_x v_y, -u v_x + u_x v_x). \tag{5.4}$$

As far as we know, the leading CAS currently lack tools to compute $\mathbf{F}$. Three questions arise:

(i) Under what conditions for f does a closed form for $\mathbf{F}$ exist?
(ii) If f is a divergence, what is it the divergence of?
(iii) Without integration by parts, can one design an algorithm to compute $\mathbf{F}$?

To answer these questions we use the following tools from the calculus of variations: the variational derivative (Euler operator), its generalizations (higher Euler operators or Lie–Euler operators), and the homotopy operator.

---

[6] We do not use *integrable* to avoid confusion with complete integrability from soliton theory.
[7] Versions 9.5 and higher of Maple can integrate such expressions as a result of our interactions with the developers.

Definition 5.4. The *variational derivative* (Euler operator), $\mathcal{L}^{(\mathbf{0})}_{\mathbf{u}(\mathbf{x})}$, is defined [33, p. 246] by

$$\mathcal{L}^{(\mathbf{0})}_{\mathbf{u}(\mathbf{x})} = \sum_{J} (-\mathrm{D})_J \frac{\partial}{\partial \mathbf{u}_J}, \tag{5.5}$$

where the sum is over all the unordered multi-indices J [33, p. 95].

For example, in the 2D case the multi-indices corresponding to second-order derivatives can be identified with $\{2x, 2y, 2z, xy, xz, yz\}$. Obviously, $(-D)_{2x} = (-\mathrm{D}_x)^2 = \mathrm{D}_x^2$, $(-D)_{xy} = (-\mathrm{D}_x)(-\mathrm{D}_y) = \mathrm{D}_x \mathrm{D}_y$, etc. For notational details see [33, pp. 95, 108, 246].

With applications in mind, we give explicit formulas for the variational derivatives in 1D, 2D, and 3D.

Example 5.5. For scalar component u they are

$$\mathcal{L}^{(0)}_{u(x)} = \sum_{k=0}^{\infty} (-\mathrm{D}_x)^k \frac{\partial}{\partial u_{kx}} = \frac{\partial}{\partial u} - \mathrm{D}_x \frac{\partial}{\partial u_x} + \mathrm{D}_x^2 \frac{\partial}{\partial u_{2x}} - \mathrm{D}_x^3 \frac{\partial}{\partial u_{3x}} + \cdots, \tag{5.6}$$

$$\begin{aligned} \mathcal{L}^{(0,0)}_{u(x,y)} &= \sum_{k_x=0}^{\infty} \sum_{k_y=0}^{\infty} (-\mathrm{D}_x)^{k_x} (-\mathrm{D}_y)^{k_y} \frac{\partial}{\partial u_{k_x x \, k_y y}} = \frac{\partial}{\partial u} - \mathrm{D}_x \frac{\partial}{\partial u_x} - \mathrm{D}_y \frac{\partial}{\partial u_y} \\ &\quad + \mathrm{D}_x^2 \frac{\partial}{\partial u_{2x}} + \mathrm{D}_x \mathrm{D}_y \frac{\partial}{\partial u_{xy}} + \mathrm{D}_y^2 \frac{\partial}{\partial u_{2y}} - \mathrm{D}_x^3 \frac{\partial}{\partial u_{3x}} - \cdots, \end{aligned} \tag{5.7}$$

and

$$\begin{aligned} \mathcal{L}^{(0,0,0)}_{u(x,y,z)} &= \sum_{k_x=0}^{\infty} \sum_{k_y=0}^{\infty} \sum_{k_z=0}^{\infty} (-\mathrm{D}_x)^{k_x} (-\mathrm{D}_y)^{k_y} (-\mathrm{D}_z)^{k_z} \frac{\partial}{\partial u_{k_x x \, k_y y \, k_z z}} \\ &= \frac{\partial}{\partial u} - \mathrm{D}_x \frac{\partial}{\partial u_x} - \mathrm{D}_y \frac{\partial}{\partial u_y} - \mathrm{D}_z \frac{\partial}{\partial u_z} + \mathrm{D}_x^2 \frac{\partial}{\partial u_{2x}} + \mathrm{D}_y^2 \frac{\partial}{\partial u_{2y}} + \mathrm{D}_z^2 \frac{\partial}{\partial u_{2z}} \\ &\quad + \mathrm{D}_x \mathrm{D}_y \frac{\partial}{\partial u_{xy}} + \mathrm{D}_x \mathrm{D}_z \frac{\partial}{\partial u_{xz}} + \mathrm{D}_y \mathrm{D}_z \frac{\partial}{\partial u_{yz}} - \mathrm{D}_x^3 \frac{\partial}{\partial u_{3x}} - \cdots. \end{aligned} \tag{5.8}$$

Note that $u_{k_x x \, k_y y}$ stands for $u_{xx \cdots x \, yy \cdots y}$ where x is repeated k_x times and y is repeated k_y times. Similar formulas hold for components v, w, etc. The first question is then answered by the following theorem [33, p. 248].

Theorem 5.6. *A necessary and sufficient condition for a function f to be a divergence, i.e. there exists an $\mathbf{F}$ so that $f = \mathrm{Div}\,\mathbf{F}$, is that $\mathcal{L}^{(0)}_{\mathbf{u}(\mathbf{x})}(f) \equiv 0$. In other words, the Euler operator annihilates divergences, just as the divergence annihilates curls, and the curl annihilates gradients.*

If, for example, $\mathbf{u} = (u, v)$ then both $\mathcal{L}^{(0)}_{u(\mathbf{x})}(f)$ and $\mathcal{L}^{(0)}_{v(\mathbf{x})}(f)$ must vanish identically. For the 1D case, the theorem says that a differential function f is exact, i.e. there exists an F so that $f = D_x F$, if and only if $\mathcal{L}^{(0)}_{\mathbf{u}(x)}(f) \equiv 0$.

Example 5.7. To test the exactness of f in (5.1) which involves just one independent variable x, we apply the zeroth Euler operator (5.5) to f for each component of $\mathbf{u} = (u, v)$ separately. For component u (of order 2), one computes

$$\begin{aligned}
\mathcal{L}^{(0)}_{u(x)}(f) &= \frac{\partial f}{\partial u} - \mathrm{D}_x \frac{\partial f}{\partial u_x} + \mathrm{D}_x^2 \frac{\partial f}{\partial u_{2x}} \\
&= 3u_x v^2 \cos u - u_x^3 \cos u + 6v\, v_x \sin u - 2u_x\, u_{2x} \sin u \\
&\quad -\mathrm{D}_x[3v^2 \sin u - 3u_x^2 \sin u + 2u_{2x} \cos u] + \mathrm{D}_x^2[2u_x \cos u] \\
&= 3u_x v^2 \cos u - u_x^3 \cos u + 6v\, v_x \sin u - 2u_x\, u_{2x} \sin u \\
&\quad -[3u_x v^2 \cos u + 6v\, v_x \sin u - 3u_x^3 \cos u - 6u\, u_{2x} \sin u \\
&\quad -2u_x\, u_{2x} \sin u + 2u_{3x} \cos u] \\
&\quad +[-2u_{3x} \cos u - 6u_x\, u_{2x} \sin u + 2u_{3x} \cos u] \\
&\equiv 0.
\end{aligned}$$

Similarly, for component v (also of order 2) one readily verifies that $\mathcal{L}^{(0)}_{v(x)}(f) \equiv 0$.

Example 5.8. As an example in 2D, one can readily verify that

$$f = u_x v_y - u_{2x} v_y - u_y v_x + u_{xy} v_x$$

from (5.3) is a divergence. Applying (5.7) to f for each component of $\mathbf{u} = (u, v)$ gives $\mathcal{L}^{(0,0)}_{u(x,y)}(f) \equiv 0$ and $\mathcal{L}^{(0,0)}_{v(x,y)}(f) \equiv 0$.

5.2. Continuous Higher Euler Operators

To compute $\mathbf{F} = \mathrm{Div}^{-1}(f)$ or, in the 1D case $F = \mathrm{D}_x^{-1}(f) = \int f\, dx$, we need higher-order versions of the variational derivative, called *higher Euler operators* [28, 33] or *Lie–Euler operators* [9]. The general formulas are given in [33, p. 367]. With applications in mind, we restrict ourselves to the 1D, 2D, and 3D cases.

Definition 5.9. The *higher Euler operators* in 1D (with variable x) are

$$\mathcal{L}^{(i)}_{\mathbf{u}(x)} = \sum_{k=i}^{\infty} \binom{k}{i} (-\mathrm{D}_x)^{k-i} \frac{\partial}{\partial \mathbf{u}_{kx}}, \tag{5.9}$$

where $\binom{k}{i}$ is the binomial coefficient.

Note that the higher Euler operator for $i = 0$ matches the variational derivative in (5.6).

Example 5.10. The explicit formulas for the first three higher Euler operators (for component u and variable x) are

$$\begin{aligned}
\mathcal{L}^{(1)}_{u(x)} &= \frac{\partial}{\partial u_x} - 2\mathrm{D}_x \frac{\partial}{\partial u_{2x}} + 3\mathrm{D}_x^2 \frac{\partial}{\partial u_{3x}} - 4\mathrm{D}_x^3 \frac{\partial}{\partial u_{4x}} + \cdots, \\
\mathcal{L}^{(2)}_{u(x)} &= \frac{\partial}{\partial u_{2x}} - 3\mathrm{D}_x \frac{\partial}{\partial u_{3x}} + 6\mathrm{D}_x^2 \frac{\partial}{\partial u_{4x}} - 10\mathrm{D}_x^3 \frac{\partial}{\partial u_{5x}} + \cdots, \\
\mathcal{L}^{(3)}_{u(x)} &= \frac{\partial}{\partial u_{3x}} - 4\mathrm{D}_x \frac{\partial}{\partial u_{4x}} + 10\mathrm{D}_x^2 \frac{\partial}{\partial u_{5x}} - 20\mathrm{D}_x^3 \frac{\partial}{\partial u_{6x}} + \cdots.
\end{aligned}$$

Definition 5.11. The *higher Euler operators* in 2D (with variables x, y) are

$$\mathcal{L}^{(i_x,i_y)}_{\mathbf{u}(x,y)} = \sum_{k_x=i_x}^{\infty} \sum_{k_y=i_y}^{\infty} \binom{k_x}{i_x}\binom{k_y}{i_y}(-\mathrm{D}_x)^{k_x-i_x}(-\mathrm{D}_y)^{k_y-i_y}\frac{\partial}{\partial \mathbf{u}_{k_x x\, k_y y}}. \tag{5.10}$$

Note that the higher Euler operator for $i_x = i_y = 0$ matches the variational derivative in (5.7).

Example 5.12. The first higher Euler operators (for component u and variables x and y) are

$$\mathcal{L}^{(1,0)}_{u(x,y)} = \frac{\partial}{\partial u_x} - 2\mathrm{D}_x\frac{\partial}{\partial u_{2x}} - \mathrm{D}_y\frac{\partial}{\partial u_{xy}} + 3\mathrm{D}_x^2\frac{\partial}{\partial u_{3x}} + 2\mathrm{D}_x\mathrm{D}_y\frac{\partial}{\partial u_{2xy}} - \cdots,$$

$$\mathcal{L}^{(0,1)}_{u(x,y)} = \frac{\partial}{\partial u_y} - 2\mathrm{D}_y\frac{\partial}{\partial u_{2y}} - \mathrm{D}_x\frac{\partial}{\partial u_{yx}} + 3\mathrm{D}_y^2\frac{\partial}{\partial u_{3y}} + 2\mathrm{D}_x\mathrm{D}_y\frac{\partial}{\partial u_{x2y}} - \cdots,$$

$$\mathcal{L}^{(1,1)}_{u(x,y)} = \frac{\partial}{\partial u_{xy}} - 2\mathrm{D}_x\frac{\partial}{\partial u_{2xy}} - 2\mathrm{D}_y\frac{\partial}{\partial u_{x2y}} + 3\mathrm{D}_x^2\frac{\partial}{\partial u_{3xy}} + 4\mathrm{D}_x\mathrm{D}_y\frac{\partial}{\partial u_{2x2y}} + \cdots,$$

$$\mathcal{L}^{(2,1)}_{u(x,y)} = \frac{\partial}{\partial u_{2xy}} - 3\mathrm{D}_x\frac{\partial}{\partial u_{3xy}} - 2\mathrm{D}_y\frac{\partial}{\partial u_{2x2y}} + 6\mathrm{D}_x^2\frac{\partial}{\partial u_{4xy}} + \mathrm{D}_y^2\frac{\partial}{\partial u_{2x3y}} - \cdots.$$

Definition 5.13. The *higher Euler operators* in 3D (with variables x, y, z) are

$$\mathcal{L}^{(i_x,i_y,i_z)}_{\mathbf{u}(x,y,z)} =$$

$$\sum_{k_x=i_x}^{\infty}\sum_{k_y=i_y}^{\infty}\sum_{k_z=i_z}^{\infty}\binom{k_x}{i_x}\binom{k_y}{i_y}\binom{k_z}{i_z}(-\mathrm{D}_x)^{k_x-i_x}(-\mathrm{D}_y)^{k_y-i_y}(-\mathrm{D}_z)^{k_z-i_z}\frac{\partial}{\partial \mathbf{u}_{k_x x\, k_y y\, k_z z}}. \tag{5.11}$$

The higher Euler operator for $i_x = i_y = i_z = 0$ matches the variational derivative given in (5.8). The higher Euler operators are useful in their own right as the following theorem [28] indicates.

Theorem 5.14. *A necessary and sufficient condition for a function f to be an rth order derivative, i.e. $\exists F$ so that $f = D_x^r F$, is that $\mathcal{L}^{(i)}_{\mathbf{u}(x)}(f) \equiv 0$ for $i = 0, 1, \ldots, r-1$.*

5.3. Continuous Homotopy Operator

We now discuss the homotopy operator which will allow us to reduce the computation of $\mathbf{F} = \mathrm{Div}^{-1}(f)$ (or in the 1D case, $F = \mathrm{D}_x^{-1}(f) = \int f\, dx$) to a single integral with respect to an auxiliary variable denoted by λ (not to be confused with λ in Section 3). Hence, the homotopy operator circumvents integration by parts and reduces the inversion of the total divergence operator, Div, to a problem of single-variable calculus.

As mentioned in Section 5.1, Div^{-1} is only defined up to a divergence-free term (a curl term). For example in 3D, Div^{-1} is represented by an equivalence class $\mathrm{Div}^{-1}(f) = \mathbf{F} + \nabla \times \mathbf{K}$ where $\mathbf{K}$ is an arbitrary vector differential function. The homotopy operator selects a particular choice of $\mathbf{K}$.

The homotopy operator is given in explicit form, which makes it easier to implement in CAS. To keep matters transparent, we present the formulas of the homotopy operator in 1D, 2D, and 3D.

Definition 5.15. The *homotopy operator* in 1D (with variable x) [33, p. 372] is

$$\mathcal{H}_{\mathbf{u}(x)}(f) = \int_0^1 \sum_{j=1}^{N} I_{u_j}(f)[\lambda \mathbf{u}] \, \frac{d\lambda}{\lambda}, \tag{5.12}$$

where u_j is the jth component of $\mathbf{u}$ and the integrand $I_{u_j}(f)$ is given by

$$I_{u_j}(f) = \sum_{i=0}^{\infty} \mathrm{D}_x^i \left(u_j \, \mathcal{L}_{u_j(x)}^{(i+1)}(f) \right). \tag{5.13}$$

The integrand involves the 1D higher Euler operators in (5.9).

In (5.12), N is the number of dependent variables and $I_{u_j}(f)[\lambda \mathbf{u}]$ means that in $I_{u_j}(f)$ one replaces $\mathbf{u}(x) \to \lambda \mathbf{u}(x)$, $\mathbf{u}_x(x) \to \lambda \mathbf{u}_x(x)$, etc.

Given an exact function f, the question how to compute $F = \mathrm{D}_x^{-1}(f) = \int f \, dx$ is then answered by the following theorem [33, p. 372].

Theorem 5.16. *For an exact function f, one has $F = \mathcal{H}_{\mathbf{u}(x)}(f)$.*

Thus, in the 1D case, applying the homotopy operator (5.12) allows one to bypass integration by parts. As an experiment, one can start from a function $\tilde{F}$, compute $f = \mathrm{D}_x \tilde{F}$, subsequently compute $F = \mathcal{H}_{\mathbf{u}(x)}(f)$, and finally verify that $F - \tilde{F}$ is a constant. Using (5.1), we show how the homotopy operator (5.12) is applied.

Example 5.17. For a system with $N = 2$ components, $\mathbf{u} = (u_1, u_2) = (u, v)$, the homotopy operator formulas are

$$\mathcal{H}_{\mathbf{u}(x)}(f) = \int_0^1 \left(I_u(f)[\lambda \mathbf{u}] + I_v(f)[\lambda \mathbf{u}] \right) \frac{d\lambda}{\lambda}, \tag{5.14}$$

with

$$I_u(f) = \sum_{i=0}^{\infty} \mathrm{D}_x^i \left(u \, \mathcal{L}_{u(x)}^{(i+1)}(f) \right) \quad \text{and} \quad I_v(f) = \sum_{i=0}^{\infty} \mathrm{D}_x^i \left(v \, \mathcal{L}_{v(x)}^{(i+1)}(f) \right). \tag{5.15}$$

These sums have only finitely many nonzero terms. For example, the sum in $I_u(f)$ terminates at $p - 1$ where p is the order of u. Take, for example,

$$f = 3 \, u_x \, v^2 \, \sin u - u_x^3 \, \sin u - 6 \, v \, v_x \, \cos u + 2 \, u_x \, u_{2x} \, \cos u + 8 \, v_x v_{2x}.$$

First, we compute

$$\begin{aligned} I_u(f) &= u \mathcal{L}_{u(x)}^{(1)}(f) + \mathrm{D}_x \left(u \mathcal{L}_{u(x)}^{(2)}(f) \right) \\ &= u \frac{\partial f}{\partial u_x} - 2u \mathrm{D}_x \left(\frac{\partial f}{\partial u_{2x}} \right) + \mathrm{D}_x \left(u \frac{\partial f}{\partial u_{2x}} \right) \\ &= 3uv^2 \sin u - u u_x^2 \sin u + 2 u_x^2 \cos u. \end{aligned}$$

Next,

$$\begin{aligned} I_v(f) &= v\mathcal{L}^{(1)}_{v(x)}(f) + \mathrm{D}_x\left(v\mathcal{L}^{(2)}_{v(x)}(f)\right) \\ &= v\frac{\partial f}{\partial v_x} - 2v\mathrm{D}_x\left(\frac{\partial f}{\partial v_{2x}}\right) + \mathrm{D}_x\left(v\frac{\partial f}{\partial v_{2x}}\right) \\ &= -6v^2\cos u + 8v_x^2. \end{aligned}$$

Formula (5.14) gives an integral with respect to λ :

$$\begin{aligned} F = \mathcal{H}_{\mathbf{u}(x)}(f) &= \int_0^1 (I_u(f)[\lambda\mathbf{u}] + I_v(f)[\lambda\mathbf{u}])\,\frac{d\lambda}{\lambda} \\ &= \int_0^1 (3\lambda^2uv^2\sin(\lambda u) - \lambda^2uu_x^2\sin(\lambda u) + 2\lambda u_x^2\cos(\lambda u) - 6\lambda v^2\cos(\lambda u) + 8\lambda v_x^2)\,d\lambda \\ &= 4v_x^2 + u_x^2\cos u - 3v^2\cos u. \end{aligned}$$

which agrees with (5.2), previously computed by hand.

We now turn to inverting the Div operator using the homotopy operator.

Definition 5.18. We define the *homotopy operator* in 2D (with variables x, y) through its two components $(\mathcal{H}^{(x)}_{\mathbf{u}(x,y)}(f), \mathcal{H}^{(y)}_{\mathbf{u}(x,y)}(f))$. The x-component of the operator is given by

$$\mathcal{H}^{(x)}_{\mathbf{u}(x,y)}(f) = \int_0^1 \sum_{j=1}^N I^{(x)}_{u_j}(f)[\lambda\mathbf{u}]\,\frac{d\lambda}{\lambda}, \tag{5.16}$$

with

$$I^{(x)}_{u_j}(f) = \sum_{i_x=0}^\infty \sum_{i_y=0}^\infty \left(\frac{1+i_x}{1+i_x+i_y}\right) \mathrm{D}_x^{i_x}\mathrm{D}_y^{i_y}\left(u_j\,\mathcal{L}^{(1+i_x,i_y)}_{u_j(x,y)}(f)\right). \tag{5.17}$$

Analogously, the y-component is given by

$$\mathcal{H}^{(y)}_{\mathbf{u}(x,y)}(f) = \int_0^1 \sum_{j=1}^N I^{(y)}_{u_j}(f)[\lambda\mathbf{u}]\,\frac{d\lambda}{\lambda}, \tag{5.18}$$

with

$$I^{(y)}_{u_j}(f) = \sum_{i_x=0}^\infty \sum_{i_y=0}^\infty \left(\frac{1+i_y}{1+i_x+i_y}\right) \mathrm{D}_x^{i_x}\mathrm{D}_y^{i_y}\left(u_j\,\mathcal{L}^{(i_x,1+i_y)}_{u_j(x,y)}(f)\right). \tag{5.19}$$

Integrands (5.17) and (5.19) involve the 2D higher Euler operators in (5.10).

After verification that f is a divergence, the question how to compute $\mathbf{F} = (F_1, F_2) = \mathrm{Div}^{-1}(f)$ is then answered by the following theorem.

Theorem 5.19. *If f is a divergence, then*

$$\mathbf{F} = (F_1, F_2) = \mathrm{Div}^{-1}(f) = (\mathcal{H}^{(x)}_{\mathbf{u}(x,y)}(f), \mathcal{H}^{(y)}_{\mathbf{u}(x,y)}(f)).$$

The superscript (x) in $\mathcal{H}^{(x)}(f)$ reminds us that we are computing the x-component of $\mathbf{F}$. As a test, one can start from any vector $\tilde{\mathbf{F}}$ and compute $f = \text{Div}\,\tilde{\mathbf{F}}$. Next, compute $\mathbf{F} = (F_1, F_2) = (\mathcal{H}^{(x)}_{\mathbf{u}(x,y)}(f), \mathcal{H}^{(y)}_{\mathbf{u}(x,y)}(f))$ and, finally, verify that $\mathbf{K} = \tilde{\mathbf{F}} - \mathbf{F}$ is divergence free.

Example 5.20. Using (5.3), we show how the application of the 2D homotopy operator leads to (5.4), up to a divergence free vector. Consider

$$f = u_x v_y - u_{2x} v_y - u_y v_x + u_{xy} v_x,$$

which is easily verified to be a divergence. In order to compute $\text{Div}^{-1}(f)$, we use (5.17) to get

$$\begin{aligned} I_u^{(x)}(f) &= u\mathcal{L}^{(1,0)}_{u(x,y)}(f) + \mathrm{D}_x\left(u\mathcal{L}^{(2,0)}_{u(x,y)}(f)\right) + \frac{1}{2}\mathrm{D}_y\left(u\mathcal{L}^{(1,1)}_{u(x,y)}(f)\right) \\ &= u\left(\frac{\partial f}{\partial u_x} - 2\mathrm{D}_x\frac{\partial f}{\partial u_{2x}} - \mathrm{D}_y\frac{\partial f}{\partial u_{xy}}\right) + \mathrm{D}_x\left(u\frac{\partial f}{\partial u_{2x}}\right) + \frac{1}{2}\mathrm{D}_y\left(u\frac{\partial f}{\partial u_{xy}}\right) \\ &= uv_y + \frac{1}{2}u_y v_x - u_x v_y + \frac{1}{2}uv_{xy}. \end{aligned}$$

Similarly, for the v component of $\mathbf{u} = (u, v)$ one gets

$$I_v^{(x)}(f) = v\mathcal{L}^{(1,0)}_{v(x,y)}(f) = v\frac{\partial f}{\partial v_x} = -u_y v + u_{xy} v.$$

Hence, using (5.16),

$$\begin{aligned} F_1 &= \mathcal{H}^{(x)}_{\mathbf{u}(x,y)}(f) = \int_0^1 \left(I_u^{(x)}(f)[\lambda\mathbf{u}] + I_v^{(x)}(f)[\lambda\mathbf{u}]\right)\frac{d\lambda}{\lambda} \\ &= \int_0^1 \lambda\left(uv_y + \frac{1}{2}u_y v_x - u_x v_y + \frac{1}{2}uv_{xy} - u_y v + u_{xy} v\right) d\lambda \\ &= \frac{1}{2}uv_y + \frac{1}{4}u_y v_x - \frac{1}{2}u_x v_y + \frac{1}{4}uv_{xy} - \frac{1}{2}u_y v + \frac{1}{2}u_{xy} v. \end{aligned}$$

Without showing the details, using (5.18) and (5.19) one computes

$$\begin{aligned} F_2 &= \mathcal{H}^{(y)}_{\mathbf{u}(x,y)}(f) = \int_0^1 \left(I_u^{(y)}(f)[\lambda\mathbf{u}] + I_v^{(y)}(f)[\lambda\mathbf{u}]\right)\frac{d\lambda}{\lambda} \\ &= \int_0^1 \left(\lambda\left(-uv_x - \frac{1}{2}uv_{2x} + \frac{1}{2}u_x v_x\right) + \lambda\left(u_x v - u_{2x} v\right)\right) d\lambda \\ &= -\frac{1}{2}uv_x - \frac{1}{4}uv_{2x} + \frac{1}{4}u_x v_x + \frac{1}{2}u_x v - \frac{1}{2}u_{2x} v. \end{aligned}$$

One can readily verify that the resulting vector

$$\mathbf{F} = \begin{pmatrix} F_1 \\ F_2 \end{pmatrix} = \begin{pmatrix} \frac{1}{2}uv_y + \frac{1}{4}u_y v_x - \frac{1}{2}u_x v_y + \frac{1}{4}uv_{xy} - \frac{1}{2}u_y v + \frac{1}{2}u_{xy} v \\ -\frac{1}{2}uv_x - \frac{1}{4}uv_{2x} + \frac{1}{4}u_x v_x + \frac{1}{2}u_x v - \frac{1}{2}u_{2x} v \end{pmatrix}$$

differs from $\tilde{\mathbf{F}} = (uv_y - u_x v_y, -uv_x + u_x v_x)$ by the divergence-free vector

$$\mathbf{K} = \tilde{\mathbf{F}} - \mathbf{F} = \begin{pmatrix} K_1 \\ K_2 \end{pmatrix}$$

$$= \begin{pmatrix} \frac{1}{2}uv_y - \frac{1}{4}u_y v_x - \frac{1}{2}u_x v_y - \frac{1}{4}uv_{xy} + \frac{1}{2}u_y v - \frac{1}{2}u_{xy}v \\ -\frac{1}{2}uv_x + \frac{1}{4}uv_{2x} + \frac{3}{4}u_x v_x - \frac{1}{2}u_x v + \frac{1}{2}u_{2x}v \end{pmatrix}.$$

As mentioned in Section 5.1, $\mathbf{K}$ can be written as $(\mathrm{D}_y\theta, -\mathrm{D}_x\theta)$, with

$$\theta = \frac{1}{2}uv - \frac{1}{4}uv_x - \frac{1}{2}u_x v.$$

The generalization of the homotopy operator to 3D is straightforward.

Definition 5.21. The *homotopy operator* in 3D (with variables x, y, z) is

$$(\mathcal{H}^{(x)}_{\mathbf{u}(x,y,z)}(f), \mathcal{H}^{(y)}_{\mathbf{u}(x,y,z)}(f), \mathcal{H}^{(z)}_{\mathbf{u}(x,y,z)}(f)).$$

By analogy with (5.16),

$$\mathcal{H}^{(x)}_{\mathbf{u}(x,y,z)}(f) = \int_0^1 \sum_{j=1}^{N} I^{(x)}_{u_j}(f)[\lambda\mathbf{u}]\,\frac{d\lambda}{\lambda},$$

with

$$I^{(x)}_{u_j}(f) = \sum_{i_x=0}^{\infty}\sum_{i_y=0}^{\infty}\sum_{i_z=0}^{\infty}\left(\frac{1+i_x}{1+i_x+i_y+i_z}\right)\mathrm{D}_x^{i_x}\mathrm{D}_y^{i_y}\mathrm{D}_z^{i_z}\left(u_j\,\mathcal{L}^{(1+i_x,i_y,i_z)}_{u_j(x,y,z)}(f)\right).$$

The y and z-operators are defined analogously. The integrands $I^{(x)}_{u_j}(f)$ involve the 3D higher Euler operators in (5.11).

By analogy with the 2D case the following theorem holds.

Theorem 5.22. *Given a divergence f one has*

$$\mathbf{F} = \mathrm{Div}^{-1}(f) = (\mathcal{H}^{(x)}_{\mathbf{u}(x,y,z)}(f), \mathcal{H}^{(y)}_{\mathbf{u}(x,y,z)}(f), \mathcal{H}^{(z)}_{\mathbf{u}(x,y,z)}(f)).$$

6. Removing Divergences and Divergence-Equivalent Terms

We present an algorithm to remove divergences and divergence-equivalent which simplifies the computation of densities.

Definition 6.1. Two scalar differential functions, $f^{(1)}$ and $f^{(2)}$, are *divergence-equivalent* if and only if they differ by the divergence of some vector $\mathbf{V}$, i.e. $f^{(1)} \sim f^{(2)}$ if and only if $f^{(1)} - f^{(2)} = \mathrm{Div}\,\mathbf{V}$. Obviously, if a scalar expression is divergence-equivalent to zero, then it is a divergence.

Example 6.2. Functions $f^{(1)} = uu_{2x}$ and $f^{(2)} = -u_x^2$ are divergence-equivalent because

$$f^{(1)} - f^{(2)} = uu_{2x} + u_x^2 = \mathrm{D}_x(uu_x).$$

Using (5.6), note that $\mathcal{L}^{(0)}_{u(x)}(uu_{2x}) = 2u_{2x}$ and $\mathcal{L}^{(0)}_{u(x)}(-u_x^2) = 2u_{2x}$ are equal. Also, $f = u_{4x} = \mathrm{D}_x(u_{3x})$ is a divergence and, as expected, $\mathcal{L}^{(0)}_{u(x)}(u_{4x}) = 0$.

Example 6.3. In the 2D case, $f^{(1)} = (u_x - u_{2x})v_y$ and $f^{(2)} = (u_y - u_{xy})v_x$ are divergence-equivalent since

$$f^{(1)} - f^{(2)} = u_x v_y - u_{2x}v_y - u_y v_x + u_{xy}v_x = \mathrm{Div}\,(uv_y - u_x v_y, -uv_x + u_x v_x).$$

Using (5.7), note that $\mathcal{L}^{(0)}_{\mathbf{u}(x,y)}(f^{(1)}) = \mathcal{L}^{(0)}_{\mathbf{u}(x,y)}(f^{(2)}) = (-v_{xy} - v_{xxy}, -u_{xy} + u_{xxy})$.

Divergences and divergence-equivalent terms can be removed with the following algorithm.

Algorithm 6.4 (for removing divergences and divergence-equivalent terms).

/* Given is a list $\mathcal{R}$ of monomial differential functions */
/* Initialize two new lists $\mathcal{S}, \mathcal{B}$ */
$\mathcal{S} \leftarrow \emptyset$
$\mathcal{B} \leftarrow \emptyset$
/* Find first member of $\mathcal{S}$ */
for each term $t_i \in \mathcal{R}$
    do $\mathbf{v}_i \leftarrow \mathcal{L}^{(0)}_{\mathbf{u}(\mathbf{x})}(t_i)$
        if $\mathbf{v}_i \neq \mathbf{0}$
            then $\mathcal{S} \leftarrow \{t_i\}$
                $\mathcal{B} \leftarrow \{\mathbf{v}_i\}$
                break
            else discard t_i and $\mathbf{v}_i$
/* Find remaining members of $\mathcal{S}$ */
for each term $t_j \in \mathcal{R} \setminus \{t_1, t_2, \ldots, t_i\}$
    do $\mathbf{v}_j \leftarrow \mathcal{L}^{(0)}_{\mathbf{u}(\mathbf{x})}(t_j)$
        if $\mathbf{v}_j \neq \mathbf{0}$
            then if $\mathbf{v}_j \notin \mathrm{Span}(\mathcal{B})$
                then $\mathcal{S} \leftarrow \mathcal{S} \cup \{t_j\}$
                    $\mathcal{B} \leftarrow \mathcal{B} \cup \{\mathbf{v}_j\}$
                else discard t_j and $\mathbf{v}_j$
return $\mathcal{S}$
/* List $\mathcal{S}$ is free of divergences and divergence-equivalent terms */

Example 6.5. Let $\mathcal{R} = \{u^3, u^2v, uv^2, v^3, u_x^2, u_xv_x, v_x^2, uu_{2x}, u_{2x}v, uv_{2x}, vv_{2x}, u_{4x}, v_{4x}\}$. We remove divergences and divergence-equivalent terms in $\mathcal{R}$ by using the above algorithm. Since $\mathbf{v}_1 = \mathcal{L}^{(0)}_{\mathbf{u}(\mathbf{x})}(u^3) = (3u^2, 0) \neq (0,0)$ we have $\mathcal{S} = \{t_1\} = \{u^3\}$ and $\mathcal{B} = \{\mathbf{v}_1\} = \{(3u^2, 0)\}$. The first for-loop is halted and the second for-loop starts. Next, $\mathbf{v}_2 = \mathcal{L}^{(0)}_{\mathbf{u}(\mathbf{x})}(u^2 v) = (2uv, u^2) \neq (0,0)$. We verify that $\mathbf{v}_1$ and $\mathbf{v}_2$ are independent and update the sets resulting in $\mathcal{S} = \{t_1, t_2\} = \{u^3, u^2 v\}$ and $\mathcal{B} = \{\mathbf{v}_1, \mathbf{v}_2\} = \{(3u^2, 0), (2uv, u^2)\}$.

Proceeding in a similar fashion, since the first seven terms are indeed independent, we have $\mathcal{S} = \{t_1, t_2, \ldots, t_7\}$ and

$$\mathcal{B} = \{\mathbf{v}_1, \mathbf{v}_2, \ldots, \mathbf{v}_7\} = \{(3u^2, 0), (2uv, u^2), \ldots, (0, -2v_{2x})\}.$$

For $t_8 = uu_{2x}$ we compute $\mathbf{v}_8 = \mathcal{L}^{(0)}_{\mathbf{u}(\mathbf{x})}(uu_{2x}) = (2u_{2x}, 0)$ and verify that $\mathbf{v}_8 = -\mathbf{v}_5$. So, $\mathbf{v}_8 \in \mathrm{Span}(\mathcal{B})$ and t_8 and $\mathbf{v}_8$ are discarded (i.e. *not* added to the respective sets). For similar reasons, t_9, t_{10}, and t_{11} as well as $\mathbf{v}_9, \mathbf{v}_{10}$, and $\mathbf{v}_{11}$ are discarded. The terms $t_{12} = u_{4x}$ and $t_{13} = v_{4x}$ are discarded because $\mathbf{v}_{12} = \mathbf{v}_{13} = (0,0)$. So, $\mathcal{R}$ is replaced by $\mathcal{S} = \{u^3, u^2 v, uv^2, v^3, u_x^2, u_x v_x, v_x^2\}$ which is free of divergences and divergence-equivalent terms.

7. Application: Conservation Laws of Nonlinear PDEs

As an application of the Euler and homotopy operators we show how to compute conserved densities and fluxes for the three PDEs introduced in Section 2. The first PDE illustrates the 1D case (one independent variable), but it involves two dependent variables $u(x)$ and $v(x)$. The second PDE (again in 1D) has a transcendental nonlinearity which complicates the computation of conserved densities and fluxes [15]. A third example illustrates the algorithm for a 2D case.

To compute conservation laws, $\mathrm{D}_t\, \rho + \mathrm{Div}\, \mathbf{J} = 0$, of polynomial systems of nonlinear PDEs, we use a direct approach. First, we build the candidate density ρ as a linear combination (with constant coefficients c_i) of terms which are uniform in rank (with respect to the scaling symmetry of the PDE). It is of paramount importance that the candidate density is free of divergences and divergence-equivalent terms. If such terms were present, their coefficients could not be determined because such terms can be moved into flux $\mathbf{J}$. We will use Algorithm 6.4 to construct a shortest density.

Second, we evaluate $\mathrm{D}_t \rho$ on solutions of the PDE, thus removing all time derivatives from the problem. The resulting expression (called E) must be a divergence (of the thus far unknown flux). Thus, we set $\mathcal{L}^{(0)}_{\mathbf{u}(\mathbf{x})}(E) \equiv 0$. Setting the coefficients of like terms to zero leads to a linear system for the undetermined coefficients c_i. In the most difficult case, such systems are parametrized by the constant parameters appearing in the given PDE. If so, a careful analysis of the eliminant (and solution branching) must be carried out. For each branch, the solution of the linear system is substituted into ρ and E.

Third, since $E = \mathrm{Div}\, \mathbf{J}$ we use the homotopy operator $\mathcal{H}_{\mathbf{u}(\mathbf{x})}$ to compute $\mathbf{J} = \mathrm{Div}^{-1}(E)$. The computations are done with our Mathematica packages [20]. Recall that $\mathbf{J}$ is only defined up a curl. Inversion of Div via the homotopy operator does not guarantee the shortest flux. Removing the curl term in $\mathbf{J}$ may lead to a shorter flux.

7.1. Conservation Laws for the Coupled KdV Equations

In (4.5) we gave the first four density-flux pairs. As an example, we will compute density $\rho^{(4)}$ and associated flux $J^{(4)}$.

Recall that the weights for the cKdV equations are $W(\partial/\partial x) = 1$ and $W(u) = W(v) = 2$. The parameter β has no weight. Hence, $\rho^{(4)}$ has rank 6. The algorithm has three steps.

Step 1: Construct the form of the density

Start from $\mathcal{V} = \{u, v\}$, i.e. the list of dependent variables with weight. Construct the set $\mathcal{M}$ which contains all monomials of selected rank 6 or less (without derivatives). Thus $\mathcal{M} = \{u^3, v^3, u^2v, uv^2, u^2, v^2, uv, u, v, 1\}$. Next, for each monomial in $\mathcal{M}$, introduce the correct number of x-derivatives so that each term has rank 6. For example,

$$\frac{\partial^2 u^2}{\partial x^2} = 2u_x^2 + 2uu_{2x}, \quad \frac{\partial^2 v^2}{\partial x^2} = 2v_x^2 + 2vv_{2x}, \quad \frac{\partial^2 (uv)}{\partial x^2} = u_{2x}v + 2u_xv_x + uv_{2x},$$
$$\frac{\partial^4 u}{\partial x^4} = u_{4x}, \quad \frac{\partial^4 v}{\partial x^4} = v_{4x}, \quad \frac{\partial^6 1}{\partial x^6} = 0. \tag{7.1}$$

Ignore the highest-order terms (typically the last terms) in each of the right-hand sides of (7.1). Augment $\mathcal{M}$ with the remaining terms, after stripping off numerical factors, to get $\mathcal{R} = \{u^3, u^2v, uv^2, v^3, u_x^2, u_xv_x, v_x^2, u_{2x}v\}$, where the 8 terms[8] are listed by increasing order

Use Algorithm 6.4 to replace $\mathcal{R}$ by $\mathcal{S} = \{u^3, u^2v, uv^2, v^3, u_x^2, u_xv_x, v_x^2\}$. Linearly combine the terms in $\mathcal{S}$ with constant coefficients to get the shortest candidate density:

$$\rho = c_1u^3 + c_2u^2v + c_3uv^2 + c_4v^3 + c_5u_x^2 + c_6u_xv_x + c_7v_x^2. \tag{7.2}$$

Step 2: Determine the constants c_i

Compute

$$\begin{aligned} E = \mathrm{D}_t\rho = \frac{\partial \rho}{\partial t} + \rho'(\mathbf{u})[\mathbf{F}] &= \frac{\partial \rho}{\partial u}u_t + \frac{\partial \rho}{\partial u_x}u_{tx} + \frac{\partial \rho}{\partial v}v_t + \frac{\partial \rho}{\partial v_x}v_{tx} \\ &= (3c_1u^2 + 2c_2uv + c_3v^2)u_t + (2c_5u_x + c_6v_x)u_{tx} \\ &\quad + (c_2u^2 + 2c_3uv + 3c_4v^2)v_t + (c_6u_x + 2c_7v_x)v_{tx}. \end{aligned}$$

Replace u_t, v_t, u_{tx} and v_{tx} from (2.2) to obtain

$$\begin{aligned} E = &(3c_1u^2 + 2c_2uv + c_3v^2)(6\beta uu_x - 6vv_x + \beta u_{3x}) \\ &+ (2c_5u_x + c_6v_x)(6\beta uu_x - 6vv_x + \beta u_{3x})_x \\ &- (c_2u^2 + 2c_3uv + 3c_4v^2)(3uv_x + v_{3x}) - (c_6u_x + 2c_7v_x)(3uv_x + v_{3x})_x. \end{aligned} \tag{7.3}$$

Since $E = \mathrm{D}_t\,\rho = -\mathrm{D}_xJ$, the expression E must be exact. Therefore, apply the variational derivative (5.6) and require that $\mathcal{L}^{(0)}_{u(x)}(E) \equiv 0$ and $\mathcal{L}^{(0)}_{v(x)}(E) \equiv 0$.

[8] Note that keeping all terms in (7.1) would have resulted in the list $\mathcal{R}$ (with 13 terms) given in the example at the end of Section 6. As shown, the algorithm would reduce $\mathcal{R}$ to 7 terms.

Group like terms and set their coefficients equal to zero to obtain the following (parametrized) linear system for the unknown coefficients c_1 through c_7:

$$
\begin{aligned}
&(3+4\beta)c_2 = 0, \quad 3c_1 + (1+\beta)c_3 = 0, \quad 4c_2 + 3c_4 = 0, \quad (1+\beta)c_3 - 6c_5 = 0,\\
&\beta(c_1 + 2c_5) = 0, \quad \beta c_2 - c_6 = 0, \quad (1+\beta)c_6 = 0, \quad c_4 + c_6 = 0,\\
&2(1+\beta)c_2 - 3(1+2\beta)c_6 = 0, \quad 2c_2 - (1+6\beta)c_6 = 0,\\
&\beta c_3 - 6c_5 - c_7 = 0, \quad c_3 + c_7 = 0.
\end{aligned}
$$

Investigate the eliminant of the system. In this example, there exists a solution for any $\beta \neq -1$. Set $c_1 = 1$ and obtain

$$
c_1 = 1, \quad c_2 = c_4 = c_6 = 0, \quad c_3 = -\frac{3}{1+\beta}, \quad c_5 = -\frac{1}{2}, \quad c_7 = \frac{3}{1+\beta}. \tag{7.4}
$$

Substitute the solution into (7.2) and multiply by $1+\beta$ to get

$$
\rho = (1+\beta)u^3 - 3uv^2 - \frac{1}{2}(1+\beta)u_x^2 + 3v_x^2, \tag{7.5}
$$

which is $\rho^{(4)}$ in (4.5).

Step 3: Compute the flux J

Compute the flux corresponding to ρ in (7.5). Substitute (7.4) into (7.3), reverse the sign and multiply by $1+\beta$, to get

$$
\begin{aligned}
E = {} & 18\beta(1+\beta)u^3u_x - 18\beta u^2vv_x - 18\beta uu_xv^2 + 18v^3v_x - 6\beta(1+\beta)u_x^3\\
& - 6\beta(1+\beta)uu_xu_{2x} + 3\beta(1+\beta)u^2u_{3x} - 3\beta v^2u_{3x} - 6v_xv_{4x} - \beta(1+\beta)u_xu_{4x}\\
& + 6uvu_{3x} + 6(\beta-2)u_xv_x^2 + 6(1+\beta)u_xvv_{2x} - 18uv_xv_{2x}.
\end{aligned} \tag{7.6}
$$

Apply (5.14) and (5.15) to (7.6) to obtain

$$
\begin{aligned}
J = {} & -\tfrac{9}{2}\beta(1+\beta)u^4 + 9\beta u^2v^2 - \tfrac{9}{2}v^4 + 6\beta(1+\beta)uu_x^2 - 3\beta(1+\beta)u^2u_{2x}\\
& + 3\beta v^2u_{2x} - \tfrac{1}{2}\beta(1+\beta)u_{2x}^2 + \beta(1+\beta)u_xu_{3x} - 6\beta vu_xv_x\\
& + 12uv_x^2 - 6uvv_{2x} - 3v_{2x}^2 + 6v_xv_{3x},
\end{aligned}
$$

which is $J^{(4)}$ in (4.5).

If $\beta = \frac{1}{2}$, the cKdV equations (2.2) are completely integrable [1, 23] and admit conserved densities at every even rank.

7.2. Conservation Laws for the Sine-Gordon Equation

Recall that the weights for the sG equation (2.3) are $W(\frac{\partial}{\partial x}) = 1$, $W(u) = 0$, $W(v) = 1$, and $W(\alpha) = 2$. The first few (of infinitely many) densities and fluxes were given in (4.6). We show how to compute densities $\rho^{(1)}$ and $\rho^{(2)}$, both of rank 2, and their associated fluxes $J^{(1)}$ and $J^{(2)}$.

In contrast to the previous example, the candidate density will no longer have *constant* undetermined coefficients c_i but *functional* coefficients $h_i(u)$ which depend on the transcendental variable u with weight zero [3]. Avoiding having to solve PDEs, we only consider examples where *one* dependent variable has weight zero.

Step 1: Construct the form of the density

Augment the list of dependent variables with α (with nonzero weight) and replace u by u_x (since $W(u) = 0$). Hence, $\mathcal{V} = \{\alpha, u_x, v\}$. Next, compute $\mathcal{R} = \{\alpha, v^2, u_{2x}, u_x v, u_x^2\}$ and remove divergences and divergence-equivalent terms to get $\mathcal{S} = \{\alpha, v^2, u_x^2, u_x v\}$. The candidate density is

$$\rho = \alpha h_1(u) + h_2(u)v^2 + h_3(u)u_x^2 + h_4(u)u_x v, \tag{7.7}$$

with undetermined functional coefficients $h_i(u)$.

Step 2: Determine the functions $h_i(u)$

Compute

$$\begin{aligned} E = \mathrm{D}_t\rho = \frac{\partial \rho}{\partial t} + \rho'(\mathbf{u})[\mathbf{F}] &= \frac{\partial \rho}{\partial u}u_t + \frac{\partial \rho}{\partial u_x}u_{tx} + \frac{\partial \rho}{\partial v}v_t \\ &= (\alpha h_1' + v^2 h_2' + u_x^2 h_3' + u_x v h_4')v + (2u_x h_3 + v h_4)v_x \\ &\quad + (2v h_2 + u_x h_4)(\alpha \sin(u) + u_{2x}). \end{aligned} \tag{7.8}$$

where h_i' means $\frac{dh_i}{du}$. Since $E = \mathrm{D}_t\,\rho = -\mathrm{D}_x J$, the expression E must be exact. Therefore, require that $\mathcal{L}^{(0)}_{u(x)}(E) \equiv 0$ and $\mathcal{L}^{(0)}_{v(x)}(E) \equiv 0$. Set the coefficients of like terms equal to zero to get a mixed linear system of algebraic and ODEs:

$$\begin{gathered} h_2(u) - h_3(u){=}0,\;\; h_2'(u){=}0,\;\; h_3'(u){=}0,\;\; h_4'(u){=}0,\;\; h_2''(u){=}0, \\ h_4''(u) = 0,\;\; 2h_2'(u) - h_3'(u) = 0,\;\; 2h_2''(u) - h_3''(u) = 0, \\ h_1'(u) + 2h_2(u)\sin u = 0,\;\; h_1''(u) + 2h_2'(u)\sin u + 2h_2(u)\cos u = 0. \end{gathered}$$

Solve the system [3] and substitute the solution

$$h_1(u) = 2c_1\cos u + c_3, \quad h_2(u) = h_3(u) = c_1, \quad h_4(u) = c_2 \tag{7.9}$$

(with arbitrary constants c_i) into (7.7) to obtain

$$\rho = c_1(2\alpha\cos u + v^2 + u_x^2) + c_2 u_x v + c_3\alpha. \tag{7.10}$$

Step 3: Compute the flux J

Compute the flux corresponding to ρ in (7.10). Substitute (7.9) into (7.8), to get

$$E = c_1(2u_{2x}v + 2u_x v_x) + c_2(v v_x + u_x u_{2x} + \alpha u_x \sin u). \tag{7.11}$$

Since $E = \mathrm{D}_t\,\rho = -\mathrm{D}_x J$, one must integrate $f = -E$. Applying (5.15) yields $I_u(f) = -2c_1 u_x v - c_2(u_x^2 + \alpha u \sin u)$ and $I_v(f) = -2c_1 u_x v - c_2 v^2$. Use formula (5.14) to obtain

$$\begin{aligned} J = \mathcal{H}_{\mathbf{u}(x)}(f) &= \int_0^1 (I_u(f)[\lambda\mathbf{u}] + I_v(f)[\lambda\mathbf{u}])\,\frac{d\lambda}{\lambda} \\ &= -\int_0^1 (4c_1\lambda u_x v + c_2(\lambda u_x^2 + \alpha u\sin(\lambda u) + \lambda v^2))\,d\lambda \\ &= -\,c_1(2u_x v) - c_2\left(\frac{1}{2}v^2 + \frac{1}{2}u_x^2 - \alpha\cos u\right). \end{aligned} \tag{7.12}$$

Finally, split density (7.10) and flux (7.12) into independent pieces (for c_1 and c_2):

$$\rho^{(1)} = 2\alpha\cos u + v^2 + u_x^2 \quad \text{and} \quad J^{(1)} = -2u_x v,$$
$$\rho^{(2)} = u_x v \quad \text{and} \quad J^{(2)} = -\frac{1}{2}v^2 - \frac{1}{2}u_x^2 + \alpha\cos u.$$

For E in (7.11), J in (7.12) can easily be computed by hand [3]. However, the computation of fluxes corresponding to densities of ranks ≥ 2 is cumbersome and requires integration with the homotopy operator.

7.3. Conservation Laws for the Shallow Water Wave Equations

In contrast to the previous two examples, (2.5) is not completely integrable as far as we know. One cannot expect to find a complete set of conserved densities and fluxes (of different ranks).

The first few densities and fluxes were given in (4.7). We show how to compute densities $\rho^{(1)}, \rho^{(3)}, \rho^{(4)}$, and $\rho^{(5)}$, which are of rank 3 under the following (choice for the) weights

$$W(\partial/\partial x)=W(\partial/\partial y)=1, W(u)=W(v)=1, W(\theta)=1, W(h)=1, W(\Omega)=2. \quad (7.13)$$

We will also compute the associated fluxes $J^{(1)}, J^{(3)}, J^{(4)}$, and $J^{(5)}$.

The fact that (2.5) is multi-uniform is advantageous. Indeed, one can use the invariance of (2.5) under one scale to construct the terms of ρ, and, subsequently, use additional scale(s) to split ρ into smaller densities. This "divide and conquer" strategy drastically reduces the complexity of the computations.

Step 1: Construct the form of the density

Start from $\mathcal{V} = \{u, v, \theta, h, \Omega\}$, i.e. the list of variables *and* parameters with weights. Use (7.13) to get

$$\mathcal{M} = \{\Omega u, \Omega v, \ldots, u^3, v^3, \ldots, u^2 v, uv^2, \ldots, u^2, v^2, \ldots, u, v, \theta, h\}$$

which has 38 monomials of rank 3 or less (without derivatives).

The terms of rank 3 in $\mathcal{M}$ are left alone. To adjust the rank, differentiate each monomial of rank 2 in $\mathcal{M}$ with respect to x ignoring the highest-order term. For example, in $\frac{du^2}{dx} = 2uu_x$, the term can be ignored since it is a total derivative. The terms $u_x v$ and $-uv_x$ are divergence-equivalent since $\frac{d(uv)}{dx} = u_x v + uv_x$. Keep $u_x v$. Likewise, differentiate each monomial of rank 2 in $\mathcal{M}$ with respect to y and ignore the highest-order term.

Produce the remaining terms for rank 3 by differentiating the monomials of rank 1 in $\mathcal{M}$ with respect to x twice, or y twice, or once with respect to x and y. Again ignore the highest-order terms. Augment the set $\mathcal{M}$ with the derivative terms of rank 3 to get

$$\mathcal{R} = \{\Omega u, \Omega v, \ldots, uv^2, u_x v, u_x\theta, u_x h, \ldots, u_y v, u_y\theta, \ldots, \theta_y h\}$$

which has 36 terms.

Instead of applying Algorithm 6.4 to $\mathcal{R}$, use the "divide and conquer" strategy to split $\mathcal{R}$ into sublists of terms of equal rank under the (general) weights

$$\begin{aligned} &W(\partial/\partial t) = W(\Omega),\ W(\partial/\partial y) = W(\partial/\partial x) = 1, \\ &W(u) = W(v) = W(\Omega) - 1, \\ &W(\theta) = 2W(\Omega) - W(h) - 2, \end{aligned} \tag{7.14}$$

where $W(\Omega)$ and $W(h)$ are arbitrary. Use (7.14) to compute the rank of each monomial in $\mathcal{R}$ and gather terms of like rank in separate lists.

For each rank R_i in Table 1, apply Algorithm 6.4 to each $\mathcal{R}_i$ to get the list $\mathcal{S}_i$. Coincidentally, in this example $\mathcal{R}_i = \mathcal{S}_i$ for all i. Linearly combine the monomials in each list $\mathcal{S}_i$ with coefficients to get the shortest candidate densities ρ_i. In Table 1, we list the 10 candidate densities and the final densities and fluxes with their ranks. These conservation laws were listed in (4.7).

i	Rank R_i	Candidate ρ_i	Final ρ_i	Final $\mathbf{J}_i$
1	$6W(\Omega)-3W(h)-6$	$c_1\theta^3$	0	0
2	$3W(h)$	c_1h^3	0	0
3	$5W(\Omega)-2W(h)-5$	$c_1u\theta^2+c_2v\theta^2$	0	0
4	$W(\Omega)+2W(h)-1$	$c_1uh^2+c_2vh^2$	0	0
5	$4W(\Omega)-W(h)-4$	$c_1u^2\theta+c_2uv\theta+c_3v^2\theta+c_4\theta^2h$	θ^2h	$\begin{pmatrix} uh\theta^2 \\ vh\theta^2 \end{pmatrix}$
6	$2W(\Omega)+W(h)-2$	$c_1u^2h+c_2uvh+c_3v^2h+c_4\theta h^2$	$u^2h+v^2h+\theta h^2$	$\mathbf{J}^{(4)}$
7	$3W(\Omega)-W(h)-2$	$c_1\Omega\theta+c_2u_y\theta+c_3v_y\theta+c_4u_x\theta$ $+c_5v_x\theta$	$2\Omega\theta-u_y\theta+v_x\theta$	$\mathbf{J}^{(5)}$
8	$W(\Omega)+W(h)$	$c_1\Omega h+c_2u_yh+c_3v_yh+c_4u_xh$ $+c_5v_xh$	Ωh	$\begin{pmatrix} \Omega uh \\ \Omega vh \end{pmatrix}$
9	$2W(\Omega)-1$	$c_1\Omega u+c_2\Omega v+c_3u_yv+c_4\theta_yh$ $+c_5u_xv+c_6\theta_xh$	0	0
10	$3W(\Omega)-3$	$c_1u^3+c_2u^2v+c_3uv^2+c_4v^3$ $+c_5u\theta h+c_6v\theta h$	0	0

TABLE 1. Candidate densities for the SWW equations

Step 2: Determine the constants c_i

For each of the densities ρ_i in Table 1 compute $E_i = \mathrm{D}_t\rho_i$ and use (2.5) to remove all time derivatives. For example, proceeding with ρ_7,

$$E_7 = \rho_7'(\mathbf{u})[\mathbf{F}] = \frac{\partial \rho_7}{\partial u_x}u_{tx} + \frac{\partial \rho_7}{\partial u_y}u_{ty} + \frac{\partial \rho_7}{\partial v_x}v_{tx} + \frac{\partial \rho_7}{\partial v_y}v_{ty} + \frac{\partial \rho_7}{\partial \theta}\theta_t$$

$$
\begin{aligned}
= & - c_4\theta(uu_x + vu_y - 2\Omega v + \tfrac{1}{2}h\theta_x + \theta h_x)_x \\
& - c_2\theta(uu_x + vu_y - 2\Omega v + \tfrac{1}{2}h\theta_x + \theta h_x)_y \\
& - c_5\theta(uv_x + vv_y + 2\Omega u + \tfrac{1}{2}h\theta_y + \theta h_y)_x \\
& - c_3\theta(uv_x + vv_y + 2\Omega u + \tfrac{1}{2}h\theta_y + \theta h_y)_y \\
& - (c_1\Omega + c_2 u_y + c_3 v_y + c_4 u_x + c_5 v_x)(u\theta_x + v\theta_y).
\end{aligned} \tag{7.15}
$$

Require that

$$
\mathcal{L}^{(0,0)}_{u(x,y)}(E_7) = \mathcal{L}^{(0,0)}_{v(x,y)}(E_7) = \mathcal{L}^{(0,0)}_{\theta(x,y)}(E_7) = \mathcal{L}^{(0,0)}_{h(x,y)}(E_7) \equiv 0,
$$

where, for example, $\mathcal{L}^{(0,0)}_{u(x,y)}$ is given in (5.7). Gather like terms. Equate their coefficients to zero to obtain

$$
c_1 + 2c_2 = 0, \quad c_3 = c_4 = 0, \quad c_1 - 2c_5 = 0, \quad c_2 + c_5 = 0.
$$

Set $c_1 = 2$. Substitute the solution

$$
c_1 = 2, \quad c_2 = -1, \quad c_3 = c_4 = 0, \quad c_5 = 1 \tag{7.16}
$$

into ρ_7 to obtain $\rho_7 = 2\Omega\theta - u_y\theta + v_x\theta$, which matches $\rho^{(5)}$ in (4.7).

Proceed in a similar way with the remaining 9 candidate densities to obtain the results given in the third column of Table 1.

Step 3: Compute the flux **J**

Compute the flux corresponding to all $\rho_i \neq 0$ in Table 1. For example, continuing with ρ_7, substitute (7.16) into (7.15) to get

$$
\begin{aligned}
E_7 = & - \theta(u_x v_x + u v_{2x} + v_x v_y + v v_{xy} + 2\Omega u_x + \tfrac{1}{2}\theta_x h_y - u_x u_y - u u_{xy} \\
& - u_y v_y - u_{2y} v + 2\Omega v_y - \tfrac{1}{2}\theta_y h_x) - (2\Omega u\theta_x + 2\Omega v\theta_y - u u_y \theta_x \\
& - u_y v\theta_y + u v_x \theta_x + v v_x \theta_y).
\end{aligned}
$$

Apply the 2D homotopy operator in (5.16)–(5.19) to $E_7 = -\mathrm{Div}\,\mathbf{J}_7$. So, compute

$$
\begin{aligned}
I_u^{(x)}(E_7) &= u\mathcal{L}^{(1,0)}_{u(x,y)}(E_7) + \mathrm{D}_x\left(u\mathcal{L}^{(2,0)}_{u(x,y)}(E_7)\right) + \frac{1}{2}\mathrm{D}_y\left(u\mathcal{L}^{(1,1)}_{u(x,y)}(E_7)\right) \\
&= u\left(\frac{\partial E_7}{\partial u_x} - 2\mathrm{D}_x\left(\frac{\partial E_7}{\partial u_{2x}}\right) - \mathrm{D}_y\left(\frac{\partial E_7}{\partial u_{xy}}\right)\right) + \mathrm{D}_x\left(u\frac{\partial E_7}{\partial u_{2x}}\right) + \frac{1}{2}\mathrm{D}_y\left(u\frac{\partial E_7}{\partial u_{xy}}\right) \\
&= -uv_x\theta - 2\Omega u\theta - \frac{1}{2}u^2\theta_y + uu_y\theta.
\end{aligned}
$$

Similarly, compute

$$
\begin{aligned}
I_v^{(x)}(E_7) &= -vv_y\theta - \frac{1}{2}v^2\theta_y - uv_x\theta, \\
I_\theta^{(x)}(E_7) &= -\frac{1}{2}\theta^2 h_y - 2\Omega u\theta + uu_y\theta - uv_x\theta, \\
I_h^{(x)}(E_7) &= \frac{1}{2}\theta\theta_y h.
\end{aligned}
$$

Next, compute

$$\begin{aligned}
J_7^{(x)}(\mathbf{u}) &= -\mathcal{H}^{(x)}_{\mathbf{u}(x,y)}(E_7) \\
&= -\int_0^1 \left(I_u^{(x)}(E_7)[\lambda\mathbf{u}] + I_v^{(x)}(E_7)[\lambda\mathbf{u}] + I_\theta^{(x)}(E_7)[\lambda\mathbf{u}] + I_h^{(x)}(E_7)[\lambda\mathbf{u}] \right) \frac{d\lambda}{\lambda} \\
&= \int_0^1 \Big(4\lambda\Omega u\theta + \lambda^2 \Big(3uv_x\theta + \frac{1}{2}u^2\theta_y - 2uu_y\theta + vv_y\theta + \frac{1}{2}v^2\theta_y \\
&\qquad + \frac{1}{2}\theta^2 h_y - \frac{1}{2}\theta\theta_y h \Big) \Big) \, d\lambda \\
&= 2\Omega u\theta - \frac{2}{3}uu_y\theta + uv_x\theta + \frac{1}{3}vv_y\theta + \frac{1}{6}u^2\theta_y + \frac{1}{6}v^2\theta_y - \frac{1}{6}h\theta\theta_y + \frac{1}{6}h_y\theta^2.
\end{aligned}$$

Analogously, compute

$$\begin{aligned}
J_7^{(y)}(\mathbf{u}) &= -\mathcal{H}^{(y)}_{\mathbf{u}(x,y)}(E_7) \\
&= 2\Omega v\theta + \frac{2}{3}vv_x\theta - vu_y\theta - \frac{1}{3}uu_x\theta - \frac{1}{6}u^2\theta_x - \frac{1}{6}v^2\theta_x + \frac{1}{6}h\theta\theta_x - \frac{1}{6}h_x\theta^2.
\end{aligned}$$

Hence,

$$\mathbf{J}_7 = \frac{1}{6} \begin{pmatrix} 12\Omega u\theta - 4uu_y\theta + 6uv_x\theta + 2vv_y\theta + u^2\theta_y + v^2\theta_y - h\theta\theta_y + h_y\theta^2 \\ 12\Omega v\theta + 4vv_x\theta - 6vu_y\theta - 2uu_x\theta - u^2\theta_x - v^2\theta_x + h\theta\theta_x - h_x\theta^2 \end{pmatrix},$$

which matches $\mathbf{J}^{(5)}$ in (4.7).

Proceed in a similar way with the remaining nonzero densities to obtain the fluxes given in the last column of Table 1.

System (2.5) has conserved densities [14, p. 294] of the form

$$\rho = hf(\theta) \quad \text{and} \quad \rho = (v_x - u_y + 2\Omega)g(\theta),$$

for any functions f and g. Our algorithm can only find f and g of the form θ^k where $k \geq 0$ is integer. A comprehensive study of all conservation laws of (2.5) is beyond the scope of this chapter.

8. Examples of Nonlinear DDEs

We consider nonlinear systems of DDEs of the form

$$\dot{\mathbf{u}}_n = \mathbf{G}(\ldots, \mathbf{u}_{n-1}, \mathbf{u}_n, \mathbf{u}_{n+1}, \ldots), \tag{8.1}$$

where $\mathbf{u}_n$ and $\mathbf{G}$ are vector-valued functions with N components. The integer n corresponds to discretization in space;[9] the dot denotes differentiation with respect to continuous time (t). For simplicity, we write $\mathbf{G}(\mathbf{u}_n)$, although $\mathbf{G}$ depends on $\mathbf{u}_n$ and a finite number of its forward and backward shifts. We assume that $\mathbf{G}$ is polynomial with constant coefficients. No restrictions are imposed on the forward or backward shifts or the degree of nonlinearity in $\mathbf{G}$. In the examples we denote

[9]We only consider DDEs with one discrete variable.

the components of $\mathbf{u}_n$ by u_n, v_n, etc. If present, parameters are denoted by lowercase Greek letters. We use the following two DDEs to illustrate the theorems and algorithms.

Example 8.1. The Kac–van Moerbeke (KvM) lattice [25],

$$\dot{u}_n = u_n(u_{n+1} - u_{n-1}), \tag{8.2}$$

arises in the study of Langmuir oscillations in plasmas, population dynamics, etc.

Example 8.2. The Toda lattice [35] in polynomial form [18],

$$\dot{u}_n = v_{n-1} - v_n, \quad \dot{v}_n = v_n(u_n - u_{n+1}), \tag{8.3}$$

models vibrations of masses in a lattice with an exponential interaction force.

9. Dilation Invariance and Uniformity in Rank for DDEs

The definitions for the discrete case are analogous to the continuous case. For brevity, we use the Toda lattice (8.3) to illustrate the definitions and concepts. (8.3) is dilation invariant under

$$(t, u_n, v_n) \rightarrow (\lambda^{-1}t, \lambda u_n, \lambda^2 v_n). \tag{9.1}$$

Definition 9.1. The *weight* W of a variable equals the exponent of the scaling parameter λ [18, 19].

Weights of dependent variables are non-negative and rational. We tacitly assume that weights are independent of n. For example, $W(u_{n-1}) = W(u_n) = W(u_{n+1})$, etc.

Example 9.2. Since t is replaced by $\frac{t}{\lambda}$ we have $W(\frac{\mathrm{d}}{\mathrm{dt}}) = W(\mathrm{D}_t) = 1$. From (9.1) we have $W(u_n) = 1$ and $W(v_n) = 2$.

Definition 9.3. The *rank* of a monomial equals the total weight of the monomial. An expression is uniform in rank if all its monomial terms have equal rank.

Ranks must be positive natural or rational numbers.

Example 9.4. The three terms in the first equation in (8.3) have rank 2; all terms in the second equation have rank 3. Each equation is uniform in rank.

Conversely, requiring uniformity in rank in (8.3) yields $W(u_n) + 1 = W(v_n)$, and $W(v_n) + 1 = W(u_n) + W(v_n)$. Hence, $W(u_n) = 1$, $W(v_n) = 2$. So, the scaling symmetry can be computed with linear algebra.

Many integrable nonlinear DDEs are scaling invariant. If not, they can be made so by extending the set of dependent variables with parameters with weights.

10. Conserved Densities and Fluxes of Nonlinear DDEs

By analogy with D_x and D_x^{-1}, we define the following operators acting on monomials m_n in u_n, v_n, etc.

Definition 10.1. D is the *up-shift operator* $\mathrm{D}\, m_n = m_{n+1}$ (also known as the forward- or right-shift operator). Its inverse, D^{-1}, is the *down-shift operator* (or backward- or left-shift operator), $\mathrm{D}^{-1}\, m_n = m_{n-1}$. The identity operator is denoted by I. Thus, $\mathrm{I}\, m_n = m_n$, and $\Delta = \mathrm{D} - \mathrm{I}$ is the *forward difference operator.* So, $\Delta\, m_n = (\mathrm{D} - \mathrm{I})\, m_n = m_{n+1} - m_n$.

Definition 10.2. A *conservation law* of (8.1),

$$\mathrm{D}_t\, \rho_n + \Delta\, J_n = 0, \tag{10.1}$$

which holds on solutions of (8.1), links a *conserved density* ρ_n to a *flux* J_n. Densities and fluxes depend on $\mathbf{u}_n$ as well as forward and backward shifts of $\mathbf{u}_n$.

To stress the analogy between one-dimensional PDEs and DDEs, we compare the defining equations in Table 2.

	Continuous case (PDE)	Semi-discrete case (DDE)
Evolution equation	$\mathbf{u}_t = \mathbf{G}(\mathbf{u}, \mathbf{u}_x, \mathbf{u}_{2x}, \ldots)$	$\dot{\mathbf{u}}_n = \mathbf{G}(\ldots, \mathbf{u}_{n-1}, \mathbf{u}_n, \mathbf{u}_{n+1}, \ldots)$
Conservation law	$\mathrm{D}_t \rho + \mathrm{D}_x J = 0$	$\mathrm{D}_t \rho_n + \Delta\, J_n = 0$

TABLE 2. Defining equations for conservation laws of PDEs and DDEs

Definition 10.3. Compositions of D and D^{-1} define an *equivalence relation* ($\equiv$) on monomials. All shifted monomials are equivalent.

Example 10.4. For example, $u_{n-1}v_{n+1} \equiv u_n v_{n+2} \equiv u_{n+1}v_{n+3} \equiv u_{n+2}v_{n+4}$. Factors in a monomial in u_n and its shifts are ordered by $u_{n+j} \prec u_{n+k}$ if $j < k$.

Definition 10.5. The *main representative* of an equivalence class is the monomial with u_n in the first position [18, 19].

Example 10.6. The main representative in $\{\ldots, u_{n-2}u_n, u_{n-1}u_{n+1}, u_n u_{n+2}, \ldots\}$ is $u_n u_{n+2}$ (not $u_{n-2}u_n$).

For monomials involving u_n, v_n, w_n, etc. and their shifts, we lexicographically order the variables, that is $u_n \prec v_n \prec w_n$, etc. Thus, for example, $u_n v_{n+2}$ but not $u_{n-2}v_n$ is the main representative of

$$\{\ldots, u_{n-2}v_n, u_{n-1}v_{n+1}, u_n v_{n+2}, u_{n+1}v_{n+3}, \ldots\}.$$

Table 3 shows the KvM and Toda lattices with their scaling invariances, weights, and a few conserved densities. Note that the conservation law "inherits" the scaling symmetry of the DDE. Indeed, all ρ_n in Table 3 are uniform in rank.

11. Discrete Euler and Homotopy Operators

11.1. Discrete Variational Derivative (Euler Operator)

Given is a scalar function f_n in discrete variables $u_n, v_n, \ldots$ and their forward and backward shifts. The goal is to find the scalar function F_n so that

$$f_n = \Delta\, F_n = F_{n+1} - F_n.$$

We illustrate the computations with the following example:

$$f_n = -u_n u_{n+1} v_n - v_n^2 + u_{n+1} u_{n+2} v_{n+1} + v_{n+1}^2 + u_{n+3} v_{n+2} - u_{n+1} v_n. \tag{11.1}$$

By hand, one readily computes

$$F_n = v_n^2 + u_n\, u_{n+1}\, v_n + u_{n+1}\, v_n + u_{n+2}\, v_{n+1}. \tag{11.2}$$

Below we will address the questions:

(i) Under what conditions for f_n does F_n exist in closed form?

(ii) How can one compute $F_n = \Delta^{-1}(f_n)$?

(iii) Can one compute $F_n = \Delta^{-1}(f_n)$ in an analogous way as in the continuous case?

Expression f_n is called *exact* if it is a total difference, i.e. there exists an F_n so that $f_n = \Delta\, F_n$. With respect to the existence of F_n in closed form, the following exactness criterion is well-known and frequently used [4, 22].

Theorem 11.1. *A necessary and sufficient condition for a function f_n, with positive shifts, to be exact is that $\mathcal{L}^{(0)}_{\mathbf{u}_n}(f_n) \equiv 0$.*

	Kac–van Moerbeke lattice	Toda lattice
Lattice	$\dot{u}_n = u_n(u_{n+1} - u_{n-1})$	$\dot{u}_n = v_{n-1} - v_n,\ \dot{v}_n = v_n(u_n - u_{n+1})$
Scaling	$(t, u_n) \rightarrow (\lambda^{-1}t, \lambda u_n)$	$(t, u_n, v_n) \rightarrow (\lambda^{-1}t, \lambda u_n, \lambda^2 v_n)$
Weights	$W(\mathrm{D}_t) = 1,\ W(u_n) = 1$	$W(\mathrm{D}_t) = 1,\ W(u_n) = 1,\ W(v_n) = 2$
Densities	$\rho_n^{(1)} = u_n,$ $\rho_n^{(2)} = \frac{1}{2}u_n^2 + u_n u_{n+1},$ $\rho_n^{(3)} = \frac{1}{3}u_n^3 + u_n u_{n+1}(u_n + u_{n+1} + u_{n+2})$	$\rho_n^{(1)} = u_n,$ $\rho_n^{(2)} = \frac{1}{2}u_n^2 + v_n,$ $\rho_n^{(3)} = \frac{1}{3}u_n^3 + u_n(v_{n-1} + v_n)$

TABLE 3. Examples of nonlinear DDEs with weights and densities

$\mathcal{L}^{(0)}_{\mathbf{u}_n}$ is the *discrete variational derivative* (discrete Euler operator of order zero) [4] defined by

$$\mathcal{L}^{(0)}_{\mathbf{u}_n} = \sum_{k=0}^{\infty} \mathrm{D}^{-k} \frac{\partial}{\partial \mathbf{u}_{n+k}} = \frac{\partial}{\partial \mathbf{u}_n} (\sum_{k=0}^{\infty} \mathrm{D}^{-k}) = \frac{\partial}{\partial \mathbf{u}_n} (\mathrm{I}+\mathrm{D}^{-1}+\mathrm{D}^{-2}+\mathrm{D}^{-3}+\cdots). \quad (11.3)$$

A proof of the theorem is given in e.g. [22]. In practice, the series in (11.3) terminates at the highest shift in the expression the operator is applied to. To verify that an expression $E(u_{n-q}, \ldots, u_n, \ldots, u_{n+p})$ involving negative shifts is a total difference, one must first remove the negative shifts by replacing E_n by $\tilde{E}_n = \mathrm{D}^q E_n$.

Example 11.2. We return to (11.1),

$$f_n = -u_n\, u_{n+1}\, v_n - v_n^2 + u_{n+1}\, u_{n+2}\, v_{n+1} + v_{n+1}^2 + u_{n+3}\, v_{n+2} - u_{n+1}\, v_n.$$

We first test that f_n is exact (i.e., the total difference of some F_n to be computed later). We then apply the discrete zeroth Euler operator to f_n for each component of $\mathbf{u}_n = (u_n, v_n)$ separately. For component u_n (with maximum shift 3) one readily verifies that

$$\mathcal{L}^{(0)}_{u_n}(f_n) = \frac{\partial}{\partial u_n} \left(\mathrm{I} + \mathrm{D}^{-1} + \mathrm{D}^{-2} + \mathrm{D}^{-3}\right)(f_n) \equiv 0.$$

Similarly, for component v_n (with maximum shift 2) one checks that $\mathcal{L}^{(0)}_{v_n}(f_n) \equiv 0$.

11.2. Discrete Higher Euler and Homotopy Operators

To compute F_n, we need higher-order versions of the discrete variational derivative. They are called *discrete higher Euler operators* or *discrete Lie–Euler operators*, $\mathcal{L}^{(i)}_{\mathbf{u}_n}$, in analogy with the continuous case [33].

In Table 4, we have put the continuous and discrete higher Euler operators side by side. Note that the discrete higher Euler operator for $i = 0$ is the discrete variational derivative.

Example 11.3. The first three higher Euler operators for component u_n from Table 4 are

$$\begin{aligned}
\mathcal{L}^{(1)}_{u_n} &= \frac{\partial}{\partial u_n} \left(\mathrm{D}^{-1} + 2\mathrm{D}^{-2} + 3\mathrm{D}^{-3} + 4\mathrm{D}^{-4} + \cdots\right), \\
\mathcal{L}^{(2)}_{u_n} &= \frac{\partial}{\partial u_n} \left(\mathrm{D}^{-2} + 3\mathrm{D}^{-3} + 6\mathrm{D}^{-4} + 10\mathrm{D}^{-5} + \cdots\right), \\
\mathcal{L}^{(3)}_{u_n} &= \frac{\partial}{\partial u_n} \left(\mathrm{D}^{-3} + 4\mathrm{D}^{-4} + 10\mathrm{D}^{-5} + 20\mathrm{D}^{-6} + \cdots\right).
\end{aligned}$$

Similar formulae hold for $\mathcal{L}^{(i)}_{v_n}$.

The discrete higher Euler operators are useful in their own right as the following theorem indicates.

Theorem 11.4. *A necessary and sufficient condition for a function f_n to be a forward difference of order r, i.e. $\exists\, F_n$ so that $f_n = \Delta^r F_n$, is that $\mathcal{L}^{(i)}_{\mathbf{u}_n}(f_n) \equiv 0$ for $i = 0, 1, \ldots, r-1$.*

Operator	Continuous case	Discrete case
Zeroth Euler	$\mathcal{L}^{(0)}_{\mathbf{u}(x)} = \sum_{k=0}^{\infty} (-\mathrm{D}_x)^k \frac{\partial}{\partial \mathbf{u}_{k\,x}}$	$\mathcal{L}^{(0)}_{\mathbf{u}_n} = \sum_{k=0}^{\infty} \mathrm{D}^{-k} \frac{\partial}{\partial \mathbf{u}_{n+k}} = \frac{\partial}{\partial \mathbf{u}_n} \sum_{k=0}^{\infty} \mathrm{D}^{-k}$
Higher Euler	$\mathcal{L}^{(i)}_{\mathbf{u}(x)} = \sum_{k=i}^{\infty} \binom{k}{i} (-\mathrm{D}_x)^{k-i} \frac{\partial}{\partial \mathbf{u}_{k\,x}}$	$\mathcal{L}^{(i)}_{\mathbf{u}_n} = \sum_{k=0}^{\infty} \binom{k}{i} \mathrm{D}^{-k} \frac{\partial}{\partial \mathbf{u}_{n+k}} = \frac{\partial}{\partial \mathbf{u}_n} \sum_{k=i}^{\infty} \binom{k}{i} \mathrm{D}^{-k}$
Homotopy	$\mathcal{H}_{\mathbf{u}(x)}(f) = \int_0^1 \sum_{j=1}^{N} I_{u_j}(f)[\lambda \mathbf{u}] \frac{d\lambda}{\lambda}$	$\mathcal{H}_{\mathbf{u}_n}(f) = \int_0^1 \sum_{j=1}^{N} I_{u_{j,n}}(f)[\lambda \mathbf{u}_n] \frac{d\lambda}{\lambda}$
Integrand	$I_{u_j}(f) = \sum_{i=0}^{\infty} \mathrm{D}_x^i \left(u_j \mathcal{L}^{(i+1)}_{u_j(x)}(f) \right)$	$I_{u_{j,n}}(f) = \sum_{i=0}^{\infty} \Delta^i \left(u_{j,n} \mathcal{L}^{(i+1)}_{u_{j,n}}(f) \right)$

TABLE 4. Continuous and discrete Euler and homotopy operators in 1D side by side

Also in Table 4, we put the formulae for the *discrete homotopy operator* $\mathcal{H}_{\mathbf{u}_n}$ and the continuous homotopy operator side by side. The integrand $I_{u_{j,n}}(f)$ of the homotopy operator involves the discrete higher Euler operators. As in the continuous case, N is the number of dependent variables $u_{j,n}$ and $I_{u_{j,n}}(f)[\lambda \mathbf{u}_n]$ means that after $I_{u_{j,n}}(f)$ is applied one replaces $\mathbf{u}_n$ by $\lambda \mathbf{u}_n$, $\mathbf{u}_{n+1}$ by $\lambda \mathbf{u}_{n+1}$, etc. To compute F_n, one can use the following theorem [21, 24, 30].

Theorem 11.5. *Given an exact function f_n, one can compute $F_n = \Delta^{-1}(f_n)$ from $F_n = \mathcal{H}_{\mathbf{u}_n}(f_n)$.*

Thus, the homotopy operator reduces the inversion of Δ (and summation by parts) to a set of differentiations and shifts followed by a single integral with respect to an auxiliary parameter λ. We present a simplified version [21] of the homotopy operator given in [24, 30], where the problem is dealt with in greater generality and where the proofs are given in the context of discrete variational complexes.

Example 11.6. For a system with components, $(u_{1,n}, u_{2,n}) = (u_n, v_n)$, the discrete homotopy operator from Table 4 is

$$\mathcal{H}_{\mathbf{u}_n}(f) = \int_0^1 \left(I_{u_n}(f)[\lambda \mathbf{u}_n] + I_{v_n}(f)[\lambda \mathbf{u}_n] \right) \frac{d\lambda}{\lambda}, \tag{11.4}$$

with

$$I_{u_n}(f) = \sum_{i=0}^{\infty} \Delta^i \left(u_n \mathcal{L}_{u_n}^{(i+1)}(f) \right) \text{ and } I_{v_n}(f) = \sum_{i=0}^{\infty} \Delta^i \left(v_n \mathcal{L}_{v_n}^{(i+1)}(f) \right). \quad (11.5)$$

Example 11.7. We return to (11.1). Using (11.5),

$$\begin{aligned}
I_{u_n}(f_n) &= u_n \mathcal{L}_{u_n}^{(1)}(f_n) + \Delta\Big(u_n \mathcal{L}_{u_n}^{(2)}(f_n)\Big) + \Delta^2\Big(u_n \mathcal{L}_{u_n}^{(3)}(f_n)\Big) \\
&= u_n \frac{\partial}{\partial u_n} \left(\mathrm{D}^{-1} + 2\mathrm{D}^{-2} + 3\mathrm{D}^{-3}\right)(f_n) + \Delta\left(u_n \frac{\partial}{\partial u_n} \left(\mathrm{D}^{-2} + 3\mathrm{D}^{-3}\right)(f_n)\right) \\
&\quad + \Delta^2\left(u_n \frac{\partial}{\partial u_n} \mathrm{D}^{-3}(f_n)\right) \\
&= 2u_n u_{n+1} v_n + u_{n+1} v_n + u_{n+2} v_{n+1},
\end{aligned}$$

and

$$\begin{aligned}
I_{v_n}(f_n) &= v_n \mathcal{L}_{v_n}^{(1)}(f_n) + \Delta\Big(v_n \mathcal{L}_{v_n}^{(2)}(f_n)\Big) \\
&= v_n \frac{\partial}{\partial v_n} \left(\mathrm{D}^{-1} + 2\mathrm{D}^{-2}\right)(f_n) + \Delta\left(v_n \frac{\partial}{\partial v_n} \mathrm{D}^{-2}(f_n)\right) \\
&= u_n u_{n+1} v_n + 2v_n^2 + u_{n+1} v_n + u_{n+2} v_{n+1}.
\end{aligned}$$

The homotopy operator (11.4) thus leads to an integral with respect to λ:

$$\begin{aligned}
F_n &= \int_0^1 \left(I_{u_n}(f_n)[\lambda \mathbf{u}_n] + I_{v_n}(f_n)[\lambda \mathbf{u}_n]\right) \frac{d\lambda}{\lambda} \\
&= \int_0^1 \left(2\lambda v_n^2 + 3\lambda^2 u_n u_{n+1} v_n + 2\lambda u_{n+1} v_n + 2\lambda u_{n+2} v_{n+1}\right) d\lambda \\
&= v_n^2 + u_n u_{n+1} v_n + u_{n+1} v_n + u_{n+2} v_{n+1},
\end{aligned}$$

which agrees with (11.2), previously computed by hand.

12. Application: Conservation Laws of Nonlinear DDEs

In [16, 22], different algorithms are presented to compute fluxes of nonlinear DDEs. In this section we show how to compute fluxes with the discrete homotopy operator. For clarity, we compute a conservation law for (8.3) in Section 8. The computations are carried out with our Mathematica packages [20]. The completely integrable Toda lattice (8.3) has infinitely many conserved densities and fluxes. As an example, we compute density $\rho_n^{(3)}$ (of rank 3) and corresponding flux $J_n^{(3)}$ (of rank 4). In this example,

$$\mathbf{G} = (G_1, G_2) = (\, v_{n-1} - v_n,\ v_n(u_n - u_{n+1})\,).$$

Assuming that the weights $W(u_n)=1$ and $W(v_n)=2$ are computed and the rank of the density is selected (say, $R=3$), our algorithm works as follows.

Step 1: Construct the form of the density

Start from $\mathcal{V} = \{u_n, v_n\}$, i.e. the list of dependent variables with weight. List all monomials in u_n and v_n of rank 3 or less: $\mathcal{M} = \{u_n^3, u_n^2, u_n v_n, u_n, v_n\}$.

Next, for each monomial in $\mathcal{M}$, introduce the correct number of t-derivatives so that each term has rank 3. Using (8.3), compute

$$\frac{d^0 u_n^3}{dt^0} = u_n^3, \quad \frac{d^0 u_n v_n}{dt^0} = u_n v_n,$$
$$\frac{du_n^2}{dt} = 2u_n \dot{u}_n = 2u_n v_{n-1} - 2u_n v_n, \quad \frac{dv_n}{dt} = \dot{v}_n = u_n v_n - u_{n+1} v_n,$$
$$\frac{d^2 u_n}{dt^2} = \frac{d\dot{u}_n}{dt} = \frac{d(v_{n-1} - v_n)}{dt} = u_{n-1}v_{n-1} - u_n v_{n-1} - u_n v_n + u_{n+1} v_n. \qquad (12.1)$$

Augment $\mathcal{M}$ with the terms from the right-hand sides of (12.1) to get $\mathcal{R} = \{u_n^3, u_n v_{n-1}, u_n v_n, u_{n-1} v_{n-1}, u_{n+1} v_n\}$.

Identify members belonging to the same equivalence classes and replace them by their main representatives. For example, $u_n v_{n-1} \equiv u_{n+1} v_n$, so the latter is replaced by $u_n v_{n-1}$. Hence, replace $\mathcal{R}$ by $\mathcal{S} = \{u_n^3, u_n v_{n-1}, u_n v_n\}$, which has the building blocks of the density. Linearly combine the monomials in $\mathcal{S}$ with coefficients c_i to get the candidate density:

$$\rho_n = c_1\, u_n^3 + c_2\, u_n v_{n-1} + c_3\, u_n v_n. \qquad (12.2)$$

Step 2: Determine the coefficients

Require that (10.1) holds. Compute $\mathrm{D}_t \rho_n$. Use (8.3) to remove $\dot{u}_n$ and $\dot{v}_n$ and their shifts. Thus,

$$E_n = \mathrm{D}_t \rho_n = (3c_1 - c_2) u_n^2 v_{n-1} + (c_3 - 3c_1) u_n^2 v_n + (c_3 - c_2) v_{n-1} v_n$$
$$+ c_2\, u_{n-1} u_n v_{n-1} + c_2 v_{n-1}^2 - c_3 u_n u_{n+1} v_n - c_3 v_n^2. \qquad (12.3)$$

To remove the negative shift $n-1$, compute $\tilde{E}_n = \mathrm{D} E_n$. Apply $\mathcal{L}_{u_n}^{(0)}$ to $\tilde{E}_n$, yielding

$$\mathcal{L}_{u_n}^{(0)}(\tilde{E}_n) = \frac{\partial}{\partial u_n}(\mathrm{I} + \mathrm{D}^{-1} + \mathrm{D}^{-2})(\tilde{E}_n)$$
$$= 2(3c_1 - c_2) u_n v_{n-1} + 2(c_3 - 3c_1) u_n v_n + (c_2 - c_3) u_{n-1} v_{n-1}$$
$$+ (c_2 - c_3) u_{n+1} v_n. \qquad (12.4)$$

Next, apply $\mathcal{L}_{v_n}^{(0)}$ to $\tilde{E}_n$, yielding

$$\mathcal{L}_{v_n}^{(0)}(\tilde{E}_n) = \frac{\partial}{\partial v_n}(\mathrm{I} + \mathrm{D}^{-1})(\tilde{E}_n)$$
$$= (3c_1 - c_2) u_{n+1}^2 + (c_3 - c_2) v_{n+1} + (c_2 - c_3) u_n u_{n+1}$$
$$+ 2(c_2 - c_3) v_n + (c_3 - 3c_1) u_n^2 + (c_3 - c_2) v_{n-1}. \qquad (12.5)$$

Both (12.4) and (12.5) must vanish identically. Solve the linear system

$$3c_1 - c_2 = 0, \quad c_3 - 3c_1 = 0, \quad c_2 - c_3 = 0.$$

Set $c_1 = \frac{1}{3}$ and substitute the solution $c_1 = \frac{1}{3}$, $c_2 = c_3 = 1$ into (12.2)

$$\rho_n = \frac{1}{3}u_n^3 + u_n(v_{n-1} + v_n). \tag{12.6}$$

Step 3: Compute the flux

In view of (10.1), one must compute $J_n = -\Delta^{-1}(E_n)$. Substitute $c_1 = \frac{1}{3}$, $c_2 = c_3 = 1$ into (12.3). Then, $\tilde{E}_n = \mathrm{D}E_n = u_n u_{n+1} v_n + v_n^2 - u_{n+1} u_{n+2} v_{n+1} - v_{n+1}^2$. Apply (11.5) to $-\tilde{E}_n$ to obtain

$$I_{u_n}(-\tilde{E}_n) = 2u_n u_{n+1} v_n, \qquad I_{v_n}(-\tilde{E}_n) = u_n u_{n+1} v_n + 2v_n^2.$$

Application of the homotopy operator (11.4) yields

$$\begin{aligned}\tilde{J}_n &= \int_0^1 \left(I_{u_n}(-\tilde{E}_n)[\lambda \mathbf{u}_n] + I_{v_n}(-\tilde{E}_n)[\lambda \mathbf{u}_n]\right) \frac{d\lambda}{\lambda} \\ &= \int_0^1 (3\lambda^2 u_n u_{n+1} v_n + 2\lambda v_n^2)\, d\lambda \\ &= u_n u_{n+1} v_n + v_n^2.\end{aligned}$$

After a backward shift, $J_n = \mathrm{D}^{-1}(\tilde{J}_n)$, we obtain J_n. With (12.6), the final result is then

$$\rho_n = \frac{1}{3}u_n^3 + u_n(v_{n-1} + v_n), \qquad J_n = u_{n-1} u_n v_{n-1} + v_{n-1}^2.$$

The above density corresponds to $\rho_n^{(3)}$ in Table 3.

13. Conclusion

Based on the concept of scaling invariance and using tools of the calculus of variations, we presented algorithms to symbolically compute conserved densities and fluxes of nonlinear polynomial and transcendental systems of PDEs in multi-spacial dimensions and DDEs in one discrete variable.

The continuous homotopy operator is a powerful, algorithmic tool to compute fluxes explicitly. Indeed, the homotopy operator handles integration by parts in multi-variables which allows us to invert the total divergence operator. Likewise, the discrete homotopy operator handles summation by parts and inverts the forward difference operator. In both cases, the problem reduces to an explicit integral from 1D calculus.

Homotopy operators have a wide range of applications in the study of PDEs, DDEs, fully discretized lattices, and beyond. We extracted the Euler and homotopy operators from their abstract setting and introduced them into applied mathematics, thereby making them readily applicable to computational problems.

We purposely avoided differential forms and abstract concepts from differential geometry and homological algebra. Our down-to-earth approach might appeal to scientists who prefer not to juggle exterior products and Lie derivatives. Our calculus-based formulas for the Euler and homotopy operators can be readily implemented in major CAS.

References

[1] M. J. Ablowitz and P. A. Clarkson, *Solitons, Nonlinear Evolution Equations and Inverse Scattering.* Cambridge University Press, Cambridge (1991).

[2] M. J. Ablowitz and H. Segur, *Solitons and the Inverse Scattering Transform.* SIAM Studies in Applied Mathematics **4**, SIAM, Philadelphia (1981).

[3] P. J. Adams, *Symbolic Computation of Conserved Densities and Fluxes for Systems of Partial Differential Equations with Transcendental Nonlinearities.* Master of Science Thesis, Department of Mathematical and Computer Sciences, Colorado School of Mines, Golden, Colorado (2003).

[4] V. E. Adler, A. B. Shabat and R. I. Yamilov, *Symmetry Approach to the Integrability Problem.* Theor. Math. Phys. **125** (2000), 1603–1661.

[5] S. C. Anco, *Conservation Laws of Scaling-invariant Field Equations.* J. Phys. A: Math. Gen. **36** (2003), 8623–8638.

[6] S. C. Anco and G. Bluman, *Direct Construction Method for Conservation Laws of Partial Differential Equations. Part I: Examples of Conservation Law Classifications.* Euro. J. Appl. Math. **13** (2002), 545–566.

[7] S. C. Anco and G. Bluman, *Direct Construction Method for Conservation Laws of Partial Differential Equations. Part II: General Treatment.* Euro. J. Appl. Math. **13** (2002), 567–585.

[8] I. M. Anderson, *Introduction to the Variational Bicomplex.* Contemp. Math. **132**. American Mathematical Society, Providence, Rhode Island (1992), pp. 51–73.

[9] I. M. Anderson, *The Variational Bicomplex.* Department of Mathematics, Utah State University, Logan, Utah (November, 2004), 318 pages. Manuscript of book is available at `http://www.math.usu.edu/~fg_mp/Publications/VB/vb.pdf`.

[10] I. M. Anderson, *The Vessiot Package*, Department of Mathematics, Utah State University, Logan, Utah (November, 2004), 253 pages. Available at `http://www.math.usu.edu/~fg_mp/Pages/SymbolicsPage/VessiotDownloads.html`.

[11] I. M. Anderson and T. Duchamp, *On the Existence of Global Variational Principles.* Amer. J. Math. **92** (1980), 781–868.

[12] A. Barone, F. Esposito, C. J. Magee, and A. C. Scott, *Theory and Applications of the Sine-Gordon Equation.* Riv. Nuovo Cim. **1** (1971), 227–267.

[13] B. Deconinck and M. Nivala, *A Maple Package for the Computation of Conservation Laws.* Available at `http://www.amath.washington.edu/~bernard/papers.html`.

[14] P. Dellar, *Common Hamiltonian Structure of the Shallow Water Equations with Horizontal Temperature Gradients and Magnetic Fields.* Phys. Fluids **15** (2003), 292–297.

[15] R. K. Dodd and R. K. Bullough, *Polynomial Conserved Densities for the Sine-Gordon Equations.* Proc. R. Soc. Lond. A **352** (1977), 481–503.

[16] H. Eklund, *Symbolic Computation of Conserved Densities and Fluxes for Nonlinear Systems of Differential-Difference Equations.* Master of Science Thesis, Department

of Mathematical and Computer Sciences, Colorado School of Mines, Golden, Colorado (2003).

[17] Ü. Göktaş and W. Hereman, *Symbolic Computation of Conserved Densities for Systems of Nonlinear Evolution Equations.* J. Symb. Comput. **24** (1997), 591–621.

[18] Ü. Göktaş and W. Hereman, *Computation of Conservation Laws for Nonlinear Lattices.* Physica D **132** (1998), 425–436.

[19] Ü. Göktaş and W. Hereman, *Algorithmic Computation of Higher-order Symmetries for Nonlinear Evolution and Lattice Equations.* Adv. Comput. Math. **11** (1999), 55–80.

[20] W. Hereman, TransPDEDensityFlux.m, PDEMultiDimDensityFlux.m, and DDEDensityFlux.m: *Mathematica* Packages for the Symbolic Computation of Conservation Laws of Partial Differential Equations and Differential-difference Equations. Available from the software section at `http://www.mines.edu/fs_home/whereman/`.

[21] W. Hereman, J. A. Sanders, J. Sayers, and J. P. Wang, *Symbolic Computation of Conserved Densities, Generalized Symmetries, and Recursion Operators for Nonlinear Differential-difference Equations.* In: *Group Theory and Numerical Methods*, CRM Proc. Lect. Ser. **39** (P. Winternitz and D. Gomez-Ullate, eds.). American Mathematical Society, Providence, Rhode Island (2005), pp. 267–282.

[22] M. Hickman and W. Hereman, *Computation of Densities and Fluxes of Nonlinear Differential-difference Equations*, Proc. Roy. Soc. Lond. A **459** (2003), 2705–2729.

[23] R. Hirota and J. Satsuma, *Soliton Solutions of a Coupled Korteweg-de Vries Equation.* Phys. Lett. **85A** (1981) 407–408.

[24] P. E. Hydon and E. L. Mansfield, *A Variational Complex for Difference Equations.* Found. Comp. Math. **4** (2004), 187–217.

[25] M. Kac and P. van Moerbeke, *On an Explicitly Soluble System of Nonlinear Differential Equations Related to Certain Toda Lattices.* Adv. Math. **16** (1975), 160–169.

[26] A. H. Kara and F. M. Mahomed, *Relationship Between Symmetries and Conservation Laws.* Int. J. Theor. Phys. **39** (2000), 23–40.

[27] I. S. Krasil'shchik and A. M. Vinogradov, *Symmetries and Conservation Laws for Differential Equations of Mathematical Physics.* American Mathematical Society, Providence, Rhode Island (1998).

[28] M. D. Kruskal, R. M. Miura, C. S. Gardner, and N. J. Zabusky, *Korteweg–de Vries Equation and Generalizations V. Uniqueness and Nonexistence of Polynomial Conservation Laws.* J. Math. Phys. **11** (1970), 952–960.

[29] G. L. Lamb, *Analytical Descriptions of Ultra Short Optical Propagation in a Resonant Medium.* Rev. Mod. Phys. **43** (1971), 99–124.

[30] E. L. Mansfield and P. E. Hydon, *On a Variational Complex for Difference Equations.* In: *The Geometrical Study of Differential Equations* (J. A. Leslie and T. P. Robart, eds.). American Mathematical Society, Providence, Rhode Island (2002), pp. 121–129.

[31] E. L. Mansfield and G. R. W. Quispel, *Towards a Variational Complex for the Finite Element Method.* In: *Group Theory and Numerical Methods*, CRM Proc. Lect. Ser. **39** (P. Winternitz and D. Gomez-Ullate, eds.). American Mathematical Society, Providence, Rhode Island (2005), pp. 207–231.

[32] J. E. Marsden and A. J. Tromba, *Vector Calculus* (3rd edn.). W. H. Freeman and Company, New York (1988).

[33] P. J. Olver, *Applications of Lie Groups to Differential Equations* (2nd edn.). Grad. Texts in Math. **107**, Springer, New York (1993).

[34] Y. B. Suris, *The Problem of Discretization: Hamiltonian Approach.* Birkhäuser, Basel (2003).

[35] M. Toda, *Theory of Nonlinear Lattices.* Springer, Berlin (1981).

[36] T. Wolf, *A Comparison of Four Approaches to the Calculation of Conservation Laws.* Euro. J. Appl. Math. **13** (2002), 129–152.

Acknowledgements and Dedication

This material is based upon work supported by the National Science Foundation (NSF) under grants nos. DMS-9732069, DMS-9912293, CCR-9901929, and FRG-DMS-0139093. Michael Nivala wishes to acknowledge support from NSF grant no. DMS-9810726 (VIGRE). Any opinions, findings, and conclusions or recommendations expressed in this material are those of the authors and do not necessarily reflect the views of NSF.

Willy Hereman is grateful for the hospitality of the Department of Mathematics and Statistics of the University of Canterbury (Christchurch, New Zealand) where he was a Visiting Erskine Fellow in Summer 2004. The Erskine Foundation is thanked for financial support. Willy Hereman thanks Jan Sanders (Free University, Amsterdam, The Netherlands) for introducing him to homotopy operators.

The authors are grateful to undergraduate students Lindsay Auble, Robert 'Scott' Danford, Ingo Kabirschke, Forrest Lundstrom, Frances Martin, Kara Namanny, and Maxine von Eye for designing Mathematica code for the project.

The research was supported in part by an Undergraduate Research Fund Award from the Center for Engineering Education awarded to Ryan Sayers. On June 16, 2003, while rock climbing in Wyoming, Ryan was killed by a lightning strike. He was 20 years old. The authors express their gratitude for the insight Ryan brought to the project. His creativity and passion for mathematics were inspiring. This chapter honors his contribution to the project.

Willy Hereman, Michael Colagrosso, Ryan Sayers, Adam Ringler
Department of Mathematical and Computer Sciences, Colorado School of Mines
Golden, CO 80401-1887, USA
e-mail: `whereman@mines.edu`

Bernard Deconinck, Michael Nivala
Department of Applied Mathematics, University of Washington
Seattle, WA 98195-2420, USA
e-mail: `bernard@amath.washington.edu`

Mark Hickman
Department of Mathematics and Statistics, University of Canterbury, Private Bag 4800
Christchurch, New Zealand
e-mail: `M.Hickman@math.canterbury.ac.nz`

Trends in Mathematics: *Differential Equations with Symbolic Computation*, 291–306

Partial and Complete Linearization of PDEs Based on Conservation Laws

Thomas Wolf

Abstract. A method based on infinite parameter conservation laws is described to factor linear differential operators out of nonlinear partial differential equations (PDEs) or out of differential consequences of nonlinear PDEs. This includes a complete linearization to an equivalent linear PDE (system) if that is possible. Infinite parameter conservation laws can be computed, for example, with the computer algebra package ConLaw.

Mathematics Subject Classification (2000). 70S10, 35-04, 35A30.

Keywords. Partial differential equations, linearization, conservation law.

1. Introduction

With the availability of computer algebra programs for the automatic computation of all conservation laws up to a given differential order of the integrating factors (as described in [5, 6]) conservation laws have been found that involve arbitrary functions, i.e. infinitely many parameters. In this paper we show how based on such conservation laws a linear differential operator can be factored out of a combination of the original nonlinear partial differential equations (PDEs) and their differential consequences. Possible outcomes include

- a complete linearization into an equivalent linear system,
- a partial linearization in the sense that a linear differential operator is factored out, splitting the problem into a linear one plus a nonlinear problem of lower order and often fewer independent variables (e.g. ordinary differential equations (ODEs)),
- the derivation of at least one linear equation from a nonlinear system (with the possibility of deriving further linear equations for the new mixed linear-nonlinear system).

An advantage of the procedure to be presented is that conservation laws need not be given explicitly in terms of the arbitrary functions. It is enough to have the conservation law determining conditions solved up to the solution of a system of consistent and necessarily linear PDEs which have arbitrary functions in their general solution.

The content of the paper is as follows. After comments are made on the computation of conservation laws in Section 3, the four computational steps of factoring out linear differential operators are illustrated using the Liouville equation in Section 4. Sufficient conditions for complete or partial linearizations are listed in Section 5, followed by a discussion of computational aspects in Section 6. A generalization involving the introduction of potentials in terms of which a linearization becomes possible is explained in Section 7. In later sections 8 and 9, an illustration is given of how the method works when nonlinear equations linearize to *inhomogeneous* equations or to *triangular linear* systems. Further examples where a complete or at least a partial linearization is possible are given in the appendix.

In this contribution we concentrate on computational aspects of the method and give examples for all of the above scenarios. An extension of the method discussing complete and partial linearizability through point and contact transformations will appear in a future publication [1], with numerous new examples and a comparison with other linearization methods found in the literature.

2. Notation

We follow the notation in [3] and denote the original nonlinear partial differential equations as $0 = \Delta_\alpha$, the dependent variables by u^β, $\alpha, \beta = 1, \ldots, q$ and the independent variables by x^i, $i = 1, \ldots, p$. In examples dealing with functions $u = u(x,t)$ or $u = u(x,y)$, partial derivatives are written as subscripts like $u_{xy} = \partial^2 u/(\partial x \partial y)$. If a formula already contains subscripts then ∂_i will be used for $\partial/\partial x^i$. The multi-indices ${}_{J},{}_{K}$ denote multiple partial derivatives like u^α_J which in our notation include u^α. With $\#J$ we denote the differential order, i.e. number of partial derivatives represented by ${}_J$. Total derivatives with respect to x^i will be denoted as D_i. We apply the convention that summation is performed over terms that involve two identical indices, one subscript and one superscript. For example, the divergence of a vector field P^i would be denoted as $D_i P^i$ $(\equiv \sum_i D_i P^i)$. The procedure to be presented repeatedly uses adjoint differential operators as follows. For given functions $f^A(x^i)$, $A = 1, \ldots, r$, let linear differential expressions H_k be defined as

$$H_k = a^J_{kA} \partial_J f^A, \quad k = 1, \ldots, s,$$

with coefficients $a^J_{kA} = a^J_{kA}(x^i)$ and summation over A and the multi-index J. The corresponding adjoint operators $\overset{*}{H}_{Ak}$ are computed for arbitrary functions $G^k(x^i)$ by repeatedly reversing the product rule of differentiation for the sum $G^k H_k$ to get

$$G^k H_k = f^A \overset{*}{H}_{Ak} G^k + D_i \bar{P}^i \tag{2.1}$$

where

$$\overset{*}{H}_{Ak} G^k = (-1)^{\#J} \partial_J \left(a^J_{kA} G^k \right). \tag{2.2}$$

and $\bar{P}^i$ are expressions resulting from integration by parts with respect to ∂_J in this computation.

3. Conservation Laws with Arbitrary Functions

Conservation laws can be formulated in different ways (see [6] for four different approaches to compute conservation laws). The form to be used in this paper is

$$D_i P^i = Q^\alpha \Delta_\alpha \tag{3.1}$$

where the components P^i of the conserved current and the so-called characteristic functions Q^α are differential expressions involving x^i, u^α_J. Other forms of conservation laws can easily be transformed into (3.1). One approach to find conservation laws for a given system of differential equations $0 = \Delta_\alpha$ is to specify a maximum differential order m of derivatives u^α_J on which P^i, Q^α may depend and then to solve condition (3.1) identically in x^i, u^α_J for the unknown functions P^i, Q^α. Due to the chain rule of differentiation in (3.1) the total derivatives D_i introduce extra derivatives u^α_K with $\#K = m+1 > m$, i.e. derivatives not occurring as variables in P^i, Q^α. Splitting with respect to these u^α_K results in an *overdetermined* and *linear* system of PDEs for P^i, Q^α.[1]

What is important in the context of this paper is that a differential Gröbner basis can be computed algorithmically and from it the dimension of the solution space can be determined, i.e. how many arbitrary functions of how many variables the general solution for P^i, Q^α depends on. In extending the capability of a program in solving condition (3.1) by not only computing a differential Gröbner basis (for linear systems) but also integrating exact PDEs (see [8]) and splitting PDEs with respect to only explicitly occurring u^α_J (which here act as independent variables), the situation does not change qualitatively. The result is still either the explicit general solution or a linear system of unsolved PDEs

$$0 = C_k(x^i, u^\alpha_J, f^A), \qquad k = 1, \ldots, r\ , \tag{3.2}$$

for some functions $f^A(x^j, u^\beta_J)$ where this system is a differential Gröbner basis and allows one to determine algorithmically the size of the solution space. The functions f^A are either the P^i, Q^α themselves or are functions arising when integrating the conservation law condition (3.1).

If the conservation law condition (3.1) is solved, i.e. P^i, Q^α are determined in terms of x^i, u^α_J, f^A_K possibly up to the solution of remaining conditions (3.2) then it is no problem to use a simple division algorithm to determine coefficients L^k satisfying

$$Q^\alpha \Delta_\alpha = D_i P^i + L^k C_k \tag{3.3}$$

identically in x^i, u^α_J, f^A_J. The coefficients L^k are necessarily free of f^A_J because (3.1) is linear and homogeneous in Q^α, P^i and this property is preserved in solving these

[1]Note that regarding (3.1) as an algebraic system for unknowns Q^α implies division through Δ_α and does therefore not produce Q^α which are regular for solutions u^α of the original system $\Delta_\alpha = 0$. For details regarding the ansatz for Q^α see [6].

conditions, so C_k are linear and homogeneous in f_J^A as well and L^k must therefore be free of f_J^A. We call relation (3.3) a *conservation law identity* because it is satisfied identically in all x^i, u_J^α and f_J^A.

4. The Procedure

The individual steps of our method are shown in detail to demonstrate that all steps are algorithmic and can be performed by computer. The REDUCE package ConLaw has the algorithm implemented and performs it whenever a conservation law computation results in a solution involving arbitrary functions possibly up to the solution of a linear system (3.2).

Input. Input to the procedure is the conservation law identity (3.3)

$$Q^\alpha \Delta_\alpha = D_i P^i + L^k C_k \tag{4.1}$$

including expressions for all its constituents Q^α, P^i, L^k, C_k in terms of x^i, u^α, f^A.

To start the procedure the functions f^A have to depend only on the variables x^i. If they depend on u_J^α then a linearization will necessarily involve a change of variables. This case is treated in [1].

Step 1. If all functions f^A depend exactly on all p independent variables x^i then proceed to step 2. Step 1 is concerned with the case that not all $f^A = f^A(x^i)$ depend on all x^i. To add the dependence of, say f^B on x^j, one has to

- compute

  $$Z := (Q^\alpha \Delta_\alpha - D_i P^i - L^k C_k)\big|_{f^B(x^i) \to f^B(x^i, x^j)}$$

  which vanishes modulo $0 = \partial_j f^B$ and therefore must have the form

  $$Z = M^J \partial_J (f^B)$$

  with suitable coefficients M^J and summation over the multi-index J,
- compute the adjoint Z_B^* as in (2.1) and (2.2) to bring Z into the form

  $$Z = D_i \bar{P}^i + Z_B^* \partial_j f^B, \tag{4.2}$$

- rename $P^i + \bar{P}^i \rightarrow P^i$ and add a new condition $C_{r+1} = \partial_j f^B$ and multiplier $L^{r+1} = Z_B^*$ to arrive at a new version of the conservation law identity $Q^\alpha \Delta_\alpha = D_i P^i + L^k C_k$ where the function f^B depends now on x^j.

This process is repeated until all f^A depend on all x^i.

Example 1. We illustrate the steps of the procedure with an investigation of the Liouville equation

$$0 = \Delta := u_{xy} - e^u. \tag{4.3}$$

Although it is not completely linearizable, we choose this equation because it involves computations in each of the first three steps.

For the Liouville equation a conservation law identity involving an arbitrary function $f(x)$ is given through

$$(f_x + fu_x)\Delta = D_x(-fe^u) + D_y(f_x u_x + fu_x^2/2), \tag{4.4}$$

i.e. $Q = f_x + fu_x$, $P^x = -fe^u$, $P^y = f_x u_x + fu_x^2/2$, $C_k = 0$. Adding a y-dependence to f requires to add to the right-hand side of our identity (4.4) the terms

$$Z = -f_{xy}u_x - f_y u_x^2/2$$

which in adjoint form (4.2) read

$$Z = D_x(-f_y u_x) + (u_{xx} - u_x^2/2)f_y,$$

giving the new conservation law identity

$$(f_x + fu_x)\Delta = D_x(-fe^u - f_y u_x) + D_y(f_x u_x + fu_x^2/2) + (u_{xx} - u_x^2/2)f_y. \tag{4.5}$$

Step 2. As the Q^α are linear homogeneous differential expressions for the f^A one can compute adjoint operators $Q_A^{\alpha *}$ as in (2.1) and (2.2) by expressing

$$Q^\alpha \Delta_\alpha = f^A Q_A^{\alpha *}\Delta_\alpha + D_i \bar{P}^i.$$

After renaming $P^i - \bar{P}^i \rightarrow P^i$ the conservation law identity takes the new form

$$f^A Q_A^{\alpha *}\Delta_\alpha = D_i P^i + L^k C_k. \tag{4.6}$$

In the case of the Liouville equation we partially integrate

$$(f_x + fu_x)\Delta - f(u_x - D_x)\Delta + D_x(f\Delta)$$

and get the conservation law identity

$$\begin{aligned} f(u_x - D_x)\Delta &= D_x(-fe^u - f_y u_x - f\Delta) + D_y(f_x u_x + fu_x^2/2) + (u_{xx} - u_x^2/2)f_y \\ &= D_x(-f_y u_x - fu_{xy}) + D_y(f_x u_x + fu_x^2/2) + (u_{xx} - u_x^2/2)f_y. \end{aligned} \tag{4.7}$$

Step 3. Because the C_k are linear homogeneous differential expressions in the f^A we can compute the adjoint form of $L^k C_k$ as in (2.1) and (2.2) by expressing

$$L^k C_k = f^A C_{Ak}^* L^k + D_i \bar{P}^i.$$

After renaming $P^i + \bar{P}^i \rightarrow P^i$ the conservation law identity takes the new form

$$f^A Q_A^{\alpha *}\Delta_\alpha = D_i P^i + f^A C_{Ak}^* L^k. \tag{4.8}$$

In our example partial integration gives

$$(u_{xx} - u_x^2/2)f_y = D_y((u_{xx} - u_x^2/2)f) - f(u_{xx} - u_x^2/2)_y$$

and substituted into (4.7) the new conservation law identity

$$\begin{aligned} f(u_x - D_x)\Delta = {} & D_x(-f_y u_x - fu_{xy}) + D_y\left(f_x u_x + fu_x^2/2 + f(u_{xx} - u_x^2/2)\right) \\ & -f(u_{xx} - u_x^2/2)_y \end{aligned} \tag{4.9}$$

after simplification.

Step 4. This step does not involve any computation; it merely completes the constructive proof how linearizations are achieved.

By bringing $f^A C^{\,*}_{Ak} L^k$ to the left-hand side of the conservation law identity (4.8) we get

$$f^A \left(Q^{\alpha *}_A \Delta_\alpha - C^{\,*}_{Ak} L^k\right) = D_i P^i \tag{4.10}$$

which still is an identity for arbitrary functions u^α, f^A. Applying the Euler operator with respect to f^A (for its definition see e.g. [3, 1]) to the left-hand side of (4.10) gives the coefficient of f^A and on the right-hand side gives zero as it is a divergence,[2] i.e. we get

$$Q^{\alpha *}_A \Delta_\alpha = C^{\,*}_{Ak} L^k \qquad \text{identically in } u^\alpha \text{ for all } A. \tag{4.11}$$

The vanishing of $D_i P^i$ on the right-hand side of (4.9) was therefore not accidental.

For the Liouville equation the identity (4.11) takes the form

$$(u_x - D_x)\Delta = -D_y L = 0 \qquad \text{with} \tag{4.12}$$

$$L = u_{xx} - u_x^2/2. \tag{4.13}$$

Integrating at first (4.12) to $L = L(x)$ leaves the Riccati ODE

$$u_{xx} - u_x^2/2 = L(x) \tag{4.14}$$

for u_x to be solved, for example, through a linearizing transformation $u(x, y) = -2\log(v(x, y))$.

Output. The results of the procedure are expressions $Q^{\alpha *}_A, C^{\,*}_{Ak}$ and L^k. The relation

$$C^{\,*}_{Ak} L^k = 0 \tag{4.15}$$

is a necessary condition which can be solved by first regarding L^k as dependent variables and then solving

$$L^k = L^k(u^\alpha_J) \tag{4.16}$$

for $u^\alpha = u^\alpha(x^i)$. The system (4.15)–(4.16) is a sufficient condition for the original system $\Delta_\alpha = 0$ if $Q^{\alpha *}_A$ is an invertible algebraic operator and it is a complete linearizing point transformation if (4.16) is purely algebraic in u^α.

5. Scope of the Procedure

The degree to which the original system $\Delta_\alpha = 0$ can be linearized depends on properties of the conservation law identity that has been computed: the number of functions f^A and the number of variables each f^A depends on, the differential order of derivatives of f^A with respect to x^i, u^α_J in C_k and in Q^α. Some properties, like the size of the solution space of remaining conditions (3.2), are essentially independent of the extent to which these conditions are solved. Other criteria, like the number of functions f^A and the number of their arguments, do depend on the

[2]To prove this statement without Euler operator we could choose the f^A to be zero outside some region R such that an integral over a volume with boundary outside R will vanish using Gauss law on the right-hand side of identity (4.10) as P^i are linear homogeneous in the f^A. Because the f^A are arbitrary inside R the coefficients of the f^A on the left-hand side must vanish identically.

extent to which conditions (3.2) were solved. The strength of the procedure to be presented is to be able to handle a wide range of situations.

The following is a list of four scenarios, sorted from most special, fully algorithmic (and most beneficial) to most general, not strictly algorithmic (and less beneficial). We refer to the computational steps described in Section 4 as "the procedure".

- If the following criteria are met:
  1. the size of the solution space of $0 = \Delta_\alpha$ is equal to the size of the solution space of $0 = C_k$,
  2. the conditions $0 = C_k$ involve q functions f^A (equal to the number of functions u^α in Δ_α) and all f^A depend on p variables (equal to the number of variables u^α depend on),
  3. the functions Q^α expressed in terms of f^A involve f^A only algebraically, i.e. no derivatives of f^A, and
  4. functions f^A do not depend on jet variables u^α_J, i.e. $f^A = f^A(x^i)$,

  then the procedure will algorithmically provide a linearizing point transformation of the system $\Delta_\alpha = 0$.

Example 2. The Burgers equation in the form

$$0 = \Delta_1 := u_t - u_{xx} - uu_x \tag{5.1}$$

for a function $u(x,t)$ cannot be linearized but in the potential form

$$0 = \Delta_2 := v_t - v_{xx} - v_x^2/2 \tag{5.2}$$

for $v(x,t)$ a conservation law identity involving a function $f(x,t)$ is given through

$$fe^{v/2}\Delta = D_t\left(2fe^{v/2}\right) + D_x\left(2f_x e^{v/2} - fe^{v/2}v_x\right) + 2e^{v/2}(-f_t - f_{xx}) \tag{5.3}$$

and the related linearization is

$$L = 2e^{v/2}$$
$$e^{v/2}\Delta = L_t - L_{xx} = 0.$$

A proof that every nonlinear PDE (system) that is linearizable through point or contact transformations can be linearized this way will be given in [1].

- If criteria 1, 2 and 3 are satisfied but not 4 then a linearization is possible but at the price of a change of variables, which will be a contact transformation if it is invertible or otherwise it will be a non-invertible transformation depending on derivatives of u^α. Furthermore, in all such cases the transformation can be derived explicitly from the conservation law identity as will be shown in [1].
- If criterion 3 is not satisfied then the partially or completely linearized equations may only be a necessary but not a sufficient condition for $\Delta_\alpha = 0$.
- If criterion 1 is satisfied but not 2 then
  - if functions f^A of fewer than p variables occur then one can add extra variable dependencies through step 1 of the procedure,

– if more than q functions f^A occur in $0 = C_k$ or functions f^A of more than p variables occur then one has to integrate more of the conditions $0 = C_k$ in order to be able to linearize the original system completely (a full treatment of this case will be given in [1]).

- If criterion 1 is not satisfied but the solution space of C_k involves at least one arbitrary function of one argument then the method will result in a differential expression for u^α_J which vanishes modulo $0 = \Delta_\alpha$ and factorizes into a linear differential operator acting on a nonlinear differential expression. Typically this leads to a PDE for u^α which is lower in differential order than Δ_α for one of the x^i. In Example 1 in Section 4 and Examples 8, 9 and 10 in the appendix an equation in one less variable results, i.e. an ODE.

The algorithmic beauty of the procedure is that the above wide range of situations is covered by one and the same 4-step algorithm.

The case that a non-local linearization exists in which the L^k depend on integrals of u^α is not covered directly as the computer algebra package ConLaw does not compute non-local conservation laws. On the other hand single conservation laws (without parametric functions) can be used to introduce potentials such that the original system re-formulated in these potentials is linearizable. This approach has been successful in all 6 linearizable evolutionary systems found in [4]. Examples given in this paper are the system (7.1)–(7.2) in Section 7, the system (9.4)–(9.5) in Section 9 and the system (10.8)–(10.9) in the appendix.

6. Computational Aspects

Given a nonlinear PDE system $0 = \Delta_\alpha$, what are possible computational hurdles to be overcome in oder to find a linearization? The method described in Section 4 is algorithmic and does not pose a problem. The formulation of the conservation law condition (3.1) and its analysis through computing a differential Gröbner basis $0 = C_k$ is algorithmic as well and could only become difficult because of a growing size of equations.

A first computational challenge lies in the fact that for linearizable systems $0 = \Delta_\alpha$ *the conservation law condition* (3.1) *has a general solution involving arbitrary functions.* It is well known that systems of equations with a large solution space are much harder to solve than systems with only few solutions or no solutions. To incorporate many solutions, algebraic Gröbner bases for algebraic systems have to be of high degree and differential Gröbner bases for differential systems have to be of sufficiently high differential order. As a consequence, the length of expressions encountered during the Gröbner basis computation is more likely to explode and exceed available resources.

The second challenge is to integrate a Gröbner basis $0 = C_k$ *sufficiently often to meet criterion* 2 *in Section* 5. Because the general solution of the conservation law conditions involves arbitrary functions, any integrations to be done can only be integrations of PDEs, not of ODEs.

The package CRACK that is used to compute the examples in this paper differs from similar other programs (as listed in [2]) in that it has a number of modules addressing the above problems. For example, the growth of expressions is lowered by a module for reducing the length of equations by replacing them through a suitable linear combination of equations as described in [7]. Integrations are handled by a module that integrates exact PDEs, that is able to introduce potentials to integrate certain generalizations of exact PDEs and that determines monomial integrating factors to achieve integration (see [8]). A relatively new module applies syzygies that result as a by-product of a differential Gröbner basis computation. This module allows to perform integrations more efficiently and to avoid a temporary explosion of the number of functions of integration generated in the process (see [9]). A module that integrates underdetermined linear ODEs with non-constant coefficients is often useful in the last stages of the computation. A description of the algorithm and its implementation is in preparation.

7. An Example Requiring the Introduction of a Potential

The following example demonstrates that a linearization of a nonlinear equation or system may only be possible if it is reformulated in terms of potentials which in turn might be found by studying conservation laws.

Example 3. The system

$$0 = \Delta_1 := u_t - u_{xx} - 2vuu_x - 2(a+u^2)v_x - v^2u^3 - bu^3 - auv^2 - cu, \quad (7.1)$$
$$0 = \Delta_2 := v_t + v_{xx} + 2uvv_x + 2(b+v^2)u_x + u^2v^3 + av^3 + bvu^2 + cv \quad (7.2)$$

with $u = u(x,t),\ v = v(x,t)$ and constants a, b, c results as one of the 15 cases of a class of generalized nonlinear Schrödinger equations [4]. This system itself does not have conservation laws involving arbitrary functions but it has the zeroth order conservation law

$$v\Delta_1 + u\Delta_2 = D_t(uv) + D_x(v_xu - u_xv + bu^2 - av^2)$$

which motivates the introduction of a function $w(x,t)$ through

$$\begin{aligned} w_x &= uv, \\ -w_t &= v_xu - u_xv + bu^2 - av^2. \end{aligned} \quad (7.3)$$

The remaining system to be solved for $r := u/v$ and w simplifies if we substitute

$$w = \frac{1}{2}\log z \quad (7.4)$$

with $z = z(x,t)$. This substitution is not essential for the following but it reduces the size of the resulting system for $r(x,t), z(x,t)$ and eases memory requirements in the computation of conservation laws of the resulting system Δ_3, Δ_4:

$$\begin{aligned} 0 = \Delta_3 :=\ & 2rr_t z_x^2 + r_x^2 z_x^2 + 2ar^2 r_x z_x^2 - 2br_x z_x^2 + 2r^2 z_x z_{xxx} \\ & - r^2 z_{xx}^2 + 2ar^3 z_x z_{xx} + 2brz_x z_{xx} + 4cr^2 z_x^2, \\ 0 = \Delta_4 :=\ & r_x z_x + rz_t - ar^2 z_x + bz_x. \end{aligned} \tag{7.5}$$

The program ConLaw finds a conservation law with integrating factors

$$\begin{aligned} Q^3 &= r^{-5/2} z_x^{-3/2}(fr + \tilde{f}), \\ Q^4 &= r^{-5/2} z_x^{-3/2} \left(-2z_x r(f_x r - \tilde{f}_x) - r_x z_x(fr + \tilde{f}) + z_{xx} r(fr - \tilde{f}) \right) \end{aligned}$$

involving two functions $f(x,t), \tilde{f}(x,t)$ that have to satisfy the conditions

$$\begin{aligned} 0 = C_1 &:= -f_t + f_{xx} + cf - 2a\tilde{f}_x, \\ 0 = C_2 &:= \tilde{f}_t + \tilde{f}_{xx} + c\tilde{f} - 2bf_x. \end{aligned}$$

The conservation law identity takes the form

$$Q^3\Delta_3 + Q^4\Delta_4 = D_t P^t + D_x P^x + L^1 C_1 + L^2 C_2 \tag{7.6}$$

with some conserved current (P^t, P^x) and coefficients L^1, L^2 of C_1, C_2

$$L^1 = 4\sqrt{z_x r}, \quad L^2 = 4\sqrt{z_x/r}. \tag{7.7}$$

Derivatives $f_x, \tilde{f}_x$ in Q^4 can be eliminated by adding total x-derivatives

$$D_x \left(r^{-5/2} z_x^{-3/2} 2z_x r(fr - \tilde{f})\Delta_4 \right)$$

to the left-hand site of the identity (7.6) and to $D_x P^x$. The modified form of the identity (7.6) is

$$\begin{aligned} 0 &= z_x^{-3/2} r^{-5/2} \left(2z_x r(fr - \tilde{f}) D_x \Delta_4 - 2r_x z_x (fr - \tilde{f})\Delta_4 + (fr + \tilde{f})\Delta_3 \right) \\ &= D_t \left(4\sqrt{z_x/r}(rf - \tilde{f}) \right) \\ &\quad + D_x \Big(2z_x^{-1/2} r^{-3/2} (-2f_x z_x r^2 - 2\tilde{f}_x z_x r + r_x z_x fr - r_x z_x \tilde{f} \\ &\qquad\qquad + z_{xx} fr^2 + z_{xx}\tilde{f}r + 4z_x a\tilde{f}r^2 + 4z_x bfr) \Big) \\ &\quad + L^1 C_1 + L^2 C_2. \end{aligned}$$

Partial integration of $L^1 C_1 + L^2 C_2$ until $f, \tilde{f}$ appear purely algebraically makes necessarily $P^t = P^x = 0$. Because $f, \tilde{f}$ are free we obtain the identities

$$0 = r^{-3/2} z_x^{-3/2} (\Delta_3 + 2rz_x D_x \Delta_4 - 2r_x z_x \Delta_4) = L_t^1 + L_{xx}^1 + cL^1 + 2bL_x^2, \tag{7.8}$$

$$0 = r^{-5/2} z_x^{-3/2} (\Delta_3 - 2rz_x D_x \Delta_4 + 2r_x z_x \Delta_4) = -L_t^2 + L_{xx}^2 + cL^2 + 2aL_x^1 \tag{7.9}$$

completing the linearization. For any solution L^1, L^2 of (7.8), (7.9), equations (7.7) provide r, z_x. With z_t from (7.5) we get z as a line integral, w from (7.4) and u, v from r and equation (7.3).

In the following section the effect of our method on PDEs is investigated that linearize to inhomogeneous equations.

8. Inhomogeneous Linear DEs

If the general solution of conservation law determining equations involves a number of free constants or free functions then individual conservation laws are obtained by setting all but one to zero. The remaining terms are homogeneous in the surviving constant or function. The question arises whether our conservation law based method is suitable to find linearizations that lead to linear but inhomogeneous equations.

Example 4. For the (ad hoc constructed) equation

$$0 = \Delta := 2uu_t + 2uu_{xx} + 2u_x^2 + 1 \tag{8.1}$$

the conservation law identity

$$\begin{aligned}(f_x + \tilde{f}_t)\Delta = {} & D_t\left(f_x u^2 + \tilde{f}_t u^2 + \tilde{f}\right) \\ & + D_x\left(-f_{xx}u^2 + 2f_x uu_x - \tilde{f}_{t,x}u^2 + 2\tilde{f}_t uu_x + f\right) \\ & + u^2(f_{t,x} - f_{xxx} - \tilde{f}_{txx} + \tilde{f}_{tt})\end{aligned}$$

involves functions $f(x,t), \tilde{f}(x,t)$ and establishes a conservation law provided $f, \tilde{f}$ satisfy

$$0 = f_{t,x} - f_{xxx} - \tilde{f}_{txx} + \tilde{f}_{tt}.$$

Our method gives the linear system

$$0 = D_x\Delta = L_{tx} + L_{xxx}, \tag{8.2}$$
$$0 = D_t\Delta = L_{tt} + L_{txx}; \tag{8.3}$$
$$L = u^2.$$

The system (8.2)–(8.3) represents the x and t derivatives of the linear equation

$$0 = L_t + L_{xx} + 1 \tag{8.4}$$

which is equivalent to equation (8.1) and is an inhomogeneous linear PDE. Although our linearization method does not quite reach (8.4), it nevertheless provides $L = u^2$ as the new unknown function which makes it easy to get to the equivalent linear equation (8.4) through a change of dependent variables in (8.1) or through an integration of (8.2), (8.3).

The way how homogeneous consequences can be derived from an inhomogeneous relation is to divide the inhomogeneous relation through the inhomogeneity, i.e. to make the inhomogeneity equal 1 and then to differentiate with respect to all independent variables and to obtain a set of linear homogeneous conditions in the same way as equations (8.2) and (8.3) are consequences of (8.4). The application in the following section leads to an inhomogeneous linear PDE with non-constant inhomogeneity.

9. An Example of a Triangular Linear System

A generalization of complete linearizability of the whole PDE system in one step is the successive linearization of one equation at a time.

Example 5. Assume a triangular system of equations, like the (ad hoc constructed) system

$$0 = \Delta_1 := u_t, \tag{9.1}$$

$$0 = \Delta_2 := vv_t - uvv_{xx} - uv_x^2 \tag{9.2}$$

with one equation (9.1) involving only one function, say $u = u(x,t)$, and this equation being linear or being linearizable and a second nonlinear equation being linear or linearizable in another function $v = v(x,t)$. How can the method in Section 4 be used to recognize that such a system can be solved by solving successively only linear equations?

In determining all conservation laws for this system with unknown functions v, u and with integrating factors of order zero we get apart from two individual conservation laws with pairs of integrating factors $(Q^1, Q^2) = (\frac{v^2}{u^2}, -\frac{2}{u})$ and $(\frac{xv^2}{u^2}, -\frac{2x}{u})$ only one with a free function $f(u,x)$:

$$f_u\Delta_1 = D_t f$$

which indicates the linearity of Δ_1 but not the linearity of Δ_2 in v once $u(x,t)$ is known.

The proper way of applying the method of Section 4 is to compute conservation laws of $0 = \Delta_2$ alone which now is regarded as an equation for $v(x,t)$ only. The function $u(x,t)$ is assumed to be parametric and given. We obtain the identity

$$2f\Delta_2 = D_x(f_x uv^2 + fu_x v^2 - 2fuvv_x) + D_t(fv^2) - v^2(f_t + uf_{xx} + 2u_x f_x + u_{xx} f)$$

which is a conservation law if f satisfies the linear condition

$$0 = f_t + uf_{xx} + 2u_x f_x + u_{xx} f. \tag{9.3}$$

This provides the linearization

$$0 = 2\Delta_2 = L_t - uL_{xx},$$
$$L = v^2.$$

The reason that now a linearization of Δ_2 is reached is that in the second try u is assumed to be known and therefore $u, u_{xx}, \ldots$ are not jet-variables and hence the condition (9.3) has solutions, otherwise not.

Examples where this triangular linearization method is successful are the systems (17) and (18) in [4]. We demonstrate the method with one of them (system (17)); the other is similar.

Example 6. The system

$$0 = \Delta_1 := u_t - u_{xx} - 4uvu_x - 4u^2v_x - 3vv_x - 2u^3v^2 - uv^3 - au, \tag{9.4}$$

$$0 = \Delta_2 := v_t + v_{xx} + 2v^2u_x + 2uvv_x + 2u^2v^3 + v^4 + av \tag{9.5}$$

involves functions $u(x,t), v(x,t)$ and the constant a. The single conservation law

$$0 = v\Delta_1 + u\Delta_2 = D_t(uv) + D_x(uv_x - u_x v - u^2v^2 - v^3)$$

motivates the introduction of a function $w(x,t)$ through

$$w_x = uv, \tag{9.6}$$

$$-w_t = uv_x - u_x v - u^2 v^2 - v^3. \tag{9.7}$$

Substitution of u from (9.6) brings equations (9.5) and (9.7) in the form

$$0 = \Delta_3 := w_t - \frac{1}{v}(-2v_x w_x + w_{xx}v + w_x^2 v + v^4), \tag{9.8}$$

$$0 = \Delta_4 := v_t + v_{xx} + 2w_{xx}v + 2w_x^2 v + av + v^4. \tag{9.9}$$

This system obeys conservation laws that involve a function $f(x,t)$ that has to satisfy $f_t = f_{xx} + af$. Our procedure provides the linearization

$$e^w(v\Delta_3 + \Delta_4) = L_t^1 + L_{xx}^1 + aL^1 = 0, \tag{9.10}$$
$$L^1 := ve^w.$$

The second linearized equation can be obtained by

- substituting $v = L^1/e^w$ into equations (9.8) and (9.9): to get the remaining condition

$$0 = \Delta_5 := w_t - w_{xx} - 3w_x^2 + 2w_x L_x^1 (L^1)^{-1} - (L^1)^3 e^{-3w}, \tag{9.11}$$

- assuming that L^1 has been solved from (9.10) and treating $L^1(x,t)$ as a parametric function when computing conservation laws for equation (9.11) which turn out to involve two functions that have to satisfy linear PDEs,
- performing the linearization method to find that the remaining equation (9.11) linearizes with $L^2 = e^{3w}$ to

$$e^{3w}\Delta_5 = L_t^2 - L_{xx}^2 + 2L_x^2 L_x^1/L^1 - 3(L^1)^3. \tag{9.12}$$

Because the condition (9.12) is inhomogeneous for L^2 due to the term $3(L^1)^3$, actually two homogeneous linear equations are generated which are the x- and t-derivative of (9.12) divided by $3(L^1)^3$ (see the previous section about linear inhomogeneous equations). But as the function $L^2 = e^{3w}$ results in this process, it is no problem to find (9.12) from (9.11) directly or from an integration of these two equations. This completes the linear triangularisation of the original problem (9.4)–(9.5) to the new system (9.10)–(9.12).

10. Summary

The paper starts with introducing conservation law identities as a natural way to formulate infinite parameter conservation laws.

Conservation law identities are the input to a four-step procedure that returns a differential consequence of the original system together with a linear differential operator that can be factored out.

Sufficient conditions on the conservation law identity which either guarantee a complete linearization or at least a partial linearization are discussed.

The possibility to find a non-local linearization arises from the application of single (finite parameter) conservation laws with the aim to introduce potentials which satisfy infinite parameter conservation laws and thus allow a linearization.

In examples it is demonstrated how the standard procedure can lead to inhomogeneous linear PDEs and how a successive linearization of one equation at a time may be possible when the whole system cannot be linearized at once.

Appendix

In this appendix we list further examples of linearizations and integrations without giving details of the calculations.

The first example of the Kadomtsev–Petviasvili equation demonstrates what our method gives when a PDE has p independent variables and the conservation law involves free functions of less than $p-1$ variables. Although the result will be less useful than in the other examples, we still include it for illustration.

Example 7. The Kadomtsev–Petviasvili equation

$$0 = \Delta = u_{tx} + u_{xxxx} + 2u_{xx}u + 2u_x^2 - u_{yy}$$

for $u(t,x,y)$ has four conservation laws with a zeroth order integrating factor and an arbitrary function $f(t)$ as given in [6]. We comment on one of these four with an integrating factor $f_t y^3 + 6fxy$ as the situation for the others is similar. Omitting the details we only give the result of our method:

$$\begin{aligned} L^1 &= y\left(u_{txxx}y^2 + 2u_{tx}uy^2 + u_{tt}y^2 + 2u_tu_xy^2\right.\\ &\qquad \left.-6u_tx - 6u_{xxx}x + 6u_{xx} - 12u_xux + 6u^2\right),\\ L^2 &= -u_{ty}y^3 + 3u_ty^2 + 6u_yxy - 6ux,\\ y(6x\Delta - y^2D_t\Delta) &= -L^1_x - L^2_y. \end{aligned}$$

The arbitrary function $f(t)$ involves only one independent variable t and the conservation law $0 = L^1_x + L^2_y$ involves two functions L^1, L^2 and has derivatives with respect to two variables x, y and is therefore not as useful as if it would be a single total derivative.

The three following equations were shown to the author first by V. Sokolov [10] who obtained their integrations earlier and independently. We add them here to demonstrate that these results can be obtained in a straight forward procedure.

Example 8. For the equation

$$0 = \Delta := u_{xy} - e^u\sqrt{u_x^2 - 4}, \qquad u = u(x,y) \tag{10.1}$$

a conservation law with an arbitrary function $f(x)$ enables to factor out D_y leaving an ODE to solve

$$(u_x - D_x)\left(\frac{\Delta}{\sqrt{u_x^2-4}}\right) = D_y\left(\frac{-u_{xx} + u_x^2 - 4}{\sqrt{u_x^2-4}}\right) = 0. \tag{10.2}$$

Another conservation law with an arbitrary function $g(y)$ gives

$$\left(u_y - \frac{u_x e^u}{\sqrt{u_x^2 - 4}} - D_y\right)\Delta = D_x\left(-u_{yy} + \frac{1}{2}u_y^2 + \frac{1}{2}e^{2u}\right) = 0. \tag{10.3}$$

Example 9. For the equation

$$0 = \Delta := u_{xy} - \left(\frac{1}{u-x} + \frac{1}{u-y}\right)u_x u_y, \quad u = u(x,y) \tag{10.4}$$

a conservation law with an arbitrary function $f(x)$ similarly to the above example provides

$$\frac{y-x}{(u-x)(u-y)}\Delta + D_x\left(\frac{\Delta}{u_x}\right) = D_y\left(\frac{u_{xx}}{u_x} - \frac{2(u_x-1)}{u-x} - \frac{u}{(u-x)x}\right) = 0. \tag{10.5}$$

A second conservation law is obtained from an arbitrary function $g(y)$ and is equivalent to (10.5) after swapping $x \leftrightarrow y$.

Example 10. For the equation

$$0 = \Delta := u_{xy} - \frac{2}{x+y}\sqrt{u_x u_y}, \quad u = u(x,y) \tag{10.6}$$

a conservation law with an arbitrary function $f(x)$ gives

$$\frac{1}{(x+y)}\left(\frac{1}{\sqrt{u_x}} - \frac{1}{\sqrt{u_y}}\right)\Delta + D_x\left(\frac{\Delta}{\sqrt{u_x}}\right) = D_y\left(\frac{u_{xx}}{\sqrt{u_x}} + \frac{2\sqrt{u_x}}{x+y}\right) = 0. \tag{10.7}$$

A second conservation law is obtained from an arbitrary function $g(y)$ and is equivalent to (10.7) after swapping $x \leftrightarrow y$.

The final example shows a linearization of a system that resulted in classifying nonlinear Schrödinger type systems in [4].

Example 11. The system

$$0 = \Delta_1 := u_t - u_{xx} - 2vu_x - 2uv_x - 2uv^2 - u^2 - au - bv - c, \tag{10.8}$$
$$0 = \Delta_2 := v_t + v_{xx} + 2vv_x + u_x \tag{10.9}$$

involves functions $u(x,t), v(x,t)$ and the constants a, b, c. The trivial conservation law

$$0 = \Delta_2 = D_t(v) + D_x(v_x + u + v^2)$$

motivates the introduction of a function $w(x,t)$ through

$$w_x = v, \tag{10.10}$$
$$-w_t = v_x + u + v^2. \tag{10.11}$$

Substitution of u, v from (10.10) and (10.11) brings equation (10.8) in the form

$$0 = \Delta_3 = w_{tt} + w_t^2 - w_t a - w_{xxxx} - 4w_{xxx}w_x - 3w_{xx}^2 - 6w_{xx}w_x^2 - w_{xx}a$$
$$- w_x^4 - w_x^2 a + w_x b + c.$$

This equation admits a conservation law identity

$$\begin{aligned} fe^w\Delta_3 = {} & D_t\left[(e^w)_t f - e^w f_t - e^w f a\right] \\ & + D_x\left[-(e^w)_{xxx} f + (e^w)_{xx} f_x - (e^w)_x f_{xx} + e^w f_{xxx} \right. \\ & \qquad \left. - (ae^w)_x f + ae^w f_x + be^w f\right] \\ & + e^w\left[f_{tt} + af_t - f_{xxxx} - af_{xx} - bf_x + cf\right]. \end{aligned}$$

From this follows the linearization

$$\begin{aligned} e^w\Delta_3 &= L_{tt} - L_t a - L_{xxxx} - L_{xx} a + L_x b + Lc = 0, \\ L &= e^w. \end{aligned}$$

References

[1] Anco, S. C., Wolf, T., Bluman, G.: *An algorithmic method for linearization of PDEs via conservation laws.* In preparation.

[2] Hereman, W.: *Symbolic software for Lie symmetry analysis.* Chapter 13 in CRC Handbook of Lie Group Analysis of Differential Equations, vol. 3 (N. H. Ibragimov, ed.), pp. 367–413. CRC Press, Boca Raton, Florida, 1996.

[3] Olver, P. J.: *Applications of Lie Groups to Differential Equations.* Grad. Texts in Math., Springer-Verlag, Berlin, 1986.

[4] Sokolov, V. V., Wolf, T.: *A symmetry test for quasilinear coupled systems.* Inverse Problems **15**, L5–L11, 1999.

[5] Wolf, T., Brand, A., Mohammadzadeh, M.: *Computer algebra algorithms and routines for the computation of conservation laws and fixing of gauge in differential expressions.* J. Symb. Comput. **27**, 221–238, 1999.

[6] Wolf, T.: *A comparison of four approaches to the calculation of conservation laws.* Euro. J. Appl. Math. **13**, part 2, 129–152, 2002.

[7] Wolf, T.: *Size reduction and partial decoupling of systems of equations.* J. Symb. Comput. **33**, no. 3, 367–383, 2002.

[8] Wolf, T.: *The symbolic integration of exact PDEs.* J. Symb. Comput. **30**, no. 5, 619–629, 2000.

[9] Wolf, T.: *The integration of systems of linear PDEs using conservation laws of syzygies.* J. Symb. Comput. **35**, no. 5, 499–526, 2003.

[10] Zhiber, A. V., Sokolov, V. V.: *Exactly integrable hyperbolic equations of Liouville type.* Uspihi Mat. Nauk **56**, no. 1, 63–106, 2001 [in Russian, translation in Russian Math. Surveys **56**, no. 1, 61–101, 2001].

Thomas Wolf
Department of Mathematics
Brock University
500 Glenridge Avenue, St. Catharines
Ontario, Canada L2S 3A1
e-mail: `TWolf@brocku.ca`

Trends in Mathematics: *Differential Equations with Symbolic Computation*, 307–325

CONSLAW: A Maple Package to Construct the Conservation Laws for Nonlinear Evolution Equations

Ruo-Xia Yao and Zhi-Bin Li

Abstract. An algorithm to explicitly compute polynomial conservation laws for nonlinear evolution equations (either uniform in rank or not) is introduced and a software package CONSLAW written in Maple to automate the computation is developed. CONSLAW can construct the polynomial conservation laws for polynomial partial differential equations automatically. Furthermore, some new integrable systems can be filtered out by analyzing the compatibility conditions which guarantee the existence of the conservation laws for given parametrized nonlinear evolution equations. The explicit forms of the conserved densities play an important role in studying the integrability, such as explicit analytical solutions, bi-Hamiltonian form, one-parameter family of Bäcklund transformation, Lax pairs, and the checking of the accuracy of numerical integration algorithm. The effectiveness of CONSLAW is illustrated by applying it to a variety of nonlinear partial differential equations.

Mathematics Subject Classification (2000). Primary 35G25; Secondary 37K05.

Keywords. Nonlinear evolution equation, scaling symmetry, conservation law, symbolic computation.

1. Introduction

Over the past two decades, the two most famous discoveries in nonlinear physics are soliton and the so-called "chaos theory", which have radically changed the thinking of scientists about the nature of nonlinearity. It was in 1895 that Korteweg and de Vries derived the equation for water waves in shallow channels, which confirmed the existence of solitary waves . The stability and particle-like behavior

This work is supported by the Research Fund for the Doctoral Program of Higher Education of China (grant no. G20020269003). Yao's research is also supported by the 2004 Ph.D. Program Scholarship Fund of East China Normal University.

of the soliton solutions can only be explained by the existence of many conservation laws, and as is well known, the obtained explicit forms of the conservation laws (CLaws) of the celebrated KdV equation lead to the discovery of the Miura transformation given by Miura and Gardner, and the finding of the Lax pair. Maybe the most important contribution of the discovery of the explicit forms of CLaws is the development of the famous inverse scattering method, which, up to now, has proved to be one of the most effective methods for solving nonlinear partial differential equation (PDEs) or systems of PDEs. The existence of an infinite number of CLaws indicates the integrability of PDEs, although the nonexistence of a sequence of CLaws does not preclude the integrability, for example the Burgers equation. Furthermore, it can make the construction of the bi-Hamiltonian form of some important nonlinear evolution equations easier. A very extensive study of CLaws may be found in [10] which includes both Lagrangian and Hamiltonian formulations. As mentioned previously, the work of the construction of the explicit forms of CLaws is meaningful, which can expedite the process of the study of nonlinear PDEs. For the first few CLaws, maybe one can obtain them by hand. However, with the increase of the degree of the conserved densities, the construction by hand becomes harder and harder and even impossible. It is exciting that the great advantage of searching for polynomial conservation laws (PCLaws) lies in that they can be found by explicit computation. As a matter of fact, possessing such PCLaws is an intrinsic and common property for most of the nonlinear evolution equations. Furthermore, from the viewpoint of CLaws, a PDE is said to be "C-integrable" if it possesses infinitely many or lots of CLaws. The most important thing therefore is how to discover them. Just so, several symbolic programs have been developed on different platforms of symbolic computation systems. Sanders and others have developed a software package in Maple and FORM [11, 13]. They use an extension of the total derivative operator to a Heisenberg algebra which allows them to invert the total derivatives on its image. Wolf has designed a REDUCE package ConLaw1-4 [14, 15]. His approach is based on solving over-determined systems of PDEs, which shows up in the computation of Lie-point symmetries and generalized symmetries. The advantages of the packages by Wolf include: (i) it can construct the explicit t, x-dependence CLaws, rational ones, as well as the PCLaws; (ii) it may give CLaws including arbitrary functions. It also can give the CLaws for PDEs with parameters. However, it needs to specify in its calls the names of parameters to be computed. Göktas and Hereman [2] have also developed a package *condens.m* in Mathematica, which is based on the scaling properties of the PDEs, and aims to search for PCLaws for nonlinear polynomial PDEs, and the package *InvariantsSymmetries* which is the successor of condens.m. The newer version of the code of InvariantsSymmetries [3] allows one to compute conserved densities (no fluxes) and generalized symmetries of nonlinear evolution equations and differential-difference equations.

This paper aims to introduce the computer algebraic algorithm and routines to search for the PCLaws for nonlinear systems of PDEs as well as nonlinear PDEs. Then a description of the software package CONSLAW to automate the calculation is given. Specially, for parametrized PDEs, CONSLAW does not need any additional processes in advance and any other specific information, and can detect all the possible parameter constraints (compatibility conditions) automatically, irrespective of whether or not which can result in a sequence of PCLaws. After some necessary substitutions and analysis, it then reports all the corresponding PCLaws, if exist, under the different parameter constraints successively and automatically. Importantly, it may lead to the finding of new integrable systems for more general PDEs. Our package starts from the invariant symmetries of PDEs and is based on a theorem given in [10]; its basic idea is that Euler operator (variational derivative) acting on an expression gives identically zero, i.e. the Euler–Lagrange equation must vanish identically, if and only if the expression is a divergence.

It is a pity that our program cannot deal with the t, x-dependence CLaws directly, which however can be constructed by utilizing the obtained PCLaws for certain type of PDEs [18]. So far, our package cannot give CLaws which may include arbitrary functions, although possessing such CLaws is not a very common property for most PDEs. In addition, our package is incapable in dealing with the case that the variables with negative weights now.

The paper is organized as follows. In the following section 2, we outline the algorithm and routines. In Section 3 several examples are given to demonstrate the effectiveness of CONSLAW and some new results are obtained. Then a summary is given in the final section 4. The details of the usage of our package will be explained in the appendix.

2. Computer Algebraic Algorithm and Routines

Before describing the general computer algebraic algorithm, let us start with the following celebrated KdV equation

$$u_t + uu_x + u_{3x} = 0, \tag{2.1}$$

where $u = u(x,t)$, x, t are independent variables and $u_{3x} = \partial^3 u/\partial x^3$, which is known to have infinitely many CLaws. Before proceeding the computation of the PCLaws, the following definitions are introduced. The notion λ-homogeneous is introduced first, which means that there exists a scaling symmetry

$$t \mapsto \lambda^{-a} t, \quad x \mapsto \lambda^{-1} x, \quad u \mapsto \lambda^{b} u, \tag{2.2}$$

which leaves (2.1) invariant. Then (2.1) is transformed to

$$\lambda^{a+b} u_t = \lambda^{2b+1} uu_x + \lambda^{b+3} u_{3x}. \tag{2.3}$$

Take $a = 3$ and $b = 2$ to make (2.1) homogeneous in λ. That is, (2.1) is 2-homogeneous of weight 3. Here, let w stand for the weight [2, 3] of a variable; then in view of (2.2), we have

$$w(u) = 2, \quad w(\partial/\partial t) = 3, \quad w(t) = -3. \tag{2.4}$$

Without loss of generality, set $w(\partial/\partial x) = 1$. Another definition of the *rank* of a monomial is useful, which is defined as the total weight of the monomial M, denoted by Rank(M). It is easy to see that all monomials in (2.1) have the same rank 5. This property is called *uniformity in rank.* Once those definitions are at hand, we can describe the algorithm in detail below. For simplicity and clarity, we will describe the algorithm in the case of nonlinear polynomial PDEs. As for systems of nonlinear PDEs, it can be carried out in the same fashion.

Consider nonlinear evolution equation

$$u_t = H(u, u_x, u_{2x}, \ldots), \tag{2.5}$$

where H is polynomial of u and u_{ix} $(i = 1, 2, \ldots)$, $u_{ix} = \partial^i u/\partial x^i$. For now, we assume that (2.5) is uniform in rank, i.e. it is invariant under some or other scaling symmetry.

A conservation law for (2.5) is a divergence expression

$$Q_t + J_x = 0, \tag{2.6}$$

where Q, J are named conserved density and conserved flux respectively, which vanishes for all solutions $u = f(x,t)$ of (2.5). As a matter of fact, for most nonlinear evolution equations, with few exceptions, the conserved density-flux pairs Q, J are polynomials in u and u_{ix}, and do not depend explicitly on x and t. Such conservation laws are called polynomial type conservation laws (PCLaws).

The computer algebraic algorithm and routines to construct the PCLaws for (2.5) are described as follows.

- *Determine the scaling symmetry, i.e. obtain the weights of the variables and parameters in* (2.1). It can be easily done by the definition of uniformity in rank. For example, the ranks of the three terms in (2.1) are $w(\partial/\partial t) + w(u), 1 + 2w(u)$ and $3 + w(u)$ respectively. The requirement of uniformity in rank leads to $w(\partial/\partial t) + w(u) = 1 + 2w(u) = 3 + w(u)$, which on solving yields $w(u) = 2, w(\partial/\partial t) = 3$. It is consistent with (2.4). However, there indeed exists such a case that maybe the equation is not uniform in rank. For instance, the terms $\alpha u u_x$ and $\beta u^2 u_x$ appearing in the KdV–mKdV equation

$$u_t + \alpha\, u u_x + \beta\, u^2 u_x + u_{3\,x} = 0, \tag{2.7}$$

  are not uniform in rank. To process such case congruously, one can treat both α and β as extra variables with (unknown) weights. Then one has

$$w(\partial/\partial t) + w(u) = 1 + w(\alpha) + 2w(u) = 1 + w(\beta) + 3w(u) = 3 + w(u),$$

  which on solving yields

$$\{w(\alpha) = w(\alpha), w(\partial/\partial t) = 3,\ w(u) = -w(\alpha) + 2,\ w(\beta) = -2 + 2w(\alpha)\}.$$

  Since all the weights should be nonnegative, one can choose $w(\alpha) = 1$ to get $w(u) = w(\alpha) = 1, w(\partial/\partial t) = 3$. That means that only α, one of the two parameters, is a weighted parameter, but one should bear in mind that it still is a parameter.
- *Determine the form of the conserved density Q, i.e. find the components (monomials) in u and its x-derivatives of the polynomial conserved densities with prescribed rank R.* Apparently, the prescribed R should be a multiple of

the smallest weight of the dependent variables and the weighted parameters. In detail, the procedure proceeds as follows.

1. Collect the dependent variables and the weighted parameters in a set V, and sort them in an ascending order by their weights.
2. Form a basis set S. The set S consists of such elements as $[M, \text{Rank}(M)]$, where M is one of the monomials of rank R or less by taking all appropriate combinations of different powers of the elements in the set V, and Rank(M) is the rank of the monomial M. Two cases arise:
   (a) If at least one of the weight values of the variables is a fraction, then some of the Rank(M)'s are inevitable fractions. In this case split S into two parts, denoted by $S1$ and $S2$ respectively. In $S1$, the value of the second element of its each member is a fraction, and then $S2$ is the complement set of S.
   (b) If all the obtained weights are integers, then it is not necessary to split S.
3. Form a set $ListH$ of all monomials in u and the x-derivatives of u with rank R. To this end, two cases should be considered. For the above case 2(a), one can proceed as follows:
   (a) If the given rank R is a fraction, one only needs to process $S1$. That is, for each element in $S1$, compute $l_i = R-\text{Rank}(M_i)$ first to form a list L. Obviously, each element of L is an integer. Compute x-derivative of M_i up to l_i to make the newly generated monomials exactly have rank R, and then gather them in a set $ListH$.
   (b) If the given rank R is an integer, then perform the same processes to $S2$.

   For the above case 2(b), likewise, compute $l_i = R-\text{Rank}(M_i)$ to form the list L. Then, compute x-derivative of M_i up to l_i to form the set $ListH$.
4. Remove the redundant monomials in the set $ListH$. If some monomials in $ListH$ belong to the same equivalence class, i.e. their conserved densities are equivalent if they only differ by a total x-derivative, then identify them. Denote the simplified set as $ListP$. It is a focal point of the whole work. In detail, we proceed:
   (a) Introduce a temporary set $TempSet$, and let $TempSet = ListH$.
   (b) Introduce three new sets and initialize them: $K1 = \{\text{Null}\}$, $K2 = \{\text{Null}\}$, $ListP = \{\text{Null}\}$.
   (c) For each element denoted by $J1$ in $TempSet$, check whether or not it is a member of the sets $K1$ and $K2$. If not, then check whether it can be rewritten as

$$M = \triangle + [N]', \tag{2.8}$$

   where $\triangle$, N are the newly generated monomials in u and its derivatives, the symbol $'$ denotes the derivative of the monomial N w.r.t.

x. If so, then rewrite it and check whether or not the newly generated monomial $\triangle$ is a member of $K1$ and $K2$. If $\triangle$ is not a member of $K1$ and $K2$, collect $\triangle$ into the sets $K2$ and $ListP$ simultaneously. If $J1$ cannot be written in the form (2.8), then collect it into $ListP$ directly. No matter how, collect the processed monomial $J1$ into the set $K1$. If $J1$ is indeed a member of the sets $K1$ and $K2$, then process the next element of $TempSet$.

(d) Check whether or not $TempSet$ equals $ListP$. If so, then terminate the loop and return the set $ListP$. If not, then go to the next step.

(e) Assign the newly generated set $ListP$ to $TempSet$, and then go to step (b).

Finally, the set $ListP$ consists of the basic components which are the building blocks of the conserved density Q.

5. Linearly combine the terms in the set $ListP$ with constant coefficients c_i's to yield the form of the polynomial type conserved density Q of rank R.

Carrying on with (2.1), we compute the form of the conserved density of rank $R = 8$. From $V = \{u\}$ we build the basis set

$$S = \{[u, 2], [u^2, 4], [u^3, 6], [u^4, 8]\}.$$

Easily, we get the list $L = [6, 4, 2, 0]$. Then compute the various x-derivatives of u, u^2 and u^3 up to $6, 4, 2$ and 0 respectively to get the newly generated monomials which exactly have rank 8. Therefore, we have

$$\begin{aligned} &\frac{d^6}{dx^6}u = u_{6x}, \quad \frac{d^4}{dx^4}(u^2) = 6u_{2x}^2 + 8u_x u_{3x} + uu_{4x}, \\ &\frac{d^2}{dx^2}(u^3) = 6uu_x^2 + 3u^2 u_{2x}, \quad \frac{d^0}{dx^0}(u^4) = u^4. \end{aligned} \tag{2.9}$$

Gather all the monomials in the right-hand sides of (2.9), we derive the set

$$ListH = \{u_{6x}, u_{2x}^2, u_x u_{3x}, uu_{4x}, uu_x^2, u^2 u_{2x}, u^4\}.$$

Since

$$\begin{aligned} u_{6x} &= \frac{d}{dx}(u_{5x}), \quad u_x u_{3x} = \frac{d}{dx}(u_x u_{2x}) - u_{2x}^2, \\ uu_{4x} &= \frac{d}{dx}(uu_{3x} - u_x u_{2x}) + u_{2x}^2, \quad u^2 u_{2x} = \frac{d}{dx}(u^2 u_x) - 2uu_x^2, \end{aligned}$$

the term u_{6x} is cancelled, $u_x u_{3x}$ and uu_{4x} are both replaced by u_{2x}^2, and $u^2 u_{2x}$ is replaced by uu_x^2. Then, we obtain the set

$$ListP = \{uu_x^2, u_{2x}^2, u^4\},$$

and the form of the conserved density of rank 8

$$Q = c_1 uu_x^2 + c_2 u_{2x}^2 + c_3 u^4. \tag{2.10}$$

- *Determine the unknown coefficients in the conserved density Q.*

1. Substitute Q into (2.6) and use (2.5) to eliminate all t-derivatives. Note that, the resulting expression E must be a total x-derivative, which means that the Euler–Lagrange equation must vanish identically [10], i.e. $L_u(E) = 0$, where L_u is the Euler operator (variational derivative) defined by
$$L_u = \sum_{n=0}^{\infty}(-1)^n D_x^n(\frac{\partial}{\partial u_{nx}}). \tag{2.11}$$
For a given nonlinear evolution equation, by means of determining the highest order of the x-derivative of the expression E, one can ensure that the expression $L_u(E)$ truncates.
2. Group the remained terms including the ones in V and all the x-derivatives of the dependent variables in V. Then set them to zero to get a linear system for the c_i's and the parameters, which on solving gives the solution sets. After removing the trivial solutions one can obtain the conserved densities of rank R.

For example, computing the t-derivative of (2.10) yields
$$Q_t = c_1 u_t u_x^2 + 2c_1 u u_x u_{xt} + 2c_2 u_{2x} u_{2xt} + 4c_3 u^3 u_t. \tag{2.12}$$
Replacement of all the t-derivatives in (2.12) by $u_t = -uu_x - u_{3x}$ yields
$$\begin{aligned} E = &-3c_1 u u_x^3 - c_1 u_x^2 u_{3x} - 2c_1 u^2 u_x u_{2x} - 2c_1 u u_x u_{4x} - 6c_2 u_{2x}^2 u_x \\ &- 2c_2 u u_{2x} u_{3x} - 2c_2 u_{2x} u_{5x} - 4c_3 u^4 u_x - 4c_3 u^3 u_{3x}. \end{aligned} \tag{2.13}$$
To compute the variational derivative, i.e. apply the Euler operator L_u to E, one has to substitute u_{ix} $(i = 1, \ldots, 5)$, not including u, u^i, by $U[i]$ first (here, to make the process convenient in Maple, we use the symbol $U[i]$), because here u and the u_{ix}'s should be viewed as independents while the operator $\partial/\partial u_{ix}$ acting on them. To this end, we give the function TransForm1(). Its algorithm is described as follows.

Input: An expression *Expr* in the dependent variables and their derivatives w.r.t. x;
Output: A symbolic expression *SymbolExpr* corresponding to *Expr*.

Begin
  SymbolExpr := 0;
  If *Expr* does not include ′+′ **Then** *Expr* := {*Expr*} **Fi**;
  Put the number of the terms in *Expr* into a variable *NumExpr*;
  For *i* **From** 1 **To** *NumExpr* **Do**
    aa := op(*i*, *Expr*);
    aaa := 1;
    If *aa* includes ′∗′ **Then**
      Put the number of the terms in *aa* into *nn*
    Else
      nn := 1
    Fi;

```
            For j From 1 To nn Do
               If nn <> 1 Then gg := op(j, aa) Else gg := aa Fi;
               Compute the degree of gg, and put it into dg;
               Compute the differential order of gg, and put it into df;
               If gg does not include x, t Then
                  aaa := aaa * gg
               Elif gg includes dependent variable u Then
                  If df = 0 Then aaa := aaa * u^dg
                  Else aaa := aaa * U[df]^dg
                  Fi
               Fi
            Od;
            SymbolExpr := SymbolExpr + aaa
         Od;
         Return(SymbolExpr)
      End
```

After the function TransForm1() is presented, we can proceed. First, compute the derivative of E w.r.t. u, and put it into a variable *summ*. Then compute the derivatives of E w.r.t. $U[1]$, …, $U[5]$ respectively, and put the results into an array *sq*, which means that $sq[i]$ contains the derivative of E w.r.t. $U[i]$. In view of the Euler operator, we must compute the ith derivatives of $sq[i]$, $i = 1, \ldots, 5$, w.r.t. x. However, u^i, $U[i]$ in $sq[i]$ are only symbolic expressions, we cannot compute their derivatives directly. Hence, we should transform u^i into function expression with the form $\widehat{U}(x,t)^i$, and $U[i]$ into $\partial^i \widehat{U}(x,t)/\partial x^i$. To this end, we give another function TransForm2(). Its algorithm is described as follows.

Input: A symbolic expression $Expr1$;
Output: A Function expression $Expr2$ corresponding to $Expr1$.

```
      Begin
         Expr2 := 0;
         If Expr1 does not include '+' Then Expr1 := Expr1 Fi;
         Put the number of Expr1 into nn;
         For k From 1 To nn Do
            aa = op(k, Expr);
            Put the number of the terms in aa into n1;
            aaa := 1;
            For i From 1 To n1 Do
               tmp := op(i, aa);
               If tmp does not contain U Then
                  If tmp does not contain u Then
                     aaa := aaa * tmp
                  Else
```

Compute the degree of tmp, and put it into n;
$aaa := aaa * \widehat{U}(x,t)^n$
Fi
Elif the degree of tmp equals 1 **Then**
♯ Fetch out the differential order of tmp.
$m :=$ op(tmp);
$ee := \widehat{U}(x,t)$;
♯ Compute the mth order x-derivative of ee.
$ee :=$ diff($ee, x\$m$);
$aaa := aaa * ee$
Else
♯ Fetch out the differential order of tmp.
$m :=$ op(op(1,tmp));
$ee := \widehat{U}(x,t)$; $ee :=$ diff($ee, x\$m$);
$n :=$ degree(tmp); $aaa := aaa * ee^n$
Fi
Od;
$Expr2 := Expr2 + aaa$
Od;
Return($Expr2$)
End

The algorithm for the computation of variational derivative is:

Begin
Determine the highest derivative of E, and put it into $HighDorder$;
$TT :=$ TransForm1(E);
Compute the derivative of TT w.r.t. the dependent variable u, and put it into $summ$;
For i **From** 1 **To** $HighDorder$ **Do**
Compute the derivatives of TT w.r.t. $U[i]$, and put them into $sq[i]$
Od;
$summ :=$ TransForm2($summ$);
For i **From** 1 **To** $HighDorder$ **Do**
$tmp :=$ TransForm2($sq[i]$);
Compute the ith order x-derivative of tmp, and put it into $tmp1$;
$summ := summ + (-1)^i * tmp1$
Od
End

After some simple computations, group the remained terms to get

$$(6c_1 + 72c_3)uu_xu_{2x} - (6c_1 + 10c_2)u_xu_{4x} - (20c_2 + 12c_1)u_{2x}u_{3x} + (2c_1 + 24c_3)u_x^2.$$

Setting the coefficients of the above expression to zero leads to

$$c_1 + 12c_3 = 0, \quad 3c_1 + 5c_2 = 0, \quad 5c_2 + 3c_1 = 0, \; c_1 + 12c_3 = 0,$$

which on solving gives

$$c_1 = -\frac{5}{3}, \quad c_2 = 1, \quad c_3 = \frac{5}{36}.$$

Then the conserved density of rank 8 is

$$Q = -\frac{5}{3}uu_x^2 + u_{2x}^2 + \frac{5}{36}u^4. \tag{2.14}$$

- *Determine the corresponding conserved flux.* Once the conserved density Q is determined, the corresponding conserved flux J can be got from (2.6) by using the idea of integrating by parts. The idea seems to be quite simple; however, when Q_t includes too many terms the computation will become considerably complicated. We process as follows.
  1. Compute $M = -Q_t$, and let $\widehat{M} = M$, $J = 0$.
  2. Initialize two new sets, i.e. $K1 = \{\text{Null}\}$ and $K2 = \{\text{Null}\}$.
  3. For each term denoted by tt of M, check whether or not it is a member of $K1$ and $K2$ first. If not, then rewrite it in the form (2.8). The processed term tt will be collected into $K1$. If the newly produced monomial $\triangle$ is not a member of $K2$ then absorb it. Compute $\widehat{M} = \widehat{M} - tt + \triangle$, and set $J = J + N$.
  4. Let $M = \widehat{M}$.
  5. Repeat steps 2, 3 and 4 until $M = 0$; then J is the conserved flux.

The process of the PCLaws as described above is completely algorithmic, but it is very tedious by hand. We present a Maple package CONSLAW that fully automates the computation.

3. Applications to Several Nonlinear PDEs

To exhibit the effectiveness of CONSLAW, PLaws for several PDEs are constructed in this section.

Example 3.1. Let us first consider the generalized 5th order KdV equation

$$u_t + au^2u_x + bu_xu_{2x} + cuu_{3x} + u_{5x} = 0, \tag{3.1}$$

with constant parameters a, b and c, which includes four well known special cases:

(1) $(a, b, c) = (10, 20, 30), (20, 40, 120), (30, 60, 270)$, the Lax equation [7];
(2) $(a, b, c) = (-15, -15, 45), (5, 5, 5), (30, 30, 30)$, the SK (Sawada–Kotera) equation [1];
(3) $(a, b, c) = (10, 25, 20), (30, 75, 180)$, the KK equation [6];
(4) $(a, b, c) = (3, 6, 2)$, the Ito equation [5].

To process the various parameters corresponding to different kinds of the 5th order class KdV equation uniformly, by introducing a simple transformation $u = u/a$, (3.1) is reduced to

$$u_t + uu_{3x} + \alpha u_xu_{2x} + \beta u^2u_x + u_{5x} = 0, \tag{3.2}$$

where $\alpha = b/a$ and $\beta = c/a^2$. Specially, the Lax equation is obtained from (3.2) for $\alpha = 2, \beta = 3/10$. The SK, KK, and Ito equations correspond to $(\alpha, \beta) = (1, 1/5)$, $(5/2, 1/5)$, $(2, 2/9)$ respectively. Next, we study the PCLaws of (3.2), which are of interest for its exact solutions, for its understanding and classification, and for supporting its numerical solutions.

For (3.2), CONSLAW reports the following information first

$$w(x) = -1, \quad w(t) = -5, \quad w(u) = 2, \quad w(\partial/\partial t) = 5.$$

Then, some of the results are listed below.

When $R = 2$, for any α and β, there exists a CLaw, namely

$$(u)_t + (u_{4x} + uu_{2x} - \frac{1}{2}u_x^2 + \frac{\alpha}{2}u_x^2 + \frac{\beta}{3}u^3)_x = 0,$$

which represents the conservation of momentum.

When $R = 4$, there also exists a CLaw if $\alpha = 2$, β is free, namely

$$(u^2)_t + (2u^2u_{2x} + \frac{\beta}{2}u^4 + 2uu_{4x} - 2u_xu_{3x} + u_{2x}^2)_x = 0,$$

which represents the conservation of energy.

From now on, as the numbers of terms of the conserved fluxes are more than 20, only the conserved densities are listed. The subscripts of Q_m refer to the rank of the corresponding conserved densities.

When $R = 6$,

$$Q_6 = u_x^2 + \frac{1-2\alpha}{15}u^3 \qquad (\beta = -\frac{1}{5}\alpha^2 + \frac{7}{10}\alpha - \frac{3}{10}, \ \alpha \text{ is free}).$$

When $R = 8$, two CLaws are obtained. The first conserved density is

$$Q_8^{(1)} = u_{2x}^2 + \left(\frac{4}{675}\alpha^2 + \frac{4}{675}\alpha + \frac{1}{675}\right)u^4 + \left(-\frac{2}{5}\alpha - \frac{1}{5}\right)uu_x^2$$
$$(\beta = -\frac{2}{45}\alpha^2 + \frac{7}{45}\alpha + \frac{4}{45}).$$

The second conserved density is

$$Q_8^{(2)} = u_{2x}^2 + \frac{1}{6}\beta u^4 - uu_x^2 \qquad (\alpha = 2).$$

When $R = 10$, only one conserved density is obtained as

$$Q_{10} = -\frac{7}{5}uu_{2x}^2 + \frac{7}{10}u^2u_x^2 + u_{3x}^2 - \frac{7}{500}u^5 \qquad (\alpha = 2, \ \beta = \frac{3}{10}).$$

When $R = 12$, CONSLAW automatically reports three branches of parameter constraints and CLaws one by one.

The first conserved density is

$$Q_{12}^{(1)} = -\frac{7}{5}uu_{3x}^2 + \frac{16}{25}u^2u_{2x}^2 - \frac{17}{75}u_x^4 + \frac{16}{15}u_{2x}^3 + u_{4x}^2 + \frac{4}{5625}u^6 - \frac{2}{15}u^3u_x^2$$
$$(\alpha = 1, \ \beta = \frac{1}{5}).$$

For the same parameter constraints, Li [8] has obtained the 2-soliton as well as the 1-soliton solutions.

The second conserved density is

$$Q_{12}^{(2)} = -2\,u u_{3\,x}^2 + \frac{34}{25}\,u^2 u_{2\,x}^2 - \frac{31}{150}\,u_x^4 + \frac{37}{15}\,u_{2\,x}^3 + u_{4\,x}^2 + \frac{16}{5625}\,u^6 - \frac{28}{75}\,u^3 u_x^2$$
$$(\alpha = \frac{5}{2},\ \beta = \frac{1}{5}).$$

For the same parameter constraints, Li has also obtained the solitary wave solutions of (3.2) [8].

The third conserved density is

$$Q_{12}^{(3)} = -\frac{9}{5}\,u u_{3\,x}^2 + \frac{63}{50}\,u^2 u_{2\,x}^2 - \frac{7}{20}\,u_x^4 + 2\,u_{2\,x}^3 + u_{4\,x}^2 + \frac{21}{5000}\,u^6 - \frac{21}{50}\,u^3 u_x^2$$
$$(\alpha = 2,\ \beta = \frac{3}{10}).$$

Some other results obtained by CONSLAW are listed in the table below.

R	P. C.	Terms	Time
2	no	1/5	1.092s
4	$\{\alpha = 2,\ \beta : \text{free}\}$	1/5	1.203s
6	$\{\beta = -\frac{1}{5}\alpha^2 + \frac{7}{10}\alpha - \frac{3}{10},\ \alpha : \text{free}\}$	2/20	1.382s
8	$\{\beta = -\frac{2}{45}\alpha^2 + \frac{7}{45}\alpha + \frac{4}{45},\ \alpha : \text{free}\}$, $\{\alpha = 2,\ \beta : \text{free}\}$	3/20	2.343s
10	$\{\alpha = 2,\ \beta = \frac{3}{10}\}$	3/24	1.442s
12	$\{\alpha = 1,\ \beta = \frac{1}{5}\}$, $\{\alpha = 2,\ \beta = \frac{3}{10}\}$, $\{\alpha = \frac{5}{2},\ \beta = \frac{1}{5}\}$	7/41	3.756s
14	$\{\alpha = 1,\ \beta = \frac{1}{5}\}$, $\{\alpha = 2,\ \beta = \frac{3}{10}\}$, $\{\alpha = \frac{5}{2}, \beta = \frac{1}{5}\}$	10/65	6.799s
16	$\{\alpha = 2,\ \beta = \frac{3}{10}\}$	14/96	9.313s
18	$\{\alpha = 1,\ \beta = \frac{1}{5}\}$, $\{\alpha = 2,\ \beta = \frac{3}{10}\}$, $\{\alpha = \frac{5}{2},\ \beta = \frac{1}{5}\}$	22/146	39.637s
20	$\{\alpha = 1,\ \beta = \frac{1}{5}\}$, $\{\alpha = 2,\ \beta = \frac{3}{10}\}$, $\{\alpha = \frac{5}{2},\ \beta = \frac{1}{5}\}$	32/218	120s
22	$\{\alpha = 2,\ \beta = \frac{3}{10}\}$	45/319	53.498s
24	$\{\alpha = 1,\ \beta = \frac{1}{5}\}$, $\{\alpha = 2,\ \beta = \frac{3}{10}\}$, $\{\alpha = \frac{5}{2},\ \beta = \frac{1}{5}\}$	67/463	134.694s
26	$\{\alpha = 1,\ \beta = \frac{1}{5}\}$, $\{\alpha = 2,\ \beta = \frac{3}{10}\}$, $\{\alpha = \frac{5}{2},\ \beta = \frac{1}{5}\}$	95/680	448.585s
28	$\{\alpha = 2,\ \beta = \frac{3}{10}\}$	134/960	1711.448s
...	...	...	...

In the above table, P. C. denotes the parameter constraints which guarantee the existence of a PCLaw. The third column shows the numbers of the conserved density-flux pairs. The forth column shows the running time. We state that we only list the running time under concrete parameter constraints when rank ≥ 22. From the above table we see that: (i) the Lax equation possesses PCLaws at every level; (ii) the SK and KK equations possess infinitely many PCLaws but with a gap. (iii) the Ito equation possesses three PCLaws at ranks 2, 4 and 8. Hence, in the sense of CLaws, the Lax, SK and KK equations are C-integrable. Furthermore, for (3.2), the package *wkptest* presented by our group [16] cannot be used to test the Painlevé property directly. However, if the parameter constraints are filtered out, *wkptest* can do it under the obtained parameter constraints respectively.

Example 3.2. Consider another classical system of Drinfel'd–Sokolov–Wilson equations [4]

$$\begin{cases} u_t + pvv_x = 0, \\ v_t + qv_{xxx} + ruv_x + svu_x = 0, \end{cases} \tag{3.3}$$

where p, q, r, s are arbitrary parameters. System (3.3) has been studied in [4]–[17]. Here, we study its PCLaws, that is its C-integrability.

For (3.3), CONSLAW first outputs

$$w(x) = -1, \quad w(t) = -3, \quad w(u) = 2, \quad w(v) = 2, \quad w(\partial/\partial t) = 3,$$

which shows that (3.3) is uniform in rank. Then we list some results as follows.

When $R = 2$, we state that without constraints on the parameters p, q, r, s,

$$(u)_t + (\frac{1}{2}pv^2)_x = 0$$

is a conservation law, and that there exists an additional CLaw if $r = s$, namely

$$(u + v)_t + (qv_{2\,x} + \frac{1}{2}\,pv^2 + svu)_x = 0,$$

which are consistent with the results obtained by using the package InvariantsSymmetries.

When $R = 4$ and 6, without constraints on the parameters p, q, r, s, system (3.3) possesses CLaws. They are shown below.

$$R = 4: \quad [v^2 + \frac{(-r + 2\,s)\,u^2}{p}]_t + (2\,sv^2u + 2\,qvv_{2\,x} - qv_x^2)_x = 0;$$

$$\begin{aligned} R = 6: \quad & [-\frac{1}{3}\,\frac{(2\,s + r)\,v^2u}{q} - \frac{2}{9}\,\frac{(2\,s^2 - sr - r^2)\,u^3}{pq} + \frac{1}{3}\,\frac{(-r + s)\,u_x^2}{p} + v_x^2]_t \\ & + [-\frac{1}{6}\,\frac{psv^4}{q} - \frac{1}{12}\,\frac{prv^4}{q} - \frac{2}{3}\,\frac{s^2v^2u^2}{q} - \frac{4}{3}\,svuv_{2\,x} + \frac{4}{3}\,ruv_x^2 + 2\,svv_xu_x \\ & + 2\,qv_xv_{3\,x} - \frac{1}{3}\,\frac{srv^2u^2}{q} - \frac{2}{3}\,rvuv_{2\,x} + \frac{2}{3}\,suv_x^2 - qv_{2\,x}^2]_x = 0. \end{aligned}$$

When $R = 8$ and $R = 10$, system (3.3) possesses CLaws if $r = 2s$. The conserved densities are

$$Q_8 = -2\,\frac{p^2v^4}{q^2} - 8\,\frac{psv^2u^2}{q^2} + \frac{16}{3}\,\frac{s^2u^4}{q^2} - 10\,\frac{pv^2u_{2\,x}}{q} + 24\,\frac{pv_x^2u}{q} - 9\,\frac{pv_{2\,x}^2}{s}$$
$$+ u_{2\,x}^2 - 12\,\frac{suu_x^2}{q},$$
$$Q_{10} = v_{3\,x}^2 + 4\,\frac{s^2u^2v_x^2}{q^2} - 4\,\frac{sv_{2\,x}^2u}{q} - \frac{2}{3}\,\frac{s^2puv^4}{q^3} + 5\,\frac{sv_x^2u_{2\,x}}{q} + \frac{7}{3}\,\frac{psv^2v_x^2}{q^2} - \frac{s^2u_x^2v^2}{q^2}$$
$$- 2\,\frac{s^2uv^2u_{2\,x}}{q^2} + 2\,\frac{svu_{3\,x}v_x}{q}.$$

Other results are no longer listed here, but we see that (3.3) possesses PCLaws at every level if $r = 2s$, which means that (3.3) is C-integrable in this case. It is interesting that (3.3) does not pass the Painlevé test by directly using the package *wkptest*, but if we choose $r = 2s$, it does. That means that a new integrable system is given for (3.3) if $r = 2s$.

Example 3.3. Consider a variant type of system of Boussinesq equations [12]

$$\begin{cases} u_t + v_x + uu_x + su_{xx} = 0, \\ v_t + (uv)_x + rv_{xx} + pu_{xxx} = 0, \end{cases} \tag{3.4}$$

where p, r, s are constant parameters. The solitary wave solutions of (3.4) have been studied in [17]. Here we also studied its PCLaws.

For (3.4), CONSLAW first outputs

$$w(x) = -1, \quad w(t) = -2, \quad w(u) = 1, \quad w(v) = 2, \quad w(\partial/\partial t) = 2.$$

Then we can get PCLaws for (3.4) in different ranks. The first four PCLaws are

$$(u)_t + (v + \frac{1}{2}\,u^2 + su_x)_t = 0,$$
$$(v)_t + (rv_x + pu_{2\,x} + uv)_x = 0,$$
$$(uv)_t + (\frac{1}{2}\,v^2 + puu_{2\,x} + svu_x + u^2v - suv_x - \frac{1}{2}\,pu_x^2)_x = 0,$$
$$(v^2 + u^2v - pu_x{}^2 + 2\,su_xv)_t + (pu^2u_{2\,x} + u^3v + 2\,uv^2 + 2\,pvu_{2\,x}$$
$$+ 4\,suvu_x - 2\,s^2u_xv_x - su^2v_x - 2\,puu_x{}^2 - 2\,pu_xv_x + 2\,s^2vu_{2\,x})_x = 0,$$

for which, with the exception of one, the parameter constraint is $r = -s$. As a matter of fact, we find that (3.4) possesses PCLaws at every level if $r = -s$, which means that (3.4) is C-integrable. Same as Example 3.2, system (3.4) cannot pass the Painlevé test if using `wkptest` directly, but if we set $r = -s$, it does. Furthermore, for this filtered integrable system, we get a dependent variable transformation as

$$u = -2\sqrt{s^2 + p}\frac{\phi_x}{\phi}, \quad v = (2s\sqrt{s^2 + p} + 2p + 2)(\ln\phi)_{xx}, \tag{3.5}$$

where $\phi = \phi(x,t)$, by which one can transform (3.4) into a bilinear form, and maybe can construct its multi-soliton solutions further. This is beyond the work of this paper.

4. Summary

We have described the computer algebraic algorithm and routines for the computation of conservation laws. A Maple package CONSLAW has been implemented to automate the computation. Using CONSLAW, one not only can obtain the scaling symmetries of PDEs, but also can construct the PCLaws of PDEs. Especially, for given parametrized nonlinear evolution equations, irrespective of whether or not the given equations are uniform in rank, CONSLAW can determine the scaling symmetries of the differentials, parameters as well as the dependent variables totally automatically. Importantly, it can detect all the possible integrable systems by analyzing the parameter constraints which may lead to the existence of a sequence of CLaws, and then output the explicit forms of the corresponding CLaws successively. As shown in the above examples a given PDE may not pass the Painlevé test by using *wkptest* directly. However, if the parameter constraints are detected, one can test its Painlevé property again. The obtained explicit forms of the conserved densities play an important role in studying the integrability of PDEs. More study on how to develop our package and how to develop the applications of the explicit forms of PCLaws is worthwhile in future.

Appendix. The Usage of CONSLAW

The package CONSLAW can work with Maple 6 and 8, or later versions. Its main procedure is denoted as CONSLAW. We take a generalized the 7th order KdV equation as an example of the use of CONSLAW.

To start, one proceeds as follows:

```
[> with(CONSLAW);
[> ConsLaw([diff(u(x,t),t)+a*u(x,t)^3*diff(u(x,t),x)+b*diff(u(x,t),x)^3
        +z*u(x,t)*diff(u(x,t),x)*diff(u(x,t),x$2)+d*u(x,t)^2*diff(u(x,t),x$3)
        +e*diff(u(x,t),x$2)*diff(u(x,t),x$3)+f*diff(u(x,t),x)*diff(u(x,t),x$4)
        +u(x,t)*diff(u(x,t),x$5)+diff(u(x,t),x$7)=0],{a,b,z,d,e,f});
```

For the main procedure CONSLAW, there are two arguments. The first one is the list of the object PDEs, and the second one is a set of parameters appearing PDEs. If the parametrized PDE is not uniform in rank, then this argument must be specified, which means that it is necessary that one or more parameters should be viewed as special weighted parameters to make it uniform in rank. However, if the parametrized PDE is uniform in rank even with no such parameters, it is the same as the case that a PDE with constant coefficients, i.e. the second argument is an empty set. Hence, for the 7th order KdV equation, we input:

[> `ConsLaw`([diff(u(x,t),t)+a*u(x,t)^3*diff(u(x,t),x)+b*diff(u(x,t),x)^3
+z*u(x,t)*diff(u(x,t),x)*diff(u(x,t),x$2)+d*u(x,t)^2*diff(u(x,t),x$3)
+e*diff(u(x,t),x$2)*diff(u(x,t),x$3)+f*diff(u(x,t),x)*diff(u(x,t),x$4)
+u(x,t)*diff(u(x,t),x$5)+diff(u(x,t),x$7)=0],{`NULL`});

`The input evolution equation is:`

$$\frac{\partial}{\partial t}u(x,t) + au(x,t)^3\frac{\partial}{\partial x}u(x,t) + b(\frac{\partial}{\partial x}u(x,t))^3 + zu(x,t)\frac{\partial}{\partial x}u(x,t)\frac{\partial^2}{\partial x^2}u(x,t)$$
$$+ du(x,t)^2\frac{\partial^3}{\partial x^3}u(x,t) + e\frac{\partial^2}{\partial x^2}u(x,t)\frac{\partial^3}{\partial x^3}u(x,t) + f\frac{\partial}{\partial x}u(x,t)\frac{\partial^4}{\partial x^4}u(x,t)$$
$$+ u(x,t)\frac{\partial^5}{\partial x^5}u(x,t) + \frac{\partial^7}{\partial x^7}u(x,t) = 0$$

`The weight of x is: -1`
`The weight of t is: -7`
`The weights of the variables is (are):`

$$[w_{u(x,t)} = 2, \; w_{\frac{\partial}{\partial t}} = 7]$$

`Input rank` R `of the conserved density, it should be a multiple of the smallest weight of the dependent variables and parameters in the above list:`

[> 12

`The number of components forming the conserved density is: 7`
`The form of the conserved density is:`

$$c_1u(x,t)(\frac{\partial^3}{\partial x^3}u(x,t))^2 + c_2(\frac{\partial^4}{\partial x^4}u(x,t))^2 + c_3(\frac{\partial^2}{\partial x^2}u(x,t))^3 + c_4(\frac{\partial}{\partial x}u(x,t))^4$$
$$+c_5(u(x,t))^6 + c_6(u(x,t))^2(\frac{\partial^2}{\partial x^2}u(x,t))^2 + c_7(u(x,t))^3(\frac{\partial}{\partial x}u(x,t)^2$$

`The solution set for the` c_i`'s are:`

$$\{\{b = \frac{15}{28}, a = 0, d = \frac{5}{28}, z = \frac{10}{7}, f = \frac{7}{2}, c_2 = c_2, c_6 = \frac{125}{196}c_2, c_3 = \frac{145}{126}c_2,$$
$$c_5 = \frac{15}{38416}c_2, c_7 = -\frac{75}{686}c_2, c_1 = -\frac{10}{7}c_2, c_4 = -\frac{25}{98}c_2, e = \frac{25}{6}\}, \{b = \frac{5}{14},$$
$$a = \frac{4}{147}, d = \frac{2}{7}, f = 72, c_2 = c_2, c_1 = -\frac{10}{7}c_2, z = \frac{9}{7}, c_4 = -\frac{31}{294}c_2, c_6 = \frac{34}{49}c_2,$$
$$c_7 = -\frac{20}{147}c_2, c_3 = \frac{37}{21}c_2, e = 6, c_5 = \frac{16}{21609}c_2\}, \{d = d, z = z, a = a, b = b,$$
$$e = e, f = f, c_7 = 0, c_1 = 0, c_4 = 0, c_2 = 0, c_6 = 0, c_3 = 0, c_5 = 0\}, \{f = -\frac{3}{2},$$
$$b = 0, e = \frac{5}{7}, a = 0, d = 0, z = 0, c_2 = c_2, c_7 = 0, c_1 = 0, c_4 = 0, c_6 = 0, c_3 = 0,$$

$$c_5 = 0\}, \{f = 3, e = 5, b = \frac{5}{14}, c_7 = -\frac{15}{98}\,c_2, c_1 = -\frac{9}{7}\,c_2, c_4 = -\frac{5}{28}\,c_2,$$
$$c_6 = \frac{9}{14}\,c_2, c_3 = \frac{10}{7}\,c_2, c_5 = \frac{3}{2744}\,c_2, a = \frac{5}{98}, z = \frac{10}{7}, d = \frac{5}{14}, c_2 = c_2\},$$
$$\{b = \frac{1}{7}, e = 3, a = \frac{4}{147}, d = \frac{2}{7}, z = \frac{6}{7}, c_5 = \frac{4}{21609}\,c_2, c_7 = -\frac{50}{1029}\,c_2,$$
$$c_1 = -c_2, c_4 = -\frac{17}{147}\,c_2, c_6 = \frac{16}{49}\,c_2, c_3 = \frac{16}{21}\,c_2, c_2 = c_2, f = 2\}\}$$

`The 1-th corresponding conserved density of rank 12 including 7 terms is:`

$$-\frac{20}{147}\,u(x,t)^3(\frac{\partial}{\partial x}u(x,t))^2 + (\frac{\partial^4}{\partial x^4}u(x,t))^2 - \frac{31}{294}\,(\frac{\partial}{\partial x}u(x,t))^4 + \frac{37}{21}\,(\frac{\partial^2}{\partial x^2}u(x,t))^3$$
$$+\frac{34}{49}\,(\frac{\partial^2}{\partial x^2}u(x,t))^2u(x,t)^2 - \frac{10}{7}\,u(x,t)(\frac{\partial^3}{\partial x^3}u(x,t))^2 + \frac{16}{21609}\,(u(x,t))^6$$

`The compatibility conditions are:`

$$a = \frac{4}{147}, e = 6, f = \frac{7}{2}, d = \frac{2}{7}, z = \frac{9}{7}, b = \frac{5}{14}$$

`If you want output the flux please input ‘y’, otherwise press any key to end.`

[> y

`The 1-th corresponding flux including 74 terms is:` (omitted)

`The 2-th corresponding conserved density of rank 12 including 7 terms is:`

$$-\frac{50}{1029}\,u(x,t)^3(\frac{\partial}{\partial x}u(x,t))^2 + (\frac{\partial^4}{\partial x^4}u(x,t))^2 - \frac{17}{147}\,(\frac{\partial}{\partial x}u(x,t))^4 + \frac{16}{21}$$
$$(\frac{\partial^2}{\partial x^2}u(x,t))^3 + \frac{16}{49}\,(\frac{\partial^2}{\partial x^2}u(x,t))^2u(x,t)^2 - u(x,t)(\frac{\partial^3}{\partial x^3}u(x,t))^2 + \frac{4}{21609}\,u(x,t)^6$$

`The compatibility conditions are:`

$$a = \frac{4}{147}, d = \frac{2}{7}, f = 2, z = \frac{6}{7}, e = 3, b = \frac{1}{7}$$

`If you want output the flux please input ‘y’, otherwise press any key to end!`

[> y

`The 2-th corresponding flux including 74 terms is:` (omitted)

`The 3-th corresponding conserved density of rank 12 including 7 terms is:`

$$-\frac{15}{98}\,u(x,t)^3(\frac{\partial}{\partial x}u(x,t))^2 + (\frac{\partial^4}{\partial x^4}u(x,t))^2 - \frac{5}{28}\,(\frac{\partial}{\partial x}u(x,t))^4 + \frac{10}{7}\,(\frac{\partial^2}{\partial x^2}u(x,t))^3$$
$$+\frac{9}{14}\,(\frac{\partial^2}{\partial x^2}u(x,t))^2u(x,t)^2 - \frac{9}{7}\,u(x,t)(\frac{\partial^3}{\partial x^3}u(x,t))^2 + \frac{3}{2744}\,u(x,t)^6$$

```
The compatibility conditions are:
```

$$b = \frac{5}{14}, a = \frac{5}{98}, d = \frac{5}{14}, f = 3, z = \frac{10}{7}, e = 5$$

```
If you want output the flux please input 'y', otherwise press any
    key to end!
[> y
The 3-th corresponding flux including 74 terms is: (omitted here)
There are 3 conserved density-flux pair(s).
Total computing time is 3.756 seconds.
Continue to compute (y/n)?
```

References

[1] R. K. Dodd, J. D. Gibbon, The prolongation structure of a higher order Korteweg–de Vries equation, *Proc. R. Soc. Lond.* A **358** (1977), 287–296.

[2] Ü. Göktas, W. Hereman, Symbolic computation of conserved densities for systems of nonlinear evolution equations, *J. Symb. Comput.* **24** (1997), 591–621.

[3] Ü. Göktas, W. Hereman, InvariantsSymmetries.m: A Mathematica integrability package for the computation of invariants and symmetries (1997). Available from MathSource (Item: 0208-932, Applications/Mathematics) via FTP: `mathsource.wolfram.com` or via Internet URL: `http://www.mathsource.com/cgi-bin/MathSource/Applications/0208-932`.

[4] R. Hirota, B. Grammaticos, A. Ramani, Soliton structure of the Drinfel'd–Sokolov–Wilson equation, *J. Math. Phys.* **27**(6) (1986), 1499–1505.

[5] M. Ito, An extension of nonlinear evolution equations of the K-dV (mK-dV) type to higher orders, *J. Phys. Soc. Jpn.* **49** (1980), 771–778.

[6] D. J. Kaup, On the inverse scattering problem for cubic eigenvalue problems of the class xxx+6Q x+6R=, *Stud. Appl. Math.* **62** (1980), 189–216.

[7] P. D. Lax, Integrals of nonlinear equations of evolution and solitary waves, *Comm. Pure Appl. Math.* **21** (1968), 467–478.

[8] Z.-B. Li, S.-Q. Pan, The soliton solutions and solitary wave solutions of a generalized 5th-order KdV equation, *Acta Phys. Sin.* **50** (2001), 402–406.

[9] Z.-B. Li, Exact solitary wave solutions of nonlinear evolution equations, in *Mathematics Mechanization and Applications*, p. 389, Academic Press, London (2000).

[10] P. J. Olver, *Applications of Lie Groups to Differential Equations*, Springer-Verlag, New York (1986).

[11] J. A. Sanders, M. Roelofs, An algorithmic approach to conservation laws using the 3-dimensional Heisenberg algebra, Tech. Rep. 2, RIACA, Amsterdam (1994).

[12] R. L. Sachs, On the integrable variant of the Boussinesq system: Painlevé property, rational solutions, a related many-body system, and equivalence with the AKNS hierarchy, *Physica* D **30** (1988), 1–27.

[13] J. A. Sanders, J. P. Wang, On the (non)existence of conservation laws for evolution equations, Tech. Rep. WS-445, Vrije Universiteit, Amsterdam (1995).

[14] T. Wolf, A. Brand, M. Mohammadzadeh, Computer algebra algorithms and routines for the computation of conservation laws and fixing of gauge in differential expressions, *J. Symb. Comput.* **27** (1999), 221–238.

[15] T. Wolf, A comparison of four approaches for the calculation of conservation laws, *Europ. J. Appl. Math.* **13** (2002), 129–152.

[16] G.-Q. Xu, Z.-B. Li, Symbolic computation of the Painlevé test for nonlinear partial differential equations using Maple, *Comp. Phys. Comm.* **161** (2004), 65–75.

[17] R.-X. Yao, Z.-B. Li, New exact solutions for three evolution equations, *Phys. Lett.* A **297** (2002), 196–204.

[18] R.-X. Yao, Z.-B. Li, On t/x-dependent conservation laws of the generalized nth-order KdV equation, *Chin. J. Phys.* **42**(4-I) (2004), 315–322.

Ruo-Xia Yao
Department of Computer Science
East China Normal University
Zhongshan North Road 3663
Shanghai 200062 and
Department of Computer Science
Shanghai Normal University
Guilin Road 100
Shanghai 200234
P. R. China
e-mail: rxyao2@hotmail.com

Zhi-Bin Li
Department of Computer Science
East China Normal University
Zhongshan North Road 3663
Shanghai, 200062
P. R. China
e-mail: lizb@cs.ecnu.edu.cn

Trends in Mathematics: *Differential Equations with Symbolic Computation*, 327–341

Generalized Differential Resultant Systems of Algebraic ODEs and Differential Elimination Theory

Giuseppa Carrá Ferro

Abstract. Generalized differential resultants of algebraic ODEs are introduced and relations between generalized differential resultant systems and differential elimination are shown.

Mathematics Subject Classification (2000). Primary 13N; Secondary 13P, 34A, 12H.

Keywords. Resultant, generalized resultant, differential resultant, generalized differential resultant, generalized differential resultant system, differential elimination.

1. Introduction

One main problem in many mathematical models is the elimination of some differential variables in a system of algebraic ordinary differential equations.

There are many differential elimination procedures, coming from different theories, for example, the characteristic set theory of Wu–Ritt [29, 24, 34, 35, 32], the Gröbner basis theory together with the characteristic set theory [1, 2, 3, 12, 13, 14, 15, 16, 17, 20, 21, 26, 28], the differential resultant theory of algebraic differential polynomials introduced by the author [8, 9] and the differential Gröbner basis theory for differential polynomial ideals introduced by the author [7] and Ollivier [27].

The algebraic multivariate resultant theory was recently studied by many people [18, 22, 23, 30, 31] and different algorithms were implemented for its calculus in the sparse case [4, 5, 6]. Since a differential resultant is a multivariate algebraic resultant, all results and algorithms can be properly used in the differential case. The differential resultant theory allows us to know some upper bounds on the orders of the differential equations in the output of a differential elimination process.

This paper was written with the support of Italian MIUR, PRIN project: Commutative and Computational Algebra 2003.

In this paper is introduced the notion of generalized differential resultants of $m+s$ ordinary algebraic differential equations $\{f_j = 0 : j = 1, \ldots, m+s\}$ in $m+r$ differential variables with $r \geq 0$, $s \geq 1$ and $\mathrm{ord}(f_j) = n_j$. Such a notion extends the analogous one in the algebraic case. In fact generalized resultants generalize the resultant of $n+1$ polynomials in n variables to the case when the number of polynomials is greater than $n+1$. The main property of the generalized resultants is that they are multiples of elements of the corresponding elimination ideal. The notion of generalized differential resultant system is introduced, as the system of the generalized differential resultants. In a similar way the notion of generalized differential resultant ideal is introduced. If $n_1 \leq n_2 \leq \cdots \leq n_{m+s}$, then the maximal orders of the differential polynomials in a generalized differential resultant system is equal to $\sum_{i=0}^{m} n_{m+s-i} + n_{m+s}$. Moreover it is proved that a multiple of a generalized differential resultant lies in the corresponding differential elimination ideal.

2. Differential Algebra Preliminaries

Throughout the paper we use the notation as in [29, 24]. Let δ be a derivation and let R be a differential ring, i.e. a commutative ring with unit and a derivation δ acting on it, such that $\mathbf{Q} \subset R$. $C_R = \{a \in R : \delta(a) = 0\}$ is the ring of constants of R and $\mathbf{N}_0 = \{0, 1, \ldots, n, \ldots\}$.

Remark 2.1. $\mathbf{Q} \subseteq C_R$ because $\mathbf{Q} \subseteq R$.

Example 2.1. $C^\infty(\mathbf{R})$, $\mathbf{Q}(t)$, $\mathbf{R}(t)$ and the field of the meromorphic functions on a domain of $\mathbf{C}$ with the usual derivation $\delta = \frac{\partial}{\partial t}$ are all differential rings. $C^m(\mathbf{R})$ for any m in $\mathbf{N}_0$ is not a differential ring.

Definition 2.1. $S = R\{X_1, \ldots, X_m\} = R[\delta^n X_i\colon i = 1, \ldots, m \text{ and } n \in \mathbf{N}_0]$ is the differential ring of the differential polynomials in the differential indeterminates $\{X_1, \ldots, X_m\}$ with coefficients in the differential ring R, where $\delta(\delta^n X_i) = \delta^{n+1} X_i$ for all $n \in \mathbf{N}_0$ and i .

If $f \in S$, then the *order* of f is $\mathrm{ord}(f) = \max\{n \in \mathbf{N}_0\colon f$ contains a power product in $X_1, \ldots, X_m, \delta X_1, \ldots, \delta X_m, \ldots, \delta^n X_1, \ldots, \delta^n X_m$ with nonzero coefficient$\}$.

Definition 2.2. An ideal I of S is a differential ideal iff $\delta^n(s) \in I$ for all $n \in \mathbf{N}_0$ and all $s \in I$. If $A \subseteq S$, then $[A] = (\delta^n(s)\colon s \in A,\ n \in \mathbf{N}_0)$ denotes the smallest differential ideal containing A.

3. Algebraic Elimination Theory

3.1. Gröbner Bases

The Gröbner basis theory is very useful in the elimination of variables.

Let K be a field. Given $k+s$ polynomials $g_1, \ldots, g_{k+s}$ in $S = K[X_1, \ldots, X_{k+r}]$ with $\deg(g_j) = d_j$ for all j we want to *eliminate* the variables $X_1, \ldots, X_k$; for example, given the ideal $I = (g_1, \ldots, g_{k+s})$ we want to find the *k-elimination ideal* $I_k = I \cap K[X_{k+1}, \ldots, X_{k+r}]$ of I. By the Gröbner basis theory it is sufficient to find a Gröbner basis of the ideal I with respect to an elimination term ordering τ_k on the set of power products P in $\{X_1, \ldots, X_{k+s}\}$ as in [10].

A general elimination term ordering τ_k is given in the following way. Given two arbitrary term orderings σ_1 on the set of power products P_1 in $\{X_1, \ldots, X_k\}$ and σ_2 on the set of power products P_2 in $\{X_{k+1}, \ldots, X_{k+s}\}$, if p, $p' \in P_1$ and q, $q' \in P_2$ then $pq <_\tau p'q'$ iff either $p <_{\sigma_1} p'$ or $p = p'$ and $q <_{\sigma_2} q'$.

If G is a Gröbner basis of I with respect to τ_k, then

$$G_k = G \cap K[X_{k+1}, \ldots, X_{k+r}]$$

is a Gröbner basis of I_k with respect to σ_2.

3.2. Multivariate Algebraic Resultant

Let R be an integral domain and let K be the quotient field of R. Given $k+1$ polynomials $g_1, \ldots, g_{k+1}$ in $S = R[X_1, \ldots, X_k]$ with $\deg(g_j) = d_j$ for all j, Macaulay [25] defines a polynomial Q only in the coefficients of $g_1, \ldots, g_{k+1}$, called the *resultant of* $g_1, \ldots, g_{k+1}$ and denoted as $Q = \mathrm{Res}(g_1, \ldots, g_{k+1})$, such that a necessary condition for the system $\{g_1 = 0, \ldots, g_{k+1} = 0\}$ to have a solution in an algebraic extension of K is $Q = 0$.

Let $D = \sum_{j=1}^{k+1} d_j - k$ and let $L = \binom{k+D}{k}$.

$$P_k^D = \left\{ X^a = X_1^{a_1} \cdots X_k^{a_k} : \sum_{i=1}^{k} a_i \leq D,\ a = (a_1, \ldots, a_k) \in \mathbf{N}_0^k \right\}$$

is the set of all power products in S of degree less than or equal to D. The set P_k^D has cardinality equal to L, while $P_k^{D-d_j}$ has cardinality equal to $L_j = \binom{k+D-d_j}{k}$ for all $j = 1, \ldots, k+1$.

P_k^D and $P_k^{D-d_j}$ are totally ordered using first the degree and then the lexicographic order with $X_1 < \cdots < X_k$.

Definition 3.1. $M = M(g_1, \ldots, g_{k+1})$ is the $(\sum_{j=1}^{k+1} L_j \times L)$-matrix defined in the following way: for each i such that $\sum_{h=1}^{j} L_h < i \leq \sum_{h=1}^{j+1} L_h$ the coefficients of the polynomial $X^a g_{j+1}$ are the entries of the ith row for each $X^a \in P_k^{D-d_{j+1}}$ that are written in decreasing order with respect to the power products in P_k^D.

Definition 3.2. Let g_j be a polynomial of degree d_j for all $j = 1, \ldots, k+1$ in $R[X_1, \ldots, X_k]$.

$$\mathrm{Res}(g_1, \ldots, g_{k+1}) = \mathrm{GCD}(\det(P)\colon P \text{ is an } L \times L\text{-submatrix of } M).$$

In [25, 11] it is shown that, in the generic case, the resultant of polynomials can be written as $\det(M_0)/\det(A)$, where M_0 is a square submatrix of M of maximal rank and A is a submatrix of M_0.

Theorem 3.1. *Let* g_j *be a polynomial of degree* d_j *for all* $j = 1, \dots, k+1$ *in* $R[X_1, \dots, X_k]$. *There exists a nonzero* r *in* R, *such that*

$$r\,\mathrm{Res}(g_1, \dots, g_{k+1}) = \sum_{j=1}^{k+1} f_j g_j,$$

where f_j *is a polynomial in* $R[X_1, \dots, X_k]$ *with* $\deg(f_j) \leq D - d_j$ *for all* $j = 1, \dots, k+1$.

Proof. It follows by definition of resultant as in [25, 11]. In the generic case $\det(A)\,\mathrm{Res}(g_1, \dots, g_{k+1}) = \det(M_0)$ and $\det(M_0) = \det(B(M_0))$, where $B(M_0)$ is obtained by substituting $X^a g_j$ to the corresponding element in the last column and in the corresponding row of M_0 for all $j = 1, \dots, k+1$ and i. Since a determinant is multilinear on the columns, $\det(M_0) = \det(B(M_0)) = \sum_{j=1}^{k+1} f_j g_j$. Here $\deg(f_j) \leq D - d_j$ for all j, because $X^a \in P_k^{D-d_j}$ in each considered polynomial $X^a g_j$ by Definition 3.1. Now if $r = \det(A) \neq 0$, then the theorem follows by definition of the matrix A. Since M_0 and A depend on a permutation $\sigma \in S_k$ as in [25] and [11, pp. 96–103], $M_0 = M_0(\sigma)$ and $A = A(\sigma)$. If $r = \det(A) = \det(A(\sigma)) = 0$, then we can choose another permutation $\sigma' \in S_k$, that gives different submatrices $M_0(\sigma')$ of M and $A(\sigma')$ with $\det(A(\sigma') \neq 0$ and $\det(A(\sigma'))\mathrm{Res}(g_1, \dots, g_{k+1}) = \det(M_0(\sigma'))$. Now since the polynomials $g_1, \dots, g_{k+1}$ have coefficients in the integral domain R, that is not specified, we have a specialization of the generic case and the theorem is proved. □

Remark 3.1. If R is a field, then $\mathrm{Res}(g_1, \dots, g_{k+1}) \in (g_1, \dots, g_{k+1})$ by Theorem 3.1.

3.3. Resultants and Elimination Ideals

The resultants can be used for the elimination of a set of variables. In fact the univariate and multivariate resultants can be used for the elimination of the set of k variables $\{X_1, \dots, X_k\}$ in a set of $k+s$ polynomials $g_1, \dots, g_{k+s}$ in $R[X_1, \dots, X_k, X_{k+1}, \dots, X_{k+r}]$ with $r \geq 0$ and $s \geq 1$ as in [19, 36, 10, 11].

Let $g_{k+1}^* = \sum_{j=k+1}^{k+s} A_j g_j$, where A_j are new indeterminates. Let us consider $\{g_1, \dots, g_k, g_{k+1}^*\}$ as a set of polynomials in

$$R[A_{k+1}, \dots, A_{k+s}, X_{k+1}, \dots, X_{k+r}][X_1, \dots, X_k].$$

$Q = \mathrm{Res}(g_1, \dots, g_k, g_{k+1}^*)$ is a polynomial in $R[A_{k+1}, \dots, A_{k+s}, X_{k+1}, \dots, X_{k+r}]$ by its own definition, and $Q = \sum_{i=1}^{p} Q_i p_i$, where the p_i's are distinct power products in $\{A_{k+1}, \dots, A_{k+s}\}$ and the Q_i's are polynomials in $\{X_{k+1}, \dots, X_{k+r}\}$. The power products p_i's have the same degree, because g_{k+1}^* is homogeneous in $\{A_{k+1}, \dots, A_{k+s}\}$ by its own definition.

Definition 3.3. Let $s \geq 1$ and let g_j and Q_i be as above for all $j = 1, \dots, k+s$ and $i = 1, \dots, p$. Then Q_i is called a *generalized resultant* of $g_1, \dots, g_k, \dots, g_{k+s}$ [10]. The system $\{Q_i = 0 : i = 1, \dots, p\}$ is called a *generalized resultant system* of $g_1, \dots, g_{k+s}$, while $E = (Q_1, \dots, Q_p)$ is called a *generalized resultant ideal* of $g_1, \dots, g_{k+s}$.

Theorem 3.2. *Let $I = (g_1, \ldots, g_{k+s})$. Let $s \geq 1$ and let g_j and Q_i be as above for all $j = 1, \ldots, k+s$ and $i = 1, \ldots, p$. Then there exists a nonzero element r in $R[X_{k+1}, \ldots, X_{k+r}]$, such that rQ_i is in the k-elimination ideal I_k of I.*

Proof. Let $Q = \operatorname{Res}(g_1, \ldots, g_k, g^*_{k+1})$. Then

$$Q \in R[X_{k+1}, \ldots, X_{k+r}][A_{k+1}, \ldots, A_{k+s}]$$

by definition. By Theorem 3.1 and by [11, exer. 7, p. 99] there exist polynomials f_j in $R[X_1, \ldots, X_{k+r}, A_{k+1}, \ldots, A_{k+s}]$ for all $j = 0, \ldots, k+1$, such that the equality

$$f_0 Q = \sum_{j=1}^{k} f_j g_j + f_{k+1} g^*_{k+1}$$

holds. Let $f_j = \sum_{h_j=1}^{r_j} q_{h_j} p_{h_j}$, where p_{h_j} is a power product in $A_{k+1}, \ldots, A_{k+s}$ and $q_{h_j} \in R[X_1, \ldots, X_{k+r}]$ for all h_j and j. We have the following equality

$$f_0 Q = \sum_{j=1}^{k} \Big(\sum_{h_j=1}^{r_j} q_{h_j} p_{h_j}\Big) g_j + \sum_{h=1}^{s} A_{k+h} \Big(\sum_{h_{k+1}=1}^{r_{k+1}} q_{h_{k+1}} p_{h_{k+1}}\Big) g_{k+h} = \sum_{i=1}^{p'} \Big(\sum_{j=1}^{k+s} q'_{ij} g_j\Big) p'_i,$$

where p'_i is again a power product in $A_{k+1}, \ldots, A_{k+s}$ and $q'_{ij} \in R[X_1, \ldots, X_{k+r}]$. Note that f_0 depends only on the coefficients of $g_1, \ldots, g_k$ by the proof of Proposition 4.6 in [11, p. 99], so it is a nonzero element of $R[X_{k+1}, \ldots, X_{k+r}]$. Furthermore $f_0 Q = \sum_{i=1}^{p} f_0 Q_i p_i$ and $p = p'$. It follows that $f_0 Q_i = \sum_{j=1}^{k+s} q'_{ij} g_j$ is in I, and thus it is in I_k, by the definition of elimination ideal for every i. □

Remark 3.2. I_k is not always generated by $Q_1, \ldots, Q_p$ as it is remarked in [10, ex. 12, p. 166]. The generalized resultants and the generalized resultant ideal depend on which k polynomials come first in the ordered sequence $\{g_1, \ldots, g_k, \ldots, g_{k+s}\}$.

Remark 3.3. Let I be as above and let $I^* = (g_1, \ldots, g_k, g^*_{k+1})$ be the corresponding ideal in $R[X_1, \ldots, X_{k+r}, A_{k+1}, \ldots, A_{k+s}]$ as polynomial ring in the variables $\{X_1, \ldots, X_k\}$ with coefficients in $T = R[X_{k+1}, \ldots, X_{k+r}, A_{k+1}, \ldots, A_{k+s}]$.

Let τ be an elimination term ordering on the set of power products in $\{X_1, \ldots, X_k\}$. By using the Gröbner basis theory let G^* be a Gröbner basis of I^* with respect to τ. Since g^*_{k+1} is homogeneous in $A_{k+1}, \ldots, A_{k+s}$, every polynomial in

$$G^* \cap R[X_{k+1}, \ldots, X_{k+r}, A_{k+1}, \ldots, A_{k+s}]$$

is homogeneous in the variables $A_{k+1}, \ldots, A_{k+s}$. The generalized resultants are in the ideal generated by the nonzero coefficients of the power products in $A_{k+1}, \ldots, A_{k+s}$, that appear in some polynomial of G^*.

The generalized resultant system and the generalized resultant ideal extend the classical notions of *resultant system* and *resultant ideal* of a set of $k+s$ polynomials $g_1, \ldots, g_{k+s}$ in $k+r$ variables.

The resultant system depends on the definite order of elimination of the variables, while the resultant ideal is independent of such order. The proof is usually given in the case $k = 1$ and then inductively by eliminating one variable at each time [19].

4. Differential Elimination Theory

4.1. Differential Gröbner Bases

Let R be a differential integral domain. Differential Gröbner bases [7, 27] can be used for the elimination of the set of m differential variables $\{X_1, \dots, X_m\}$ in a set of $m+s$ differential polynomials $f_1, \dots, f_{m+s}$ in $R\{X_1, \dots, X_m, X_{m+1}, \dots, X_{m+r}\}$ with $r \geq 0$ and $s \geq 1$.

In other words, given the differential ideal $I = [f_1, \dots, f_{m+s}]$ we want to find the *differential m-elimination ideal* $I_m = I \cap R\{X_{m+1}, \dots, X_{m+r}\}$. By the Gröbner basis theory it is sufficient to a find a differential Gröbner basis of the differential ideal I with respect to a differential elimination term ordering.

Unfortunately there are only partial results. Even differential term orderings on the set of differential power products, i.e. power products in the variables and in their derivatives, are not completely classified [33].

Let $S = R\{X_1, \dots, X_{m+r}\}$ and let

$$S_t = R[X_1, \dots, X_{m+r}, \dots, \delta^t X_1, \dots, \delta^t X_{m+r}]$$

for all $t \in \mathbf{N}_0$. The following are useful examples of differential term ordering.

Example 4.1. Let σ_t be the lexicographic term ordering on the set of power products P_t in $X_1, \dots, X_{m+r}, \dots, \delta^t X_1, \dots, \delta^t X_{m+r}$ with

$$\delta^t X_1 >_{\sigma_t} \cdots >_{\sigma_t} \delta^t X_{m+r} >_{\sigma_t} \cdots >_{\sigma_t} X_1 >_{\sigma_t} \cdots >_{\sigma_t} X_{m+r}$$

for every $t \in \mathbf{N}_0$. The set $\{\sigma_t : t \in \mathbf{N}_0\}$ defines a differential term ordering σ on the set of power products in $\{X_1, \dots, X_{m+r}\}$ and their derivatives by taking its restriction to P_t equal to σ_t. σ is not a differential elimination term ordering.

Example 4.2. Let τ_t be the lexicographic term ordering on P_t with

$$\delta^t X_1 >_{\tau_t} \cdots >_{\tau_t} X_1 >_{\tau_t} \cdots >_{\tau_t} \delta^t X_{m+r} >_{\tau_t} \cdots >_{\tau_t} X_{m+r}$$

for every $t \in \mathbf{N}_0$. The set $\{\tau_t : t \in \mathbf{N}_0\}$ defines a differential term ordering τ on the set of power products in $\{X_1, \dots, X_{m+r}\}$ and their derivatives by taking its restriction to P_t equal to τ_t. τ is a differential elimination term ordering.

Example 4.3. Let ρ_t be the lexicographic term ordering on P_t with

$$\delta^t X_1 >_{\rho_t} \cdots >_{\rho_t} \delta^t X_m >_{\rho_t} \cdots >_{\rho_t} X_1 >_{\rho_t} \cdots >_{\rho_t} X_m >_{\rho_t} \delta^t X_{m+1}$$
$$>_{\rho_t} \cdots >_{\rho_t} X_{m+1} >_{\rho_t} \delta^t X_{m+r} >_{\rho_t} \cdots >_{\rho_t} X_{m+r}$$

for every $t \in \mathbf{N}_0$. The set $\{\rho_t : t \in \mathbf{N}_0\}$ defines a differential term ordering ρ on the set of power products in $\{X_1, \dots, X_{m+r}\}$ and their derivatives by taking its restriction to P_t equal to ρ_t. ρ is a differential elimination term ordering, that allows to eliminate the differential variables $X_1, \dots, X_m$.

4.2. Multivariate Differential Resultant

Let R be a differential integral domain and let K be the differential quotient field of R. K is a differential field by defining

$$\delta(\frac{r}{s}) = \frac{\delta(r)s - r\delta(s)}{s^2} \quad \text{for all } \frac{r}{s} \in K.$$

Let f_j be a differential polynomial in $R\{X_1, \dots, X_m\}$ of order n_j and degree d_j for each $j = 1, \dots, s$. Then $f_j \in R[X_1, \dots, X_m, \dots, \delta^{n_j} X_1, \dots, \delta^{n_j} X_m]$, which is a polynomial ring in $(n_j + 1)m$ variables.

It is necessary to find an s-tuple of non-negative integers $(r_1, \dots, r_s)$, such that $f_j, \delta(f_j), \dots, \delta^{r_j}(f_j)$, $j = 1, \dots, s$ are

$$\sum_{j=1}^{s}(r_j + 1) = (\sum_{j=1}^{s} r_j) + s$$

polynomials in $(\sum_{j=1}^{s} r_j) + s - 1$ variables. So the number of variables is

$$(n_j + r_j + 1)m = (n_{j+1} + r_{j+1} + 1)m$$

for all $j = 1, \dots, s - 1$ and the number of polynomials is

$$(\sum_{j=1}^{s} r_j) + s = (n_j + r_j + 1)m + 1$$

for some j. If $s = m + 1$, then $N = \sum_{j=1}^{m+1} n_j$ and

$$(r_1, \dots, r_{m+1}) = (N - n_1, \dots, N - n_{m+1})$$

by [9]. The *differential resultant* $\delta\mathrm{Res}(f_1, \dots, f_{m+1})$ of $f_1, \dots, f_{m+1}$ is the Macaulay algebraic resultant of the $mN + m + 1$ polynomials

$$\delta^{N-n_1}(f_1), \dots, \delta(f_1), f_1, \dots, \delta^{N-n_{m+1}}(f_{m+1}), \dots, \delta(f_{m+1}), f_{m+1}$$

in $S_{mN+m} = R[X_1, \dots, X_m, \dots, \delta^N X_1, \dots, \delta^N X_m]$, which is a polynomial ring in $mN + m$ variables by [9].

Example 4.4. Let

$$f_j = \delta X_j - g_j(X_1, \dots, X_m), \quad j = 1, \dots, m,$$

where g_j is in $R[X_1, \dots, X_m]$ and let f_{m+1} be a differential polynomial of order 1 in $R\{X_1, \dots, X_m\}$. Then $\mathrm{ord}(f_j) = 1$ for all j, $N = m + 1$, and the differential resultant of $f_1, \dots, f_{m+1}$ is the Macaulay algebraic resultant of the $m^2 + 2m + 1$ polynomials

$$\delta^m(f_1), \dots, \delta(f_1), f_1, \dots, \delta^m(f_{m+1}), \dots, \delta(f_{m+1}), f_{m+1}$$

in $S_{m^2+2m} = R[X_1, \dots, X_m, \dots, \delta^{m+1} X_1, \dots, \delta^{m+1} X_m]$, which is a polynomial ring in $m^2 + 2m$ variables.

Example 4.5. Let $f_j = \delta X_j - g_j(X_1, \ldots, X_{m+1})$, $j = 1, \ldots, m+1$, where g_j is in $R[X_1, \ldots, X_{m+1}]$. The variables $\{X_1, \ldots, X_m\}$ can be eliminated. Note that $\mathrm{ord}(f_j) = 1$ for all $j = 1, \ldots, m$, while $\mathrm{ord}(f_{m+1}) = 0$ as differential polynomials in $R\{X_{m+1}\}\{X_1, \ldots, X_m\}$. Now $N = m$, and the differential resultant of $f_1, \ldots, f_{m+1}$ is the Macaulay algebraic resultant of the $m^2 + m + 1$ polynomials

$$\delta^{m-1}(f_1), \ldots, \delta(f_1), f_1, \ldots, \delta^{m-1}(f_m), \ldots, \delta(f_m), f_m, \delta^m(f_{m+1}), \ldots, \delta(f_{m+1}), f_{m+1}$$

in $R[X_{m+1}, \delta X_{m+1}, \ldots, \delta^{m+1} X_{m+1}][X_1, \ldots, X_m, \ldots, \delta^m X_1, \ldots, \delta^m X_m]$.

Remark 4.1. The differential resultant can be used for the elimination of the set of m differential variables $\{X_1, \ldots, X_m\}$ in a set of $m+1$ differential polynomials $f_1, \ldots, f_{m+1}$ in

$$R\{X_1, \ldots, X_m, X_{m+1}, \ldots, X_{m+r}\} = R\{X_{m+1}, \ldots, X_{m+r}\}\{X_1, \ldots, X_m\}.$$

In this case the differential resultant is in $R\{X_{m+1}, \ldots, X_{m+r}\}$.

4.3. Differential Resultants and Differential Elimination

The differential resultant can be used for the elimination of the set of m differential variables $\{X_1, \ldots, X_m\}$ in a set of $m+s$ differential polynomials $f_1, \ldots, f_{m+s}$ in $R\{X_1, \ldots, X_m, X_{m+1}, \ldots, X_{m+r}\}$ with $r \geq 0$ and $s \geq 1$.

Let $f^*_{m+1} = \sum_{j=m+1}^{m+s} A_j f_j$, where $\{A_{m+1}, \ldots, A_{m+s}\}$ is a set of new differential indeterminates. Let us consider $\{f_1, \ldots, f_m, f^*_{m+1}\}$ as a set of differential polynomials in the differential polynomial ring

$$R\{A_{m+1}, \ldots, A_{m+s}, X_{m+1}, \ldots, X_{m+r}\}\{X_1, \ldots, X_m\}.$$

By definition

$$Q = \delta\mathrm{Res}(f_1, \ldots, f^*_{m+1}) \in R\{A_1, \ldots, A_{m+s}, X_{m+1}, \ldots, X_{m+r}\};$$

$Q = \sum_{i=1}^{p} Q_i m_i$, where the m_i's are distinct power products in $\{A_{m+1}, \ldots, A_{m+s}\}$ and their derivatives and the Q_i's are differential polynomials in the differential variables $\{X_{m+1}, \ldots, X_{m+r}\}$. The power products m_i's have the same degree, because g^*_{m+1} and its derivatives are homogeneous in $\{A_{m+1}, \ldots, A_{m+s}\}$ and their derivatives by its own definition.

Definition 4.1. Let $s \geq 1$ and let f_j and Q_i be as above for all $j = 1, \ldots, m+s$ and $i = 1, \ldots, p$. Q_i is called a *generalized differential resultant* of $f_1, \ldots, f_m, \ldots, f_{m+s}$. The system $\{Q_i = 0,\ i = 1, \ldots, p\}$ is called a *generalized differential resultant system* of $g_1, \ldots, g_{k+s}$, while $E = [Q_1, \ldots, Q_p]$ is called a *generalized differential resultant ideal* of $f_1, \ldots, f_{m+s}$.

Theorem 4.1. *Let $s \geq 1$ and let g_j and Q_i be as above for all $j = 1, \ldots, m+s$ and $i = 1, \ldots, p$. Then there exists a nonzero element r in $R\{X_{m+1}, \ldots, X_{m+r}\}$, such that rQ_i is in the differential m-elimination ideal I_m of $I = [f_1, \ldots, f_{m+s}]$.*

Proof. Let $n_j = \mathrm{ord}(f_j)$ for all $j = 1, \ldots, m+s$. Let $n'_j = \mathrm{ord}(f_j)$ for all $j = 1, \ldots, m+s$ as differential polynomials in the differential variables $\{X_1, \ldots, X_m\}$.

Of course $n'_j \leq n_j$ for all j. Let $n^*_{m+1} = \text{ord}(f^*_{m+1})$ as a differential polynomial in the differential variables $\{X_1, \dots, X_m\}$. We have

$$n^*_{m+1} = \max\{n'_j : j = m+1, \dots, m+s\} \leq \max\{n_j : j = m+1, \dots, m+s\}.$$

Let $N^* = \sum_{j=1}^{m} n'_j + n^*_{m+1}$ and let $Q = \delta\text{Res}(f_1, \dots, f_m, f^*_{m+1})$. We have

$$Q = \text{Res}(\delta^{N^*-n'_1}(f_1), \dots, f_1, \dots, \delta^{N^*-n'_m}(f_m), \dots, f_m, \delta^{N^*-n^*_{m+1}}(f^*_{m+1}), \dots, f^*_{m+1})$$

by [9]. So Q is the multivariate algebraic resultant of $mN^* + m + 1$ polynomials in

$$T_{mN^*+m} = T[X_1, \dots, X_m, \dots, \delta^{N^*}X_1, \dots, \delta^{N^*}X_m],$$

which is a polynomial ring in $mN^* + m$ variables with coefficients in the polynomial ring

$$T = R[X_{m+b}, A_{m+b}, \dots, \delta^{N^*}X_{m+b}, \delta^{N^*}A_{m+b} : b = 1, \dots, s].$$

Q is in $R\{X_{m+1}, \dots, X_{m+r}\}\{A_{m+1}, \dots, A_{m+s}\}$ by its own definition.

By Proposition 3.2 in [11, p. 89], we have

$$Q' = \text{Res}(\delta^{N^*-n'_1}(f_j), \dots, \delta(f_1), f_1, \dots, \delta^{N^*-n'_m}(f_m), \dots,$$
$$\delta(f_m), f_m, f^*_{m+1}, \delta^{N^*-n^*_{m+1}-1}(f^*_{m+1}), \dots, \delta(f^*_{m+1}), \delta^{N^*-n^*_{m+1}}(f^*_{m+1}))$$

equal to Q up to the sign. By Theorem 3.1, by Definition 10 and Proposition 12 in [9], and by exer. 7 in [11, p. 99], there exist differential polynomials F_0, F_{h_j} in $R\{X_1, \dots, X_{m+r}, A_{m+1}, \dots, A_{m+s}\}$ for all $h_j = 0, 1, \dots, N^* - n'_j$, $j = 1, \dots, m$ and for all $h_{m+1} = 0, 1, \dots, N^* - n^*_{m+1}$, such that we have the equality

$$F_0 Q' = \sum_{j=1}^{m} \Big(\sum_{h_j=0}^{N^*-n'_j} F_{h_j} \delta^{h_j}(f_j) \Big) + \sum_{h_{m+1}=0}^{N^*-n^*_{m+1}} F_{h_{m+1}} \delta^{h_{m+1}}(f^*_{m+1}).$$

Let $F_{h_j} = \sum_{a_{h_j}=1}^{p_{h_j}} q_{a_{h_j}} m_{a_{h_j}}$, where $q_{a_{h_j}} \in R\{X_1, \dots, X_{m+r}\}$ and $m_{a_{h_j}}$ is a power product in

$$\{A_{m+1}, \dots, A_{m+s}, \dots, \delta^{N^*-n^*_{m+1}} A_{m+1}, \dots, \delta^{N^*-n^*_{m+1}} A_{m+s}\}$$

for all a_{h_j}. We have

$$F_0 Q' = \sum_{j=1}^{m} \Big(\sum_{h_j=0}^{N^*-n'_j} \Big(\sum_{a_{h_j}=1}^{p_{h_j}} q_{a_{h_j}} m_{a_{h_j}} \Big) \delta^{h_j}(f_j) \Big)$$
$$+ \sum_{b=1}^{s} \Big(\sum_{h_{m+1}=0}^{N^*-n^*_{m+1}} \Big(\sum_{a_{h_{m+1}}=1}^{p_{h_{m+1}}} q_{a_{h_{m+1}}} m_{a_{h_{m+1}}} \Big) \Big) \delta^{h_{m+1}}(A_{m+b} f_{m+b})$$
$$= \sum_{i=1}^{p'} \Big(\sum_{j=1}^{m} \Big(\sum_{h_j=0}^{N^*-n'_j} q'_{h_{i_j}} \delta^{h_j}(f_j) \Big) \Big) + \sum_{j=m+1}^{m+s} \Big(\sum_{h_j=0}^{N^*-n^*_{m+1}} q'_{h_{i_j}} \Big) \delta^{h_j}(f_j) m'_i.$$

F_0 does not depend on the coefficients of $\delta^{N^*-n^*_{m+1}}(f^*_{m+1})$ by Proposition 4.6 in [11, p. 99]. By the definition of derivative of a polynomial the coefficients of f^*_{m+1} and their derivatives up to order $N^* - n^*_{m+1}$ are among the coefficients of $\delta^{N^*-n^*_{m+1}}(f^*_{m+1})$. It follows that F_0 does not depend on the coefficients of any derivative of f^*_{m+1} up to order $N^* - n^*_{m+1}$, i.e. F_0 is a nonzero element of $R\{X_{m+1}, \ldots, X_{m+r}\}$. Furthermore $F_0Q' = \sum_{i=1}^{p} F_0Q_i m_i$ and $p = p'$. So $F_0Q_i = \sum_{j=1}^{m+s} q'_{i_j} f_j$ is in I, and thus it is in I_m, by the definition of elimination ideal for every i. □

Remark 4.2. I_m could be generated not always by $Q_1, \ldots, Q_p$ as in the algebraic case and in the following example. In a similar way the generalized differential resultants and the differential generalized resultant ideal depend on which m polynomials come first in the ordered sequence $\{f_1, \ldots, f_m, \ldots, f_{m+s}\}$.

Example 4.6. Let $S = R\{X_1, X_2\}$,

$$f_1 = \delta X_1 - 1, \quad f_2 = X_2X_1 - 1, \quad f_3 = X_2^2X_1 + X_2X_1 + 1$$

and let $I = [f_1, f_2, f_3]$. We want to eliminate the variable $\{X_1\}$.

$$f_2^* = (A_2X_2 + A_3X_2^2 + A_3X_2)X_1 + (-A_2 + A_3).$$

$\delta\mathrm{Res}(f_1, f_2^*) = \mathrm{Res}(f_1, \delta(f_2^*), f_2^*))$, because f_1 and f_2^* have order respectively 1 and 0 as differential polynomials in $S = R\{X_2\}\{X_1\}$. The matrix $M = M_0$, because f_1 and f_2^* have degree 1 as differential polynomials in $S = R\{X_2\}\{X_1\}$.

$$\begin{aligned}\delta\mathrm{Res}(f_1, f_2^*) = {} & (X_2^2 + 2X_2)\delta A_2A_3 - (X_2^2 + 2X_2)A_2\delta A_3 - (X_2^2 + \delta X_2)A_2^2 \\ & - (2X_2^2 + 2X_2^3 + 2X_2\delta X_2)A_2A_3 \\ & + (2X_2\delta X_2 + \delta X_2 - X_2^4 - 2X_2^3 - X_2^2)A_3^2.\end{aligned}$$

The generalized differential resultants are

$$\begin{aligned}&Q_1 = X_2^2 + 2X_2, \quad Q_2 = -(X_2^2 + 2X_2), \quad Q_3 = -(X_2^2 + \delta X_2), \\ &Q_4 = -(2X_2^2 + 2X_2^3 + 2X_2\delta X_2), \quad Q_5 = (2X_2\delta X_2 + \delta X_2 - X_2^4 - 2X_2^3 - X_2^2)\end{aligned}$$

and the generalized differential resultant ideal $E = [Q_1, Q_2, Q_3, Q_4, Q_5] = [X_2]$. The differential ideal $I = [1]$. So $I_1 = [1]$ and $E \subset I_1$.

Corollary 4.2. *Let $s \geq 1$ and let g_j and Q_i be as above for all $j = 1, \ldots, m+s$ and $i = 1, \ldots, p$. Suppose that $n_1 \leq n_2 \leq \cdots \leq n_{m+s}$. Then $\mathrm{ord}(Q_i) \leq \sum_{j=1}^{m} n_j + n_{m+s}$ for all i, once the m polynomials $\{f_1, \ldots, f_m\}$ come first in the ordered sequence $\{f_1, \ldots, f_m, \ldots, f_{m+s}\}$.*

Proof. By hypothesis on n_j for all $j = 1, \ldots, m+s$ and by the proof of Theorem 4.1

$$N^* = \sum_{j=1}^{m} n'_j + n^*_{m+1} \leq \sum_{j=1}^{m} n_j + n^*_{m+1}.$$

Since $\mathrm{ord}(f^*_{m+1}) \leq n_{m+s}$, the order of every coefficient of f^*_{m+1} is either less than or equal to n_{m+s}. In a similar way, since $\mathrm{ord}(\delta^i(f^*_{m+1})) \leq n_{m+s} + i$ for all

$i = 1, \ldots, N^* - n^*_{m+1}$, the order of every coefficient of $\delta^i(f^*_{m+1})$ is either less than or equal to $N^* - n^*_{m+1} + n_{m+s}$. Let

$$Q = \delta\text{Res}(f_1, \ldots, f_m, f^*_{m+1}) = \sum_{i=1}^{p} Q_i m_i.$$

Then $\text{ord}(Q_i) \leq \text{ord}(F_0 Q_i)$ by the definition of differential resultant as the GCD of determinants. $\text{ord}(F_0 Q_i)$ is either less than or equal to the maximal order of the coefficients of $\delta^i(f_j)$ for all $j = 1, \ldots, m$, $i = 0, 1, \ldots, N^* - n'_j$ and the coefficients of $\delta^i(f^*_{m+1})$ for all $i = 0, 1, \ldots, N^* - n^*_{m+1}$, which is less than or equal to

$$\sum_{j=1}^{m} n_j + n_{m+s}.$$

□

Corollary 4.3. *Let f_j be a differential polynomial in $R\{X_1, \ldots, X_{m+r}\}$ of order n_j for each $j = 1, \ldots, m+s$ and let $s \geq 1$. Let Q_i be a generalized differential resultant of $f_1, \ldots, f_m, \ldots, f_{m+s}$. If $n_1 \leq n_2 \leq \cdots \leq n_{m+s}$, then*

$$\text{ord}(Q_i) \leq \sum_{j=0}^{m} n_{m+s-j} + n_{m+s}$$

for all i and for every ordering of the sequence $\{f_1, \ldots, f_m, \ldots, f_{m+s}\}$.

Proof. Let $J_{m+1} = \{j_1, \ldots, j_{m+1}\} \subset \{1, \ldots, m+s\}$ with $|J_{m+1}| = m+1$ and let

$$N^*_{J_{m+1}} = \sum_{h=1}^{m+1} n_{j_h}.$$

Let $N = \max\{N^*_{J_{m+1}} : |J_{m+1}| = m+1\}$. We have $N = \sum_{j=0}^{m} n_{m+s-j}$, and $N^*_{J_{m+1}} \leq N$ for every ordering of the sequence $\{f_1, \ldots, f_m, \ldots, f_{m+s}\}$. By hypothesis on n_j and by the proofs of Theorem 4.1 and Corollary 4.2

$$\text{ord}(Q_i) \leq \max\{N^*_{J_{m+1}} - n'_{j_h} + n_{j_h} : h = 1, \ldots, m+1\} \leq N + n_{m+s}$$

for all i . □

Corollary 4.4. *Let f_j be a differential polynomial in $R\{X_1, \ldots, X_m, X_{m+1}, \ldots, X_{m+r}\}$ of order n_j and let $s \geq 1$. Let $I = [f_1, \ldots, f_m, \ldots, f_{m+s}]$ and let I_m be the differential m-elimination ideal of I. If $n_1 \leq n_2 \leq \cdots \leq n_{m+s}$, then there are differential polynomials of order either less than or equal to $\sum_{j=0}^{m} n_{m+s-j} + n_{m+s}$ among the generators of I_m.*

Proof. By the proof of Theorem 4.1 the differential polynomials $F_0 Q_i \in I_m$ for all $i = 1, \ldots, p$. Now by the proof of Corollary 4.3

$$\text{ord}(Q_i) \leq \text{ord}(F_0 Q_i) \leq \sum_{j=0}^{m} n_{m+s-j} + n_{m+s} - n_1$$

for all i. □

4.4. Examples

Example 4.7. Let $f_j = \delta X_j - g_j(X_1, \ldots, X_{m+r})$ for all $j = 1, \ldots, m+r$, where g_j is a polynomial in $\{X_1, \ldots, X_m, X_{m+1}, \ldots, X_{m+r}\}$ for all j.

We want to eliminate the variables $\{X_1, \ldots, X_m\}$. $n_j = \mathrm{ord}(f_j) = 1$ for all $j = 1, \ldots, m$, while $n_j = \mathrm{ord}(f_j) = 0$ for all $j = m+1, \ldots, m+r$ as differential polynomials in the differential variables $\{X_1, \ldots, X_m\}$. Let Q_i be a generalized differential resultant of $f_1, \ldots, f_m, f_{m+1}, \ldots, f_{m+r}$ for all $i = 1, \ldots, p$. By Corollary 4.2, Q_i is a differential polynomial in $\{X_{m+1}, \ldots, X_{m+r}\}$ with $\mathrm{ord}(Q_i) \leq m+1$ for all $i = 1, \ldots, p$.

If $r = 1$, then there is only one generalized differential resultant and its order is either less than or equal to $m+1$.

If $r = m+1$, then the generalized resultant $Q' = \mathrm{Res}(f_{m+1}, \ldots, f_{2m+1})$ of $f_{m+1}, \ldots, f_{2m+1}$ is a polynomial of order either less than or equal to 1 in $\{X_{m+1}, \ldots, X_{2m+1}\}$.

If $r > m+1$, then the generalized resultants Q'_h of $f_{m+1}, \ldots, f_{m+r}$ are polynomials of order either less than or equal to 1 in $\{X_{m+1}, \ldots, X_{m+r}\}$ for all $h = 1, \ldots, q$.

The following example shows the use of algebraic and differential resultants in differential elimination.

Example 4.8 (n-strain model [28]). Let

$$f_j = \delta X_j - g_j(X_1, \ldots, X_{2^n-1}, Y_1, \ldots, Y_{n2^{n-1}}, Z_1, \ldots, Z_n)$$

when $j = 1, \ldots, 2^n - 1$,

$$f_j = \delta Y_j - g_j(X_1, \ldots, X_{2^n-1}, Y_1, \ldots, Y_{n2^n-1}, Z_1, \ldots, Z_n)$$

when $j = 2^n, \ldots, n2^{n-1} + 2^n - 1$, and

$$f_j = \delta Z_j - g_j(X_1, \ldots, X_{2^n-1}, Y_1, \ldots, Y_{n2^n-1}, Z_1, \ldots, Z_n)$$

when $j = n2^{n-1}+2^n, \ldots, n2^{n-1}+2^n+n-1$, where g_j are polynomials. It is possible to eliminate the variables X_j's and Y_j's. $\mathrm{ord}(f_j) = n_j = 1$ for all $j = 1, \ldots, n2^{n-1} + 2^n - 1$, while $\mathrm{ord}(f_j) = n_j = 0$ for all $j = n2^{n-1} + 2^n, \ldots, n2^{n-1} + 2^n - 1 + n$ as differential polynomials in

$$R\{Z_1, \ldots, Z_n\}\{X_1, \ldots, X_{2^n-1}, Y_1, \ldots, Y_{n2^{n-1}}\}.$$

So $N = n2^{n-1} + 2^n - 1$.

If $n = 1$, it is sufficient to take the algebraic resultant of the polynomials $\delta(f_1), f_1, \delta(f_2), f_2, \delta^2(f_3), \delta(f_3), f_3$ in the polynomial ring

$$R[Z, \delta Z, \delta^2 Z, \delta^3 Z][X, \delta X, \delta^2 X, Y, \delta Y, \delta^2 Y],$$

which is in $R[Z, \delta Z, \delta^2 Z, \delta^3 Z]$.

If $n > 1$, then let

$$f^*_{n2^{n-1}+2^n} = \sum_{j=n2^{n-1}+2^n}^{n2^{n-1}+2^n+n-1} A_j g_j,$$

where A_j are arbitrary differential variables.

Now it is possible to eliminate the variables X_j's and Y_j's from the system

$$\{f_1 = 0, \ldots, f_{n2^{n-1}+2^n-1} = 0, f^*_{n2^{n-1}+2^n} = 0\}$$

by using the differential resultants and the generalized differential resultants. As in the example above the order of each generalized differential resultant is either less than or equal to $n2^{n-1} + 2^n$.

The following example shows a procedure for the elimination of differential variables by using algebraic and differential Gröbner bases.

Example 4.9 (n-strain model [28]). Let

$$f_j = \delta X_j - g_j(X_1, \ldots, X_{2^n-1}, Y_1, \ldots, Y_{n2^{n-1}}, Z_1, \ldots, Z_n)$$

when $j = 1, \ldots, 2^n - 1$,

$$f_j = \delta Y_j - g_j(X_1, \ldots, X_{2^n-1}, Y_1, \ldots, Y_{n2^n-1}, Z_1, \ldots, Z_n)$$

when $j = 2^n, \ldots, n2^{n-1} + 2^n - 1$, and

$$f_j = \delta Z_j - g_j(X_1, \ldots, X_{2^n-1}, Y_1, \ldots, Y_{n2^n-1}, Z_1, \ldots, Z_n)$$

when $j = n2^{n-1} + 2^n, \ldots, n2^{n-1} + 2^n + n - 1$, where the g_j's are polynomials.

$$\{f_j : j = 1, \ldots, n2^{n-1} + 2^n + n - 1\}$$

is a differential Gröbner basis of the differential ideal

$$I = [f_j : j = 1, \ldots, n2^{n-1} + 2^n + n - 1]$$

with respect to σ, but it is not in general a differential Gröbner basis of I with respect to τ.

We want to eliminate the X_j's and the Y_j's.

Let G_{01} be the Gröbner basis of $I = I_{01}$ as an ideal in S_1 with respect to τ_1 and let $G_{01}(Z) = G_{01} \cap R\{Z_1, \ldots, Z_n\}$. Let $I_{02} = (g, \delta(g) : g \in I_{01})$ in the ring S_2.

Let G_{02} be the Gröbner basis of I_{02} with respect to τ_2 and let

$$G_{02}(Z) = G_{02} \cap R\{Z_1, \ldots, Z_n\}.$$

If $G_{02}(Z)$ contains only the differential polynomials in $G_{01}(Z)$ and their derivatives, then stop. Otherwise, let $I_{03} = (g, \delta(g) : g \in I_{02})$ in the ring S_3.

Let G_{03} be the Gröbner basis of I_{03} with respect to τ_3 and let

$$G_{03}(Z) = G_{03} \cap R\{Z_1, \ldots, Z_n\}.$$

If $G_{03}(Z)$ contains only the differential polynomials in $G_{02}(Z)$ and their derivatives, then stop. Otherwise, let $I_{04} = (g, \delta(g) : g \in I_{03})$ in the ring S_4 and so on.

If N is the upper bound given by Macaulay's differential resultant theory, then $N + 1$ is an upper bound for the number of steps.

The same procedure can be used if we take ρ instead of τ.

References

[1] Boulier, F., Lazard, D., Ollivier, F. and Petitot, M.: Representation for the radical of a finitely generated differential ideal. In: *Proc. ISSAC '95*, pp. 158–166. ACM Press, New York (1995).

[2] Boulier, F.: Etude et implantation de quelques algorithmes en algebre differentielle. Ph.D. thesis, Universite de Lille, France (1994).

[3] Boulier, F.: Some improvements of a lemma of Rosenfeld. Preprint, Universite de Lille, France (1996).

[4] Canny, J. and Emiris, J.: An efficient algorithm for the sparse mixed resultant. In: *Applied Algebra, Algebraic Algorithms and Error-correcting Codes* (AAECC-10) (G. Cohen, T. Mora and O. Moreno, eds.), Lecture Notes in Computer Science **673**, pp. 89–104. Springer-Verlag, Berlin Heidelberg (1993).

[5] Canny, J. and Emiris, J.: A subdivision-based algorithm for the sparse resultant. *J. ACM* **47** (2000), 417–451.

[6] Canny, J. and Manocha, D.: Multipolynomial resultant algorithms. *J. Symb. Comput.* **15** (1993), 99–122.

[7] Carrá Ferro, G.: Gröbner bases and differential algebra. In: *Lecture Notes in Computer Science* **356**, pp. 129–140. Springer-Verlag, Berlin Heidelberg (1987).

[8] Carrá Ferro, G.: A resultant theory for the systems of two ordinary algebraic differential equations. *Appl. Alg. Eng. Comm. Comp.* (1997), 539–560.

[9] Carrá Ferro, G.: A resultant theory for ordinary algebraic differential equations. In: *Lecture Notes in Computer Science* **1255**, pp. 55–65. Springer-Verlag, Berlin Heidelberg (1997).

[10] Cox, D., Little, J. and O'Shea D.: *Ideals, Varieties and Algorithms.* Springer-Verlag, New York (1992).

[11] Cox, D., Little, J. and O'Shea D.: *Using Algebraic Geometry.* Springer-Verlag, New York (1998).

[12] Diop, S. and Fliess, M.: On nonlinear observability. In: *Proc. First Europ. Control Conf.*, pp. 152–157. Hermes, Paris (1991).

[13] Diop, S.: Elimination in control theory. *Math. Control Signal Syst.* **4** (1991), 17–32.

[14] Diop, S.: Finite morphisms of differential algebraic varieties and elimination theory. In: *Analysis of Controlled Dynamical Systems — Proc. Conf. 1990*, pp. 193–200. Birkhäuser, Basel Boston (1991).

[15] Diop, S.: Differential algebraic decision methods and some application to system theory. *Theoret. Comput. Sci.* **98** (1992), 137–161.

[16] Fliess, M.: Automatique et corps differentiels. *Forum Math.* **1** (1989), 227–238.

[17] Fliess, M.: Generalized controller canonical forms for linear and nonlinear dynamics. *IEEE Trans. Autom. Control* **35** (1990), 994–1001.

[18] G'elfand, I. M., Kapranov, M. M. and Zelevinski, A. V.: Discriminants, resultants and multidimensional determinants. In: *Mathematics: Theory and Applications.* Birkhäuser, Boston (1994).

[19] Hodge, W. V. D. and Pedoe, D.: *Methods of Algebraic Geometry*, vol. 1. Cambridge University Press, Cambridge (1947).

[20] Hubert, E.: Factorization-free decomposition algorithms in differential algebra. *J. Symb. Comput.* **29** (2000), 641–622.

[21] Hubert, E.: Notes on triangular sets and triangular decomposition algorithms II. In: *Lecture Notes in Computer Science* **2630**, pp. 40–87. Springer-Verlag, Berlin Heidelberg (2003).

[22] Jouanolou, J.-P.: Le formalisme du résultant. *Adv. Math.* **90** (1991), 117–263.

[23] Jouanolou, J.-P.: Formes d'inertieet résultant: un formulaire. *Adv. Math.* **126** (1997), 119–250.

[24] Kolchin, E.: *Differential Algebra and Algebraic Groups.* Academic Press, New York (1973).

[25] Macaulay, F. S.: *The Algebraic Theory of Modular Systems.* Cambridge University Press, Cambridge (1916).

[26] Margaria, G., Riccomagno, E. and White, L. J.: Structural identifiability analysis of some highly structured families of statespace models using differential algebra. *J. Math. Biol.* **49** (2004), 433–454.

[27] Ollivier, F.: Standard bases of differential ideals. In: *Lecture Notes in Computer Science* **508**, pp. 304–321. Springer-Verlag, Berlin Heidelberg (1990).

[28] Riccomagno, E. and White, L. J.: Multi-strain species modeling via differential algebra reduction. Preprint (1997).

[29] Ritt, J. F.: *Differential Algebra*, AMS Coll. Publ. **33**. AMS, New York (1950).

[30] Sturmfels, B.: On the Newton polytope of the resultant. *J. Algebraic Comb.* **3** (1994), 207–236.

[31] Sturmfels, B.: Sparse elimination theory. In: *Computational Algebraic Geometry and Commutative Algebra* (D. Eisenbud and L. Robbiano, eds.), pp. 264–298. Cambridge University Press, Cambridge (1993).

[32] Wang, D.: Geometric theorem proving with existing technology. MEDLAR II Deliverable DII.41P (1993). In: *Proc. ATCM '95* (Singapore, December 18–21, 1995), pp. 561–570.

[33] Weispfenning, V.: Differential-term orders. In: *Proc. ISAAC '93*, pp. 245–253. ACM Press, New York (1993).

[34] Wu, W.-t.: Mechanical derivation of Newton's gravitational laws from Kepler's laws. *Mathematics Mechanization Research Reprints* **1** (1987), 53–61.

[35] Wu, W.-t.: On the foundation of algebraic differential geometry. *Mathematics Mechanization Research Reprints* **3** (1989), 1–26. *Syst. Sci. Math. Sci.* **2**, 289–312.

[36] Van der Waerden, B. L.: *Modern Algebra II* (3rd edn.). F. Ungar Publ. Co., New York (1950).

Giuseppa Carrá Ferro
Department of Mathematics and Computer Science
University of Catania
Viale Andrea Doria 6
95125 Catania, Italy
e-mail: `carra@dmi.unict.it`

Trends in Mathematics: *Differential Equations with Symbolic Computation*, 343–350

On "Good" Bases of Algebraico-Differential Ideals

Wen-tsün Wu

Abstract. The characteristic set method of polynomial equations-solving is naturally extended to the differential case, which gives rise to an algorithmic method for solving arbitrary systems of algebrico-differential equations. The existence of "good bases" of the associated algebrico-differential ideals is also studied in this way. As an illustration of the method, the Devil problem of Pommaret is studied in detail.

Mathematics Subject Classification (2000). Primary 13P10; Secondary 13N10, 35G05.

Keywords. Algebrico-differential equations, (differential) zero-decomposition theorem, integrability differential polynomial, compatibility differential polynomial, Pommaret's Devil problem, "good basis".

1. Introduction

In the seminar DESC held in Beijing, April 14–16, 2004, the present author gave a talk bearing the title "On 'Good Bases' of Polynomial Ideals" [10]. The present paper is an extended form of that talk in extending the notion of "good bases" of polynomial ideals to that of algebrico-differential ideals.

To begin with, let us consider a finite polynomial set PS in the polynomial ring $R = K[x_1, \ldots, x_n]$, K being a coefficient field of characteristic 0. Then there are two important problems to be studied, viz:

Problem P1. Determine the totality of solutions of $PS = 0$ in all conceivable extension fields of K, to be denoted by $\mathrm{Zero}(PS)$ in what follows.

Problem P2. For the ideal $\mathrm{Ideal}(PS)$ with basis PS, determine some kind of *good* basis which will enjoy some *good* properties to be made precise.

We shall show how to solve Problem 1 in the polynomial case in Section 2, explain how to extend the solution to the algebrico-differential case in Section 3, and solve Problem 2 in the polynomial case in Section 4 by using the method developed in Section 3. In Section 5 we shall study Problem 2 in the algebrico-differential case and introduce the notion of probably existing "good basis" for certain algebrico-differential ideals. Finally in Section 6 we shall provide a solution to the Devil problem of Pommaret as an illustrative example.

2. Problem 1 in the Polynomial Case

For Problem P1 the present author has given a method for determining Zero(PS) completely, which may be described briefly as follows.

Arrange the variables $x_1, \ldots, x_n$ in the natural order; then any non-constant polynomial $P \in R$ may be written in the canonical form

$$P = I_0 x_c^d + I_1 x_c^{d-1} + \cdots + I_d,$$

in which all the I_j are either constants or polynomials in $x_1, \ldots, x_{c-1}$ alone with *initial* $I_0 \neq 0$. With respect to *class* c and *degree* d, we may introduce a partial ordering $\prec$ for all non-zero polynomials in R, with non-zero constant polynomials in the lowest ordering. Consider now some polynomial set, which either consists of a single non-zero constant polynomial, or in which the polynomials may be arranged with classes all positive and steadily increasing. We call such polynomial sets *ascending sets*. Then we may introduce a partial ordering $\prec$ among all such ascending sets, with the trivial ones consisting of single non-zero constant polynomials in the lowest ordering. For a finite polynomial set consisting of non-zero polynomials, any ascending set wholly contained in it and of lowest ordering is called a *basic set* of the given polynomial set. A partial ordering among all finite polynomial sets may then be unambiguously introduced according to their basic sets.

For any finite polynomial set $PS \subset R$, consider now the scheme

$$\begin{array}{lllllll} PS = PS^0 & PS^1 & \cdots & PS^i & \cdots & PS^m & \\ \phantom{PS = {}} BS^0 & BS^1 & \cdots & BS^i & \cdots & BS^m = CS & \\ \phantom{PS = {}} RS^0 & RS^1 & \cdots & RS^i & \cdots & RS^m = \emptyset. & \end{array} \tag{S}$$

In this scheme, each BS^i is a basic set of PS^i, each RS^i is the set of non-zero remainders, if any, of the polynomials in $PS^i \setminus BS^i$ with respect to BS^i, and $PS^{i+1} = PS \cup BS^i \cup RS^i$ if RS^i is non-empty. It may be easily proved that the sequences in the scheme should terminate at certain stage m with $RS^m = \emptyset$. The corresponding basic set $BS^m = CS$ is then called a *characteristic set* (abbreviated *char-set*) of the given polynomial set PS. The *zero set* of PS, Zero(PS), which is the collection of common zeros of all the polynomials in PS, is closely connected with that of CS by the *well-ordering principle* in the form

$$\mathrm{Zero}(PS) = \mathrm{Zero}(CS/IP) \cup \mathrm{Zero}(PS \cup \{IP\}),$$

in which IP is the product of all initials of the polynomials in CS and

$$\mathrm{Zero}(CS/IP) = \mathrm{Zero}(CS) \setminus \mathrm{Zero}(IP).$$

Now $PS \cup \{IP\}$ is easily seen to be a polynomial set of lower ordering than PS. If we apply the well-ordering principle to $PS \cup \{IP\}$ and proceed further and further in the same way, we should stop in a finite number of steps and arrive at the following

Zero-Decomposition Theorem. There is an algorithm which may compute, from any finite polynomial set PS and in a finite number of steps, a finite set of ascending sets CS^s with initial-product IP^s such that

$$\text{Zero}(PS) = \bigcup_s \text{Zero}(CS^s/IP^s). \tag{Z}$$

Now all CS^s are ascending sets. Hence all the zero sets Zero(CS^s) and all Zero(CS^s/IP^s) may be considered as well-determined in some natural sense. The formula (Z) gives thus actually an explicit determination of Zero(PS) for all finite polynomial sets PS, which serves for the solving of arbitrary systems of polynomial equations. This solves Problem 1 in the polynomial case.

3. Extension to Algebrico-Differential Systems

The above method of solving arbitrary systems of polynomial equations has been extended to arbitrary systems of *algebrico-differential* equations, either ordinary or partial ones, which will be explained below.

Let $y, u_j, j \in J$, be infinitely differentiable functions in independent variables $X = \{x_1, \ldots, x_n\}$. A polynomial in various derivatives of y and u_j with respect to x_k with coefficients in the differential field of rational functions of X will be called an *algebrico-differential polynomial.* Suppose that we are given a finite set of such polynomials $DPS = \{DP_i \mid i \in I\}$. Let us consider the associated system of partial differential equations of y with u_j supposed known:

$$DPS = 0, \quad \text{or} \quad DP_i = 0, \ i \in I.$$

Our problem is to determine the integrability conditions in terms of x_k, u_j for y to be solvable and in the affirmative case to determine the set of all possible *formal* solutions of y.

Criteria and even algorithmic methods for solving the above problem in some sense were known in quite remote times, for which we may cite in particular the work of C. H. Riquier, M. Janet, and E. Cartan. The method of Riquier and Janet was reformulated by J. F. Ritt in his books [5, 6]. In recent years, J. F. Pommaret has given a systematic *formal intrinsic* way of treatment and published several voluminous treatises. On the other hand, the present author has given an alternative method in following essentially the steps of Riquier and Janet as reformulated by Ritt [7]. The method consists in first extending naturally the notions of ascending sets, basic sets, remainders, etc. in the ordinary case to the present algebrico-differential case. Orderings among all derivatives and then partial orderings may then be successively introduced among all algebrico-differential polynomials, all differential-ascending sets, and finally all systems of algebrico-differential polynomial sets, somewhat analogous to the ordinary case.

For any system *DPS* of algebrico-differential polynomials, we may then form a scheme (dS) analogous to the scheme (S) in the ordinary case as shown below:

$$\begin{array}{lllllllll} DPS = DPS^0 & DPS^1 & \cdots & DPS^i & \cdots & DPS^m & & \\ \ DBS^0 & DBS^1 & \cdots & DBS^i & \cdots & DBS^m & = & DCS \\ \ DRIS^0 & DRIS^1 & \cdots & DRIS^i & \cdots & DRIS^m & = & \emptyset \\ \ DCPS^0 \cup & DCPS^1 \cup & \cdots \cup & DCPS^i \cup & \cdots \cup & DCPS^m & = & DCPS. \end{array} \tag{dS}$$

In the scheme (dS), *DPS* is the given algebrico-differential polynomial set. For each i, DBS^i is a differential basic set of DPS^i. The set $DRIS^i$ is the union of two parts: one is the set of all possible non-zero differential remainders in the sense of Ritt formed from differential polynomials in $DPS^i \setminus DBS^i$ with respect to DBS^i, while the other is the set of integrability differential polynomials formed from certain pairs of differential polynomials in DPS^i, so far they contain actually y or its derivatives. Such pairs may be determined by the notions of *multiplicativity* and *non-multiplicativity* due to Riquier and Janet. On the other hand, those containing no y or its derivatives but containing possibly u_j or their derivatives form a set of *compatibility differential polynomials* whose vanishing gives the *compatibility conditions* under which the given system of equations $DPS = 0$ has solutions. In case $DRIS^i$ is non-empty, the union $DPS \cup DBS^i \cup DRIS^i$ forms the next differential polynomial set DPS^{i+1}.

As in the ordinary case the sequences will terminate at a certain stage m with $DRIS^m = \emptyset$. The corresponding differential basic set $DBS^m = DCS$ is then called a *differential characteristic set* (abbreviated *d-char-set*) of the given differential polynomial set *DPS*. The union *DCPS* of all sets $DCPS^i, i = 1, \ldots, m$, will form the totality of all possible compatibility differential polynomials whose vanishing forms the compatibility conditions to guarantee the existence of solutions of the system of partial differential equations $DPS = 0$.

As in the ordinary case the above will lead finally to the formation of the totality of formal solutions of the given system of algebrico-differential equations under suitable initial data for which we refer to the paper [9].

4. Problem 2 in the Polynomial Case

Let us now consider the particular case for which the differential polynomials in *DPS* are all *linear* with *constant coefficients.* For each tuple of non-negative integers $\mu = (i_1, \ldots, i_n)$, let us write $\|\mu\|$ for $i_1 + \cdots + i_n$ and make the correspondence

$$\text{Partial derivative } \frac{\partial^{\|\mu\|}}{\partial x_1^{i_1} \cdots \partial x_n^{i_n}} \longleftrightarrow \text{Monomial } x_1^{i_1} \cdots x_n^{i_n}.$$

Then the partial differentiation of a derivative with respect to some x_j will correspond to the multiplication of the corresponding monomial with the variable x_j.

In this way a differential polynomial set DPS consisting of only linear differential polynomials with constant coefficients will become, under the above correspondence, a polynomial set PS in the ordinary sense. The scheme (dS) will then be turned into some scheme (W) for PS somewhat of the following form:

$$\begin{array}{llllllll} PS = & PS^0 & PS^1 & \cdots & PS^i & \cdots & PS^m & \\ & WS^0 & WS^1 & \cdots & WS^i & \cdots & WS^m & = WS \\ & IS^0 & IS^1 & \cdots & IS^i & \cdots & IS^m & = \emptyset. \end{array} \tag{W}$$

In the above scheme the WS^i are certain subsets of PS^i enjoying some well-arranged properties and each IS^i consists of remainders of the polynomials in $PS^i \setminus WS^i$ with respect to WS^i as well as those determined from certain pairs of polynomials in WS^i determined by the notions of multiplicativity and non-multiplicativity of Riquier and Janet. The union of WS^i, IS^i and eventually PS^0 will then be PS^{i+1} so far $IS^i \neq \emptyset$. It turns out that the final set WS is a basis of the given ideal Ideal(PS) and possesses many nice properties. It turns out too that this basis WS is just the well-known *Gröbner basis* of the given ideal Ideal(PS), which may now be found in some way different from the original one of B. Buchberger. Moreover, many known properties connected with the Gröbner basis which are dispersed in the literature have been proved in some simple and unanimous manner. We refer to the paper [8] for details. In particular, we have the following nice property of Gröbner bases which solves the important *membership problem.*

MP. A polynomial P in R belongs to the ideal Ideal(PS) if and only if the remainder of P with respect to the Gröbner basis of PS is 0.

It turns out that the Russian mathematician V. P. Gerdt has also found the Gröbner basis of a polynomial ideal essentially in the same way as above. He has used an alternative name of *involutive basis* and has given also a detailed analysis of various possible notions of *multiplicativity* and *non-multiplicativity* due to Riquier, Janet, Thomas, and Gerdt himself. For more details we refer to the paper [2] by Gerdt. At this point the author would like to express his hearty thanks to D. Wang who pointed out to the author the above-mentioned work of Gerdt.

5. Problem 2 in the Algebrico-Differential Case

Let us consider now Problem 2 of algebrico-differential systems in the general case. Let DPS be an arbitrary finite algebrico-differential polynomial set as before. The problem is to find some *finite* differential basis of the differential ideal dIdeal(DPS) that enjoys some nice properties as the Gröbner basis in the polynomial case and solves in particular the corresponding membership problem. It is natural to extend the method of Buchberger in the polynomial case to the present algebrico-differential case. Unfortunately, in 1986 G. Carrá Ferro showed in a well-known remarkable paper [1] that such a *finite* differential Gröbner basis does not exist in general. In later years the possibility of existence of such *finite* differential Gröbner bases was widely studied, notably by F. Ollivier (see [3]).

Now let us try to deal with this problem by our method explained in Section 3. Consider again the diagram (dS). We suppose naturally that all the compatibility conditions are verified. It is clear from the constructions that

$$\mathrm{dIdeal}(DPS) = \mathrm{dIdeal}(DPS^0) = \mathrm{dIdeal}(DPS^1) = \cdots = \mathrm{dIdeal}(DPS^m).$$

Suppose that for the final d-char-set *DCS* the following condition GC is verified.

GC. The initials and separants of the algebrico-differential polynomials in *DCS* are all constants.

By the differential remainder theorem of Ritt, it is readily seen that *DCS* is a differential basis of dIdeal(*DPS*) and a differential polynomial *DP* belongs to dIdeal(*DPS*) if and only if the differential remainder of *DP* with respect to *DCS* is 0.

It is thus seen that under the condition GC the final d-char-set *DCS* will serve as a *finite* differential basis of dIdeal(*DPS*), which solves the membership problem in a simple way. The condition GC is clearly less stringent than the condition of linearity and coefficients-constancy, which leads to the usual Gröbner basis in the polynomial case. On the other hand, the verification of the condition GC can be seen only after lengthy computations of d-char-set. In any way we may lay down the following definition.

Definition. An algebrico-differential polynomial set *DPS* verifying condition GC is called a *good set* and the corresponding differential basis formed by the final d-char-set is called a *good basis* of dIdeal(*DPS*).

In view of the significance and also the weakness of the above notion of *good basis*, we suggest now some problems for further study.

Problem 3. Try to find some *intrinsic* conditions for an algebrico-differential polynomial set to be "good" directly from the given set without passing to the final d-char-set.

Problem 4. Try to weaken the condition GC such that the differential ideal generated by the given algebrico-differential polynomials still has a *finite* differential basis that verifies some simple membership condition.

Problem 5. Compare our condition GC with other known conditions introduced by Ollivier and other authors.

6. Example: Pommaret's Devil Problem

To illustrate our treatment of algebrico-differential polynomial sets, let us consider the *Devil problem* of Pommaret, given for example in his paper [4]. We shall treat this Devil problem in detail by our method as exhibited in Section 3. For this purpose let us reproduce the statement of the Devil problem below.

Devil Problem. Let u, v, y be three functions of the Cartesian coordinates x_1, x_2, x_3 in Euclidean spaces related by the following two partial differential equations

$$\begin{aligned} DP_1 &= \partial_{200}y - x_2\partial_{002}y - u = 0, \\ DP_2 &= \partial_{020}y - v = 0, \end{aligned} \tag{D}$$

with the corresponding algebrico-differential polynomial set $DPS = \{DP_1, DP_2\}$. Note that here and below we use the notation $\partial_{i_3i_2i_1}$ for the partial derivative $\dfrac{\partial^{i_3+i_2+i_1}}{\partial x_3^{i_3} x_2^{i_2} x_1^{i_1}}$.

The functions u, v are supposed to be known. The problem consists in finding the compatibility conditions to be satisfied by u and v in order to insure the existence of solutions for y and to see whether the given algebrico-differential polynomial set DPS is a *good* one or not.

It turns out that our procedure ends at the stage $m = 2$ so that the scheme (dS) in the present case becomes

$$\begin{array}{llllll} DPS = DPS^0 & & DPS^1 & & DPS^2 & \\ DBS^0 & & DBS^1 & & DBS^2 & = DCS \\ DRIS^0 & & DRIS^1 & & DRIS^2 & = \emptyset \\ DCPS^0 & \cup & DCPS^1 & \cup & DCPS^2 & = DCPS. \end{array} \tag{dS$'$}$$

The final d-char-set DCS is found to consist of 4 algebrico-differential polynomials

$$\begin{aligned} G_1 &= \partial_{004}y - z, \\ G_2 &= \partial_{012}y - w, \\ G_3 &= \partial_{020}y - v, \\ G_4 &= \partial_{200}y - x_2\partial_{002}y - u. \end{aligned} \tag{G}$$

In (G), w and z are given by

$$\begin{aligned} w &= \tfrac{1}{2}(\partial_{200}v - x_2\partial_{002}v - \partial_{020}u), \\ z &= \partial_{200}w - \partial_{012}u - x_2\partial_{002}w. \end{aligned}$$

The *compatibility conditions* are found to be $A = 0$ and $B = 0$ with

$$\begin{aligned} \tfrac{1}{2}A &= \partial_{010}w - \partial_{002}v, \\ B &= \partial_{400}w - 2x_2\partial_{202}w + x_2^2\partial_{004}w - \partial_{212}u + x_2\partial_{014}u - \partial_{004}u. \end{aligned} \tag{CC}$$

It may also be shown further that the two compatibility conditions $A = 0$ and $B = 0$ are not independent of each other. They are in fact connected by the differential identity

$$\partial_{400}A - 2x_2\partial_{202}A + x_2^2\partial_{004}A - 2\partial_{010}B = 0.$$

Naturally all the above were found by Pommaret by his method and in his notations, which are different from ours.

Now we see that the d-char-set DCS consisting of the 4 algebrico-differential polynomials $G_1, \ldots, G_4$ verifies the condition GC so that the given algebrico-differential polynomial set DPS is a *good* one with a *good* basis for the corresponding dIdeal(DPS).

References

[1] G. Carrá Ferro: Gröbner Bases and Differential Ideals. In: *Proc. AAECC-5*, Menorca, Spain, pp. 129–140. Springer-Verlag, Berlin Heidelberg (1987).

[2] V. P. Gerdt, Y. A. Blinkov: Minimal Involutive Bases. *Math. Comput. Simul.* **45**: 543–560 (1998).

[3] F. Ollivier: Standard Bases of Differential Ideals. In: *Proc. AAECC-8*, Tokyo, Japan, LNCS **508**, pp. 304–321. Springer-Verlag, Berlin Heidelberg (1990).

[4] J. F. Pommaret, A. Haddack: Effective Methods for Systems of Algebraic Partial Differential Equations. In: *Effective Methods in Algebraic Geometry* (T. Mora, C. Traverso, eds.), pp. 441–426. Birkhäuser, Basel Boston (1991).

[5] J. F. Ritt: *Differential Equations from the Algebraic Standpoint.* Amer. Math. Soc., New York (1932).

[6] J. F. Ritt: *Differential Algebra.* Amer. Math. Soc., New York (1950).

[7] W.-t. Wu: On the Foundation of Algebraic Differential Geometry. *Syst. Sci. Math. Sci.* **2**: 289–312 (1989).

[8] W.-t. Wu: On the Construction of Gröbner Basis of a Polynomial Ideal Based on Riquier–Janet Theory. *Syst. Sci. Math. Sci.* **4**: 193–207 (1991) [reprinted in this volume].

[9] W.-t. Wu: On Algebrico-Differential Equations-Solving. *J. Syst. Sci. Complex.* **17**(2): 153–163 (2004).

[10] W.-t. Wu: On "Good Bases" of Polynomial Ideals. Presented at the DESC seminar, Beijing, April 14–16, 2004.

Wen-tsün Wu
Key Laboratory of Mathematics Mechanization
Academy of Mathematics and System Sciences
Academia Sinica
Beijing 100080, China
e-mail: wtwu@mmrc.iss.ac.cn

Trends in Mathematics: *Differential Equations with Symbolic Computation*, 351–368

On the Construction of Groebner Basis of a Polynomial Ideal Based on Riquier–Janet Theory

Wenjun Wu (Wen-tsün Wu)

Abstract. As a consequence of a previous study of algebraic differential geometry ([see WU1]) there may be associated to certain special kinds of differential ideals some well-behaved basis enjoying some well-behaved properties. If the differential ideals are further specialized so that they correspond to ordinary polynomial ideals then such a well-behaved basis will become the usual Groebner basis of the polynomial ideals while the latter is not known for differential ideals.

Mathematics Subject Classification (2000). Primary 13P10; Secondary 13N10.

Keywords. Polynomial ideal, differential ideal, well-behaved basis, Groebner basis.

0. Introduction

Riquier and Janet have created a theory of PDE which has been further developed by Ritt and Thomas and is closely related to the corresponding theory of E. Cartan. Based on such a theory the author has shown in a previous paper [WU1] how to construct a d-char-set *DCS* of a d-polset *DPS* for which their d-zero-sets are closely connected according to the following decomposition formula:

$$\text{d-Zero}(DPS) = \text{d-Zero}(DCS/J) + \text{SUM}_i\, \text{d-Zero}(DPS_i).$$

In the formula J is the product of all initials and separants of the d-pols in *DCS*, and DPS_i are the enlarged d-polsets of *DPS* in adjoining to it one of such initials

The present paper is partially supported by NSFC Grant JI85312. It is reprinted, with permission and with minor editorial changes, from *Systems Science and Mathematical Sciences* 4/3 (1991): 193–207, originally received on September 4, 1990.

or separants. In the particular case for which all these initials and separants are non-zero constants in the basic d-field, the above formula becomes simply

$$\text{d-Zero}(DPS) = \text{d-Zero}(DCS).$$

Moreover, denoting the differential ideal with *DPS* as a basis by d-Ideal(*DPS*), we see from the construction of *DCS* and theorems proved in that paper that *DCS* is also a basis of this ideal, or

$$\text{d-Ideal}(DPS) = \text{d-Ideal}(DCS).$$

Furthermore, this basis *DCS* possesses the following well-behaved property:

A d-pol *DP* will belong to the differential ideal d-Ideal(*DPS*) if and only if the d-remainder of *DP* w.r.t. the basis *DCS* is 0:

$$\text{d-Remdr}(DP/DCS) = 0.$$

For this reason we shall call the corresponding d-char-set *DCS*, in the above particular case, a *well-behaved basis* of the differential ideal d-Ideal(*DPS*) with *DPS* as a given basis.

Let us consider now the further specialized d-polset *DPS* with the following restrictions:

1. The basic d-field is one with trivial differentiations so that it is just an ordinary field of characteristic 0.

2. The independent variables are still $X_1, \ldots, X_n$ while there is only one dependent Y.

3. Each d-pol *DP* in the d-polset *DPS* is of the form

$$DP = \text{SUM}_t C_t * \text{DER}_t Y,$$

in which t runs over a finite set of n-tuples of nonnegative integers and C_t are non-zero constants in the basic field.

We are thus in the situation of a system $DPS = 0$ of linear PDE with constant coefficients. Now to each partial derivative $\text{DER}_t Y$ we may make a corresponding monomial $X^t = X_1^{i_1} * \cdots * X_n^{i_n}$ in which $(i_1, \ldots, i_n)$ is the n-tuple t. Under the correspondence the d-pols will then be turned into ordinary pols in $X_1, \ldots, X_n$ with coefficients in an ordinary field of characteristic 0. The above theory will then give a well-behaved basis of an ordinary polset *PS*. It turns out that this well-behaved basis is, in the present non-differential case, just the usual Groebner basis of the corresponding polynomial ideal Ideal(*PS*). This offers thus an alternative method of constructing a Groebner basis of a polynomial ideal different from that of Buchberger.

In the present paper we shall consider the last case of ordinary polsets alone. Our exposition will be so given that it is independent of the Riquier-Janet theory and the previous paper [WU1], though it will follow closely the steps exhibited in that paper. In studying the properties of the well-behaved basis of a polynomial ideal introduced in this way it will follow that this basis is just the Groebner basis of that ideal. We prove now several well-known beautiful properties of the

Groebner basis in a way along the line of the thoughts of the previous paper based on Riquier-Janet theory. The proofs are thus somewhat different from the known ones scattered in the literature. These proofs may in fact be carried over to the differential case as stated above for the well-behaved basis, while the Groebner basis is undefined in that case. We remark in passing that our theory will give a unique expression for an arbitrary pol w.r.t. such a basis of a polynomial ideal, while for the usual theory of the Groebner basis such an expression is unique only modulo the basis in some way. Finally, we give a concrete example for which the Groebner basis is determined by the present method in using the REDUCE implemented in our machine SUN3/140. Further examples are yet to be studied and a complexity study of the present method is required.

1. Tuples of Integers

Let n be a positive integer fixed throughout the present paper.

DEF. An ordered sequence of n non-negative integers

$$t = (I_1, \dots, I_n)$$

is called an n-tuple or simply a *tuple*. I_i is then called the i-th *coordinate* of t, to be denoted by

$$\mathrm{COOR}_i(t) = I_i.$$

DEF. The particular tuple with all coordinates $= 0$ will be called the 0-*tuple*, to be denoted as 0.

Notation. For any tuple u and any integer $i \geq 1$ and $\leq n$, the tuple u' with

$$\mathrm{COOR}_i(u') = \mathrm{COOR}_i(u) + 1,$$
$$\mathrm{COOR}_j(u') = \mathrm{COOR}_j(u), \quad j \neq i,$$

will be denoted by ui or iu.

DEF. For any two tuples u and v, we say u is a *multiple* of v or v is a *divisor* of u, if

$$\mathrm{COOR}_i(u) \geq \mathrm{COOR}_i(v), \quad i = 1, \dots, n.$$

We write then

$$u \gg v \ \text{ or } \ v \ll u.$$

DEF. For any two tuples u and v, their *product* $uv = vu$ is the tuple with

$$\mathrm{COOR}_i(uv) = \mathrm{COOR}_i(u) + \mathrm{COOR}_i(v), \quad i = 1, \dots, n.$$

We introduce now an ordering among all the n-tuples according to the following

DEF. For any two tuples u and v we say that u is *higher than* v or v is *lower than* u if there is some $k > 0$ and $\leq n$ such that

$$\mathrm{COOR}_i(u) = \mathrm{COOR}_i(v), \quad i > k,$$
$$\mathrm{COOR}_k(u) > \mathrm{COOR}_k(v).$$

We write then

$$u > v \text{ or } v < u.$$

DEF. A set of tuples T is said to be *autoreduced* if no t in T is a multiple of anther t' in T.

The following two lemmas are already known or easily deduced from known results.

Lemma 1. *Any sequence of tuples steadily decreasing in order is finite.*

Lemma 2. *Any autoreduced set of tuples is finite.*

DEF. For any finite set of tuples T, the *maximum* of T, to be denoted by Max(T), is the tuple defined by

$$\begin{aligned}&\mathrm{Max}(T) = n\text{-tuple}(\mathrm{MAX}_1(T), \ldots, \mathrm{MAX}_n(T)), \quad \text{with}\\ &\mathrm{MAX}_i(T) = \mathrm{Max}\{\mathrm{COOR}_i(t)/\ t \in T\}.\end{aligned}$$

DEF. For any finite set of tuples T, the *completion* of T, to be denoted by Comp(T), is the set of tuples defined by

$$\mathrm{Comp}(T) = \{u/\ u \ll \mathrm{Max}(T) \text{ and } u \gg t \text{ for some } t \text{ in } T\}.$$

DEF. For any finite set of tuples T and any tuple $t \ll \mathrm{Max}(T)$, the integer $i\,(\geq 1, \leq n)$ is called a *multiplier* of t w.r.t. T if

$$\mathrm{COOR}_i(t) = \mathrm{MAX}_i(T).$$

Otherwise i is called a *non-multiplier* of t w.r.t. T. In that case we have

$$\mathrm{COOR}_i(t) < \mathrm{MAX}_i(T).$$

Notation. For any finite set of tuples T and any tuple $t \ll \mathrm{Max}(T)$, we shall set

$$\begin{aligned}&\mathrm{Mult}(t/T) = \text{set of all multipliers of } t \text{ w.r.t. } T,\\ &\mathrm{Nult}(t/T) = \text{set of all non-multipliers of } t \text{ w.r.t. } T.\end{aligned}$$

DEF. For $t \ll \mathrm{Max}(T)$, the set of all multiples tu of t with

$$\mathrm{COOR}_i(u) = 0 \text{ for } i \text{ in } \mathrm{Nult}(t/T)$$

is called the *total multiple set* of t w.r.t. T, to be denoted by

$$\mathrm{TMU}(t/T) = \{tu/\ \mathrm{COOR}_i(u) = 0 \text{ for } i \text{ in } \mathrm{Nult}(t/T)\}.$$

Lemma 3. Let T be a finite set of tuples. For any tuple v there is a unique tuple $t \ll \mathrm{Max}(T)$ such that v is in TMU(t/T). Moreover, if v is a multiple of some tuple in T, then t is in Comp(T).

Proof. t is determined as $\mathrm{COOR}_i(t) = \mathrm{Min}(\mathrm{COOR}_i(v), \mathrm{MAX}_i(T))$. □

Tuple-decomposition Theorem. *For any finite tuple set T the totality of tuples each of which is a multiple of some tuple in T is the disjoint union of sets* TMU(t/T) *with t running over* Comp(T).

Proof. This follows directly from Lemma 3. □

We shall now introduce an ordering in the totality of autoreduced sets as follows.

Let us consider any two autoreduced sets AS and AS' with tuples arranged in increasing order:

$$(AS): \quad t_1 < t_2 < \cdots < t_r,$$
$$(AS)': \quad t'_1 < t'_2 < \cdots < t'_s.$$

DEF. The autoreduced set (AS) is said to be *higher than* the autoreduced set $(AS)'$, or $(AS)'$ *lower than* (AS), if either of the two following cases holds true:

(a) There is some $k \leq r$ and $\leq s$ such that

$$t_i = t'_i \text{ for } i < k, \text{ while } t_k > t'_k.$$

(b) $r < s$ and $t_i = t'_i$ for $i \leq r$.

In notation, we shall set then

$$(AS) > (AS)', \text{ or } (AS)' < (AS).$$

Lemma 4. *Any sequence of autoreduced sets steadily decreasing in order is finite.*

Proof. Let the sequence be

$$(S): \quad AS_1 > AS_2 > \cdots$$

and suppose the contrary that it is infinite. For each autoreduced set AS_i let its tuples be arranged in increasing order. By Lemma 1 the sequence $as_{11}, as_{21}, \ldots, as_{i1}, \ldots$ of which as_{i1} is the first tuple of AS_i should consist of the same tuple, say t_1, from a certain stage onwards. Denote the corresponding infinite sequence of autoreduced sets from that stage onwards by

$$(S1): \quad AS_{11} > AS_{12} > \cdots.$$

Again by Lemma 1 the sequence of second tuples in AS_{1i}, should consist of the same tuple, say t_2, from a certain stage onwards. Denote the corresponding infinite sequence of autoreduced sets from that stage onwards by

$$(S2): \quad AS_{21} > AS_{22} > \cdots.$$

The above reasoning can be repeated indefinitely so that we get an infinite sequence of tuples

$$(T): \quad t_1 < t_2 < \cdots,$$

which is clearly an autoreduced set. This contradicts however Lemma 2 and hence (S) is finite. □

From the very definition of the ordering we have also

Lemma 5. *Let T be an autoreduced set and u be a tuple which is not a multiple of any tuple in T. Let T' be the autoreduced set obtained by adjoining u to T and then removing all tuples in T which are multiples of u. Then T' is of lower order than T.*

2. Well-Arranged Basis of a Polynomial Ideal

Henceforth throughout the paper there will be fixed an integer n, a set of variables $X_1, \ldots, X_n$, and a field K of characteristic 0. By a *pol* will then be meant, unless otherwise stated, a polynomial in $K[X_1, \ldots, X_n]$.

By a *monom* is meant a power-product in X_i of the form

$$X^t = X_n^{i_n} * \cdots * X_j^{i_j} * \cdots * X_1^{i_1},$$

in which the tuple $t = (i_1, \ldots, i_n)$ will be called the *degree-tuple* of the monom X^t. Any non-zero pol P can then be written in the unique *normal form*

$$P = A_1 * X^{t_1} + A_2 * X^{t_2} + \cdots + A_r * X^{t_r},$$

with A_i non-zero in K, and the degree-tuples t_i in decreasing order, viz.

$$t_1 > t_2 > \cdots > t_r.$$

We call $A_1 * X^{t_1}, X^{t_1}, A_1$, and t_1 resp. the *leading term*, the *leading monom*, the *leading coefficient*, and the *leading degree-tuple*, to be denoted resp. by

$$\text{Lterm}(P),\ \text{Lmonom}(P),\ \text{Lcoef}(P),\ \text{and Ldeg}(P).$$

DEF. For two non-zero pols P_1 and P_2, P_1 is said to be *higher than, lower than, or incomparable to* P_2 according as whether the leading degree-tuple of P_1 is *higher than, lower than, or identical to* that of P_2. In notation, we shall write resp.

$$P_1 > P_2,\ P_1 < P_2,\ \text{and } P_1 <> P_2.$$

DEF. For a finite polset *PS* of non-zero pols the set of leading degree-tuples of pols in *PS* will be called the *degree-tuple-set* of *PS* to be denoted by DTS(*PS*).

DEF. A finite polset *PS* of non-zero pols is said to be *autoreduced* if its degree-tuple-set is autoreduced.

DEF. Let *AS* be an autoreduced polset of non-zero pols and $T = \text{DTS}(AS)$ be its degree-tuple-set. A non-zero pol P is said to be *reduced* w.r.t. *AS* if for each term in P, the corresponding degree-tuple is not a multiple of any tuple in T. The autoreduced *AS* itself is said to be *reduced* if each pol of *AS* is reduced w.r.t. the autoreduced polset formed from *AS* by removing that pol.

For any autoreduced polset *AS* consisting of non-zero pols F_i there may be different ways of putting P into a form

$$P = \text{SUM}_i E_i * F_i + R, \tag{2.1}$$

in which E_i, R are pols and R, if not zero, is reduced w.r.t. *AS*. We shall now proceed in the following way to get a unique R from P as follows. Write P in the normal form. Let $c * X^t$, c in K, be the non-zero term in P, if it exists, such that t is of highest order with t a multiple of some tuple u in T, u being chosen to be the highest one in T. Write $t = uv$ and let the pol in *AS* having u as its leading-degree-tuple be

$$F_i = a * X^u + F_i',$$

with $a * X^u$ as its leading term, $a \neq 0$ being in K. Set

$$P_1 = P/c - X^v * F_i/a.$$

Then P_1 is such a pol that the term of highest order in P_1 having its degree-tuple t_1 as a multiple of some tuple in T, if it exists, is of lower order than t above. We can then apply the same procedure to P_1 as above to get a pol P_2. The procedure can be continued until we get a pol P_s which is reduced w.r.t. AS. This pol P_s will then be the R required.

DEF. The unique pol R reduced w.r.t. the given autoreduced set AS got from P in the above manner will be called the *rest* of P w.r.t. AS , to be denoted by

$$R = \text{Rest}(P/AS).$$

DEF. The autoreduced polset AS is said to be *higher than* the autoreduced polset AS', or AS' *lower than* AS if

$$T = \text{DTS}(AS) > T' = \text{DTS}(AS').$$

Given an arbitrary finite polset PS of non-zero pols let us form now a scheme (SA) below:

$$\begin{array}{llll} PS = PS_0 & PS_1 & \cdots & PS_r \\ AS_0 & AS_1 & \cdots & AS_r \\ RS_0 & RS_1 & \cdots & RS_r = \text{Empty}. \end{array} \tag{SA}$$

The scheme is formed in the following manner:

For each i AS_i is an autoreduced polset with pols chosen from PS_i such that the degree-tuple of any remaining pol in PS_i is a multiple of the degree-tuple of some pol in AS_i. Each RS_i is then the polset of all non-zero rests, if it exists, of the pols in $PS_i - AS_i$ w.r.t. AS_i. The polset PS_{i+1} is just the union of the previous AS_i and RS_i:

$$PS_{i+1} = AS_i + RS_i.$$

From the construction we see by Lemma 5 of Sect. 1 that the autoreduced sets AS_i are steadily decreasing in order:

$$AS_0 > AS_1 > \cdots .$$

By Lemma 4 of Sect. 1 the sequence is finite so that the procedure has to stop at a certain stage with its corresponding rest-set RS_r = Empty as shown in the diagram (SA).

Theorem 1. *The final autoreduced polset* AS_r *in the scheme* (SA) *forms a basis for the ideal* Ideal(PS) *with* PS *as a basis. In other words,*

$$\text{Ideal}(PS) = \text{Ideal}(AS_r).$$

Proof. Let AS_0 consist of pols P_i and the other pols in PS_0 be Q_j so that Ideal(PS) has a basis consisting of pols P_i and Q_j. Let $R_j = \text{Rest}(Q_j/AS_0)$. Then by definition of rest, it is clear that the ideal Ideal(PS_0) has also a basis consisting of pols P_i and those R_j which are non-zero, or

$$\text{Ideal}(PS) = \text{Ideal}(PS_0) = \text{Ideal}(AS_0 + RS_0) = \text{Ideal}(PS_1).$$

In the same way we have

$$\text{Ideal}(PS_1) = \text{Ideal}(PS_2) = \cdots = \text{Ideal}(PS_r).$$

Hence Ideal(PS) = Ideal(PS_r) = Ideal(AS_r) as to be proved. □

DEF. The final autoreduced polset AS_r in the scheme (SA) will be called a *well-arranged basis* of the ideal Ideal(PS).

3. Well-Behaved Basis of a Polynomial Ideal

Let AS be an autoreduced polset with degree-tuple set T. For any tuple u in Comp(T) let $u = tv$ with t the highest tuple in T which is a divisor of u. Let F_t be the pol in AS with t as its degree-tuple and let us set $H_u = X^v * F_t$. In particular, if u is itself in T, then $u = t$ and v is the 0-tuple so that H_u is just the pol F_t of AS.

DEF. The pol H_u defined above will be called the *completed pol* of AS *relative to* u. The polset consisting of all such completed pols will be called the *completed polset* of AS.

DEF. A product of the form $M * H_u$ in which H_u is the completed pol of AS relative to u in Comp(T), and M a monom X^w for which each i with $\text{COOR}_i(w) \neq 0$ is a multiplier of u will be called an *M-product* of AS.

DEF. A finite linear combination of M-products of AS with coefficients in K will be called an *M-pol* of AS.

Theorem 2. *Any pol P can be written uniquely in the form*

$$P = MP + N, \tag{3.1}$$

in which MP is an M-pol of AS and N is reduced w.r.t. AS.

Proof. Suppose that P is not reduced w.r.t. AS. Then in P there will be a term $a * X^u$ of highest order with u a multiple of some tuple in T, $a \neq 0$ being in K. By Lemma 3 of Sect. 1 there is a unique t in Comp(T) with $u{=}vt$ such that each i with $\text{COOR}_i(v) \neq 0$ is a multiplier of t. Let H_t be the completed pol of AS relative to t with leading term $\text{Lterm}(H_t) = b * X^t$. Set

$$P_1 = P/a - X^v * H_t/b,$$

or

$$P = c_1 * MP_1 + b_1 * P_1, \quad (c_1 = a/b,\, b_1 = a)$$

with $MP_1 = X^v * H_t$ an M-product. If P_1 is not reduced w.r.t. AS, then there will be a term $a_1 * X^{u_1}$ of highest order in P_1 with u_1 a multiple of some tuple in T and u_1 is of lower order than u. Apply now the preceding procedure to P_1 and we get a pol P_2 so that $P_1 = c_2 * MP_2 + b_2 * P_2$, with b_2, c_2 in K and MP_2 an M-product of lower order than MP_1. The procedure can be continued to get pols P_3, etc. until we arrive at some pol P_r which is reduced w.r.t. AS. We may then write P in the form (3.1) with MP an M-pol and $N = b_r * P_r$ reduced w.r.t. AS as required. That the decomposition of form (3.1) is unique follows also easily from Lemma 3 of Sect. 1. □

DEF. The pols MP and N in (3.1) will be called resp. the *M-part* and the *N-part* of the pol P w.r.t. AS.

Consider now any u in Comp(T) with corresponding completed pol H_u and any non-multiplier i of u. Then $ui = v$ is also in Comp(T) and the decomposition of $X_i * H_u$ into the M- and N-parts can be put in the form

$$X_i * H_u = a * H_v + MP_{ui} + N_{ui}, \tag{3.2}$$

in which $a \neq 0$ is in K, MP_{ui} is an M-pol with each M-product in it of lower order than H_v or $X_i * H_u$, and N_{ui} is the N-part of $X_i * H_u$. Note that N_{ui} is reduced w.r.t. AS. Owing to its importance we shall lay down the following

DEF. The N-part N_{ui} of pol $X_i * H_u$ in (3.2) will be called the *N-pol* of AS relative to the tuple u in Comp(T) and the non-multiplier i of u.

Consider now a finite polset PS and let us form the scheme (SB) below:

$$\begin{array}{llll} PS = PS_0 & PS_1 & & PS_s \\ WS_0 & WS_1 & \cdots & WS_s \\ NS_0 & NS_1 & \cdots & NS_s = \text{Empty}. \end{array} \tag{SB}$$

The scheme is formed in the following way:

For each i WS_i is a well-arranged basis of the ideal Ideal(PS_i), determined from PS_i as in Sect. 2 with scheme (SA) applied to PS_i, and NS_i is the set of all non-zero N-pols of WS_i, if it exists. Finally, the polset PS_{i+1} is the union of the preceding sets WS_i and NS_i, or

$$PS_{i+1} = WS_i + NS_i.$$

As in the case of scheme (SA), the sequence of autoreduced sets WS_i is steadily decreasing in order so that the above procedure will end in a certain stage with corresponding NS_s = Empty as shown in the diagram (SB).

Theorem 3. *The final polset WS_s in the scheme* (SB) *is a basis of the ideal* Ideal(PS), *or*

$$\text{Ideal}(PS) = \text{Ideal}(WS_s).$$

Proof. By Theorem 1 of Sect. 2 we have Ideal(PS_0) = Ideal(WS_0). Now each pol N in NS_0 is the N-part of some pol $X_i * H_u$ with H_u the completed pol of WS_0 relative to the tuple u in Comp(T_0) where T_0 is the degree-tuple-set of WS_0 and i a non-multiplier of u so that $X_i * H_u = MP + N$ with MP an M-pol of WS_0. As both H_u and MP are clearly pols in the ideal Ideal(WS_0), the same is for N. Hence

$$\text{Ideal}(PS_0) = \text{Ideal}(WS_0) = \text{Ideal}(WS_0 + NS_0) = \text{Ideal}(PS_1).$$

Proceeding further in the same way we get then successively

$$\text{Ideal}(PS) = \text{Ideal}(PS_1) = \cdots = \text{Ideal}(PS_s) = \text{Ideal}(WS_s),$$

as to be proved. □

DEF. The final autoreduced polset WS_s in the scheme (SB) will be called a *well-behaved basis* of the ideal Ideal(PS).

In the next section it will be shown that the notion of well-behaved set coincides with the usual notion of Groebner basis.

4. Identification of Well-Behaved Basis with Groebner Basis

Consider any ideal ID for which the well-behaved basis, say WB, has been determined as in Sect. 3 so that ID = Ideal(WB).

Theorem 4. *Any pol in the ideal ID is an M-pol of its well-behaved basis WB, or the N-part of any such pol is* 0.

Proof. Let T be the degree-tuple set of WB. For any u in Comp(T) let H_u be the corresponding completed pol. It is enough to prove that any product of the form $M * H_u$ with M a monom and u in Comp(T) is an M-pol. We shall prove this by induction on the order of $M * H_u$ as well as on the number of X's in the monom M as follows.

If each i for which X_i appears in the monom M is a multiplier of u, then $M * H_u$ is already an M-pol and nothing is to be proved. Suppose therefore $M = M' * X_i$ with i a non-multiplier of u. As WB is the well-behaved basis of the ideal, the N-pol relative to u and i is 0 so that (3.2) of Sect. 3 may be written as

$$X_i * H_u = a * H_v + MP, \tag{4.1}$$

in which $v = ui$, and MP is an M-pol of lower order than H_v or $X_i * H_u$. It follows that $M * H_u = a * M' * H_v + M' * MP$, of which $M' * MP$ is of lower order than $M * H_u$ and M' has a smaller number of X's than M. By induction $M' * H_v$ and each term in $M' * MP$ are M-pols and so is $M * H_u$. The theorem is thus proved. □

Theorem 5. *The rest of any pol P w.r.t. the well-behaved basis WB coincides with the N-part of P w.r.t. WB.*

Proof. The rest is determined as the pol R in $P = \mathrm{SUM}_k C_k * W_k + R$, in which W_k are the pols in WB, C_k are pols too, and R is reduced w.r.t. WB. By Theorem 4 $\mathrm{SUM}_k C_k * W_k$ is an M-pol so that R is the N-part of P, as to be proved. □

From Theorems 4 and 5 we get the following

Theorem 6. *A pol P belongs to an ideal ID if and only if its rest w.r.t. the well-behaved basis WB of ID is* 0*:*

$$P \in ID \iff \mathrm{Rest}(P/WB) = 0.$$

The previous results may be further put into a strengthened form as follows.

Theorem 7. *A well-behaved basis WB with degree-tuple-set T of an ideal ID possesses the following well-behaved property:*

Any pol P in $K[X_1, \dots X_n]$ has a unique expression

$$P = \mathrm{SUM}_u a_u * M_u * H_u + N, \tag{4.2}$$

in which H_u are completed H-pols with u running over the completion Comp*(T) of T, M_u are monoms in these X_i with each i a multiplier of u, a_u are constants in K, and N is reduced w.r.t. WB. Moreover, P is in the ideal ID if and only if $N = 0$.*

From the unique expression (4.2) for any pol in *ID* w.r.t. *WB* we get immediately the following theorem due to Macaulay, cf. [M]:

Theorem 8. *The Hilbert function of an ideal is completely determined by the degree-tuple-set of a well-behaved basis of the ideal.*

Theorem 9. *Let the well-behaved basis WB of an ideal ID consist of the pols $W_1, \dots, W_r$. For any completed pol H_u of WB and any non-multiplier i of WB w.r.t. u let us rewrite* (4.1) *in the form*

$$\mathrm{SUM}_i S_{uij} * W_j = 0. \tag{4.3}$$

Then the sets $S_{ui} = (S_{ui1}, \dots, S_{uir})$ form a basis of the linear space of possible solutions $(S_1, \dots, S_r)$ in pols for the syzygy equation

$$\mathrm{SUM}_j S_j * W_j = 0. \tag{4.4}$$

Proof. Consider any solution of equation (4.4) in pols S_j. Denote the left-hand side of (4.4) by S. Then S is a pol belonging to the ideal *ID* with a well-behaved basis *WB*. From the proof of Theorem 4 we see that S can be shown to be 0 by successive reductions in the form of (4.1) or (4.3). Hence $S = 0$ is a consequence of equations (4.3) or S is a linear combination of S_{ui} with pols as coefficients, as to be proved. We remark only that the solutions S_{ui} are not necessarily independent ones.

If the ideal *ID* is given a basis $F_1, \dots F_m$, then each F_i is a linear combination with pol-coefficients of W_j in the well-behaved basis *WB* and vice versa, which can be explicitly determined by means of the constructions in schemes (SA) and

(SB). Hence the above will furnish a method of deriving a basis of the solutions $(S_1, \ldots S_m)$ of the syzygy equation

$$\mathrm{SUM}_k S_k * F_k = 0. \qquad \square$$

Theorem 10. *The reduced well-behaved basis WB of an ideal ID* = Ideal(*PS*) *with polset PS as a basis is uniquely determined up to constant multiples by the following two properties:*

(*a*) *WB is a reduced autoreduced basis of ID.*

(*b*) *Let T be the degree-tuple-set of WB. Then for any tuple u in* Comp(T) *with completed pol H_u and any non-multiplier i of WB w.r.t. u, the N-part of $X_i * H_u$ is* 0.

Proof. We have shown how to determine from *PS* by schemes (SA) and (SB), by a further reduction if necessary, a well-behaved basis *WB* of *ID* verifying properties (*a*) and (*b*). From the proofs of Theorems 5 and 6 we see that there will follow also the following property (*c*).

(*c*) The rest of any pol P in the ideal *ID* w.r.t. *WB* is 0.

Consider now any polset WB' verifying the analogous properties $(a)'$, $(b)'$ and hence also $(c)'$. There is no loss of generality in assuming that all the pols in *WB* and WB' have been normalized to have their leading coefficients = 1. We are to prove that WB' coincides with *WB*.

To see this let us arrange the pols in *WB* and WB' both in decreasing order, viz.

$$(WB): \quad W_1 > W_2 > \cdots > W_r,$$
$$(WB)': \quad W_1' > W_2' > \cdots > W_s'.$$

By (c) we have $\mathrm{Rest}(W_1'/WB) = 0$ and by the corresponding rest formula we see that the leading degree-tuple of W_1' should be a multiple of the leading-degree-tuple of some pol in *WB* , say W_i. In the same way, by $(c)'$ the leading-degree-tuple of W_i should be a multiple of the leading-degree-tuple of some W_j' of WB'. As WB' is autoreduced it will only be possible that W_j' coincides with W_1'. Then W_1' will have the same leading-monom as W_i. Applying the same reasoning to W_1 we see that W_1 should have the same leading monom as some W_k' of WB'. This is only possible when $W_i = W_1, W_k' = W_1'$ and W_1, W_1' have the same leading-monoms.

Applying now the same reasoning to W_2 and W_2' we see that they should have the same leading monoms. Continuing we see then *WB* and WB' should have the same number of pols or $r = s$ and each pair W_i and W_i' should have the same leading monoms.

Consider now the last two pols W_r and W_r' in *WB* and WB'. As W_r' has the same leading monom as W_r and W_r' has rest 0 w.r.t. *WB* we see that W_r' should be identical to W_r. Let us consider the pair W_{r-1} and W_{r-1}'. As the rest of W_{r-1}' w.r.t. *WB* is 0 we should have an identity of the form

$$W_{r-1}' = W_{r-1} + M_r,$$

in which M_r is an M-pol constructed from W_r. Now W'_{r-1} and W_{r-1} have the same leading monoms and no other monoms in W'_{r-1} and W_{r-1} can be multiples of the leading monom of W_r. It follows from the Tuple Decomposition Theorem that this will be possible only when $M_r = 0$ or W'_{r-1} is identical to W_{r-1}. Applying now the same reasoning to the other pairs of pols in WB and WB' successively in the reverse order we see that all the pairs should be identical to each other. The theorem is thus proved. □

Consider now an ideal ID with a reduced well-behaved basis WB. For $m < n$ let ID' be the ideal of all pols in ID in $X_1, \ldots, X_m$ alone. Let WB' be the autoreduced polset consisting of such pols in WB in $X_1, \ldots, X_m$ alone too. Then we have the following.

Theorem 11. *Let WB be a reduced well-behaved basis of an ideal*

$$ID \subset K[X_1, \ldots, X_n].$$

Then the autoreduced polset

$$WB' = WB \cap K[X_1, \ldots, X_m]$$

is a reduced well-behaved basis of the ideal

$$ID' = ID \cap K[X_1, \ldots, X_m].$$

Proof. Let T be the degree-tuple-set of WB and T' that of WB'. Consider now any pol P in ID'. Let us consider P as a pol in ID and write it in the form (4.2). By Theorem 6 we have $N = 0$. By the Tuple Decomposition Theorem we see that in (4.2) for each term in H_u we should have $\mathrm{COOR}_k(u) = 0$ for $k > m$. Let $\mathrm{Max}'(T)$ be the m-tuple got from $\mathrm{Max}(T)$ by deleting the last $n - m$ coordinates. It is clear that $\mathrm{Max}(T') \ll \mathrm{Max}'(T)$. It follows that for each H_u in (4.2) for which $\mathrm{COOR}_k(u) = 0$ for $k > m$, each i with X_i occurring in M_u which is a multiplier of u w.r.t. WB should also be a multiplier u w.r.t. WB'. Hence the N-part of P, considered as a pol in ID', is 0 too w.r.t. WB'. This implies in particular property (b) in Theorem 10 corresponding to WB' of ID'. By Theorem 10 again WB' is thus a reduced well-behaved basis of the ideal ID'. This completes the proof of the theorem. □

Finally, in comparing with the usual definition of Groebner basis of a polynomial ideal we see readily from Theorem 6 the following.

Theorem 12. *Any well-behaved basis of a polynomial ideal ID is a Groebner basis of ID. If the well-behaved basis is reduced and the leading coefficient of each pol in the basis is normalized to* 1, *then the basis is coincident with the reduced Groebner basis of ID.*

The well-behaved basis of a polynomial ideal, being nothing else but the usual Groebner basis of ID, will enjoy the various already well-known properties of Groebner basis. Some of such properties have been restated and reproved in the form of well-behaved basis as given above. The treatments and proofs are however done along the line of the thoughts of previous sections, giving thus alternative proofs of these known theorems about Groebner basis different from the known ones. Moreover, the proofs are given in order that they may be readily transferred to the differential case as described in the Introduction for which the corresponding notion of Groebner basis is non-existent. Furthermore, the above treatment shows that any pol in $K[X_1, \ldots, X_m]$ will have a *unique* expression w.r.t. a well-behaved basis, i.e. a Groebner basis, of a polynomial ideal in the form of equation (4.2), which is a property more precise than the corresponding known one for a Groebner basis under the usual known treatment.

5. An Example

The schemes (SB) and (SA) in the previous sections give an algorithm for the determination of a well-behaved basis, i.e. a Groebner basis of an ideal Ideal(PS) with a given basis PS. As an illustrative example (Problem 9(b) in [CG]) let us consider the following polset $PS = \{P_1, P_2, P_3\}$ with

$$\begin{cases} P_1 = X^2 + Y * Z + D * X + 1, \\ P_2 = Y^2 + Z * X + E * Y + 1, \\ P_3 = Z^2 + X * Y + F * Z + 1. \end{cases}$$

Introduce now an ordering among the various indeterminates by

$$Z > Y > X > D > E > F.$$

This amounts to equating these indeterminates to X_i such that $X_i > X_j$ if and only if $i > j$. We shall retain however the usual notations of Z, etc. as it will not cause misunderstandings.

According to the scheme (SB) we form first the well-behaved set (in decreasing order) WS_0 consisting of W_1, W_2, W_3 with

$$W_1 = P_3, \quad W_2 = P_1, \quad W_3 = P_2.$$

The leading-degree-tuple set of WS_0 is

$$T = \{(0,0,2), (0,1,1), (1,0,1)\}$$

so that $\mathrm{Max}(T) = (1,1,2)$. The completed H-pols arranged in descending order are thus 7 in number, viz.

$$H_1 = Y * X * W_1, \quad H_2 = Y * W_1, \quad H_3 = X * W_1,$$
$$H_4 = W_1, \quad H_5 = X * W_2, \quad H_6 = W_2, \quad H_7 = W_3.$$

Let the N-part of an H-pol H w.r.t. a non-multiplier i be denoted by $N(H/X_i)$. Most of the N-parts may be directly seen to be 0 by definition and the only non-zero N-parts are readily found to be the following ones:

$$\begin{aligned}
N_1 &= -N(H_5/Z) \\
&= -Z*H_5+H_1-F*H_5+X^2*H_7+D*X*H_7+H_7 \\
&= 2*Y^2*X^2+Y^2*X*D+Y^2+Y*X^2*E+Y*X*D*E \\
&\quad +Y*X+Y*E-X^3*F-X^2*D*F+X^2+X*D-X*F+1, \\
N_2 &= N(H_6/Z) \\
&= Z*H_6-H_2+F*H_6-X*H_7-D*H_7 \\
&= Z-2*Y^2*X-Y^2*D-Y*X*E-Y*D*E-Y \\
&\quad +X^2*F+X*D*F-X-D+F, \\
N_3 &= N(H_7/Y) \\
&= Y*H_7-H_5 \\
&= Y^3+Y^2*E+Y-X^3-X^2*D-X, \\
N_4 &= N(H_7/Z) \\
&= Z*H_7-H_3-Y*H_6-E*H_6+F*H_7 \\
&= Z+Y^2*F-2*Y*X^2-Y*X*D+Y*E*F-Y \\
&\quad -X^2*E-X*D*E-X-E+F.
\end{aligned}$$

The polset $PS_1 = WS_0 + NS_0$ thus consists of 7 pols, W_i and N_j. We proceed to form a well-arranged basis WS_1 of Ideal(PS_1) according to scheme (SA) in starting from $QS_0 = PS_1$, viz.

$$\begin{array}{llll}
QS_0 & QS_1 & \dots & QS_r, \\
AS_0 & AS_1 & \dots & AS_r, \\
RS_0 & RS_1 & \dots & RS_r.
\end{array}$$

It is found that for $r = 6$ the polset QS_6 consists of 4 pols Q_i below:

$$\begin{aligned}
Q_1 &= Z+Y^2*F-2*Y*X^2-Y*X*D+Y*E*F-Y-X^2*E \\
&\quad -X*D*E-X-E+F, \\
Q_2 &= Y^2*D*F+Y^2*F^2+2*Y^2+\cdots, \\
Q_3 &= Y*X^2*G_3+\cdots, \\
Q_4 &= Y*X*G_4+\cdots,
\end{aligned}$$

in which G_3, G_4 are pols in D, E, and F alone. The number of the terms of Q_3 and Q_4 are resp. 90 and 314 and Q_4 is non-factorizable. To make computations not too complicated we shall consider the special case of $F = E$ which will not influence the computations already done. It turns out that in this case of $F = E$

the pols Q_i will be simplified to the following ones:

$$\begin{aligned}
Q_1 &= Z + Y^2 * E - 2 * Y * X^2 - Y * X * D + Y * E^2 - Y - X^2 * E \\
&\quad - X * D * E - X, \\
Q_2 &= Y^2 * D * E + Y^2 * E^2 + 2 * Y^2 + 4 * Y * X^3 + 2 * Y * X^2 * D \\
&\quad - 2 * Y * X^2 * E - Y * X * D * E - Y * X * E^2 + 2 * Y * X \\
&\quad + Y * D * E^2 + Y * E^3 + 2 * Y * E + 2 * X^3 * E + 2 * X^2 * D * E \\
&\quad - 2 * X^2 * E^2 + 2 * X^2 - 2 * X * D * E^2 + D * E - E^2 + 2, \\
Q_3 &= F_1 * F_2, \\
Q_4 &= F_1 * F_3.
\end{aligned}$$

The pols Q_3 and Q_4 split into factors F_i with

$$\begin{aligned}
F_1 &= 2 * X^2 + X * D - X * E + 2, \\
F_2 &= Y * D * E - Y * E^2 - 2 * X^3 - 3 * X^2 * D - X * D^2 - X - E, \\
F_3 &= 2 * X^4 + 3 * X^3 * D + 2 * X^3 * E + X^2 * D^2 + 2 * X^2 * D * E \\
&\quad + X^2 * E^2 + X^2 + X * D * E^2 + 2 * X * E + E^2.
\end{aligned}$$

The polset QS_6 is now already an autoreduced one and may be taken as the corresponding well-arranged set AS_6. Let us denote the completed H-pols by

$$\begin{aligned}
H_{1ij} &= X^i * Y^j * Q_1, \quad i \le 6, j \le 2, \\
H_{2i} &= X^i * Q_2, \quad i \le 6, \\
H_{3i} &= X^i * Q_3, \quad i \le 4, \\
H_4 &= Q_4.
\end{aligned}$$

The variables corresponding to the non-multipliers are then resp. at most

$$\begin{aligned}
&X, Y \text{ for } H_{1ij}, \\
&X, Z \text{ for } H_{2i}, \\
&X, Y, Z \text{ for } H_{3i}, \\
&Y, Z \text{ for } H_4.
\end{aligned}$$

To determine the N-pols let us consider first $N(H_{30}/Y)$ where $H_{30} = Q_3$. By direct computation we find

$$(D*E+E^2+2)*E*(D-E)*Y*H_{30} = P_2*Q_4+P_3*Q_3+E^2*(D-E)^2*F_1*Q_2,$$

in which P_2, P_3 are pols in X of degree 2 and 3 resp. As all terms $X^i * Q_j$ occurring in the right-hand side of the above equation are M-products, it follows that

$$N(H_{30}/Y) = 0.$$

Consider now $N(H_4/Y)$ where $H_4 = Q_4$. Write F_2 in the form $F_2 = Y * E * (D - E) - P$ with P a pol in X of degree 3. Then by simply rewriting we get

$$E * (D - E) * Y * H_4 = F_3 * Q_3 + P * Q_4,$$

in which all terms $X^i * Q_j$ on the right-hand side are M-products. We thus again have

$$N(H_4/Y) = 0.$$

That other N-pols are all 0 may be deduced from the above ones or directly by rewriting in a similar way almost without computation. In conclusion it follows that QS_6 is already a well-behaved basis or a Groebner basis *GB* of the ideal Ideal(*PS*).

The zero-set Zero(*PS*) = Zero(*GB*) may be determined as follows. As F_1, F_3 are easily seen to be prime to each other (D, E are independent indeterminates) so $Q_4 = 0$ has 6 zeros of X, 2 from $F_1 = 0$ and 4 from $F_3 = 0$. For each zero of $F_1 = 0, Q_3$ will be 0 too and $Q_2 = 0$, $Q_1 = 0$ will give 2 zeros of *GB*. On the other hand, for each zero of $F_3 = 0$ we have $F_1 \neq 0$ and the resultant of Q_2 and F_2 is found to be 0, so such a zero of F_3 will be extended to only one zero of *GB* determined by $F_2 = 0$ and $Q_1 = 0$. In all we have 8 zeros of *GB* or *PS*. We remark that in the present case each zero of $Q_4 = 0$ can be extended to at least one zero of *GB*. This is however not the case in general. Cf. e.g. [WU2] and [L].

References

[BU1] Buchberger, B., Ein algorithmisches Kriterium fuer die Loesbarkeit eines algebraishes Gleichungs-systems, *Aeq. Math.*, **4** (1970), 374–383.

[BU2] Buchberger, B., Some properties of Groebner bases for polynomial ideals, *ACM SIGSAM. Bull.*, **10** (1976), 19–24.

[BU3] Buchberger, B., An algorithmic method in polynomial ideal theory, in N.K. Bose (Ed.), Recent trends in multidimensional systems theory, Reidel, 184–232, 1985.

[CAR] Cartan, E., Les systèmes différentiels extérieurs et leurs applications géometriques, Paris, 1945.

[J] Janet, M., Leçons sur les systèmes d'équations aux dérivées partielles, Gauthier Villars, 1920.

[K] Kolchin, E. R., Differential algebra and algebraic groups, Academic Press, 1973.

[L] Li, Ziming, Automatic implicitization of parametric objects, MM-Res. Preprints, No. 4 (1989), 54–62.

[M] Macaulay, F. S., Some properties of enumeration in the theory of modular systems. *Proc. London Math. Soc.*, **26** (1927), 531–555.

[RQ] Riquier, C. H., La méthode des fonctions majorantes et les systèmes d'équations aux dérivées partielles, Gauthier Villars, 1928.

[RITT1] Ritt, J. F., Differential equations from the algebraic standpoint, Amer. Math. Soc., 1932.

[RITT2] Ritt, J. F., Differential algebra, Amer. Math. Soc., 1950.

[TH] Thomas, J. M., Differential systems, Amer. Math. Soc., 1937.

[TR] Trinks, W., On Buchberger's method for solving systems of algebraic equations, *J. Number Theory*, **10** (1978), 475–488.

[WA] Wall, B., On the computation of syzygies, *ACM SIGSAM Bull*, **23** (1989), 5–14.

[WU1] Wu, Wen-tsün, On the foundation of algebraic differential geometry, MM-Res. Preprints, No. 3 (1989), 1–26, also in *Sys. Sci. & Math Scis.*, **2** (1989), 289–312.

[WU2] Wu, Wen-tsün, On a projection theorem of quasi-varieties in elimination theory, MM-Res. Preprints, No. 4 (1989), 40–48, also in *Chin. Ann. of Math.*, **11** B:2 (1990), 220–226.

Wenjun Wu (Wen-tsün Wu)
Institute of Systems Science
Academia Sinica
Beijing 100080, China

Index

GPSR Compliance

The European Union's (EU) General Product Safety Regulation (GPSR) is a set of rules that requires consumer products to be safe and our obligations to ensure this.

If you have any concerns about our products, you can contact us on ProductSafety@springernature.com

In case Publisher is established outside the EU, the EU authorized representative is:

Springer Nature Customer Service Center GmbH
Europaplatz 3
69115 Heidelberg, Germany

Batch number: 10732470

Printed by Printforce, the Netherlands